Fish Chemoreception

Fish Chemoreception

Edited by

Toshiaki J. Hara

Department of Fisheries and Oceans
Freshwater Institute
and
Department of Zoology
University of Manitoba
Winnipeg
Canada

CHAPMAN & HALL
London · Glasgow · New York · Tokyo · Melbourne · Madras

Published by Chapman & Hall, 2–6 Boundary Row, London SE1 8HN

Chapman & Hall, 2–6 Boundary Row, London SE1 8HN, UK

Blackie Academic & Professional, Wester Cleddens Road, Bishopbriggs, Glasgow G64 2NZ, UK

Van Nostrand Reinhold Inc., 115 5th Avenue, New York NY10003, USA

Chapman & Hall Japan, Thomson Publishing Japan, Hirakawacho Nemoto Building, 7F, 1-7-11 Hirakawa-cho, Chiyoda-ku, Tokyo 102, Japan

Chapman & Hall Australia, Thomas Nelson Australia, 102 Dodds Street, South Melbourne, Victoria 3205, Australia

Chapman & Hall India, R. Seshadri, 32 Second Main Road, CIT East, Madras 600 035, India

First edition 1992

© 1992. Chapman & Hall

Typeset in 10/12pt Photina by Interprint Ltd., Malta
Printed in Great Britain at the University Press, Cambridge

ISBN 0 412 35140 4   0 442 31534 1 (USA)

Apart from any fair dealing for the purposes of research or private study, or criticism or review, as permitted under the UK Copyright Designs and Patents Act, 1988, this publication may not be reproduced, stored, or transmitted, in any form or by any means, without the prior permission in writing of the publishers, or in the case of reprographic reproduction only in accordance with the terms of the licences issued by the Copyright Licensing Agency in the UK, or in accordance with the terms of licences issued by the appropriate Reproduction Rights Organization outside the UK. Enquiries concerning reproduction outside the terms stated here should be sent to the publishers at the London address printed on this page.

The publisher makes no representation, express or implied, with regard to the accuracy of the information contained in this book and cannot accept any legal responsibility or liability for any errors or omissions that may be made.

A catalogue record for this book is available from the British Library

Library of Congress Cataloging-in-Publication data available

Printed on permanent acid-free text paper, manufactured in accordance with the proposed ANSI/NISO Z 39.48–199X and ANSI Z 39.48–1984

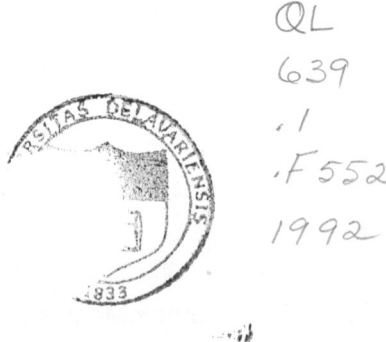

Contents

Contributors	x
Preface	xii
Series foreword T.J. Pitcher	xiii

1 Overview and introduction 1
  T.J. Hara

   1.1 Chemical stimuli in the aquatic environment 1
   1.2 Characteristic features of fish chemosensory organs 2
   1.3 Signal transduction and transmission 4
   1.4 Receptor sensitivity and specificity 5
   1.5 Chemoreception in behaviour 6
   1.6 Chemoreception and water pollution 9
   1.7 Conclusion 10
   References 10

2 Structure, development, and evolutionary aspects of the peripheral olfactory system 13
  E. Zeiske, B. Theisen and H. Breucker

   2.1 Introduction 13
   2.2 Gross morphology 13
   2.3 Histology 22
   2.4 Development 29
   2.5 Adaptation and evolution 30
   2.6 Summary and conclusion 32
   Acknowledgement 32
   References 32

3 Synaptic organization of the olfactory bulb and its central projection 40
  M. Satou

   3.1 Introduction 40
   3.2 Olfactory bulb 41
   3.3 Telencephalic hemisphere and secondary olfactory areas 50
   Acknowledgements 53

	Abbreviations	54
	References	54
4	Structure of the peripheral gustatory organ, represented by the siluroid fish *Plotosus lineatus* (Thunberg) K. Reutter	60
	4.1 Introduction	60
	4.2 Distribution and gross anatomy of taste buds	61
	4.3 Fine structure of taste buds	63
	4.4 Discussion	72
	Acknowledgements	75
	Abbreviations	75
	References	76
5	Central representation and projections of gustatory systems J.S. Kanwal and T.E. Finger	79
	5.1 Introduction	79
	5.2 Primary gustatory centres	83
	5.3 Medullary networks and descending gustatory projections	88
	5.4 Supramedullary gustatory pathways	91
	5.5 Summary	97
	Acknowledgement	98
	Abbreviations	98
	References	100
6	Solitary chemosensory cells M. Whitear	103
	6.1 Introduction	103
	6.2 Distribution	104
	6.3 Cytology	109
	6.4 Innervation	113
	6.5 Physiology	114
	6.6 The common chemical sense	116
	6.7 Paraneurones	119
	6.8 Conclusion	120
	Acknowledgements	121
	References	121
7	Molecular mechanisms of chemosensory transduction: gustation and olfaction J.G. Brand and R.C. Bruch	126
	7.1 Introduction	126

	7.2 Receptor events in chemoreception	127
	7.3 Receptor-mediated chemosensory transduction	133
	7.4 Conclusion	144
	Acknowledgements	145
	References	145
8	Mechanisms of olfaction *T.J. Hara*	150
	8.1 Introduction	150
	8.2 Sensory transduction	151
	8.3 Signal transmission	154
	8.4 Sensitivity and specificity	156
	8.5 Neural coding	164
	8.6 Summary and conclusion	166
	Acknowledgements	166
	References	166
9	Teleost gustation *T. Marui and J. Caprio*	171
	9.1 Introduction	171
	9.2 Historical background of gustatory peripheral nerve physiology	172
	9.3 Response features for simple chemical compounds	177
	9.4 Receptor site types for amino acids	182
	9.5 Taste responses of single facial taste fibres	185
	9.6 Enhanced taste activity	187
	9.7 Behaviour to chemical stimuli	188
	9.8 Tactile sensitivity of peripheral neurones	191
	9.9 Summary	191
	Acknowledgements	192
	References	192
10	Hormones, pheromones and chemoreception *P.W. Sorensen*	199
	10.1 Introduction	199
	10.2 History of the definition of pheromone	199
	10.3 Origins of the hypothesis that fish use hormonal sex pheromones	200
	10.4 Current status of the hypothesis and future considerations	215
	10.5 Neural responsiveness to sex pheromones	217
	10.6 General conclusions	220

	Acknowledgements	221
	References	221
11	**Kin recognition in fish mediated by chemical cues** *K.H. Olsén*	**229**
	11.1 Introduction	229
	11.2 Salmonids	230
	11.3 Studies of nonsalmonid fishes	243
	Acknowledgement	244
	References	244
12	**Olfactory control of homing behaviour in salmonids** *O.B. Stabell*	**249**
	12.1 Introduction	249
	12.2 Ecological framework	250
	12.3 Olfaction and homing behaviour	255
	12.4 Olfactory hypotheses in salmonid homing	259
	12.5 Possible mechanisms of olfactory control	263
	12.6 A model system and its ecological implications	264
	12.7 Summary and conclusions	265
	References	266
13	**Gustation and nutrition in fishes: application to aquaculture** *M. Takeda and K. Takii*	**271**
	13.1 Introduction	271
	13.2 Identification of feeding stimulants	271
	13.3 Relevance of feeding stimulants to nutrition and feeding	272
	13.4 Dietary application of feeding stimulants	273
	13.5 Feeding stimulant effects on nutrition	277
	13.6 Conclusions and prospects	283
	Acknowledgements	284
	References	285
14	**Food search behaviour in fish and the use of chemical lures in commercial and sports fishing** *K.A. Jones*	**288**
	14.1 Introduction	288
	14.2 Food search behaviour	289
	14.3 Modifiers of food search	297
	14.4 Effectiveness of natural materials as food search inducers	301
	14.5 Chemical nature of food search inducers	302
	14.6 Chemical lures	307

	Acknowledgements	314
	References	314
15	Environmental contaminants and chemoreception in fishes *D.A. Klaprat, R.E. Evans and T.J. Hara*	321
	15.1 Introduction	321
	15.2 Histological effects of contaminants	322
	15.3 Physiological and biochemical effects of contaminants	326
	15.4 Behavioural responses to contaminants	330
	15.5 Future research	334
	Ackowledgements	335
	References	335
	Author index	343
	Species index	356
	Subject index	363

Contributors

Joseph G. Brand
Monell Chemical Senses Center, 3500 Market Street,
and
Department of Biochemistry, School of Dental Medicine, University of Pennsylvania,
and
Veterans Administration Medical Center, Philadelphia, Pennyslvania 19104–3308, USA

Haide Breucker
Anatomical Institute, University of Hamburg, D–2000 Hamburg 20, Germany

Richard C. Bruch
Department of Neurobiology and Physiology, Northwestern University, Evanston, Illinois 60208, USA

John Caprio
Department of Zoology and Physiology, Louisiana State University, Baton Rouge, Louisiana 70803–1725, USA

Robert E. Evans
Department of Fisheries and Oceans, Freshwater Institute, Winnipeg, Manitoba, Canada R3T 2N6

Thomas E. Finger
Department of Cellular and Structural Biology, University of Colorado Medical School, Denver, Colorado 80262, USA

Toshiaki J. Hara
Department of Fisheries and Oceans, Freshwater Institute,
and
Department of Zoology, University of Manitoba, Winnipeg, Manitoba, Canada R3T 2N6

Keith A. Jones
Research and Development, Outdoor Technologies Group, One Berkley Drive, Spirit Lake, Iowa 51360, USA

Jagmeet S. Kanwal
Department of Biology, Washington University, St Louis, Missouri 63130, USA

Dorthy A. Klaprat
Department of Fisheries and Oceans, Freshwater Institute, Winnipeg, Manitoba, Canada R3T 2N6

Takayuki Marui
Department of Oral Physiology, Kagoshima University Dental School, Kagoshima 890, Japan

K. Håkan Olsén
Department of Zoophysiology, Uppsala University, S–75122 Uppsala, Sweden

Klaus Reutter
Department of Anatomy, University of Tübingen, D–7400 Tübingen, Germany

Masahiko Satou
Division of Information, Science Graduate School of Integrated Science, Yokohama City University, Yokohama 236, Japan

Peter W. Sorensen
Department of Fisheries and Wildlife, University of Minnesota, St Paul, Minnesota 55108, USA

Ole B. Stabell
Center of Veterinary Medicine, N–9001, Tromsø, Norway

Masahiko Takeda
Professor Emeritus, Laboratory of Fish Nutrition, Faculty of Agriculture, Kochi University, Nankoku, Kochi 783, Japan

Kenji Takii
Laboratory of Fisheries, Kinki University, Nachikatsuura, Wakayama 649–51, Japan

Birgit Theisen
Institute of Cell Biology and Anatomy, University of Copenhagen, DK–2100 Copenhagen, Denmark

Mary Whitear
Department of Zoology, University College London, London WC1E 6BT, United Kingdom

Eckart Zeiske
Zoological Institute and Zoological Museum, University of Hamburg, D–2000 Hamburg 13, Germany

Preface

This book describes in general how the chemosensory systems of fish function at various levels. In many ways, fish are typical vertebrates differing only slightly from other vertebrates including humans. In other ways, their aquatic environment imposes strict requirements or offers unique opportunities which have resulted in some unusual functions having no counterpart in higher vertebrates.

This new volume is necessitated by advances in many vital areas as the field of chemical senses continues to grow at a rapid pace. Most significant is the application of the contemporary electrophysiological technique of patch-clamping, recognition of a second messenger system in chemosensory transduction processes and the identification of hormonal pheromones in fish reproductive behaviour. The last major synthesis of our knowledge about fish chemoreception, *Chemoreception in Fishes*, was published ten years ago (Elsevier, Amsterdam, 1982). In that volume four aspects of fish chemoreception, i.e. morphology of the peripheral chemoreceptors, primary sensory processes, roles in behaviour, and its interactions with environment, were discussed.

This book is intended to be helpful to students, scientists and aquaculturists not only as a source book but also as a textbook on chemical senses. My intent in editing the volume has been twofold: first, to review recent advances in fundamental structure and function of the fish chemical senses and their relationships to the environment, and second, to illustrate future directions for research. Studies on the fish chemoreceptor systems will contribute to the solution of broad problems of olfaction and gustation in general. One book, of course, cannot be all-encompassing in the area of fish chemoreception. The alarm substance-fright reaction system has not been included nor has thorough coverage of orientation mechanisms.

In assembling this book, I have been greatly helped and encouraged by the diligence and enthusiasm of expert contributors. The book owes its good features to their cooperation. I would like to thank Dr Thomas P. Quinn for help and refereeing and Dr Charles Hollingworth for his thorough reading of the manuscripts. Finally, I would like to thank my colleagues, Scott Brown, Robert Evans, Dorthy Klaprat, Diane Malley, Torarinn Sveinsson and Chunbo Zhang, for providing advice, support and encouragement throughout the gestation of this book. Carol Catt provided excellent secretarial services. Jason Duncan gave invaluable help in compiling the index.

Toshiaki J. Hara
Winnipeg

Series foreword

Among the fishes, a remarkably wide range of biological adaptations to diverse habitats has evolved. As well as living in the conventional habitats of lakes, ponds, rivers, rock pools and the open sea, fish have solved the problems of life in deserts, in the deep sea, in the cold antarctic, and in warm waters of high alkalinity or of low oxygen. Along with these adaptations, we find the most impressive specializations of morphology, physiology and behaviour. For example we can marvel at the high-speed swimming of the marlins, sailfish and warm-blooded tunas, air-breathing in catfish and lungfish, parental care in the mouth-brooding cichlids and viviparity in many sharks and toothcarps.

Moreover, fish are of considerable importance to the survival of the human species in the form of nutritious and delicious food of numerous kinds. Rational exploitation and management of our global stocks of fishes must rely upon a detailed and precise insight of their biology.

The Chapman and Hall *Fish and Fisheries Series* aims to present timely volumes reviewing important aspects of fish biology. Most volumes will be of interest to research workers in biology, zoology, ecology and physiology but an additional aim is for the books to be accessible to a wide spectrum of non-specialist readers ranging from undergraduates and postgraduates to those with an interest in industrial and commercial aspects of fish and fisheries.

Toshiaki Hara (Winnipeg) has assembled an impressive raft of 21 international experts in this synoptic edited volume reviewing fish chemoreception, the sixth book in the *Fish and Fisheries Series*. The chemical senses are the oldest of the vertebrate sensory systems; revealing their ancient heritage, alone among vertebrate neurones, the olfactory nerves feed information directly into the fore-brain. It is not surprising therefore that chemical signals are critical to the essential functions of feeding and breeding.

Unlike air-borne stimuli, soluble nonvolatile compounds of low molecular weight are the most important in fish chemoreception. The first eight chapters, forming the core of the book, present the essential details of the functional neuroanatomy, neurophysiology and electrophysiology of the chemosensory transducers. Fish have three systems that transduce binding of these molecules at ciliary-derived membrane sites into coded nervous impulses: smell, taste and the recently-discovered solitary cells. Neuroethology of pheromone-mediated responses in social behaviour, kin recognition and reproductive homing, recently recognized as a key feature of teleost evolutionary success, are discussed in three subsequent chapters.

The book concludes with three chapters which provide a thorough review of applied aspects of fish chemoreception by examining the role of chemical appetite stimulation and attractants in aquaculture, commercial fishing, sport fishing and environmental pollution.

As with other books in this new series, I am confident that this book will soon assume the role of a major reference volume in the field, providing an interesting and significant contribution to the study of fish chemoreception.

Dr Tony J. Pitcher
Editor, Chapman and Hall Fish and Fisheries Series
Special Research Fellow, Imperial College, London

Note: Throughout this book the spellings of Doving, Döving and Døving are used interchangeably.

Chapter one

Overview and introduction

T.J. Hara

1.1 CHEMICAL STIMULI IN THE AQUATIC ENVIRONMENT

Fish are immersed in their physical and chemical environment, and their sensory systems are in continuous interaction with environmental perturbations. The aquatic environment is similar to the terrestrial environment in that it contains a multitude of chemical *mélanges*. However, the aquatic environment differs from the terrestrial environment in the ways in which chemical compounds can be distributed: (1) molecules need to be in solution rather than in the gaseous phase to be transported, and (2) water is a slower carrier medium, both for diffusion and for currents (Atema, 1980). Thus, solubility rather than volatility determines the type of compounds that can be found and utilized as chemical signals. Consequently, nonvolatile compounds with rather small molecular weights, such as amino acids and some steroids, are prominent compounds of fish olfaction and gustation and have been implicated in various behavioural roles (Table 7.1).

Chemical stimuli have three important properties: (1) extreme specificity, (2) persistence, and (3) non-directionality. Specificity of a chemical signal contained in its molecular structure, including stereoisometric configuration, has been demonstrated. Chemical signals, unlike visual, mechanical, sound, or electrical signals, last beyond the moment of production. This situation is actively employed in territorial and shelter marking by various animal species. In close proximity to the chemical signal source the distribution pattern of the molecules will be determined by both the diffusion characteristics of the substance and the degree of turbulence of the water (Kleerekoper, 1982). The concentration gradient will also be dependent on these two variables. Under these conditions two kinds of locomotor responses by fish are anticipated on perception of the chemical signals: (1) undirected arousal responses, and (2) orientated responses.

1.2 CHARACTERISTIC FEATURES OF FISH CHEMOSENSORY ORGANS

The chemical senses are the most ancient of sensory systems, having evolved 500 million years ago. They are involved in mediating the functions most basic to the survival of the individual and the species: feeding and reproduction. Fish detect chemical stimuli through at least two different channels of chemoreception – olfaction (smell) and gustation (taste). The olfactory organ in teleosts originates in an anlage formed by the ectoderm, and the whole organ remains ectodermal throughout its formation. The taste buds, in contrast, are endodermal in origin, although the external taste buds are claimed to be of ectodermal origin (Kapoor et al., 1975). The distinction between these two senses in fish is not always as clear as in terrestrial, air-breathing vertebrates, mainly because in fish both olfaction and gustation are mediated by molecules dissolved in water. In air or water, however, all chemoreception can be considered an aquatic phenomenon, because the receptor is covered with fluid materials where the initial sensory transduction process takes place. This fact may prove extremely useful to the elucidation of common principles for the mechanism by which chemoreceptor cells of all organisms function.

The peripheral olfactory organ in fish is variable in morphology, reflecting the enormous diversity of species adapted to different environmental conditions and multiple evolutionary pathways (Zeiske et al., Chapter 2 herein). Different ecostructural adaptations may have been achieved by selection pressure to optimize the properties relevant to hydrodynamic and other demands. Generally, two morphologically distinct receptor cell (olfactory neurone) types, ciliated and microvillar, exist in teleosts. The olfactory neurones, listed fifth of the seven wonders of the modern world by Thomas (1984), are the only neurones whose axons carry external information directly to the brain, and unlike any other neurones of the vertebrate central nervous system, olfactory neurones are replaced every few weeks. The microvillar receptor cells were once considered precursors of the ciliated receptor cells. However, ontogenetic as well as degeneration and regeneration studies have now established that both cell types represent independent entities (Zielinski and Hara, 1988; Zeiske et al., Chaper 2 herein). Ultrastructural and ontogenetic characteristics of the microvillar receptor cells in fish show remarkable similarities to those of the vomeronasal organ in higher vertebrates (Wysocki and Meredith, 1987). These cells may also have homology in their functional characteristics by playing a major role in the detection of chemical stimuli of a social nature. Functional differentiation of the two olfactory neurone types has yet to be determined.

The olfactory nerve fibres, unmyelinated axons of the receptor neurones, course to the olfactory bulb, where they make synaptic contact with the second-order bulbar neurones in the form of glomeruli. The accepted view is that the organization of the teleost olfactory bulb is basically similar to that in higher vertebrates (Satou, Chapter 3 herein). Furthermore, the olfactory system of teleosts, as in higher vertebrates, is segregated into two functionally distinct subdivisions – lateral and medial. In goldfish, *Carassius auratus*, stimulation by amino acids is primarily mediated by the lateral olfactory tract, whereas stimulation by pheromones (Sorensen, Chapter 10 herein) is primarily mediated by the medial tract.

The taste bud constitutes the structural basis of the gustatory organ, the other major chemosensory channel of vertebrates. In fish, taste buds are located on the gills, barbels, fins, oral cavity and pharynx, as well as over the entire body surface of some species. They are not usually found on the tongue, in contrast to higher vertebrates. Although there is general agreement on the gross structure of the taste bud, nomenclature and interpretation of morphology and function of some cell types are not always consistent. At least three distinct cell types have been identified: light and dark microvillar cells and basal cells. The light cell is generally interpreted as the receptor cell. The dark cell, considered by most to be a supporting cell, contains numerous, small microvilli. Reutter (1982; Chapter 4 herein) describes two types of elongated cells, light and dark, which differ in electron density. Basal cells in fishes are thought to be interneurones, and their resemblance to Merkel cells suggests a mechanoreceptive function. The basal cells of fish taste buds are thus different from those of mammalian taste buds, which are stem cells or regenerative cells of receptor cells. Jakubowski and Whitear (1990) assert that this light/dark terminology is misleading and should be abandoned. They favour the terminology of 'gustatory' and 'supporting' cells, respectively. The variation in overall electron density of gustatory cells is likely related to ageing of the individual cells. Both dark and light cell processes are linked to the basal cell by desmosomes, but direct, synaptic cell-to-cell communication within a taste bud is not established (Jakubowski and Whitear, 1990). A large number of nerve fibres concentrate beneath the basal lamina against which the taste bud rests. These fibres, often entering the bud and forming an intragemmal plexus, terminate on the surface of the receptor cells. These synapses are marked by membrane densities but not always by transmitter vesicles (Jakubowski and Whitear, 1990).

The gustatory receptor cell is a secondary sensory cell innervated by facial (cranial nerve VII), glossopharyngeal (IX), or vagal (X) nerves. Generally, all cutaneous taste buds are supplied by branches of the facial nerve, whereas the taste buds of the oropharyngeal cavity and palate are innervated by glossopharyngeal and vagal nerves. The central organization of the gustatory system is highly specialized for stimulus localization in some fish species. This is why fish have historically served as a vertebrate model for the study of the gustatory system (e.g. Herrick, 1901). The facial, glossopharyngeal and vagal inputs are distributed and encoded in parallel at the primary sensory centre in the medulla (Kanwal and Finger, Chapter 5 herein). Thus, the taste system is automatically divisible into two subsystems, facial and vagal, each serving separate phases of feeding behaviour. Kanwal and Finger (Chapter 5) examine the topographical representation of the overall organization within the medulla as well as the ascending projections from the facial and vagal fibres to higher centres in the silurids and cyprinids, where the gustatory system is highly developed.

The third chemosensory system is the 'common chemical sense', a term first used by Parker (1912) based on his observations on fish behaviour. In the common chemical sense, the epidermal free nerve endings, fibres whose cell body lies in a deep-seated ganglion and whose proximal end is embedded in the central organ, are stimulated by the action of irritant chemicals. Unlike olfactory and gustatory systems, however, this system is not precisely defined; Silver (1987), for instance, writes, '... perhaps a more general definition of the common chemical sense would be the sensation due to the stimulation of epithelial or mucosal free nerve endings by chemicals....'

Uncertain that the common chemical sense is an entity, and suggesting that the nomenclature of a basically unknown system is not worthy of argument, Whitear (Chapter 6 herein) describes solitary chemosensory cells based on electron microscopic studies of diverse fish species. She puts forward a hypothesis that secondary sensory cells capable of responding to chemical stimuli differentiate in the epithelia of primary aquatic vertebrates. Some of these chemosensory cells are incorporated in discrete taste buds while others, the solitary chemosensory cells, do not form a special association with surrounding epithelial cells. Further studies are needed to elucidate the functions of all non-olfactory and non-gustatory chemosenses, including fin-ray chemoreceptors of sea robins, *Prionotus carolinus* (Silver and Finger, 1984) and the vibratile fin of rocklings, *Ciliata mustela* (Peters *et al.*, 1991), as well as solitary chemosensory cells.

1.3 SIGNAL TRANSDUCTION AND TRANSMISSION

Sensory information about the chemical environment is transmitted to the brain by olfactory receptor neurones through a series of molecular, membranous and neural processes. Odorant molecules bind to receptor proteins in the ciliary plasma membrane, enabling them to activate a G protein. The activated G protein then activates adenylate cyclase. The resulting increase in second messenger cAMP concentration opens ion channels in the membrane, or translocates ions directly, causing membrane depolarization (Brand and Bruch, Chapter 7 herein). This eventually leads to the generation of impulses by which the sensory information is transmitted to the olfactory bulb, the first relay station in the olfactory system. The axons of the olfactory bulb neurones, in turn, project to the brain, where higher-level processing of olfactory information allows the discrimination of odours by the brain (Chapters 3 and 8 herein).

Since Adrian and Ludwig's (1938) first electrophysiological work on the olfactory system, fishes have been a valuable model for advancement of chemoreception studies in general. Fishes often possess specific, sensitive and readily accessible chemosensory organs, and the use of classical receptor binding techniques has permitted a critical evaluation of the transduction sequences following the initial ligand binding (Cagan and Zeiger, 1978; Novoselov et al., 1980; Brown and Hara, 1981). Subsequent comprehensive studies on amino acid receptors and their associated transduction sequences in ictalurid catfish suggest that there exist multiple mechanisms for the transduction of chemosensory information in teleosts (Brand and Bruch, Chapter 7 herein). Recent progress in characterizing transduction sequences in fish chemosensory systems stimulated the search for similar mechanisms in mammals, and eventually led to a major discovery of a large family of genes that encodes an array of olfactory receptor proteins (Buck and Axel, 1991). This study suggests the presence of a large number of distinct odorant receptors.

1.4 RECEPTOR SENSITIVITY AND SPECIFICITY

Since Sutterlin and Sutterlin's (1971) and Suzuki and Tucker's (1971) initial works on amino acid stimulation of the olfactory receptors in Atlantic salmon, *Salmo salar*, and white catfish, *Ictalurus catus*, respectively, research on fish olfaction has centred on characterizing amino acid receptors in a variety of species (reviews: Hara, 1982, 1986; Caprio, 1984). Recently, new groups of chemicals, steroids and prostaglandins, have been identified and their extreme olfactory stimulatory characteristics for fish have been investigated (Hara, Chapter 8 herein; Sorensen, Chapter 10 herein).

For the best-studied amino acids, electrophysiological thresholds lie between 10^{-7} and 10^{-9} M and the concentration–response relationship exhibits a broad dynamic range of sensitivity, covering over 6–7 log units. This wide dynamic range, coupled with data from binding experiments (Cagan and Zeiger, 1978), cross-adaptation experiments (Caprio and Byrd, 1984) and kinetic analyses (Sveinsson and Hara, 1990), suggests the existence of multiple receptor sites for amino acids in fish olfaction. A structure–activity relationship study established that the most effective stimuli are unsubstituted L-α-amino acids, containing unbranched and uncharged side chains (Hara, 1973). This finding led to one hypothetical receptor model which involves two charged subsites, one anionic and one cationic, capable of interacting with ionized amino and carboxyl groups of stimulant amino acid molecules (Hara, 1977). Subsequent binding and cross-adaptation experiments led to the general postulate that a multiplicity of receptor types or transduction mechanisms exist in the fish olfactory system (Cagan and Zeiger, 1978; Caprio and Byrd, 1984).

Unlike amino acids, bile salts and sex steroids as well as prostaglandins seem to be more specific in terms both of species and of receptor interactions. Bile salts have been found highly stimulatory in salmonids, while goldfish preovulatory pheromone 17α,20β-dihydroxy-4-pregnen-3-one and related sex steroids seem stimulatory only in cyprinids. Prostaglandins are active in both cyprinids and salmonids. Minor alterations in molecules dramatically reduce their activities in all chemicals. Receptor site specificity is further demonstrated by cross-adaptation experiments, where exposure to high concentrations of any one of the four groups of chemicals maintains responses to the others (Hara, Chapter 8 herein; Sorensen, Chapter 10 herein).

Contrary to olfaction, gustatory responses to amino acids vary greatly among species, in terms both of sensitivity and of specificity (Marui and Caprio, Chapter 9 herein). Two fish groups can be identified: (1) those species that possess several amino acid receptors and respond to a wide range of amino acids, and (2) those species that possess one or two amino acid receptors and respond to a limited number of amino acids (Hara and Zielinski, 1989). Channel catfish, *Ictalurus punctatus*, and several marine species studied may be typical of those fish with a broad response range, while some charrs (*Salvelinus* spp.) may represent the simplest form of the latter, responding only to L-proline and related chemicals. Generally, L-amino acids are more stimulatory than their stereoisomers, but recent electrophysiological data indicate the existence of D-amino acid, especially D-alanine, gustatory receptors in some catfish (Marui and Caprio, Chapter 9 herein). In fact, D-amino acids are more abundant than their L-isomers in some bivalve molluscs.

Some of the bile salts stimulatory for the olfactory system also stimulate gustatory receptors in some species. In addition, the gustatory receptors are sensitive to CO_2 and are capable of distinguishing between CO_2 and H^+ (Yamashita *et al.*, 1989). Although the gustatory responses to CO_2 are interpreted as a behavioural strategy functioning primarily in preventing respiratory distress, they may have a direct influence on central respiratory regulating mechanisms. The sensitive, specific gustatory receptors for marine toxins such as tetrodotoxin and saxitoxin suggest the existence of a mechanism for avoiding poisonous prey organisms that has adaptive advantage to the receiver (Yamamori *et al.*, 1988).

1.5 CHEMORECEPTION IN BEHAVIOUR

The encoded chemosensory information is transmitted and integrated into behavioural patterns through spatially separated neuroanatomical substrates within the central nervous system. The following three principal areas of interface between chemosensory and fish behaviour deserve special attention: (1) chemical signals—pheromones, (2) homing migration in salmon, and (3) feeding.

Chemical signals – pheromones

Our understanding of the role of chemosensory and chemical signalling systems has been limited, but dramatic advances in the study of sex pheromones have been made since 1980. The most comprehensive study describes a hormonal pheromone system in goldfish. Sorensen (Chapter 10 herein) reviews these advances from a historical perspective emphasizing the goldfish as a model. These studies strongly suggest that sex hormones and their metabolites function as sex pheromones with distinct, fundamental roles controlling the reproductive physiology and behaviour of goldfish. Evidence is rapidly accumulating that pheromonal and endocrinological functions are closely intertwined in teleost fish. It seems ironic that the original definition of a pheromone states that, '... unlike hormones ... the substance is not secreted into the blood but outside the body; it does not serve humoral correlation within the organism but communication between individuals' (Karlson and Lüscher, 1959). It now appears that hormones have dual functions in general, at least in some teleosts. Although hormonal pheromones meet some of the characteristic features of true chemical communication, it is uncertain whether they have a communication role in the strict sense (Liley, 1982).

Social organisms frequently distinguish relatives from non-relatives and determine their degree of relatedness. Odours often play an important role in this kin recognition. Recent studies on fish present evidence for sibling recognition and the existence of population-specific chemical cues in salmonids (Olsén, Chapter 11 herein). Their roles in social behaviour and mate choice as well as in homing behaviour are implicated (Stabell, Chapter 12 herein). An alarm substance is released when the epidermis of an ostariophysan fish is damaged. The development and maintenance of the alarm substance and fright reaction can be interpreted as animal altruism maintained by kin recognition (Smith, 1992).

Homing migration in salmon

Olfaction seems essential for homing migration of salmonids, at least during their final riverine stage. The imprinting hypothesis, i.e. learning of environmental chemical cues during a sensitive period of early development, and the pheromone hypothesis, i.e. innate responses to pheromones specific for local populations, are the two prevailing hypotheses. Wisby and Hasler's (1954) classical experiment on olfactory occlusion in coho salmon, *Oncorhynchus kisutch*, was verified by artificial imprinting experiments using the chemical morpholine (Cooper and Hirsch, 1982; Hasler and Scholz, 1983).

Recent studies present evidence indicating that imprinting is under control of thyroid hormones. In Atlantic salmon a sensitive period for olfactory learning and imprinting exists during smoltification, which is correlated with increased activity of thyroid hormones (Morin et al., 1989).

Lack of evidence for existence of odorants specific for stream waters led Stabell to conclude that 'the pheromone theory should be given priority' (Chapter 12 herein). The pheromone hypothesis states that (1) populations or races of salmon in different streams emit pheromones that serve to identify fish distinctly from each other, (2) the memory of this population-specific pheromone is inherited, and (3) homing adults follow pheromone trails released by juveniles residing in the stream, i.e. the juveniles provide a constant source of population odour (Nordeng, 1977). Stabell considers that a specific homing to a native spawning site is under genetic influence, and that possible genetic contamination of pheromones resulting from hatchery escapes or random stocking programmes might seriously interfere with homing performance and population structures. He stresses the importance of the concept of chemical ecology in management practices of salmonid fisheries.

Chemoreception, feeding and nutrition

When exposed to chemical stimuli associated with food, fish initiate food-search behaviour. Feeding behaviour is a stereotyped sequence of behavioural components which can normally be differentiated into several phases: (1) arousal, (2) search, and (3) uptake and ingestion (consummatory). However, they are in reality a continuum without necessarily distinct transition (Jones, Chapter 14 herein). Fish rely upon information received through all sensory channels. The relative importance of individual sense organs differs in various species and is determined by their ecological niches, feeding strategy, motivation, and other biotic and abiotic environmental factors (Pavlov and Kasumyan, 1990). A wide variety of fish species are known to utilize chemical signals in search, location, and ingestion of food; attempts to identify and isolate active components of food extracts have often been frustrated by a loss of potency as the fractionation stages progress. Although some generalization can be made as to the chemical nature of feeding stimulants, i.e. they are of low molecular weight, non-volatile, nitrogenous and amphoteric, a question still remains as to how and what chemical components fish use in food selection and ingestion. This uncertainty is due partly to differences in methodologies employed, and partly to simultaneous involvement of both olfactory and gustatory systems. The same stimulus often exerts different behavioural effects depending upon whether it is received by olfactory or by gustatory organs. In conventional behavioural studies, the distinction of senses involved is not always clear. Some chemical stimulants may act as attractants via olfaction, and others may act as promoters or enhancers of food intake or ingestion.

Recently, increased attention has been given to the effects on chemical feeding stimulants on fish nutrition, with special emphasis on their possible application to aquaculture (Takeda and Takii, Chapter 13 herein). A diet supplemented with feeding stimulant amino acids and nucleotides stimulates feeding activity, leading to enhanced growth performance, in Japanese eels, *Anguilla japonica*. These effects are thought to be attributed primarily to improved food intake, digestion and absorption. Chemosensory stimulation by dietary feeding stimulants induces the cephalic reflex, which serves to increase the activities of digestion, absorption and metabolism at early postprandial states. In view of the importance of diet development in aquaculture, more research is encouraged on dietary supplementation with feeding stimulants to improve the palatability and nutritional value of diets.

The overt response of fish to food odours and flavours has not escaped Man's commercial eye, and for centuries commercial and sports fishermen alike have employed chemical lures in diverse ways to increase their catch (Jones, Chapter 14 herein). Although chemical feeding lures have been based almost exclusively on natural materials, the potential for the development of wholly synthetic lures appears considerable. Their greatest potential lies in their high specificity and potency. Furthermore, if the structure–activity relationships of the active compounds were sufficiently defined in olfactory and gustatory responses, it would be theoretically possible to design chemical stimulants which accentuate those molecular features necessary for activity. The end results could be chemical lures having potencies beyond those attained by natural materials.

1.6 CHEMORECEPTION AND WATER POLLUTION

Increased inputs of anthropogenic contaminants, and other stresses such as the destruction of habitats, have brought about drastic changes in aquatic ecosystems. Fish chemoreceptive membranes directly exposed to the aquatic environment are susceptible to deleterious effects of water-borne chemicals. Toxic substances may alter chemosensory perception via several modes of action: (1) through cellular uptake and damage to organelles and enzyme systems (Chapters 2 and 4), (2) by direct interactions with membrane receptor sites (Chapter 7), or (3) by masking biologically important chemical signals through competitive binding (Chapters 7, 10–14). Klaprat *et al.* (Chapter 15 herein) review developments in the study of effects of aquatic contamination on fish chemoreception since the work of Hara *et al.* (1983). Widely used behavioural responses of fishes to adverse conditions are variable and are not always related to the toxicity of pollutants. Physiological, biochemical and histochemical methods that are sensitive and specific to a chemical or group of chemicals should be developed to serve as early warning indicators. Aquatic toxicology in general has undergone significant changes since 1980. Studies have expanded to areas involving metabolic conversions of carcinogens, modification of DNA, and other distinctly biochemical processes. Recent advances in biochemical studies of chemosensory transduction mechanisms (Chapter 7) should help us to understand the nature and mode of action of toxicants. The olfactory epithelium of a number of fish species contains high levels of the cytochrome P-450 monooxygenase (mixed-function oxidase, MFO) system, which can be activated by exposure to toxicants. However, the basic biochemical and physiological functions of MFO systems must be understood before they can be used to predict the effect of contaminant-induced changes.

1.7 CONCLUSION

High chemosensitivity is ubiquitous throughout fish fauna. Fish have provided a valuable model for advancing the study of chemoreception. This book is devoted to three main chemosensory channels: olfaction, gustation and solitary chemoreceptors. As new techniques are adopted (patch clamping, binding and DNA recombinant), underlying mechanisms of sensory transduction and discrimination are gradually being unveiled. Although dramatic advancements have been recorded in the study of hormonal pheromones, the specific roles of the majority of chemical signals still remain undefined.

REFERENCES

Adrian, E.D. and Ludwig, C. (1938) Nervous discharges from the olfactory organs of fish. *J. Physiol., Lond.*, **94**, 441–60.

Atema, J. (1980) Chemical senses, chemical signals and feeding behavior in fishes, in *Fish Behavior and its Use in the Capture and Culture of Fishes* (eds J.E. Bardach, J.J. Magnuson, R.C. May and J.M. Reinhart), ICLARM, Manila, pp. 57–101.

Brown, S.B. and Hara, T.J. (1981) Accumulation of chemostimulatory amino acids by a sedimentable fraction isolated from olfactory rosettes of rainbow trout (*Salmo gairdneri*). *Biochim. biophys. Acta*, **675**, 149–62.

Buck, L. and Axel, R. (1991) A novel multigene family may encode odorant receptors – a molecular basis for odor recognition. *Cell*, **65**, 175–87.

Cagan, R.H. and Zeiger, W.N. (1978) Biochemical studies of olfaction: binding specificity of radioactively labeled stimuli to an isolated olfactory preparation from rainbow trout (*Salmo gairdneri*). *Proc. natn. Acad. Sci. U.S.A.*, **75**, 4679–83.

Caprio, J. (1984) Olfaction and taste in fish, in *Comparative Physiology of Sensory Systems* (eds L. Bolis, R.D. Keynes and S.H.P. Madrell), Cambridge University Press, Cambridge, pp. 257–83.

Caprio, J. and Byrd, R.P. jun. (1984). Electrophysiological evidence for acidic, basic, and neutral amino acid olfactory receptor sites in the catfish. *J. gen. Physiol.*, **84**, 403–22.

Cooper, J.C. and Hirsch, P.J. (1982) The role of chemoreception in salmonid homing, in *Chemoreception in Fishes* (ed. T.J. Hara), Elsevier, Amsterdam, pp. 343–62.

Hara, T.J. (1973) Olfactory responses to amino acids in rainbow trout, *Salmo gairdneri*. *Comp. Biochem. Physiol.*, **44A**, 407–16.

Hara, T.J. (1977) Further studies on the structure–activity relationships of amino acids in fish olfaction. *Comp. Biochem. Physiol.*, **56A**, 559–65.

Hara, T.J. (ed.) (1982) *Chemoreception in Fishes*, Elsevier, Amsterdam, 333 pp.

Hara, T.J. (1986) Role of olfaction in fish behaviour, in *The Behaviour of Teleost Fishes* (ed. T.J. Pitcher), Croom Helm, London, pp. 152–76.

Hara, T.J., Brown, S.B. and Evans, R.E. (1983) Pollutants and chemoreception in aquatic organisms, in *Aquatic Toxicology* (ed. J.O. Nriagu), Wiley, New York, pp. 247–306.

Hara, T.J. and Zielinski, B. (1989) Structural and functional development of the olfactory organ in teleosts. *Trans. Am. Fish. Soc.*, **118**, 183–94.

Hasler, A.D. and Scholz, A.T. (1983) *Olfactory Imprinting and Homing in Salmon*, Springer-Verlag, Berlin, 134 pp.

Herrick, C.J. (1901) The cranial nerves and cutaneous sense organs of the North American silurid fishes. *J. comp. Neurol. Psychol.*, **11**, 177–249.

Jakubowski, M. and Whitear, M. (1990) Comparative morphology and cytology of taste buds in teleosts. *Z. mikrosk.-anat. Forsch.*, **104**, 529–60.

Kapoor, B.G., Evans, H.E. and Pevzner, R.A. (1975) The gustatory system in fish. *Adv. mar. Biol.*, **13**, 53–108.

Karlson, P. and Lüscher, M. (1959) 'Pheromones': a new term for a class of biologically active substances. *Nature, Lond.*, **183**, 55–6.

Kleerekoper, H. (1982) Research in olfaction in fishes: historical aspects, in *Chemoreception in Fishes* (ed. T.J. Hara), Elsevier, Amsterdam, pp. 1–14.

Liley, N.R. (1982) Chemical communication in fish. *Can. J. Fish. Aquat. Sci.*, **39**, 22–35.

Morin, P.-P., Dodson, J.J. and Doré, F.Y. (1989) Thyroid activity concomitant with olfactory learning and heart rate changes in Atlantic salmon, *Salmo salar*, during smoltification. *Can. J. Fish. Aquat. Sci.*, **46**, 131–6.

Nordeng, H. (1977) A pheromone hypothesis for homeward migration in anadromous salmonids. *Oikos*, **28**, 155–9.

Novoselov, V.I., Krapivinskaya, L.D. and Fesenko, E.E. (1980) Molecular mechanisms of odor sensing. V. some biochemical characteristics of the alanineous receptor from the olfactor epithelium of the skate *Dasyatis pastinaca*. *Chem. Senses*, **5**, 195–203.

Parker, G.H. (1912) The reactions of smell, taste, and the common chemical sense in vertebrates. *Proc. Acad. nat. Sci. Philad.*, **15**, 221–34.

Pavlov, D.S. and Kasumyan, A.O. (1990) Sensory principles of the feeding behavior of fishes. *J. Ichthyol.*, **30**, 77–93.

Peters, R.C., Kotrschal, K. and Krautgartner, W.-D. (1991) Solitary chemoreceptor cells of *Ciliata mustela* (Gadidae, Teleostei) are tuned to mucoid stimuli. *Chem. Senses*, **16**, 31–42.

Reutter, K. (1982) Taste organ in the barbel of the bullhead, in *Chemoreception in Fishes* (ed. T.J. Hara), Elsevier, Amsterdam, pp. 77–91.

Silver, W.L. (1987) The common chemical sense, in *Neurobiology of Taste and Smell* (eds T.E. Finger and W.L. Silver), Wiley, New York, pp. 65–87.

Silver, W.L. and Finger, T.E. (1984) Electrophysiological examination of a non-olfactory, non-gustatory chemosense in the searobin, *Prionotus carolinus*. *J. comp. Physiol.*, **154A**, 167–74.

Smith, R.J.F. (1982) The adaptive significance of the alarm substance–fright reaction system, in *Chemoreception in Fishes* (ed. T.J. Hara), Elsevier, Amsterdam, pp. 327–42.

Smith, R.J.F. (1992) Alarm signals in fishes. *Rev. Fish Biol. Fish.*, **2**, in press.

Sutterlin, A.M. and Sutterlin, N. (1971) Electrical responses of the olfactory epithelium of Atlantic salmon (*Salmo salar*). *J. Fish. Res. Bd Can.*, **28**, 565–72.

Suzuki, N. and Tucker, D. (1971) Amino acids as olfactory stimuli in freshwater catfish, *Ictalurus catus* (Linn.). *Comp. Biochem. Physiol.*, **40A**, 399–404.

Sveinsson, T. and Hara, T.J. (1990) Analysis of olfactory responses to amino acids in Arctic char (*Salvelinus alpinus*) using a linear multiple-receptor model. *Comp. Biochem. Physiol.*, **97A**, 279–87.

Thomas, L. (1984) *Late Night Thoughts of Listening to Mahler's Ninth Symphony*, Bantam Books, New York, p. 168.

Wisby, W.J. and Hasler, A.D. (1954) Effect of olfactory occlusion in migrating silver salmon (*Oncorhynchus kisutch*). *J. Fish. Res. Bd Can.*, **11**, 472–8.

Wysocki, C.J. and Meredith, M. (1987) The vomeronasal system, in *Neurobiology of Taste and Smell* (eds T.E. Finger and W.L. Silver), Wiley-Interscience, New York, pp. 125–50.

Yamamori, K., Nakamura, M., Matsui, T. and Hara, T.J. (1988) Gustatory responses to tetrodotoxin and saxitoxin in fish: a possible mechanism for avoiding marine toxins. *Can. J. Fish. Aquat. Sci.*, **45**, 2182–6.

Yamashita, S., Evans, R.E. and Hara, T.J. (1989) Specificity of the gustatory chemoreceptors for CO_2 and H^+ in rainbow trout (*Oncorhynchus mykiss*). *Can. J. Fish. Aquat. Sci.*, **46**, 1730–4.

Zielinski, B. and Hara, T.J. (1988) Morphological and physiological development of olfactory receptor cells in rainbow trout (*Salmo gairdneri*) embryos. *J. Comp. Neurol.* **27**, 300–11.

Chapter two

Structure, development, and evolutionary aspects of the peripheral olfactory system

E. Zeiske, B. Theisen and H. Breucker

2.1 INTRODUCTION

The morphology of the olfactory organ in fish has fascinated scientists for a long time. At the turn of the 18th to the 19th century, investigators were already attracted by the formation of olfactory lamellae in fish. From the second half of the 19th century, extensive comparative investigations followed (comprehensive historical review: Kleerekoper, 1969, 1982). Schultze (1863) was the pioneer of studies on the histology of the olfactory epithelium of vertebrates, including fish. He found two types of cells, epithelial and sensory, the latter of which seemed to be the peripheral end of the olfactory nerve. A new epoch of histology began with the introduction of the electron microscope. Using this technique the first results from the olfactory epithelium of fish were obtained by Trujillo-Cenoz (1961). The last review of the morphology of the peripheral olfactory organ in fish (Yamamoto, 1982) was focused on teleosts. This chapter deals also with non-teleost fish. Aspects of ontogeny, growth, adaptation, and evolution of the peripheral olfactory organ are included.

2.2 GROSS MORPHOLOGY

Olfactory organs in fish are basically paired structures situated in the snout. Each olfactory organ consists of an olfactory chamber which connects with

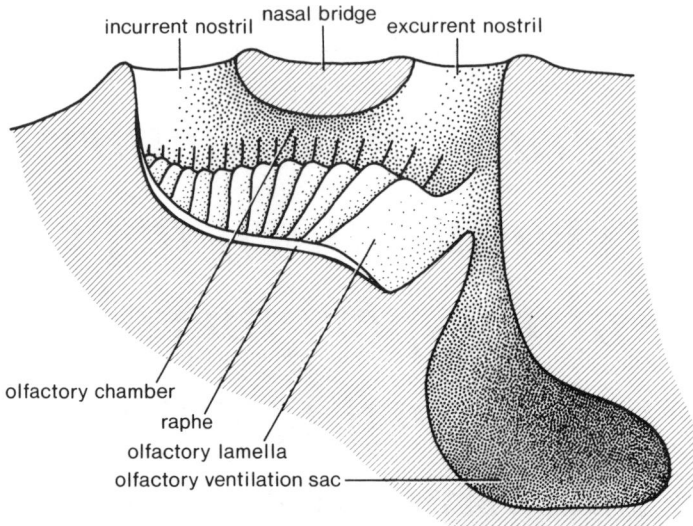

Fig. 2.1 Diagrammatic median section of the olfactory organ of a generalized fish.

the exterior through two openings, the incurrent and excurrent nostrils, separated by a nasal bridge (Fig. 2.1). An olfactory rosette is situated on the floor of the olfactory chamber and comprises a longitudinal ridge, the raphe, with two rows of olfactory lamellae. The olfactory epithelium with the olfactory receptor cells covers the olfactory lamellae. An extension from the olfactory chamber may form an olfactory ventilation sac. Great variations of this scheme are found in different fish groups as will be revealed below.

The cyclostomes (hagfish and lampreys) are unique in possessing only a single olfactory (monorhinal) organ with a single nostril. From the nostril, a nasal duct leads to the olfactory chamber and continues below and behind this chamber as a nasopharyngeal duct, which opens into the pharynx in hagfish (Fig. 2.2(A)), whereas in lampreys the nasal duct ends blindly above the region of the anterior gill openings as a nasopharyngeal pouch. A thin valve without muscle fibres is attached to the wall of the nasal duct just in front of the olfactory chamber (Bütschli, 1921; Kleerekoper and van Erkel, 1960; Ross, 1963; Theisen, 1973). In hagfish the olfactory rosette is composed of about seven free lamellae arranged longitudinally and parallel to one another (Fig. 2.2(B)), whereas in lampreys it consists of a number of longitudinally arranged lamellae radiating from the wall of the olfactory chamber. Ventilation of the olfactory organs is created by a pumping mechanism associated with the respiratory movements. In hagfish the water current is produced by the movements of a muscular velum in the pharynx. The water enters through the nostril, passes to the olfactory organ, and leaves through the gill opening (Fig. 2.2(A); Johansen and Strahan, 1963). In lampreys the pharyngeal pouch is activated by the pressure changes in

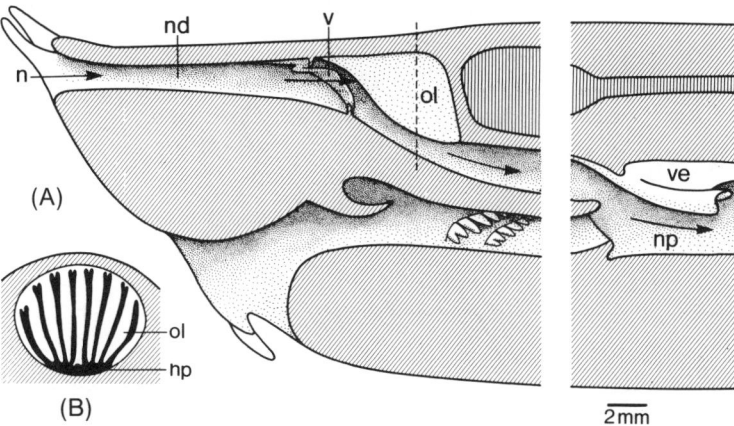

Fig. 2.2 Olfactory organ of hagfish, *Myxine glutinosa*. (A) Paramedian section of the head; (B) cross-section of the olfactory organ. The broken line in (A) indicates the section in (B). The arrows in (A) indicate the direction of the water current. Abbreviations: n, nostril; nd, nasal duct; np, nasopharyngeal duct; ol, olfactory lamella; v, valve; ve, velum. (Redrawn from Theisen (1973) and from a preserved specimen.)

the respiratory tube, which create a water current in and out of the nostril. In both cases the valve in front of the olfactory chamber directs the incoming water into the spaces between the olfactory lamellae (Kleerekoper and van Erkel, 1960; Theisen, 1976).

The olfactory organs in elasmobranchs (sharks and rays) are in cartilaginous capsules situated ventrally in the snout. The incurrent and excurrent nostrils of each organ open ventrally and are incompletely separated by nasal flaps. The nostrils vary greatly in size, shape and position (Tester, 1963; Meng and Yin, 1981a,b). They lie obliquely relative to the long axis of the body, with the incurrent nostril positioned lateral to the excurrent nostril (Fig. 2.3). The olfactory chamber is filled with a rosette composed of two rows of olfactory lamellae on each side of a transverse raphe.

In chimaeras the two olfactory organs lie close together, separated only by a median cartilaginous septum, and the incurrent nostrils are situated medially, just above the mouth. The excurrent nostrils are connected with the corner of the mouth.

In elasmobranchs the ventilatory flow of water through the olfactory organ is created by pressure differential between the incurrent and excurrent nostrils. The shape of the nostrils indicates that forward motion of the animal creates a water current over the nostrils, which establishes this pressure differential. In benthic elasmobranchs, where the olfactory organs are connected with the corners of the mouth, the water flow is supported by the respiratory activity (Theisen *et al.*, 1986; Zeiske *et al.*, 1989). In chimaeras, the respiratory activity is assumed to create the

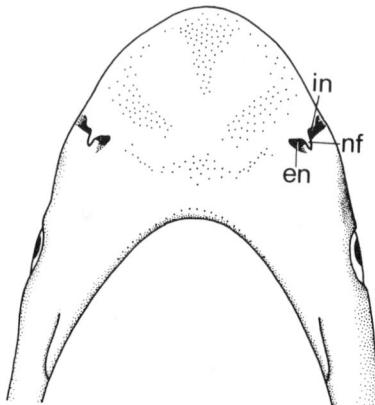

Fig. 2.3 Nostrils of Caribbean sharpnose shark, *Rhizoprionodon porosus*. Abbreviations: en, excurrent nostril; in, incurrent nostril; nf, nasal flap. (Drawn from a preserved specimen.)

ventilatory flow, but forward motion of the animal may also take part (Holl, 1973a).

Lungfish (Dipneustei) have large olfactory organs situated ventrally in the snout. The incurrent nostril of each side is found at the entrance of the mouth (Huxley, 1876), and the excurrent nostril opens posteriorly into the oral cavity (Fig. 2.4(A) and (B)). The large olfactory chambers are elongated and curved (Fig. 2.4(C)). The olfactory rosette consists of a raphe with two rows of olfactory lamellae which hang down from the roof and side walls of the olfactory chamber (Fig. 2.4(D); Pfeiffer, 1969; Theisen, 1972; Derivot, 1984). The two longitudinal flaps in front of the olfactory lamellae may aid in regulating the water current into the olfactory chamber, or may protect the olfactory lamellae against mud particles (Bertmar, 1965). In water, ventilation of the olfactory organs is provided by pumping mechanisms. In air, the olfactory organs are believed not to be used (Atz, 1952; Bertmar, 1965). During dry periods both *Protopterus* and *Lepidosiren* may aestivate. *Protopterus* makes a cocoon of mud with a tube for breathing, but the nostrils become closed by mucus (Atz, 1952; Bone and Marshall, 1982).

The olfactory organs of *Latimeria*, the only extant member of the coelacanths, have a complicated structure (Millot and Anthony, 1965; Pfeiffer, 1969). The nostrils of each organ are widely separated. The incurrent nostril lies at the end of a nasal tube just above the mouth, and the excurrent nostril is a narrow slit close to the eye. No choanae are present. The cartilaginous and bony nasal capsule contains the ovoid olfactory chamber and adipose tissue. The olfactory rosette appears as five conical sections with many ramifications, lodging the olfactory lamellae. The morphology of the olfactory organ and the presence of nonsensory cilia (Millot and Anthony, 1965) indicate that these cilia may provide the water current through the organ.

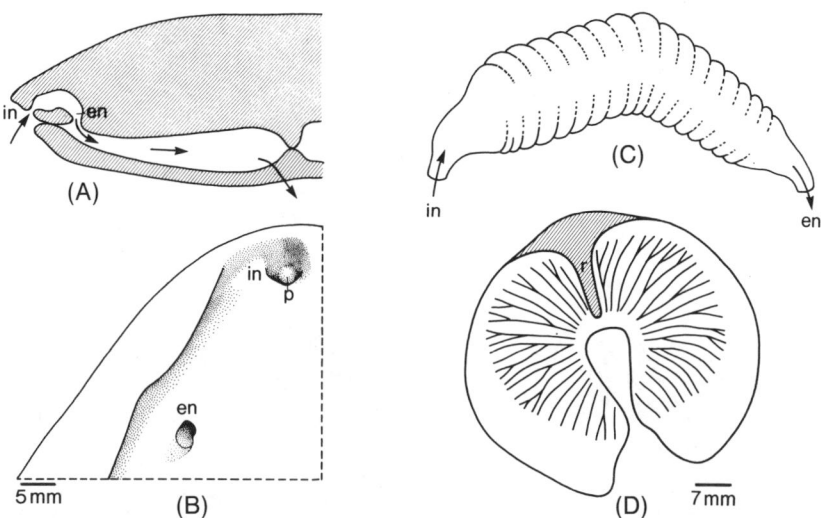

Fig. 2.4 Olfactory organs of lungfish. (A) Paramedian section through the head of *Protopterus annectens*; arrows indicate the olfactory water current during olfaction; (B) *P. annectens*, magnification of part of the roof of the mouth with nostrils; (C) *P. annectens*, olfactory organ; (D) *Neoceratodus forsteri*, a pair of olfactory lamellae. Abbreviations: en, excurrent nostril; in, incurrent nostril; p, pad of tissue; r, raphe. ((A) and (C) based on Derivot *et al.* (1979a) and Derivot (1984); (B) and (D) drawn from preserved specimens.)

In short-finned fish (Brachiopterygii) the olfactory organs are large and have a complicated structure, unique among fish (Pfeiffer, 1968; Theisen, 1970; Schulte and Holl, 1971). The incurrent nostril is found at the end of a nasal tube and the valvular excurrent nostril lies just in front of the eye (Fig. 2.5(A) and (B)). Each of the paired olfactory chambers is nearly fully surrounded by a nasal capsule. The olfactory chamber comprises a main and a small chamber (Fig. 2.5(A)). The main chamber has a compact central axis with five septa dividing it into sections (Fig. 2.5(C)). Both the small chamber and each section of the main chamber lodge two rows of olfactory lamellae separated by a mushroom-shaped lumen. Each section may be compared with the olfactory chamber of the general fish type, with the septa of each section corresponding to the walls of the olfactory chamber, and the central floor of the lumen may be compared with a low raphe (Fig. 2.5(C)). Beating of numerous kinocilia present on the lamellae and parts of the walls (Pfeiffer, 1968; Schulte and Holl, 1971) undoubtedly produces the water current through the olfactory organs.

Olfactory organs of ray-finned fish (Actinopterygii) are generally located dorsally in the snout. The structural variation is great and correlated with the diversity among teleostean species (Burne, 1909; Derscheid, 1924; Pipping, 1926, 1927; Matthes, 1934; Teichmann, 1954; Holl, 1965; Zeiske,

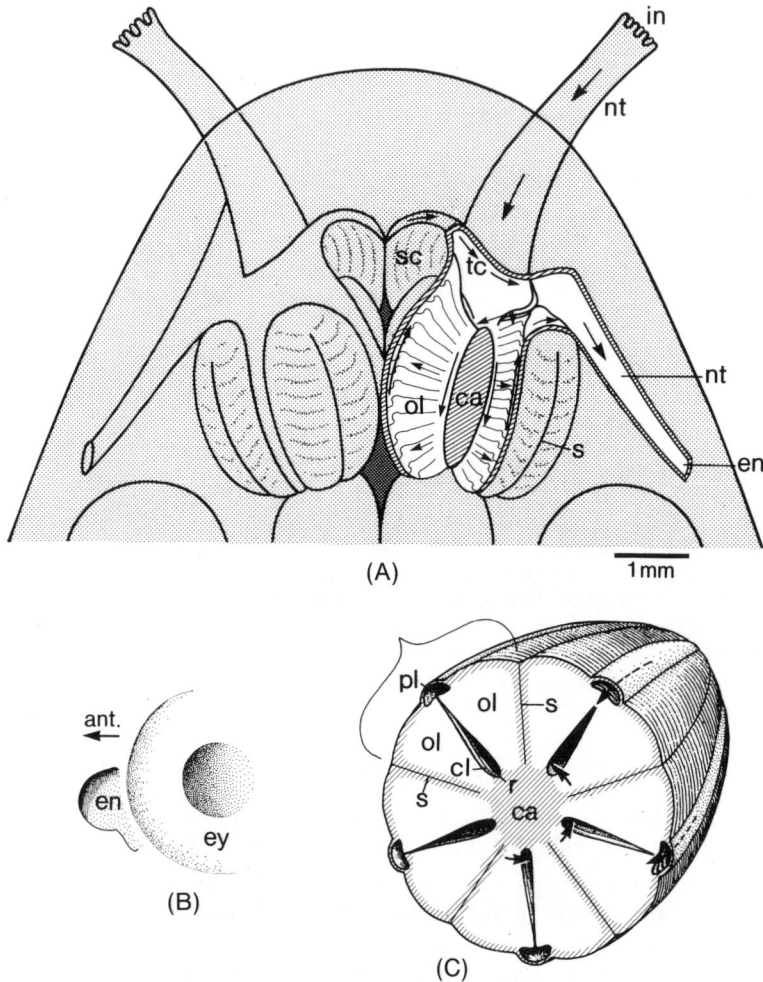

Fig. 2.5 Olfactory organs of *Erpetoichthys* (syn. *Calamoichthys*) *calabaricus*. (A) The olfactory organs in ventral view; the part shown in white has been opened and two rows of olfactory lamellae and a septum have been removed; (B) excurrent nostril (in front of the eye), lateral view; (C) posterior half of a main chamber, schematized; arrows indicate water currents; brace encloses a section comparable with the olfactory chamber of a general fish. Abbreviations: ant., anterior; ca, central axis; cl, central lumen; en, excurrent nostril; ey, eye; in, incurrent nostril; nt, nasal tube; ol, olfactory lamella; pl, peripheral lumen; r, raphe; s, septum; sc, small chamber; tc, transverse canal. (Modified from Theisen (1970) and drawn from preserved specimens.)

1973, 1974; Hara, 1975; Yamamoto and Ueda, 1979c; Yamamoto, 1982). In general each olfactory organ is provided with two nostrils with varying distances between them (Fig. 2.6). In many forms the incurrent nostril is situated at the end of a tubule of varying length (Fig. 2.6 (B), (D), (E), (G)),

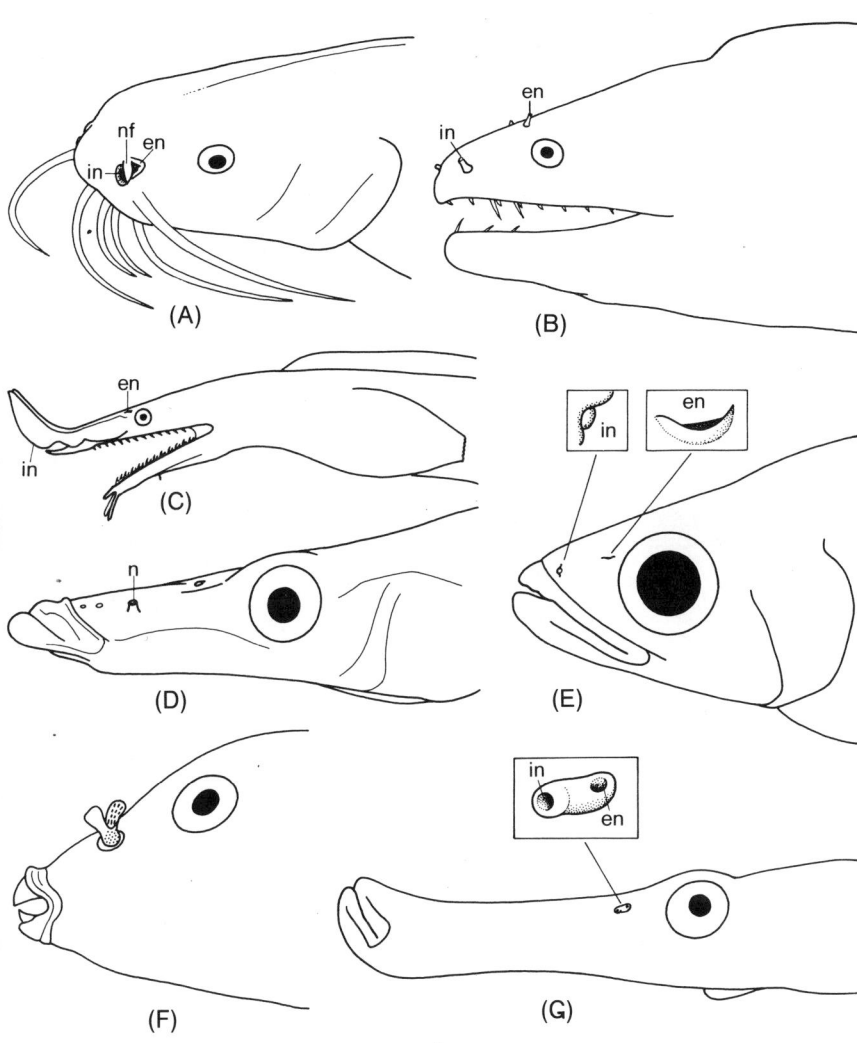

Fig. 2.6 Examples of nostrils of ray-finned fish showing different shapes and positions. Not drawn to scale. (A) Hardhead sea catfish, *Arius felis*; (B) giant moray, *Lycodontis javanicus*, from a photograph by Ingvar Bundgaard Jensen; (C) moray eel, *Rhinomuraena ambonensis*, based on Holl et al. (1970); (D) fifteenspine stickleback, *Spinachia spinachia*, redrawn from Theisen (1982); (E) silverside, *Bedotia geayi*, redrawn from Melinkat and Zeiske (1979); (F) parrotfish, *Arothron (Tetraodon) nigropunctatus*, based on Wiedersheim (1887); (G) pipefish, *Siphonostoma typhle*, from a preserved specimen. Abbreviations: en, excurrent nostril; in, incurrent nostril; n, nostril; nf, nasal flap.

but it can also be funnel-shaped (Fig. 2.6(A) and (C)) or just a simple opening (e.g. some hatchetfish; Derscheid, 1924). In certain groups the posterior border of this nostril is prolonged to form an upright nasal flap (e.g. Cyprinidae; Burne, 1909) or an inward extension. The excurrent nostril

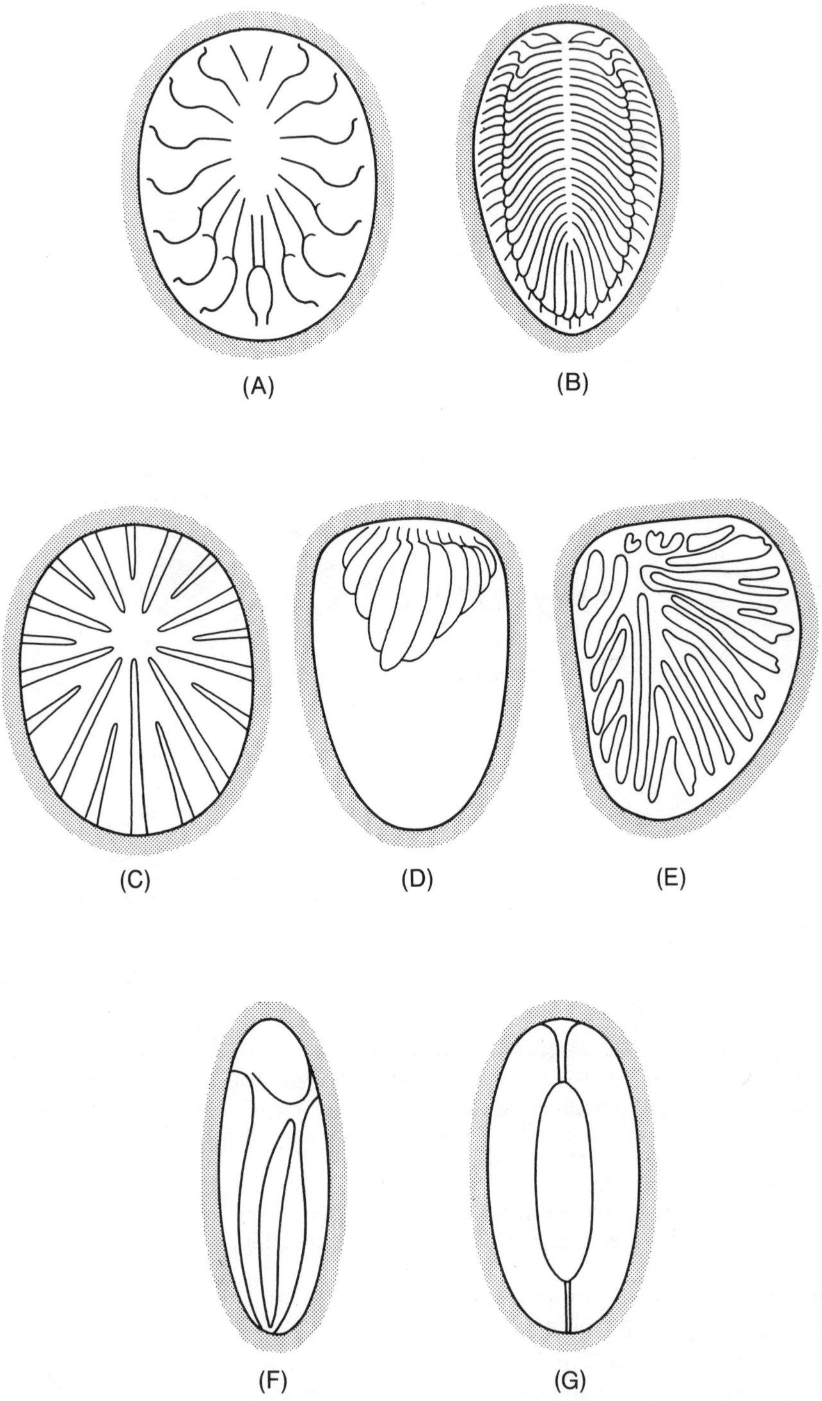

20 Structure, development, and evolutionary aspects

(A)

(B)

(C)

(D)

(E)

(F)

(G)

mostly lies flush with the skin of the snout. A valve may be present, mostly in forms possessing an olfactory ventilation sac (Fig. 2.6(C) and (E)). In a few cases this nostril is tubular (Fig. 2.6(B)) or provided with a nasal flap (Fig. 2.6(A)). In some groups each olfactory organ possesses a single nostril (Fig. 2.6(D)). The olfactory chamber may also be totally open to the exterior, with an olfactory papilla extending from the floor of the chamber (Exocoetoidei; Theisen et al., 1980). In some tetraodontiform fish, nostrils are lacking and each organ may appear as a bifid (Fig. 2.6(F)) or solid tentacle-like structure (Su and Li, 1982). Besides external nostrils, a few benthic teleosts (stargazers, *Astroscopus, Ichthyoscopus*, Dahlgren, 1908, 1927; Mees, 1962; *Gymnodraco acuticeps*, Jakubowski, 1975) also possess internal nostrils, connecting the olfactory chambers to the oral cavity.

The olfactory rosette shows great variability in the arrangement, shape and number of olfactory lamellae. In the majority of teleosts, the rosette is oval and consists of a row of lamellae on each side of a midline raphe (Fig. 2.7(A) and (B)). In some forms the lamellae are arranged in a circle (Fig. 2.7(C)), in a semicircle (snakefish, *Trachinocephalus myops*; Yamamoto and Ueda, 1978c), lie parallel to one another (Fig. 2.7(D)) or form irregular patterns (Fig. 2.7(E)). Rosettes may have few (Fig. 2.7(F)) to many lamellae, and lamellae may also be totally absent, as in the pipefish, *Siphonostoma typhle* (Liermann, 1933) or in clingfish (Yamamoto and Ueda, 1978d). The sand eel, *Ammodytes*, possesses a kidney-shaped bulge (Fig. 2.7(G)). Although in some forms the lamellae are only slightly raised folds, in most species they are true lamellae of different shape which to a certain extent is correlated with the different types of ventilation. Olfactory rosettes grow by increasing number and size of their lamellae.

A sexual dimorphism of the olfactory organs is found in some ceratioid angler fish (Bertelsen, 1951, 1980), stomiiform fish (Marshall, 1967), lophiid angler fish (Caruso, 1975), gulper eels (Saccopharyngidae; Nielsen and Bertelsen, 1985), and in the deep-sea eel family Monognathidae (Bertelsen and Nielsen, 1987). The olfactory organs are weakly developed in females, whereas they are very large in males of these families. This is thought to be an adaptation enabling the male to find the female.

In many teleosts the olfactory chamber extends to form one or more olfactory ventilation sacs (Burne, 1909; Liermann, 1933), which are primarily a device for ventilating the olfactory organs. They have also been

---

Fig. 2.7 Examples of olfactory rosettes in ray-finned fish. Not drawn to scale. (A) Carp, *Carassius auratus*, based on Breiphol et al. (1974a); (B) catfish, *Silurus glanis*, based on Jakubowski and Kunysz (1979); (C) pike, *Esox lucius*, based on Holl (1965); (D) striped eel catfish, *Plotosus lineatus*, from a preserved specimen; (E) swordtail, *Xiphophorus helleri*, based on Zeiske (1973); (F) fifteenspine stickleback, *Spinachia spinachia*, based on Theisen (1982); (G) sand eel, *Ammodytes personatus*, based on Yamamoto and Ueda (1979c).

proposed to have a secretory function (Kyle, 1900; Burne, 1909), contributing to the efficiency of the olfactory organs (Jakubowski and Kunysz, 1979) or helping in the removing of debris (Kapoor and Ojha, 1972; Døving et al., 1977; Theisen, 1982). The ventilation of the olfactory organs in the ray-finned fish is achieved by different mechanisms (Burne, 1909; Pipping, 1926, 1927; Berghe, 1929; Liermann, 1933; Kleerekoper, 1969; Oelschläger, 1976; Døving and Thommesen, 1977; Døving et al., 1977; Melinkat and Zeiske, 1979; Kux et al., 1988; Nevitt, 1991).

In many forms the water current through the olfactory organs is created by the action of cilia (isosmates; Døving et al., 1977). In other forms the water current is created by expansion and compression of one or more olfactory ventilation sacs (cyclosmates; Døving et al., 1977). The ventilation sacs are not provided with muscles, and the pulsatile movements are created by transmission of respiratory pressure changes in the oral cavity to the sac or by movements of cranial bones which activate the sac. A unidirectional water current is secured by some kind of valve mechanism. In some pelagic forms, forward motion of the fish creates the water current through or along the olfactory organ. Some are ditrematous, with large nostrils and an upright nasal flap (e.g. hardhead sea catfish, *Arius felis*, Fig. 2.6(A)), whereas others have an open cavity with a protruding olfactory papilla (e.g. needlefish, *Belone belone*; Theisen el al., 1980). In many sharks the water current is considered to be created by forward motion, although the olfactory lamellae are provided with kinocilia (Theisen et al., 1986; Zeiske et al., 1987). The hardhead sea catfish also has kinocilia, which may aid in creating the water current (Theisen et al., 1989). The tench, *Tinca tinca*, has kinocilia, but swimming may increase the flow of water through the olfactory organ (Døving et al., 1977). In the catfish, *Silurus glanis*, ciliary action causes the water current, but small olfactory ventilation sacs are present (Døving et al., 1977). Thus, although different ventilatory types can be identified, combined use of the mechanisms generally facilitates the water flow over the olfactory epithelium.

2.3 HISTOLOGY

Sensory and nonsensory epithelium

The olfactory lamella is lined with sensory and nonsensory epithelium; these epithelia show great variability in their arrangement and distribution. Both epithelia are covered with a mucous layer and rest on a basal lamina and the underlying lamina propria.

Sensory epithelium

The sensory epithelium is a columnar pseudostratified epithelium, up to 110 µm thick in the pike, *Esox lucius* (Holl, 1965), and consists of three main components: receptor, supporting and basal cells (Fig. 2.8). The receptor cells are bipolar neurones with their cell bodies (perikarya) arranged in some layers throughout the broad midregion of the epithelium. They send a single dendrite to the surface and a thin axon in proximal direction. While passing through the basal part of the epithelium, axons aggregate to form larger bundles before they leave the epithelium and enter the lamina propria. The supporting cells extend from the epithelial surface down to the basal lamina. Their nuclei are found in the lower third of the epithelium below the nuclear region of the receptor cells. Reverse stratification of the nuclei of receptor and supporting cells is reported from some fish species (Schulte, 1972; Schulte and Riehl, 1978). The basal cells are the smallest cells of the sensory epithelium and are scattered among the supporting cells at the base of the epithelium.

The receptor cells are separated from each other by supporting cells. However, receptor cell dendrites may also juxtapose each other by appositions or cell projections called spines (Breipohl, 1974; Breipohl et al., 1974b; Breucker et al., 1979; Zeiske et al., 1979; Theisen et al., 1980). Interreceptor contacts are also present between the perikarya. Junctional complexes consisting of tight junction, adherent junction and desmosome extend down from the epithelial surface along the cellular membranes. The junctions between supporting and receptor cells may differ from those of supporting cells by their asymmetrical configuration (Bannister, 1965; Zeiske et al., 1976b). Below the junctional complex, supporting cells interdigitate with supporting, basal and receptor cells, and with the basal lamina.

Nonsensory epithelium

The term 'indifferent epithelium' is frequently used, but should be avoided, because, contrary to long-lasting assumption, the nonsensory epithelium does not differentiate into the sensory. Nonsensory epithelium exhibits ciliated and nonciliated surface areas (Fig. 2.8). The epithelial cell, the basic component of the nonsensory epithelium, contains numerous filaments, some of which are associated with desmosomes (Henrikson and Matoltsy, 1968; Hawkes, 1974; Whitear, 1986). Nonciliated areas form a stratified squamous epithelium. Its general structure is identical to that of the external eipdermis from which it is derived. The surface has microridges arranged in fingerprint-like patterns (Yamada, 1968; Uehara et al., 1988, 1991), which are covered with a glycocalyx. These are also characteristic of the free surface of the fish epidermis, especially in teleosts. Several possible functions of the microridges, such as imparting epithelial rigidity or plasticity or adhesion of the mucous layer, have been suggested. The ciliated epithelial cells are scattered throughout the nonsensory and/or sensory epithelium as single cells or patchy clusters, probably facilitating olfactory ventilatory processes.

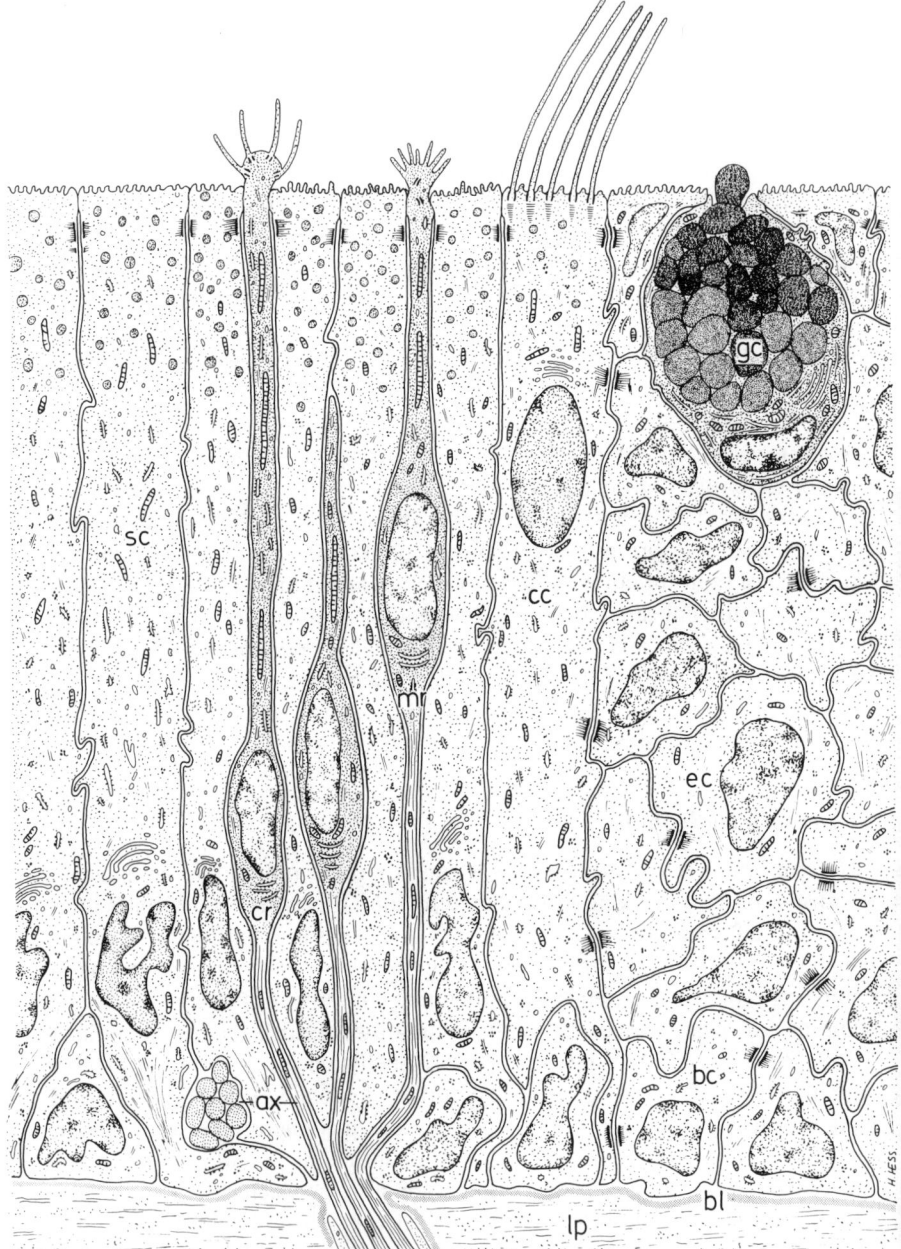

Fig. 2.8 Simplified diagram of the olfactory epithelium of teleosts, showing sensory and nonsensory epithelium. Abbreviations: ax, axon; bc, basal cell; bl, basal lamina; cc, ciliated nonsensory cell; cr, cilated receptor cell; ec, epithelial cell; gc, goblet cell; lp, lamina propria; mr, microvillous receptor cell; sc, supporting cell.

Sensory and nonsensory cells

Receptor cells

Olfactory receptor cells are bipolar neurones, either ciliated or microvillous, as demonstrated by means of degeneration and regeneration techniques (Ichikawa and Ueda, 1977; Cancalon, 1982). Microvillous receptor cells have been considered precursors of the ciliated receptor type (Bannister, 1965; Bakhtin, 1976, 1977). However, ontogenetic investigations, as well as studies on degeneration and regeneration, have shown that both cell types represent independent entities (Pyatkina, 1976; Ichikawa and Ueda, 1977; Breucker et al., 1979; Jakubowski, 1981; Cancalon, 1982; Evans et al., 1982; Zielinski and Hara, 1988; Hara and Zielinski, 1989). Ciliated and microvillous olfactory neurones are present in almost all teleosts (Yamamoto, 1982; for exceptions see Rowley and Moran, 1983; Fisher et al., 1984). Elasmobranchs exhibit the microvillous type only (Reese and Brightman, 1970; Holl, 1973b; Bronshtein, 1976; Theisen et al., 1986; Zeiske et al., 1987). While both receptor types have been recognized in hagfish (Theisen, 1973, 1976) and the African lungfish, *Protopterus annectens* (Derivot et al., 1979b), the river lamprey, *Lampetra fluviatilis*, shows only the ciliated type (Thornhill, 1967), and the Australian lungfish, *Neoceratodus forsteri*, only the microvillous receptor type (Theisen, 1972).

Both cell types have many cytological features in common: a granular endoplasmic reticulum arranged in parallel arrays (ergastoplasm) in close proximity to the nucleus, a Golgi complex above the ergastoplasm, and free ribosomes scattered throughout the perikaryon. The axon contains smooth endoplasmic reticulum and microtubules. Characteristics of the dendrite are longitudinally orientated microtubules and long mitochondria. At the level of the terminal web in the surrounding supporting cells, the dendrite forms a terminal swelling, the olfactory knob, protruding slightly above the epithelial surface. Intramembrane particles are found on fracture faces of the membrane of ciliated and microvillous olfactory dendrites. These particles are considered integral membrane proteins and possible candidates for different olfactory receptor molecules (Hernádi and Röhlich, 1988).

Ciliated receptor cells have cilia up to 10 μm long projecting laterally from the knob (Fig. 2.9(A) and (B)). Olfactory cilia normally contain a $9+2$ axonemal complex (Yamamoto, 1982), or a variation of a $9+0$ set in some cyprinodontiform fish including killifish, rainbow fish, *Melanotaenia* (syn. *Nematocentris*) *maccullochi*, and needlefish (Zeiske et al., 1976b, 1979; Theisen et al., 1980), and in hagfish (Theisen, 1973, 1976). Dynein arms are usually present in the cilia of the $9+2$ type, whereas they are missing in the $9+0$ type. Ciliary rootlets are seldom observed and appear less developed than in nonsensory cells (Schulte and Holl, 1971; Schulte, 1972; Zeiske et al.,

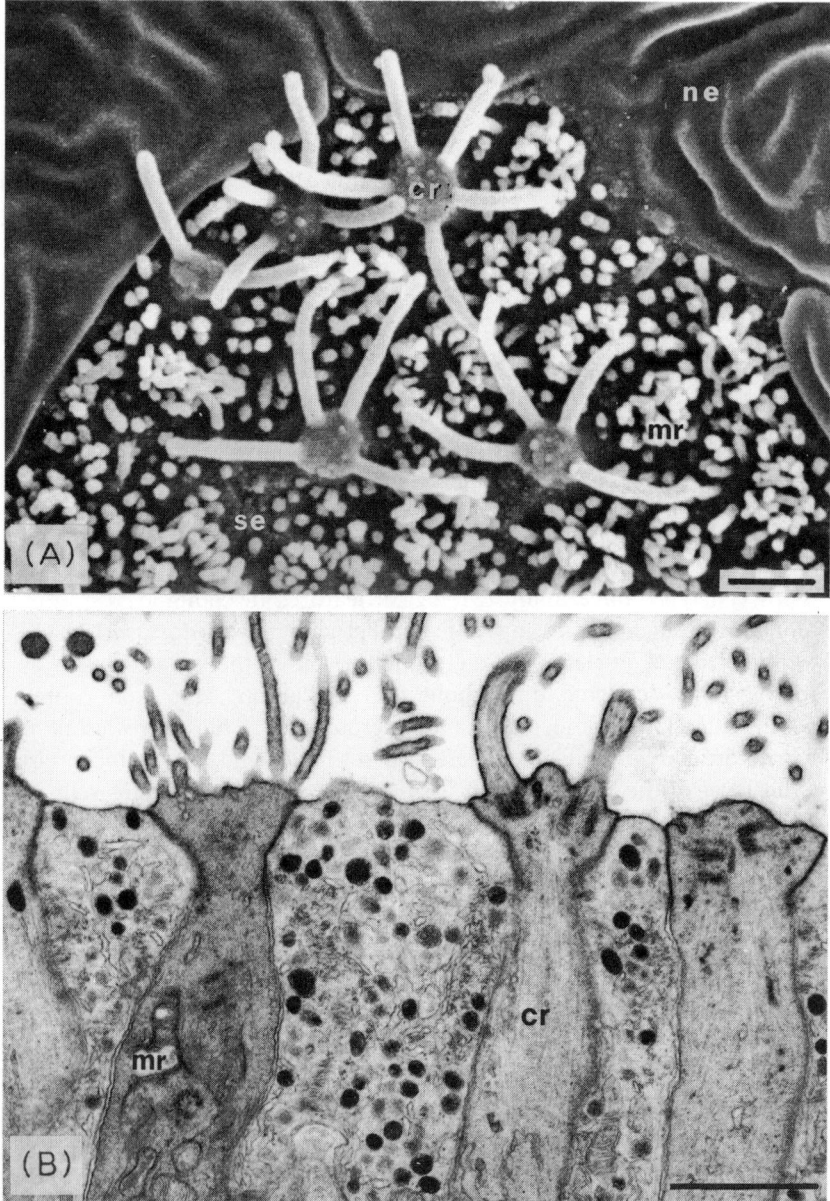

Fig. 2.9 Ciliated (cr) and microvillous (mr) receptor cells. (A) Scanning and (B) transmission electron micrographs; ne, nonsensory epithelium; se, sensory epithelium. Scale bars, 1 μm.

1979). There is still controversy concerning whether the receptor-cell cilia are actively motile.

Microvillous receptor cells have a flattened free surface and bear up to 80 microvilli (Fig. 2.9(A) and (B)). The microvilli are up to 5 μm long and possess a number of centrioles buried at various depths in the cytoplasm of the dendrite, an observation which led to the assumption that this cell type was an immature form of the ciliated receptor cell (Bannister, 1965). Some investigators suggest that the microvillous receptor type was derived phylogenetically from the ciliated one.

Cells with a rod-shaped protrusion have been observed and sometimes considered a distinct receptor cell type. Rod cells containing bundles of closely packed filaments (Bannister, 1965; Schulte, 1972) were found to remain unaffected by olfactory nerve-bundle transection, disproving that this cell type could be an olfactory neurone (Ichikawa and Ueda, 1977). Rod-shaped protrusions containing one or more basal bodies, each of which emits a set of an axonemal complex or randomly arranged microtubules (giant cilia, according to Theisen, 1973) (Yamamoto and Ueda, 1977; Theisen et al., 1980; Delfino et al., 1981; Rhein et al., 1981; Yamamoto, 1982; Muller and Marc, 1984), are considered artefacts of ciliated receptor-cell endings produced by degeneration or severe impact such as deleterious effects of water pollution or fixation (Rowley and Moran, 1985).

Ciliated receptor cells dominate in the olfactory epithelium of migratory sturgeons, whereas in stationary species microvillous receptor cells are relatively abundant (Pyatkina, 1976). Also, differential distribution of the ciliated and microvillous receptor cells has been reported from the olfactory lamellae of the eel catfish, *Plotosus lineatus* (Yamamoto and Ueda, 1978b; Theisen et al., 1991), channel catfish, *Ictalurus punctatus* (Cancalon, 1982; Erickson and Caprio, 1984) and salmonids (Thommesen, 1982, 1983). Ontogenetic studies on the olfactory organ of sturgeon, rainbow fish, and rainbow trout, *Oncorhynchus mykiss* (formerly *Salmo gairdneri*), revealed that ciliated receptor cells appear prior to microvillous ones (Pyatkina, 1976; Breucker et al., 1979; Evans et al., 1982; Zielinski and Hara, 1988).

Available data indicate remarkably high frequencies of olfactory receptor cells (400 000–500 000 mm^{-2}) in two species of killifish (cyprinodonts) (Zeiske et al., 1976b; Yamamoto, 1982). There are other fish species with lower receptor cell density (e.g. 200 000–250 000 mm^{-2} in rainbow fish, Zeiske et al., 1979; 24 000 mm^{-2} in salmonids, Thommesen, 1983). A correlation between olfactory acuity and receptor cell number has not yet been established.

Supporting cells

Supporting cells are generally restricted to the sensory epithelium. At their free surface, supporting cells give off microvilli or microvillous-like protrusions. In the river lamprey and the reedfish, *Erpetoichthys* (syn. *Calamoichthys*) *calabaricus*, supporting cells are found bearing cilia in addition to microvilli (Thornhill, 1967; Schulte and Holl, 1971). Bertmar (1973) described ciliated and nonciliated supporting cells in Baltic sea trout, *Salmo trutta trutta*. Most of the cells described as ciliated supporting cells, however, are identical to ciliated nonsensory cells in structure. Characteristic internal features of the supporting cells are bundles of tonofilaments, generally longitudinally arranged. But bundles perpendicular to these are also present, interconnecting the desmosomes of different junctional complex. At the level of the intermediate junctions, there are also horizontally-running filaments which form a terminal web slightly below the free surface. In the uppermost region of the cell, vesicles may be accumulated, which vary in size and electron density and arise from a prominent Golgi complex. Adjacent to the Golgi complex, profiles of granular endoplasmic reticulum, free ribosomes, and mitochondria are recognizable. Single centrioles may occur. The nucleus is oval or lobed, sometimes to a high degree. Several functions of the supporting cells have been suggested: secretory, absorptive and glia function (Dogiel, 1887; Moulton and Beidler, 1967; Thornhill, 1967; Schulte and Holl, 1971; Theisen, 1972, 1973; Bertmar, 1973; Breipohl et al., 1976; Zeiske et al., 1976b; Yamamoto and Ueda, 1978a; Theisen et al., 1980; Delfino et al., 1981).

Basal cells

These relatively small cells have a round-to-oval nucleus and contain tonofilaments. Basal cells may be stem cells for regeneration of lost or damaged ciliated nonsensory and mucous cells (Möller et al., 1989). Ongoing cell dynamics and regeneration after degeneration, indicated by mitotic activities of basal cells, are found in the olfactory epithelium of fish (Thornhill, 1967; Zeiske et al., 1976b, 1979; Evans et al., 1982). Thus, basal cells (blastema cells) may also be precursors of receptor cells or, at a lower rate, of supporting cells (further discussion: Thornhill, 1970; Graziadei, 1971; Graziadei and Monti Graziadei, 1976, 1978; Cancalon, 1982; Breipohl et al., 1986; Shepherd, 1988; Yamagishi et al., 1989; Suzuki and Takeda, 1991).

Goblet cells

Goblet cells are abundant in the stratified squamous epithelium and are mostly absent from the sensory epithelium (Fig. 2.8). They have a flattened nucleus, a prominent Golgi complex, and a granular endoplasmic reticulum. Mature goblet cells are packed with positive Alcian blue and periodic-acid-Schiff-reaction (PAS) secretory vesicles, indicating the presence of acidic mucopolysaccharides. The secretory droplets are released through the stomatum of the cell at the surface of the epithelium.

Ciliated nonsensory cells

Ciliated nonsensory cells are cylindrical and extend from the base to the free surface of the epithelium, where they bear numerous cilia and microvilli (Fig. 2.8). The cilia are typical kinocilia, containing the characteristic $9+2$ axonemal structure; here each A-subfibre of the nine outer doublets is provided with dynein arms. The complex rootlet system connecting basal apparatuses indicates a deep anchorage of each cilium for generating effective strokes and possibly coordination transmitted by the connecting fibres (Holley, 1984; González Santander et al., 1984). It is assumed that these cilia propel both water and mucus (Sleigh et al., 1988; Sleigh, 1989).

2.4 DEVELOPMENT

The development of the olfactory organ begins with the appearance of an epithelial thickening, the olfactory placode, in the ectoderm on each side of the head in front of the eye, except in monorhinal cyclostomes, where a single anlage occurs (Kleerekoper, 1969). The olfactory placode starts developing very early in ontogeny. About 24 h after fertilization in zebrafish, *Brachydanio rerio* (Wilson et al., 1990), the olfactory placode is connected to the brain by axons of the olfactory nerve. The following step is the development of an olfactory pit.

The formation of the nostrils varies in different groups. In salmonids (Gawrilenko, 1910; Reinke, 1937; Teichmann, 1964; Jahn, 1972; Elston et al., 1981; Zielinski and Hara, 1988; Hara and Zielinski, 1989), the primary olfactory groove orientated in the anterior–posterior axis of the head elongates further, as its midregion extensions arise from the superficial epithelial borders (Fig. 2.10). The extensions form an isthmus constricting the primary opening in a dumb-bell shape (Fig. 2.10, lower right). Finally, an epithelial bridge separates two apertures at the extreme ends of the primary groove, the incurrent and excurrent nostrils. The salmonid type of nostril formation is widespread in various fish groups and is considered to be phylogenetically the primary type (Reinke, 1937; Breucker et al., 1979).

A second type of nostril formation, characterized by separate development of the two apertures a priori, is found in killifish (cyprinodonts) (Reinke, 1937; Zeiske et al., 1976a). In a third type, found in the needlefish, *Belone belone* (Sewertzoff, 1931; Theisen et al., 1980), the olfactory anlage gradually elongates into a triangular groove. As the juvenile grows, a small spherical elevation appears on the floor of the groove and develops into a larger, oval projection, the papilla, orientated transversely to the swimming direction. The olfactory cavity always remains open.

It has been shown for sturgeon (Pyatkina, 1976), rainbow fish (Breucker et al., 1979), and rainbow trout (Evans et al., 1982; Zielinski and Hara,

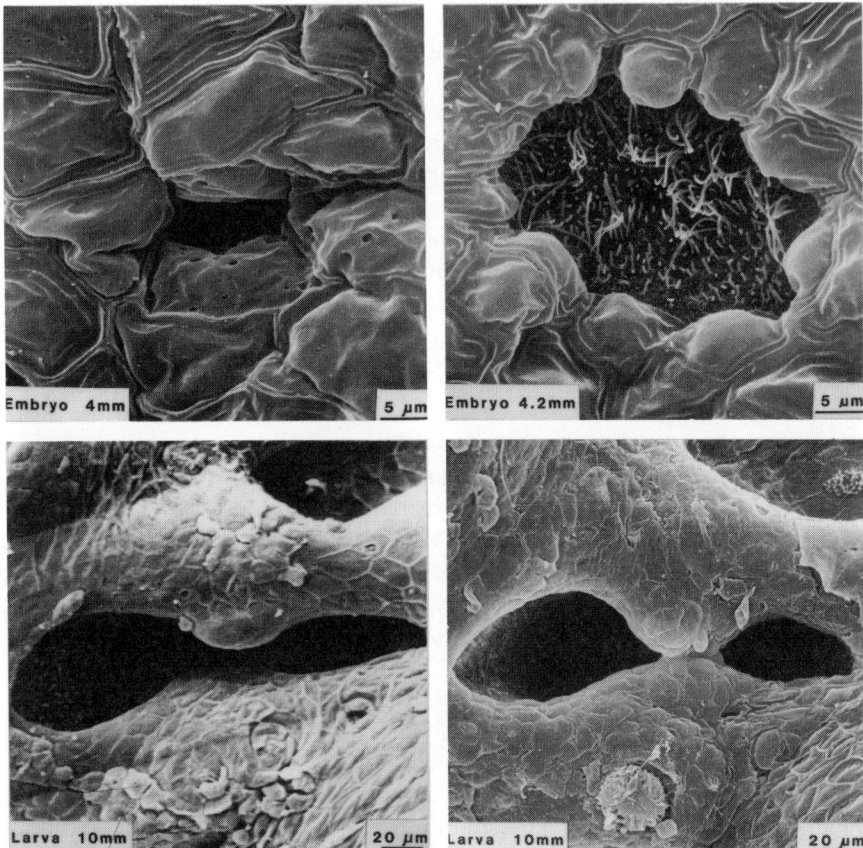

Fig. 2.10 Nostril formation (salmonid type). See text for explanation.

1988; Hara and Zielinski, 1989) that during early larval stages ciliated receptor cells appear slightly ahead of microvillous receptors. These ontogenetic studies unequivocally support the hypothesis that ciliated and microvillous receptor cells are distinct types of olfactory neurones.

2.5 ADAPTATION AND EVOLUTION

The hypothesis that the olfactory organ is a branchial sense organ, phylogenetically derived from a pair of anterior gill clefts, and adjoined epibranchial placodes (Bjerring, 1989), can be traced back to Beard (1886), Marshall (1879) and Dohrn (1875). This idea is associated with theories on the segmentation of the head of vertebrates. Other investigators doubt that the olfactory organs of vertebrates were originally related to visceral openings. According to the Haeckelian biogenetic law, they suggest that the olfactory placode is an ontogenetic recapitulation of an ancestral sense organ (Lubosch, 1905), or in other words that the olfactory organ of ancestral vertebrates had a superficial position because it is characteristic of early ontogenetic development in recent vertebrates (Bertmar, 1969; Holl, 1973a).

Special attention has been paid to the olfactory openings in fish with respect to the origin of the choana and, thus, the origin of tetrapods (Panchen, 1967; Schmalhausen, 1968; Bertmar, 1969; Rosen et al., 1981). Conflicting statements exist on the choanate or pseudochoanate character of the posterior nostrils of lungfish.

The olfactory organ in recent fish demonstrates a great variety in its morphological characteristics. The variety is correlated with the enormous diversity among fish species adapted to different environmental conditions, and is a consequence of multiple evolutionary pathways. A considerable diversity in morphological structures of the olfactory organ can be found even in smaller systematic entities, for example in tetraodonts, perciform and atheriniform fish (Zeiske, 1973, 1974; Zeiske et al., 1976b, 1979; Yamamoto and Ueda, 1979a,b,c; Theisen et al., 1980; Su and Li, 1982). Different ecostructural adaptations in these fish were achieved by selection pressure to optimize properties relevant to hydrodynamics and other demands. The arrangements of the olfactory epithelium demonstrates the effective use of the space available in the olfactory chamber, and therefore represents adaptation which maximizes the sensory area under the given geometrical restrictions (Bertmar, 1972). This ensures that the largest possible number of odour molecules will come into contact with the sensory epithelium.

Fish with exposed and widely opened olfactory organs, and those with organs that are located in a roofed cavity with incurrent and excurrent nostrils, have been grouped together in a single order on the basis of osteological evidence (Atheriniformes *sensu* Rosen, 1964). A classification based, however, on the morphology of the olfactory organ would support a systematic revision (for critical remarks see also Yamamoto and Ueda, 1979a). The use of general form and structure of the olfactory organ as a systematic character of fish is not new. It has been employed for example in classification of flatfish (Norman, 1934) and identification of tunas (Iwai and Nakamura, 1964). Pfeiffer (1969) discusses relationships between Elasmobranchii, Actinopterygii, Brachiopterygii, Crossopterygii, and Dipnoi on the basis of the structure of their olfactory organs. However, sexual dimorphism of the olfactory organ (Marshall, 1967; Caruso, 1975) and convergences (multiple reduction of the olfactory rosette, widespread occurrence of ventilatory nasal sacs and monotrematous olfactory organs) limit drastically the systematic significance of the olfactory organ in fish.

2.6 SUMMARY AND CONCLUSION

This chapter describes the anatomy and histology of the olfactory organ in fish, including cyclostomes. The olfactory organs exhibit great variety in their morphological characteristics. This variety is correlated with the diversity found among fish. The olfactory epithelium has a basic histological structure consisting of receptor cells (bipolar neurones), supporting cells and basal cells, but shows considerable group-specific differences in composition and arrangement. Aspects of development, growth, adaptation and evolution of the olfactory organ are considered. Advanced studies on functional morphology and development of the olfactory organ in fish are still necessary to improve our understanding of adaptation and evolution of this chemical sense organ.

ACKNOWLEDGEMENT

For skilful help with the drawings we are grateful to Mrs Beth Beyerholm, Copenhagen.

REFERENCES

Atz, J.W. (1952) Narial breathing in fishes and the evolution of internal nares. *Q. Rev. Biol.*, **27**, 366–77.

Bakhtin, E.K. (1976) The morphology of the olfactory organ in some fish species and its functional interpretation. *Vop. Ikhtiol.*, **16**, 867–88. (In Russian)

Bakhtin, E.K. (1977) Peculiarities of the fine structure of the olfactory organ of *Squalus acanthias*. *Tsitologiya*, **19**, 725–31. (In Russian)

Bannister, L.H. (1965) The fine structure of the olfactory surface of teleostean fishes. *Q.J. microsc. Sci.*, **106**, 333–42.

Beard, J. (1886) The system of branchial sense organs and their associated ganglia in Ichthyopsida. A contribution to the ancestral history of vertebrates. *Q.J. microsc. Sci.*, **26**, 95–156.

Bertelsen, E. (1951) The ceratoid fishes. *Dana Rep.*, **39**, 1–276.

Bertelsen, E. (1980) Notes on Linophrynidae V: a revision of the deepsea anglerfishes of the *Linophryne arbrifera*-group (Pisces, Ceratioidei). *Steenstrupia*, **6**, 29–70.

Bertelsen, E. and Nielsen, J.G. (1987) The deep sea eel family Monognathidae (Pisces, Anguilliformes). *Steenstrupia*, **13**, 141–98.

Bertmar, G. (1965) The olfactory organ and upper lips in Dipnoi, an embryological study. *Acta zool., Stockh.*, **46**, 1–40.

Bertmar, G. (1969) The vertebrate nose, remarks on its structural and functional adaptation and evolution. *Evolution*, **23**, 131–52.

Bertmar, G. (1972) Ecostructural studies on olfactory organ in young and adult sea trout (Osteichthyes, Salmonidae). *Z. Morph. Tiere*, **72**, 307–30.

Bertmar, G. (1973) Ultrastructure of the olfactory mucosa in the homing Baltic sea trout *Salmo trutta trutta*. *Mar. Biol.*, **19**, 74–88.

Bjerring, H.C. (1989) Apertures of craniate olfactory organs. *Acta zool., Stockh.*, **70**, 71–85.

Bone, Q. and Marshall, N.B. (1982) *Biology of Fishes*, Blackie, London, 262 pp.

Breipohl, W. (1974) Licht- und elektronenmikroskopische Befunde zur Struktur der Regio olfactoria des Goldfisches. *Verh. anat. Ges. Jena*, **68**, 479–85.
Breipohl, W., Bijvank, G.J. and Pfefferkorn, G.E. (1974a) Scanning electron microscopy of various sensory receptor cells in different vertebrates, in *Scanning Electron Microscopy*. Vol. VII (eds O. Johari and J. Gorvin), JJT-Research Institute, Chicago, pp. 557–64.
Breipohl, W., Laugwitz, H.J. and Bornfeld, N. (1974b) Topological relations between the dendrites of olfactory sensory cells and sustentacular cells in different vertebrates. An ultrastructural study. *J. Anat.*, **117**, 89–94.
Breipohl, W., Zippel, H.P., Rückert, K. and Oggolter, H. (1976) Morphologische und elektrophysiologische Studien zur Struktur und Funktion des olfaktorischen Systems beim Goldfisch unter normalen und experimentellen Bedingungen. *Beitr. Elektronenmikrosk. Direktabb. Oberfl.*, **9**, 561–84.
Breipohl, W., Mackay-Sim, A., Grandt, D., Rehn, B. and Darrelmann, C. (1986) Neurogenesis in the vertebrate main olfactory epithelium, in *Ontogeny of Olfaction* (ed. W. Breipohl) Springer-Verlag, Berlin, pp. 21–33.
Breucker, H., Zeiske, E. and Melinkat, R. (1979) Development of the olfactory organ in the rainbow fish *Nematocentris maccullochi* (Atheriniformes, Melanotaeniidae). *Cell Tissue Res.*, **200**, 53–68.
Bronshtein, A.A. (1976) Some peculiarities of fine structure of the olfactory organ in elasmobranchs. *Zh. evol. Biokhim. Fiziol.*, **12**, 63–7. (In Russian)
Burne, R.H. (1909) The anatomy of the olfactory organs in teleostean fishes. *Proc. zool. Soc. Lond.*, **1909**, 610–63.
Bütschli, O. (1921) *Vorlesungen über vergleichende Anatomie.* Vol. I, Springer-Verlag, Berlin, 946 pp.
Cancalon, P. (1982) Degeneration and regeneration of olfactory cells induced by $ZnSO_4$ and other chemicals. *Tissue Cell*, **14**, 717–33.
Caruso, J.H. (1975) Sexual dimorphism of the olfactory organs of lophiids. *Copeia*, **1975**, 380–1.
Dahlgren, U. (1908) The oral opening of the nasal cavity in *Astroscopus*. *Science, Wash., D.C.*, **27**, 993–4.
Dahlgren, U. (1927) The life history of the fish *Astroscopus* (the 'stargazer'). *Sci. Mon.*, **24**, 348–65.
Delfino, G., Bianchi, S. and Ercolini, A. (1981) On the olfactory epithelium in cyprinids: a comparison between hypogean and epigean species. *Monitore zool. ital. (N.S.), Suppl.*, **14**, 153–80.
Derivot, J.H. (1984) Functional anatomy of the peripheral olfactory system of the African lungfish *Protopterus annectens* Owen: macroscopic, microscopic, and morphometric aspects. *Am. J. Anat.*, **169**, 177–92.
Derivot, J.H., Dupé, M. and Godet, R. (1979a) Anatomie functionelle de l'organe olfatif de *Protopterus annectens* Owen (Dipneustes): contribution à la connaissance du mécanisme d'irrigation de l'organe olfactif. *Acta zool., Stockh.*, **60**, 251–7.
Derivot, J.H., Mattei, X., Godet, R. and Dupé, M. (1979b) Etude ultrastructurale de la région apicale des cellules de l'épithélium olfactif de *Protopterus annectens* Owen (Dipneustes). *J. Ultrastruct. Res.*, **66**, 22–31.
Derscheid, J.M. (1924) Contribution à la morphologie céphalique des vertébrés. A: Structure de l'organe olfactif chez les poissons. *Annls Soc. r. zool. Belg.*, **54**, 79–162.
Dogiel, A. (1887) Ueber den Bau des Geruchsorganes bei Ganoiden, Knochenfischen und Amphibien. *Arch. mikrosk. Anat. EntwMech.*, **29**, 74–139.
Dohrn, A. (1875) *Der Ursprung der Wirbeltiere und das Princip des Functionswechsels*, Engelmann, Leipzig, 102 pp.

Døving, K.B. and Thommesen, G. (1977) Some properties of the fish olfactory system, in *Olfaction and Taste. VI* (eds J. Le Magnen and P. McLeod), Information Retrieval, London, pp. 175–83.

Døving, K.B., Dubois-Dauphin, M., Holley, A. and Jourdan, F. (1977) Functional anatomy in the olfactory organ of fish and the ciliary mechanism of water transport. *Acta zool., Stockh.,* **58**, 245–55.

Elston, R., Corazza, L. and Nickum, J.G. (1981) Morphology and development of the olfactory organ in larval walleye (*Stizostedion vitreum*). *Copeia,* **1981**, 890–3.

Erickson, J.R. and Caprio, J. (1984) The spatial distribution of ciliated and microvillous olfactory receptor neurons in the channel catfish is not matched by a differential specificity to amino acid and bile salt stimuli. *Chem. Senses,* **9**, 127–41.

Evans, R.E., Zielinski, B. and Hara, T.J. (1982) Development and regeneration of the olfactory organ in rainbow trout, in *Chemoreception in Fishes* (ed. T.J. Hara), Elsevier, Amsterdam, pp. 15–37.

Fisher, J.W., Mattie, D.R., Paulos, L.M. and Helton, C.D. (1984) Surface morphology of bluegill olfactory lamellae. *Trans. Am. microsc. Soc.,* **103**, 93–7.

Gans, C. and Northcutt, R.G. (1983) Neural crest and the origin of vertebrates: a new head. *Science, Wash. DC,* **220**, 268–74.

Gawrilenko, A. (1910) Die Entwicklung des Geruchsorgans bei *Salmo salar. Anat. Anz.,* **36**, 411–27.

González Santander, R., Martínez Cuadrado, G., Rubio Sáez, M., Bujan Varela, J. and Laraña Solé, A. (1984) Ultrastructural basis of the coordination of ciliary movement. *Acta anat.,* **118**, 82–90.

Graziadei, P.P.C. (1971) The olfactory mucosa of vertebrates, in *Handbook of Sensory Physiology.* Vol. IV (1) (ed. L.M. Beidler), Springer-Verlag, Berlin, pp. 29–58.

Graziadei, P.P.C. and Monti Graziadei, G.A. (1976) Olfactory epithelium of *Necturus maculosus* and *Ambystoma tigrinum. J. Neurocytol.,* **5**, 11–32.

Graziadei, P.P.C. and Monti Graziadei, G.A. (1978) Continuous nerve cell renewal in the olfactory system, in *Handbook of Sensory Physiology.* Vol. IX (ed. M. Jacobson), Springer-Verlag, Berlin, pp. 55–82.

Hara, T.J. (1975) Olfaction in fish. *Progr. Neurobiol.,* **5**, 271–335.

Hara, T.J. and Zielinski, B. (1989) Structural and functional development of the olfactory organ in teleosts. *Trans. Am. Fish. Soc.,* **118**, 183–94.

Hawkes, J.W. (1974) The structure of fish skin. I. General organization. *Cell Tissue Res.,* **149**, 147–58.

Henrikson, R.C. and Matoltsy, A.G. (1968) The fine structure of teleost epidermis. I. Introduction and filament-containing cells. *J. Ultrastruct. Res.,* **21**, 194–212.

Hernádi, L. and Röhlich, P. (1988) Freeze–fracture study of the receptor membranes in the olfactory organ of *Alburnus alburnus* (Teleostei). *Zoomorphology,* **108**, 41–6.

Holl, A. (1965) Vergleichende morphologische und histologische Untersuchungen am Geruchsorgan der Knochenfische. *Z. Morphol. Ökol. Tiere,* **54**, 707–82.

Holl, A. (1973a) Funktionsmorphologie der Nase von *Chimaera monstrosa* (Holocephali). *Z. Morphol. Tiere,* **74**, 271–96.

Holl, A. (1973b) Feinstruktur des Riechepithels von *Chimaera monstrosa* (Holocephali). *Mar. Biol.,* **23**, 59–72.

Holl, A., Schulte, E. and Meinel, M. (1970) Funktionelle Morphologie des Geruchsorgans und Histologie der Kopfanhänge der Nasenmuräne *Rhinomuraena ambonensis* (Teleostei, Anguilliformes). *Helgoländer wiss. Meeresunters.,* **21**, 103–23.

Holley, M.C. (1984) The ciliary basal apparatus is adapted to the structure and mechanics of the epithelium. *Tissue Cell,* **16**, 287–310.

Huxley, T.H. (1876) On *Ceratodus forsteri*, with observations on the classification of fishes. *Proc. zool. Soc. Lond.,* **1876**, 23–59.

Ichikawa, M. and Ueda, K. (1977) Fine structure of the olfactory epithelium in the goldfish, *Carassius auratus*. A study of retrograde degeneration. *Cell Tissue Res.*, **183**, 445–55.

Iwai, T. and Nakamura, I. (1964) Olfactory organs of tunas with special reference to their systematic significance. *Bull. Misaki mar. biol. Inst. Kyoto Univ.*, **7**, 1–8.

Jahn, L.A. (1972) Development of the olfactory apparatus of the cutthroat trout. *Trans. Am. Fish. Soc.*, **101**, 284–9.

Jakubowski, M. (1975) Anatomical structure of olfactory organs provided with internal nares in the Antarctic fish *Gymnodraco acuticeps* Boul. (Bathydraconidae). *Bull. Acad. Pol. Sci., Sér. Sci., Biol.*, **23**, 115–20.

Jakubowski, M. (1981) Ultrastructure (SEM, TEM) of the olfactory epithelium in the wels, *Siluris glanis* L. (Siluridae, Pisces). *Z. mikrosk.-anat. Forsch.*, **95**, 337–52.

Jakubowski, M. and Kunysz, E. (1979) Anatomy and morphometry of the olfactory organ of the wels *Silurus glanis* L. (Siluridae, Pisces). *Z. Mikrosk.-anat. Forsch.*, **93**, 728–35.

Jinxiang, S. and Wanduan, L. (1982) A study of patterns of the olfactory organ of tetraodontiform fishes in China. *Acta zool. sin.*, **28**, 389–98. (In Chinese)

Johansen, K. and Strahan, R. (1963) The respiratory system of *Myxine glutinosa* L., in *The Biology of Myxine* (eds A. Brodal and R. Fänge), Universitets-forlaget, Oslo, pp. 352–71.

Kapoor, A.S. and Ojha, P.P. (1972) Studies on ventilation of the olfactory chambers of fishes with a critical reevaluation of the role of the accessory nasal sacs. *Arch. Biol.*, **83**, 167–78.

Kleerekoper, H. (1969) *Olfaction in Fishes*, Indiana University Press, Bloomington, 238 pp.

Kleerekoper, H. (1982) Research in olfaction in fishes: historical aspects, in *Chemoreception in fishes* (ed. T.J. Hara), Elsevier, Amsterdam, pp. 1–14.

Kleerekoper, H. and van Erkel, G.A. (1960) The olfactory apparatus of *Petromyzon marinus* L. *Can. J. Zool.*, **38**, 209–23.

Kux, J., Zeiske, E. and Osawa, Y. (1988) Laser Doppler velocimetry measurement in the model flow of a fish olfactory organ. *Chem. Senses.*, **13**, 257–65.

Kyle, H.M. (1900) On the presence of nasal secretory sacs and a nasopharyngeal communication in Teleostei, with special reference to *Cynoglossus semilaeris* Gthr. *J. Linn. Soc.*, **27**, 541–56.

Liermann, K. (1933) Über den Bau des Geruchsorgans der Teleostier. *Z. Anat. EntwGesch.*, **100**, 1–39.

Lubosch, W. (1905) Die Entwicklung und Metamorphose des Geruchsorganes von *Petromyzon* und seine Bedeutung für die vergleichende Anatomie des Geruchsorganes. *Jena Z. Naturwiss.*, **40**, 95–148.

Marshall, A.M. (1879) The morphology of the vertebrate olfactory organ. *Q.J. microsc. Sci.*, **14**, 300–40.

Marshall, N.B. (1967) The olfactory organs of bathypelagic fishes. *Symp. zool. Soc. Lond.*, **19**, 57–70.

Matthes, E. (1934) Geruchsorgan, in *Handbuch der vergleichenden Anatomie der Wirbeltiere*. Vol. II, Pt. 2 (eds. L. Bolte, E. Göppert, E. Kallius and W. Lubosch), Urban und Schwarzenberg, Berlin, pp. 879–948.

Mees, F. (1962) Occurrence of internal nares in the genus *Ichthyoscopus* (Pisces, Uranoscopidae). *Copeia*, **1962**, 462.

Melinkat, R. and Zeiske, E. (1979) Functional morphology of ventilation of the olfactory organ in *Bedotia geayi* Pellegrin 1909 (Teleostei, Atherinidae). *Zool. Anz.*, **203**, 354–68.

Meng, Q. and Yin, M. (1981a) A study of the olfactory organ in the sharks. *Trans. Chin. Ichthyol. Soc.*, **2**, 1–24. (In Chinese)

Meng, Q. and Yin, M. (1981b) A study on the olfactory organ of sharks, rays and chimaeras. *J. Fish. Chin.*, **5**, 209–28. (In Chinese)

Millot, J. and Anthony, J. (1965) *Anatomie de* Latimeria chalumnae. II. *Système Nerveux et Organes des Sens*, CNRS, Paris, 206 pp.

Möller, P.C., Partridge, L.R., Cox, R.A., Pellegrini, V. and Ritchie, D.G. (1989) The development of ciliated and mucus cells from basal cells in hamster tracheal epithelial cell cultures, *Tissue Cell*, **21**, 195–8.

Moulton, D.G. and Beidler, L.M. (1967) Structure and function in the peripheral olfactory system. *Physiol. Rev.*, **47**, 1–52.

Muller, J.F. and Marc, R.E. (1984) Three distinct morphological classes of receptors in fish olfactory organs. *J. comp. Neurol.*, **222**, 482–95.

Nevitt, G.A. (1991) Do fish sniff? A new mechanism of olfactory sampling in pleuronectid flounders. *J. exp. Biol.*, **157**, 1–18.

Nielsen, J.G. and Bertelsen, E. (1985) The gulper-eel family Saccopharyngidae (Pisces, Anguilliformes). *Steenstrupia*, **11**, 157–206.

Norman, J.R. (1934) *A Systematic Monograph of the Flatfishes* (Heterosomata). Vol. I. *Psettodidae, Bothidae, Pleuronectidae*, British Museum (Nat. Hist.), London, 468 pp.

Northcutt, R.G. and Gans, C. (1983) The genesis of neural crest and epidermal placodes: a reinterpretation of vertebrate origins. *Q. Rev. Biol.*, **58**, 1–28.

Oelschläger, H.A. (1976) Morphologisch-funktionelle Untersuchungen am Geruchsorgan von *Lampris guttatus* (Brünnich 1788) (Teleostei: Allotriognathi). *Zoomorphologie*, **85**, 89–110.

Panchen, A.L. (1967) The nostrils of choanate fishes and early tetrapods. *Biol. Rev.*, **42**, 374–420.

Pfeiffer, W. (1968) Das Geruchsorgan der Polypteridae (Pisces, Brachiopterygii). *Z. Morph. Tiere*, **63**, 75–110.

Pfeiffer, W. (1969) Das Geruchsorgan der rezenten Actinistia und Dipnoi (Pisces). *Z. Morph. Tiere*, **64**, 309–37.

Pipping, M. (1926) Der Geruchssinn der Fische mit besonderer Berücksichtigung seiner Bedeutung für das Aufsuchen des Futters. *Soc. Sci. Fennica, Commentat. Biol.*, **2**(4), 1–28.

Pipping, M. (1927) Ergänzende Beobachtungen über den Geruchssinn der Fische mit besonderer Berücksichtigung seiner Bedeutung für das Aufsuchen des Futters. *Soc. Sci. Fennica, Commentat. Biol.*, **2**(9), 1–10.

Pyatkina, G.A. (1976) Receptor cells of various types and their proportional interrelation in the olfactory organ of larvae and adults of acipenserid fishes. *Tsitologiya*, **18**, 1444–9. (In Russian)

Reese, T.S. and Brightman, M.W. (1970) Olfactory surface and central olfactory connections in some vertebrates, in *Taste and Smell in Vertebrates* (eds G.E.W. Wolstenholme and J. Knight), Ciba Found. Symp., Churchill, London, pp. 115–43.

Reinke, W. (1937) Zur Ontogenie und Anatomie des Geruchsorgans der Knochenfische. *Z. Anat. EntwGesch.*, **106**, 600–24.

Rhein, L.D., Cagan, R.H., Orkand, P.M. and Dolack, M.K. (1981) Surface specializations of the olfactory epithelium of rainbow trout, *Salmo gairdneri*. *Tissue Cell*, **13**, 577–87.

Rosen, D.E. (1964) The relationships and taxonomic position of the halfbeaks, killifishes, silversides, and their relatives. *Bull. Am. Mus. nat. Hist.*, **127**, 217–68.

Rosen, D.E., Forey, P.L., Gardiner, B.G. and Patterson, C. (1981) Lungfishes, tetrapods, paleontology, and plesiomorphy. *Bull. Am. Mus. nat. Hist.*, **167**, 159–276.

Ross, D.M. (1963) The sense organs of *Myxine glutinosa* L., in *The Biology of Myxine* (eds A. Brodal and R. Fänge), Universitetsforlaget, Oslo, pp. 150–60.

Rowley, J.C., III and Moran, D.T. (1983) Comparative ultrastructure of olfactory epithelia from three species of marine fish. V. Ann. Meet. Assoc. Chemorecept. Sci., 27 April–1 May 1983, Sarasota, Florida, USA. Abstract.

Rowley, J.C., III and Moran, D.T. (1985) HRP applied to transected trout olfactory nerves fills ciliated receptors, microvillar receptors, and some basal cells. *Chem. Senses*, **10**, 392–3.

Schmalhausen, I.I. (1968) *The Origin of Terrestrial Vertebrates*, Academic Press, London, 336 pp.

Schulte, E. (1972) Untersuchungen an der Regio olfactoria des Aals, *Anguilla anguilla* L. I. Feinstruktur des Riechepithels. *Z. Zellforsch.*, **125**, 210–28.

Schulte, E. and Holl, A. (1971) Feinstruktur des Riechepithels von *Calamoichthys calabaricus* J.A. Smith (Pisces, Brachiopterygii). *Z. Zellforsch*, **120**, 261–79.

Schulte, E. and Riehl, R. (1978) Feinstruktur der Regio olfactoria vom Piranha, *Serrasalmus natteri* (Kner, 1860) (Teleostei, Characidae). *Zool. Anz.*, **200**, 119–31.

Schultze, M. (1863) Untersuchungen über den Bau der Nasenschleimhaut, namentlich die Stuctur und Endigungsweise der Geruchsnerven bei dem Menschen und den Wirbelthieren. *Abh. Naturforsch. Ges. Halle*, **7**, 1–100.

Sewertzoff, A.N. (1931) *Morphologische Gesetzmäßigkeiten der Evolution*, Fisher-Verlag, Jena, 386 pp.

Shepherd, G.M. (1988) Studies of development and plasticity in the olfactory sensory neuron. *J. Physiol., Paris*, **83**, 240–5.

Sleigh, M.A. (1989) Adaptations of ciliary systems for the propulsion of water and mucus. *Comp. Biochem. Physiol.*, **94A**, 359–64.

Sleigh, M.A., Blake, J.R. and Liron, N. (1988) The propulsion of mucus by cilia. *Am. Rev. Respir. Dis.*, **137**, 726–46.

Suzuki, Y. and Takeda, M. (1991) Keratins in the developing olfactory epithelia. *Dev. Brain Res.*, **59**, 171–8.

Teichmann, H. (1954) Vergleichende Untersuchungen an der Nase der Fische. *Z. Morph. Ökol. Tiere*, **43**, 171–212.

Teichmann, H. (1964) Experimente zur Nasenentwicklung der Regenbogenforelle (*Salmo irideus* W. Gibb.) *Wilhelm Roux Arch. EntwMech. Org.*, **155**, 129–43.

Tester, A.L. (1963) Olfaction, gustation, and the common chemical sense in sharks, in *Sharks and Survival* (ed. P.W. Gilbert), O.C. Heath and Company, Boston, Mass., pp. 255–82.

Theisen, B. (1970) The morphology and vascularization of the olfactory organ in *Calamoichthys calabaricus* (Pisces, Polypteridae). *Vidensk. Meddr dansk. naturh. Foren.*, **133**, 31–50.

Theisen, B. (1972) Ultrastructure of the olfactory epithelium in the Australian lungfish *Neoceratodus forsteri*. *Acta zool., Stockh.*, **53**, 205–18.

Theisen, B. (1973) The olfactory system in the hagfish *Myxine glutinosa*. I. Fine structure of the apical part of the olfactory epithelium. *Acta zool., Stockh.*, **54**, 271–84.

Theisen, B. (1976) The olfactory system in the Pacific hagfish *Eptatretus stoutii*, *Eptatretus deani*, and *Myxine circifrons*. *Acta zool., Stockh.*, **57**, 167–73.

Theisen, B. (1982) Functional morphology of the olfactory organ in *Spinachia spinachia* (L.) (Teleostei, Gasterosteidae). *Acta zool., Stockh.*, **63**, 247–54.

Theisen, B., Breucker, H., Zeiske, E. and Melinkat, R. (1980) Structure and development of the olfactory organ in the garfish *Belone belone* (L.) (Teleostei, Atheriniformes). *Acta zool., Stockh.*, **61**, 161–70.

Theisen, B., Zeiske, E. and Breucker, H. (1986) Functional morphology of the olfactory organs in the spiny dogfish (*Squalus acanthias* L.) and the small-spotted catshark (*Scyliorhinus canicula* (L.)). *Acta zool., Stockh.*, **67**, 73–86.

Theisen, B., Zeiske, E. and Breucker, H. (1989) Olfactory organs in two marine catfish species. A comparison. *Annls Soc. r. zool. Belg.*, **119**, (Suppl. 1), 84.

Theisen, B., Zeiske, E., Silver, W.L., Marui, T. and Caprio, J. (1991) Morphological and physiological studies on the olfactory organ of the striped eel catfish, *Plotosus lineatus*. *Mar. Biol.*, **110**, 127–35.

Thommesen, G. (1982) Specificity and distribution of receptor cells in the olfactory mucosa of char (*Salmo alpinus* L.). *Acta physiol. scand.*, **115**, 47–56.

Thommesen, G. (1983) Morphology, distribution, and specificity of olfactory receptor cells in salmonid fishes. *Acta physiol. scand.*, **117**, 241–50.

Thornhill, R.A. (1967) The ultrastructure of the olfactory epithelium of the lamprey *Lampetra fluviatilis*. *J. Cell Sci.*, **2**, 591–602.

Thornhill, R.A. (1970) Cell division in the olfactory epithelium of the lamprey, *Lampetra fluviatilis*. *Z. Zellforsch.*, **109**, 147–57.

Trujillo-Cenoz, O. (1961) Electron microscope observations on chemo- and mechano-receptor cells of fishes. *Z. Zellforsch.*, **54**, 654–76.

Uehara, K., Miyoshi, M. and Miyoshi, S. (1988) Microridges of oral mucosal epithelium in carp, *Cyprinus carpio*. *Cell Tissue Res.*, **251**, 547–53.

Uehara, K., Miyoshi, M. and Miyoshi, S. (1991) Cytoskeleton in microridges of the oral mucosal epithelium in the carp, *Cyprinus carpio*. *Anat. Rec.*, **230**, 164–8.

van den Berghe, L. (1929) Observations sur l'olfaction et le mécanisme des courants olfactifs chez quelques téléostéens. *Bull. Acad. R. Belg. Sci. Lettr. Beaux-Arts, Cl. Sci.* (5. Sér.) **15**, 278–305.

Whitear, M. (1986) The skin of fishes including cyclostomes. 2. Epidermis, in *Biology of the Integument*. Vol. II, *Vertebrates* (eds J. Bereiter-Hahn and A.G. Matoltsy), Springer-Verlag, Berlin, pp. 8–38.

Wiedersheim, R. (1887) Über rudimentäre Fischnasen. *Anat. Anz.*, **2**, 652–7.

Wilson, S.W., Ross, L.S., Parrett, T. and Easter, S.S., jun. (1990) The development of a simple scaffold of axon tracts in the brain of the embryonic zebrafish, *Brachydanio rerio*. *Development*, **108**, 121–45.

Yamada, J. (1968) A study on the structure of surface cell layers in the epidermis of some teleosts. *Annot. Zool. Jpn.*, **41**, 1–8.

Yamagishi, M., Nakamura, H., Takahashi, S., Nakano, Y. and Iwanaga, T. (1989) Olfactory receptor cells: immunocytochemistry for nervous system-specific proteins and re-evaluation of their precursor cells. *Arch. Histol. Cytol.*, **52** (Suppl.), 375–81.

Yamamoto, M. (1982) Comparative morphology of the peripheral olfactory organ in teleosts, in *Chemoreception in Fishes* (ed. T.J. Hara), Elsevier, Amsterdam, pp. 39–59.

Yamamoto, M. and Ueda, K. (1977) Comparative morphology of fish olfactory epithelium. I. Salmoniformes. *Bull. Jap. Soc. scient. Fish.*, **43**, 1163–74.

Yamamoto, M. and Ueda, K. (1978a) Comparative morphology of fish olfactory epithelium. II. Clupeiformes. *Bull. Jap. Soc. scient. Fish.*, **44**, 855–9.

Yamamoto, M. and Ueda, K. (1978b) Comparative morphology of fish olfactory epithelium. IV. Anguilliformes and Myctophiformes. *Bull. Jap. Soc. scient. Fish.*, **44**, 1207–12.

Yamamoto, M. and Ueda, K. (1978c) Comparative morphology of fish olfactory epithelium. V. Gasterosteiformes, Channiformes and Synbranchiformes. *Bull. Jap. Soc. scient., Fish.*, **44**, 1309–14.

Yamamoto, M. and Ueda, K. (1978d) Comparative morphology of fish olfactory epithelium. VII. Gadiformes, Lophiiformes and Gobiesociformes. *J. Fac. Sci. Tokyo Univ. Sect. 4*, **14**, 115–25.

References

Yamamoto, M. and Ueda, K. (1979a) Comparative morphology of fish olfactory epithelium. VIII. Atheriniformes. *Zool. Mag. (Tokyo),* **88**, 155–64.

Yamamoto, M. and Ueda, K. (1979b) Comparative morphology of fish olfactory epithelium. IX. Tetraodontiformes. *Zool. Mag. (Tokyo),* **88**, 210–18.

Yamamoto, M. and Ueda, K. (1979c) Comparative morphology of fish olfactory epithelium. X. Perciformes, Beryciformes, Scorpaeniformes and Pleuronectiformes. *J. Fac. Sci. Tokyo Univ. Sect. 4,* **14**, 273–97.

Zeiske, E. (1973) Morphologische Untersuchungen am Geruchsorgan von Zahnkarpfen (Pisces, Cyprinodontoidea). *Z. Morph. Tiere,* **74**, 1–16.

Zeiske, E. (1974) Morphologische und morphometrische Untersuchungen am Geruchsorgan oviparer Zahnkarpfen (Pisces). *Z. Morph. Tiere,* **77**, 19–50.

Zeiske, E., Kux, J. and Melinkat, R. (1976a) Development of the olfactory organ of oviparous and viviparous cyprinodonts (Teleostei). *Z. Zool. Syst. Evol.–forsch.,* **14**, 34–40.

Zeiske, E., Melinkat, R., Breucker, H. and Kux, J. (1976b) Ultrastructural studies on the epithelia of the olfactory organ of cyprinodonts (Teleostei, Cyprinodontoidea). *Cell Tissue Res.,* **172**, 245–68.

Zeiske, E., Breucker, H. and Melinkat, R. (1979) Gross morphology and fine structure of the olfactory organ of rainbow fish (Atheriniformes, Melanotaeniidae). *Acta zool., Stockh.,* **60**, 173–86.

Zeiske, E., Theisen, B. and Gruber, S.H. (1987) Functional morphology of the olfactory organ of two carcharhinid shark species. *Can. J. Zool.,* **65**, 2406–12.

Zeiske, E., Theisen, B. and Breucker, H. (1989) Olfactory organs in pelagic and benthic elasmobranchs, in *Progress in Zoology.* Vol. 35, *Trends in Vertebrate Morphology* (eds H. Splechtna, and H. Hilgers), Fischer-Verlag, Stuttgart, pp. 370–2.

Zielinski, B. and Hara, T.J. (1988) Morphological and physiological development of olfactory receptor cells in rainbow trout (*Salmo gairdneri*) embryos. *J. comp. Neurol.,* **271**, 300–11.

Chapter three

Synaptic organization of the olfactory bulb and its central projection

Masahiko Satou

3.1 INTRODUCTION

Olfaction plays an important role in teleost behaviour, such as feeding, social interaction, and migration (reviews: Kleerekoper, 1969; Hara, 1975, 1982, 1986; Liley, 1982; Döving, 1986; see also Chapters 11–13). Our understanding of the teleost olfactory system has increased rapidly during the last 10–15 years, through investigation using physiological, anatomical and behavioural methods (reviews: Hara, 1975, 1982, 1986; Döving, 1986; Satou, 1990). Recent findings are concerned with (1) the organization of the olfactory bulb (Ichikawa, 1976; Kosaka and Hama, 1979, 1980, 1981, 1982; Kosaka, 1980; Oka, 1983; Satou et al., 1983; Fujita et al., 1988) and of the telencephalic olfactory regions (Scalia and Ebbesson, 1971; Finger, 1975; Ichikawa, 1975; Oka, 1980; Bass, 1981a,b; Davis et al., 1981; Ebbesson et al., 1981; Murakami et al., 1983; Bartheld et al., 1984; Fujita et al., 1984; Levine and Dethier, 1985), and (2) specific odorants (pheromones) related to the reproductive functions and their effects on the olfactory system (Dulka et al., 1987; Sorensen et al., 1987, 1988; Stacey et al., 1987; Fujita et al., 1991). This chapter deals with structural and functional organization of the olfactory bulb, and its central projection. For reviews from a comparative neuro-anatomical viewpoint, see Allison (1953), Nieuwenhuys (1963, 1967), Andres (1970), Northcutt and Braford (1980), and Northcutt and Davis (1983).

3.2. OLFACTORY BULB

The olfactory bulb is the first relay station receiving the primary olfactory nerve inputs and sending its output signals to several target areas in the telencephalon and diencephalon (gross brain anatomy: Fig. 5.8 herein). The organization of the teleost olfactory bulb is basically similar to that of other vertebrate classes including mammals (Allison, 1953; Nieuwenhuys, 1967; Andres, 1970; Northcutt and Braford, 1980), in spite of various degrees of differentiation of neuronal elements and development of intrabulbar local circuitry. Hence, in an evolutionary sense the olfactory bulb is one of the most conservative regions of the vertebrate central nervous system. This view has been supported through electron microscopic studies (Andres, 1970; Ichikawa, 1976; Kosaka and Hama, 1982; Oka, 1983) as well as electrophysiological studies of bulbar synaptic organization (MacLeod, 1976; MacLeod and Lowe, 1976; Satou et al., 1983; Fujita et al., 1988), although the uniqueness of the teleost olfactory bulb has been emphasized by Kosaka and Hama (1982–1983).

Anatomy

The teleost olfactory bulb is a concentrically laminated structure similar to that in higher vertebrates. It consists of four layers, from superficial to deep: (1) olfactory nerve layer (ONL), (2) glomerular layer (GL), (3) mitral cell layer (MCL), and (4) internal cell layer (ICL).*

The carp olfactory nerve can be divided into the lateral and medial olfactory nerve bundles (Sheldon, 1912). Fibres in the lateral bundle originate from the receptor cells in the caudal part of the olfactory epithelium and project to the lateral part of the olfactory bulb. Those in the medial bundle originate from the receptor cells in the rostral part of the olfactory epithelium and project to the medial part of the bulb. Thus, a relatively clear-cut topographical relation among the olfactory epithelium, nerve and bulb can be recognized at the gross anatomical level. Two types of receptor cells, ciliated and microvillar (Ichikawa and Ueda, 1977; Zeiske et al., Chapter 2 herein) send unmyelinated axons which terminate in the olfactory glomeruli.

Golgi staining (Sheldon, 1912; Holmgren, 1920; Kosaka and Hama, 1982; Oka, 1983) and intracellular horseradish peroxidase (HRP) staining (Fujita et al., 1988) studies showed that teleost mitral cells consist of large neurones sparsely distributed throughout the MCL (Fig. 3.1). The somata are fusiform, elongated, oval, or irregular. They emit several thick dendrites with glomerular tufts ending in the glomerular region and send single axons to the olfactory tract. In the carp, Cyprinus carpio, two to five thick dendrites

---

*Abbreviations are listed on p. 54.

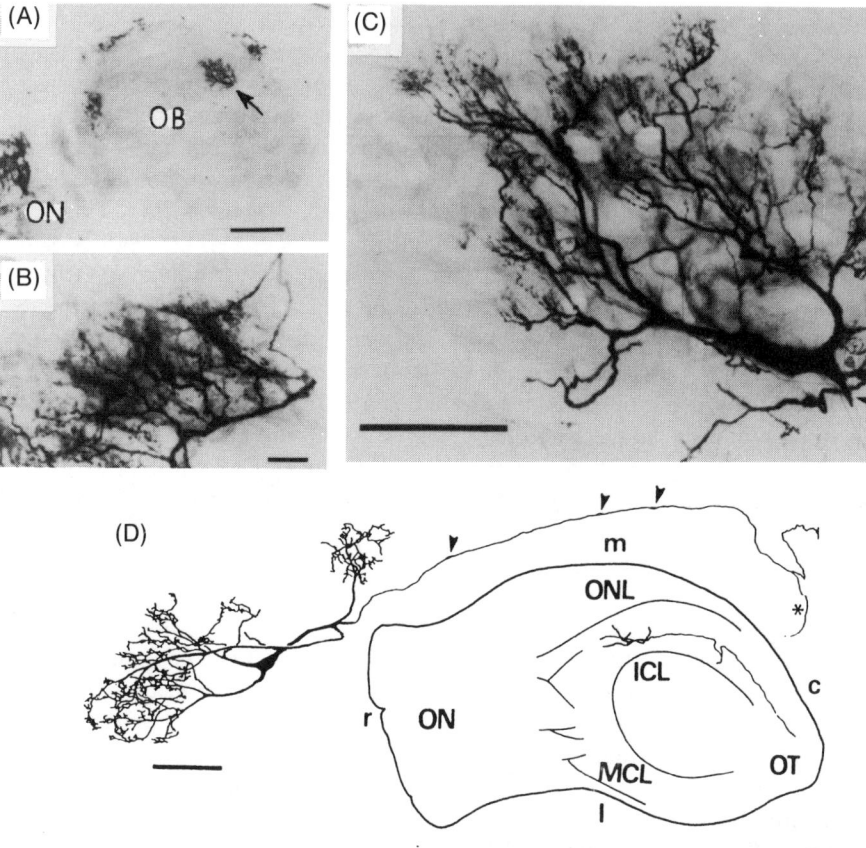

Fig. 3.1 Morphology of intracellularly stained carp, *Cyprinus carpio*, mitral cells. (A) Photomicrograph of an HRP-labelled mitral cell (arrow) at low magnification, showing its location and extent of its dendrites. Horizontal section. Upward and right are medial and caudal directions, respectively. Scale bar, 500 μm. (B) Photomicrograph showing glomerular tufts of dendrites. Scale bar, 20 μm. (C) Photomicrograph of an HRP-labelled mitral cell at higher magnification. Arrow indicates initial portion of the axon, arising from the first-order dendrite. Same cell as shown in (A). Scale bar, 100 μm. (D) Reconstruction of an HRP-labelled mitral cell from serial horizontal sections, showing its soma/dendritic morphology and axonal trajectory. Arrowheads and asterisk indicate elliptical swellings and an intrabulbar axon collateral, respectively. Scale bars, 100 μm, 1 mm (inset). Abbreviations: ICL, internal cell layer; MCL, mitral cell layer; OB, olfactory bulb; ON, olfactory nerve; ONL, olfactory nerve layer; OT, olfactory tract; m,l,r,c, medial, lateral, rostral and caudal directions, respectively. (After Fujita *et al.*, 1988.)

arising from the somata are directed toward the olfactory bulb surface. The dendrites emit successive branches that end in glomerular tufts with many irregular, varicose, or spiny appendages. Axons arising from the dendrites or somata run caudally through the MCL or the most superficial part of the ICL toward the olfactory tract.

Electron microscopic studies (Andres, 1970; Ichikawa, 1976; Kosaka and Hama, 1982; Oka, 1983) showed that mitral cell dendrites receive numerous synaptic inputs from the olfactory nerve on their spiny appendages, where they form excitatory, asymmetrical synapses. The dendritic shafts and the somata of the mitral cell make numerous dendro–dendritic reciprocal synapses with the peripheral dendrites of the presumed granule cells. The direction of the excitatory, asymmetrical synapses is from mitral to granule cells, while that of the inhibitory, symmetrical synapses is from granule to mitral cells. Thus, a functional segregation of dendritic regions can be recognized: (1) the distal dendritic tufts where the olfactory nerve inputs converge, and (2) the proximal dendrites where the dendro–dendritic reciprocal synaptic interaction with the granule cells occurs. Such a functional segregation is more distinct in mammalian mitral cells. They emit (1) single tuft-bearing 'primary' dendrites that integrate the excitatory olfactory nerve inputs, and (2) several non-tuft-bearing 'secondary' dendrites that integrate the inhibitory granule cell inputs (reviews: Shepherd, 1972, 1977; Mori, 1987).

In goldfish, *Carassius auratus*, catfish, *Parasilurus asotus*, and sea eel, *Conger myriaster*, another type of neurone (ruffed cell) exists which has many pedunculated protrusions on the initial unmyelinated portion of the axon (Kosaka and Hama, 1979, 1980, 1981, 1982–1983; Kosaka, 1980). Kosaka and Hama (1980) suggested that the ruffed cells may be present in all teleost species. The dendrites of the ruffed cells make few synaptic contacts with other neurones, including the olfactory nerve terminals, and surround the mitral cell dendrites resembling glial cell processes (Kosaka, 1980; Kosaka and Hama, 1981, 1982–1983). The pedunculated protrusions of initial segments of the axon make numerous reciprocal synapses with the presumed granule cells (Kosaka and Hama, 1979, 1980, 1981, 1982–1983; Kosaka, 1980). These morphological features of the ruffed cells suggest their distinct function.

The ICL is the major source of local neurones contributing to the intrabulbar local integration. Analyses of the bulbar field potentials (MacLeod, 1976; MacLeod and Lowe, 1976; Satou *et al.*, 1983) and mitral cell intracellular potentials (Satou *et al.*, 1983) showed that local inhibitory neurones in the ICL project their processes to the MCL and interact with mitral cell dendrites through the dendro–dendritic reciprocal synapses (see below, Electrophysiology). These inhibitory interneurones are thought to be a functional equivalent of the mammalian granule cells. However, the teleost ICL cells are morphologically different from the mammalian granule cells. The dendrites of the Golgi-stained teleost ICL cells are smooth (Sheldon, 1912; Holmgren, 1920; Oka *et al.*, 1982), but those of the mammalian granule cells have many spines ('gemmules') (Cajal, 1911; Price and Powell, 1970; Mori *et al.*, 1983).

Centrifugal fibres from the telencephalon terminate mainly in the ICL, making excitatory, asymmetrical synapses with the deep dendrites or somata of the presumed granule cells (Ichikawa, 1976). Within the ICL, a segregation of the terminal fields of the centrifugal fibres from different origins (pages 52–3) can be recognized (Bass, 1981b; Bartheld et al., 1984). Figure 3.2 schematically illustrates the synaptic organization of the teleost olfactory bulb, based on anatomical and electrophysiological knowledge.

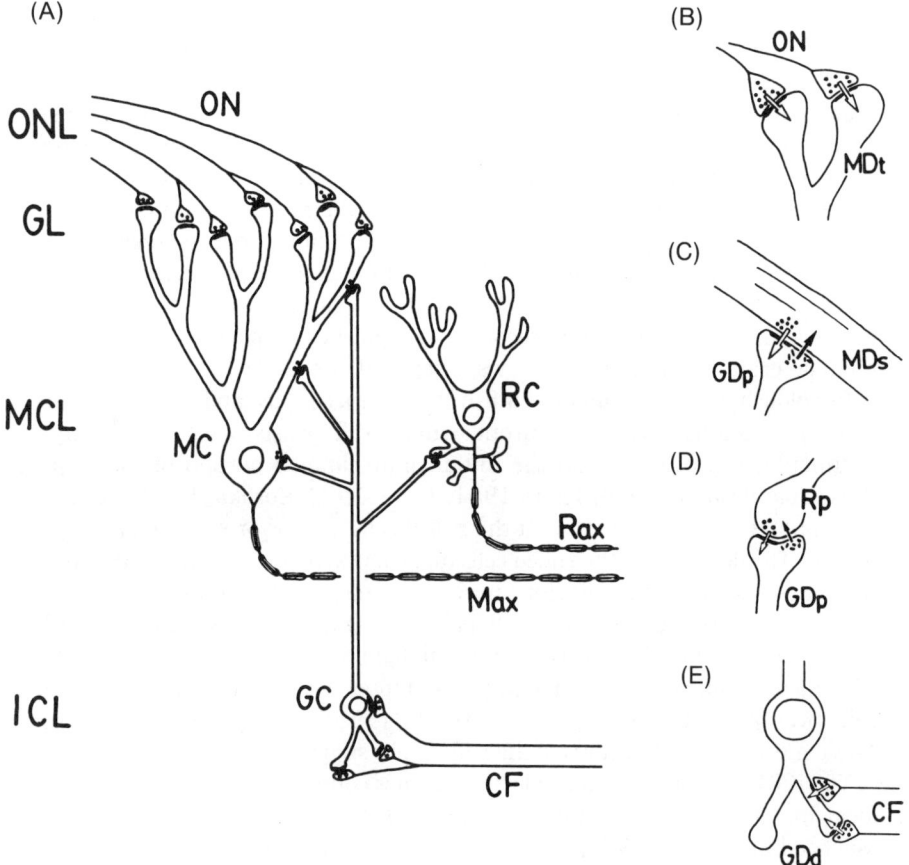

Fig. 3.2 Schematic diagram of synaptic organization of the teleost olfactory bulb. (A) Distribution of synapses on mitral cell, ruffed cell and granule cell; (B) synaptic inputs from olfactory nerve fibres to mitral cell dendritic tufts; (C) dendro–dendritic reciprocal synapse between mitral cell dendritic shaft and granule cell peripheral dendrite; (D) reciprocal synapse between the pedunculated protrusion of the initial segment of ruffed cell axon and granule cell peripheral dendrite; (E) synaptic inputs from centrifugal fibres to granule cell deep dendrite. Abbreviations: CF, centrifugal fibre, GC, granule cell; GDd, granule cell deep dendrite; GDp, granule cell peripheral dendrite; GL, glomerular layer; ICL, internal cell layer; Max, mitral cell axon; MC, mitral cell; MCL, mitral cell layer; MDs, mitral cell dendritic shaft; MDt, mitral cell dendritic tuft; ON, olfactory nerve; ONL, olfactory nerve layer; Rax, ruffed cell axon; RC, ruffed cell; Rp, pedunculated protrusion of the initial segment of ruffed cell axon. Open arrows indicate direction of asymmetrical synapses, filled arrows indicate direction of symmetrical synapses. (Reproduced with permission from Satou, 1990.)

Another notable neuronal element of the teleost olfactory bulb is the ganglion cells of the terminal nerve (TN) (Sheldon, 1909, 1912). The TN runs along the olfactory system of all vertebrates (Demski and Schwanzel-Fukuda, 1987), and has been suggested to have a chemosensory function related to reproduction (Demski and Northcutt, 1983; Wirsig and Leonard, 1987) (see p. 50, Electrophysiology). The teleost TN ganglion cells are located within the olfactory nerve rostromedially in the olfactory bulb or in the transitional zone between the caudoventral olfactory bulb and ventrorostral telencephalon (Sheldon, 1909, 1912; Münz et al., 1982; Demski and Northcutt, 1983; Springer, 1983; Stell et al., 1984; Levine and Dethier, 1985; Bartheld and Meyer, 1986; Münz and Claas, 1987). They project to various areas in the telencephalon, diencephalon, mesencephalon and the retina (Sheldon, 1909, 1912; Münz et al., 1982; Springer, 1983; Stell et al., 1984; Levine and Dethier, 1985; Kah et al., 1986; Bartheld and Meyer, 1986; Grober et al., 1987). The TN cells show immunoreactivity to luteinizing hormone-releasing hormone (LHRH), molluscan cardioexcitatory peptide (FMRFamide), and substance P (Münz et al., 1981, 1982; Halpern-Sebold and Schreibman, 1983; Stell et al., 1984, 1985).

The olfactory tract comprises two distinct fibre bundles, the lateral olfactory tract (LOT) and medial olfactory tract (MOT) (Sheldon, 1912; Holmgren, 1920; Ichikawa and Ueda, 1979; Oka, 1980; Bartheld et al., 1984). Axons from the mitral cells in the lateral part of the olfactory bulb course mainly through the LOT, while those in the medial part of the bulb course mainly through the MOT. Centrifugal fibres to the olfactory bulb run through the lateral part of the MOT. Axons from the TN cells run through the ventromedial part of the MOT.

Electrophysiology

The teleost olfactory bulb has a concentric laminar structure and synaptic organization basically similar to that of higher vertebrates. This was confirmed electrophysiologically by laminar analysis of the field potentials (MacLeod, 1976; MacLeod and Lowe, 1976; Satou et al., 1983), intracellular analysis of the mitral cells (Guselnikova and Guselnikov, 1975; Satou et al., 1983), and extracellular study of the presumed granule cells (Satou, 1990). Electrophysiological studies further revealed a functional separation of the olfactory bulb into lateral and medial subdivisions, each of which has input and output pathways and inhibitory interneurones (Satou et al., 1979, 1983; Fujita et al., 1988).

The topographical relationship between the mitral cells and the olfactory tract was examined using antidromic activation methods in the carp (Satou et al., 1979, 1983) and the tench, *Tinca tinca* L. (Dubois-Dauphin et al., 1980). In agreement with anatomical results, the antidromically activated units (presumed mitral cells) are diffusely arranged in a concentric layer (MCL). Units activated antidromically by the LOT and MOT volleys are located mainly laterally and medially in the olfactory bulb, respectively. Axonal conduction velocity of carp mitral cells ranges from 0.14 to 3.89 ms^{-1} (Satou et al., 1979), indicating that the size of mitral cell axons varies considerably.

In the carp olfactory bulb, four types of field potentials (component 1 (C_1)–component 4 (C_4) waves) are evoked by volleys from the olfactory nerve (ON) or tract (Satou et al., 1983). The C_1 wave reflects the compound action potentials from the mitral cell somata. Laminar analysis of the C_2–C_4 waves indicated that the extracellular currents flow across the MCL from deep to superficial during the period of the C_2 wave, and from superficial to deep during the period of the C_3 and C_4 waves. The granule cells in the carp olfactory bulb have the necessary anatomical arrangement to provide an intracellular return pathway for such currents (Sheldon, 1912). Therefore, it was concluded that the C_2 wave reflects synaptic depolarization at the peripheral dendrites of the granule cells through synapses from the mitral cell dendrites. In contrast, the C_3 and C_4 waves reflect synaptic depolarization at the deep dendrites and somata of the granule cells through synapses from the centrifugal fibres to the granule cells (Satou et al., 1983).

The C_2–C_4 waves proved to be useful to assess the spatial distribution of excitatory synaptic inputs to the granule cell population (Satou et al., 1983). In the lateral part of the olfactory bulb, the LOT-evoked C_2 wave is negative at ONL and GL, reversed in polarity at MCL, and positive at ICL. In the medial part of the bulb, the LOT-evoked C_2 wave remains positive irrespective of the layers. The distribution of the MOT-evoked C_2 wave shows a mirror image to that of the LOT-evoked C_2 wave. However, the MOT-evoked C_3 and C_4 waves are negative at ICL and generally positive or small in size at ONL and GL in both the lateral and medial parts of the olfactory bulb (Fig. 3.3).

Thus, the patterns of excitatory synaptic inputs to peripheral dendrites of the granule cell population differ between the lateral and medial parts of the olfactory bulb. However, synaptic inputs to deep dendrites and somata of the granule cell population are similar in the two parts of the bulb. These findings suggest that a functional segregation of the olfactory bulb into lateral and medial parts also exists at the level of inhibitory interneurones.

Intracellular recordings from mitral cells in the carp revealed basic properties of synaptic responses which were interpreted in terms of the bulbar local neuronal circuitry (Guselnikova and Guselnikov, 1975; Satou et al., 1983). The ON volleys elicit excitatory postsynaptic potentials (EPSPs) on which one or more action potentials are superimposed. The EPSPs are followed by long-lasting (up to 4 s) inhibitory postsynaptic potentials (IPSPs) during which the spontaneously occurring action potentials are inhibited (Fig. 3.4(A)). The MOT volleys elicit long-lasting IPSPs in the mitral cells in the medial and lateral parts of the olfactory bulb. The LOT volleys elicit long-lasting IPSPs in the mitral cells located in the lateral, but not medial, olfactory bulb (Fig. 3.4(B) and (C)). Such a distribution of mitral cells showing

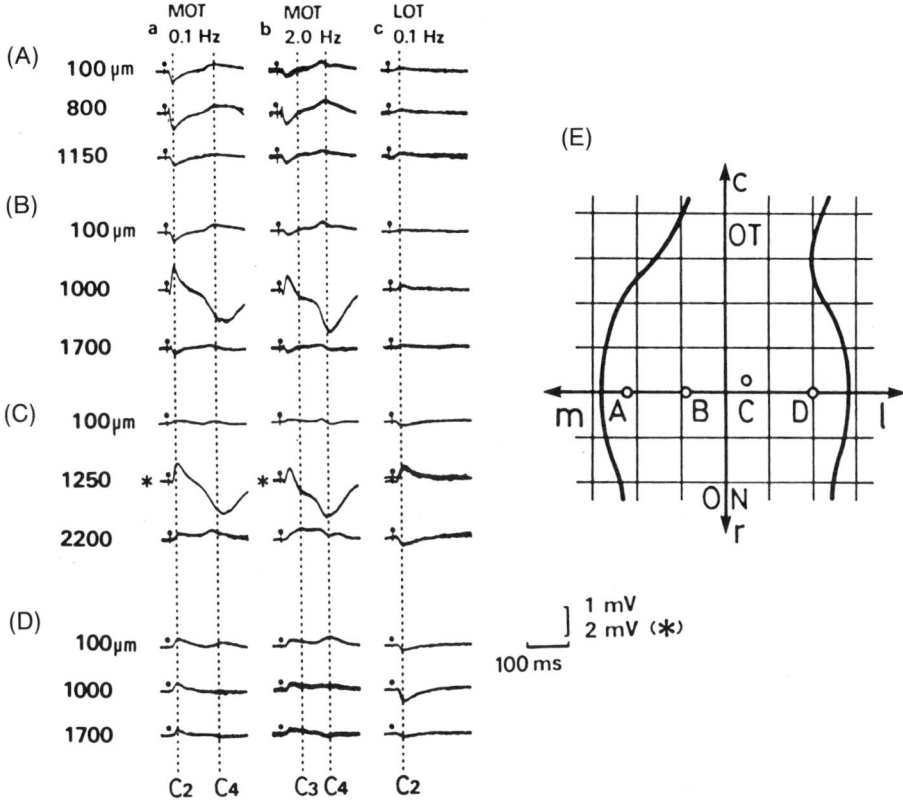

Fig. 3.3 Spatial distribution of the field potentials over the olfactory bulb in the carp. Traces in (A) to (D) show field potentials recorded at three depths along four electrode tracks illustrated in (E). (Aa–Da) Responses to medial olfactory tract (MOT) volleys at a low stimulation frequency (0.1 Hz); (Ab–Db) responses to MOT volleys at a higher stimulus frequency (2.0 Hz); (Ac–Dc) responses to lateral olfactory tract (LOT) volleys at a low stimulation frequency (0.1 Hz). Recorded depths from the olfactory bulb surface are shown at left of records. Note that traces marked with asterisks have different voltage scale (2 mV) from other traces (1 mV). (E) Schematic dorsal view of the left olfactory bulb, with locations of electrode insertion A–D. Interval of the lattice is 500 μm. Directions: m,l,r,c, medial, lateral, rostral and caudal, respectively. (Reproduced with permission from Satou et al., 1983.)

IPSPs over the olfactory bulb agrees with the distribution of C_2–C_4 waves, and further supports the view of latero–medial separation of the olfactory bulb in terms of local inhibition (Satou et al., 1983).

Extracellular recordings showed that the presumed granule cells do not discharge spikes spontaneously, but respond synaptically mainly by the ON and MOT volleys. These ON- and MOT-evoked spikes are superimposed on the peak to trailing phase of the C_2–C_4 waves (Satou, 1990). The granule cells discharge spikes in response to the MOT or ON volleys after the mitral cell

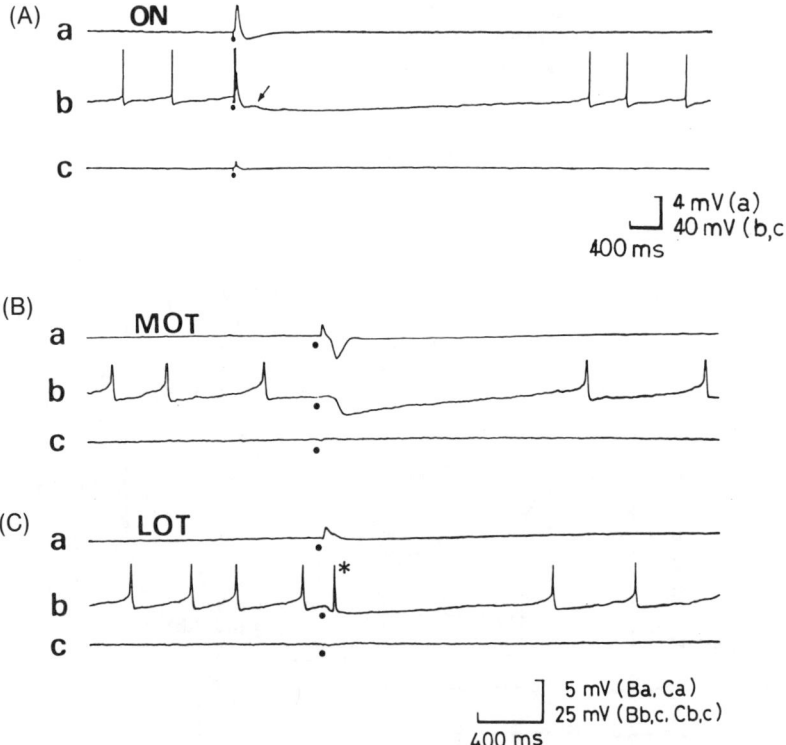

Fig. 3.4 Intracellular responses (DC records) from carp mitral cells. (A) Intracellular response from a mitral cell located at the middle part of the olfactory bulb to (Ab) electrical stimulation of the middle part of the olfactory nerve, and (Aa) field potential response recorded simultaneously in the internal cell layer (ICL); (Ac) extracellular field potential just outside the impaled cell; arrow indicates a depolarizing wave that appeared on the rising phase of the IPSPs. (B) and (C) Intracellular responses from a mitral cell located in the lateral part of the olfactory bulb to electrical stimulation of the MOT (Bb) and the LOT (Cb); (Ba and Ca) simultaneously recorded field potentials in ICL; (Bc and Cc) extracellular field potentials just outside the impaled cell; *indicates the antidromic action potential. Recorded by 2M K-citrate electrode. LOT, lateral olfactory tract; MOT, medial olfactory tract; ON, olfactory nerve. (After Satou et al., 1983.)

IPSPs begin, and they rarely discharge spikes in response to the LOT volleys. This suggests that the synaptic inhibition in the mitral cells is mainly mediated by local, graded depolarizing responses in the granule cells. The spikes generated in the granule cell somata (or proximal dendrites) may function as a booster mechanism so that the EPSPs at the granule cell somata or deep dendrites elicited by the centrifugal fibre volleys can activate remote inhibitory synapses located on the granule cell peripheral dendrites. The basic neuronal circuit within the carp olfactory bulb, illustrating its functional segregation into lateral and medial subdivisions, is shown in Fig. 3.5.

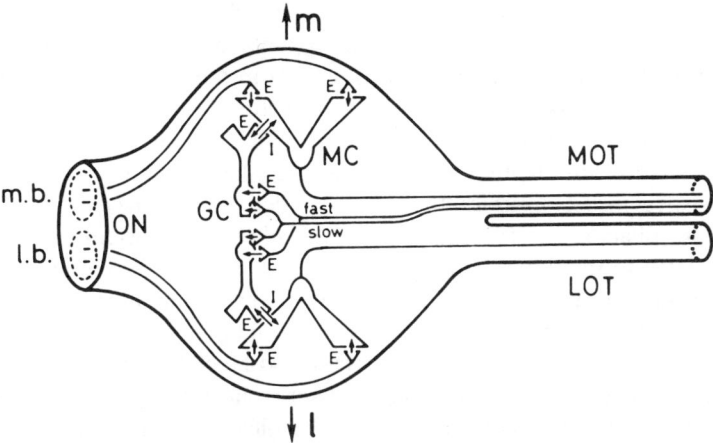

Fig. 3.5 Neuronal circuit diagram of the carp olfactory bulb schematically illustrating its functional segregation into the lateral and medial subdivisions. Abbreviations: E, excitatory synapse; GC, granule cell; I, inhibitory synapse; l, lateral direction; l.b., lateral bundle of the olfactory nerve; LOT, lateral olfactory tract; m, medial direction; m.b., medial bundle of the olfactory nerve; MC, mitral cell; MOT, medial olfactory tract; ON, olfactory nerve; fast, 'fast' conducting centrifugal fibre; slow, 'slow' conducting centrifugal fibre. (Reproduced with permission from Satou et al., 1983.)

Several studies in which natural odour stimuli were used also support the functional latero-medial separation of the teleost olfactory bulb. In salmonid fishes the lateral part of the olfactory bulb responds mainly to amino acids, whereas the medial part responds mainly to bile acids (Thommesen, 1978; Döving et al., 1980). Recently identified teleost sex pheromones ($17\alpha,20\beta$-dihydroxy-4-pregnen-3-one (17,20P) and F-series prostaglandins (PGFs)) (Sorensen, Chapter 11 herein) stimulate the olfactory receptors at extremely low concentrations (10^{-13} M for 17,20P, 10^{-9} M for $PGF_{2\alpha}$, 10^{-11} M for 15K-$PGF_{2\alpha}$) (Sorensen et al., 1987, 1988), and elicit excitatory or inhibitory responses in the mitral cells located at the medial part of the olfactory bulb (Fujita et al., 1991).

Behavioural studies in several teleosts also suggest that the lateral and medial parts of the olfactory bulb have different functions. In freely swimming cod, Gadus morhua, different behaviour is evoked when different bundlets separated from the LOT or MOT are electrically stimulated (Döving and Selset, 1980). In the male goldfish, courtship behaviour is impaired after the MOT is sectioned, but not after the LOT is sectioned (Stacey and Kyle, 1983). Feeding behaviour is more reduced by the LOT section than by the MOT section (Stacey and Kyle, 1983).

These results suggest that sex-pheromone-related information is processed in the medial part of the olfactory bulb, and thus, the quality of the odorants is represented in a spatial pattern of activated olfactory bulb neurones (Adrian, 1953; Mozell, 1958; Moulton, 1967, 1976). Such a spatial pattern of odour-induced activity likely exists to some degree within the olfactory epithelium, since a gross topographical relationship among the olfactory epithelium, nerve and bulb is present in the carp (Sheldon, 1912). However, the dimensionality of the olfactory stimulus representation in the olfactory epithelium is not clear (Shepherd, 1985). Since the signals arising from each of the lateral and medial parts of the olfactory bulb are transmitted to respective target areas in the telencephalon and diencephalon (p. 51, Secondary olfactory areas), certain types of olfactory information, such as those related to reproduction or to feeding, may be processed independently through the two distinct 'channels' or 'subsystems', the lateral and medial olfactory systems (Satou et al., 1979, 1983; Fujita et al., 1984, 1988). The olfactory system of higher vertebrates can also be divided into functionally distinct subsystems, the main and accessory olfactory systems (Raisman, 1972). It is interesting to note that the medial olfactory system in teleosts and the accessory olfactory system in higher vertebrates share a function related to reproduction (Powers and Winans, 1975), although the homology of these two subsystems is still unclear.

Some of the electrophysiological characteristics of the TN cells were investigated intracellularly in the carp (Fujita et al., 1985). The MOT volleys elicit long-lasting IPSPs in all TN cells tested, and antidromic action potentials in about half of them, whereas the LOT volleys elicit no response. The ON volleys elicit long-lasting IPSPs in about three-quarters of the TN cells. However, the relationship between the TN cells and the olfactory system proper remains unclear. It has been proposed that the TN cells mediate responses to sex pheromones in the goldfish (Demski and Northcutt, 1983; Stell et al., 1984). However, recent experiments in which the effects of several identified sex pheromones ($17,20P$, $PGF_{2\alpha}$, $15K\text{-}PGF_{2\alpha}$) on TN cell activity were examined showed that goldfish TN cells did not respond to any of the pheromones tested (Fujita et al., 1991). This raises a question as to the hypothesis that the TN system is involved in sex-pheromone-mediated responses in the goldfish. As mentioned above, information related to the sex pheromones is supposed to be processed not through the TN system, but through the medial olfactory system.

3.3 TELENCEPHALIC HEMISPHERE AND SECONDARY OLFACTORY AREAS

The telencephalic hemisphere of teleosts has been extensively studied from a comparative neurological viewpoint to understand the evolution of the telencephalon throughout the vertebrates (Sheldon, 1912; Holmgren, 1920; Nieuwenhuys, 1963, 1967; Northcutt and Braford, 1980; Northcutt and Davis, 1983). The unique developmental process of the teleost telencephalon, eversion, contrasts to the process of inversion of the telencephalon of land vertebrates. Attempts by researchers to establish homology of brain areas are controversial (Nieuwenhuys, 1967; Northcutt and Braford, 1980; Echteler and Saidel, 1981; Murakami et al., 1983). The evolution of the telencephalon is undoubtedly a subject of great interest, but beyond the scope of this review. This section describes the secondary olfactory areas, origins of the centrifugal fibres to the olfactory bulb, and telencephalic targets of other sensory inputs.

Secondary olfactory areas

The secondary olfactory areas where the axons of the olfactory bulb neurones terminate have been examined in several teleost species by means of anatomical (Scalia and Ebbesson, 1971; Ito, 1973; Finger, 1975; Ichikawa, 1975; Ichikawa and Ueda, 1978; Bass, 1981a; Davis et al., 1981; Ebbesson et al., 1981; Murakami et al., 1983; Bartheld et al., 1984; Levine and Dethier, 1985) and electrophysiological methods (Fujita et al., 1984). These studies showed that the LOT and MOT fibres project to several restricted areas in the telencephalon and diencephalon. Figure 3.6(A) schematically illustrates the pathways and terminal fields of the LOT and MOT fibres in the carp, based on electrophysiological results (Fujita et al., 1984). In the telencephalon, three terminal regions can be recognized bilaterally: (1) the precommissural ventrolateral region (Dlv, Vl), (2) the precommissural ventromedial region (Vv–1, Vv–2), and (3) the postcommissural region (Dlp, Dp, Dc–3, NT, Vi). A terminal region (POA) also exists in the diencephalon. These areas correspond in location to the lateral (LTF), medial (MTF), and posterior or posterior-central (PTF) terminal fields, respectively, as described in other teleosts (Scalia and Ebbesson, 1971; Finger, 1975; Ichikawa, 1975; Bass, 1981a; Bartheld et al., 1984). Thus, a basically similar projection pattern can be recognized among the different teleost species examined (moray eel, *Gymnothorax funebris*, goldfish, bullhead catfish, *Ictalurus nebulosus*, channel catfish, *I. punctatus*, piranha, *Serrasalmus nattereri* and *Macropodus opercularis*), although some minor differences exist. For example, the LOT fibres also project to the ipsilateral MTF in the channel catfish (Bass, 1981a), but the MOT fibres do not project to the contralateral LTF in the channel catfish (Bass, 1981a) and the goldfish (Bartheld et al., 1984). Contralateral projection from the LOT fibres through the habenular commissure in addition to the anterior commissure is known in the bullhead catfish (Finger, 1975), channel catfish (Bass, 1981a), and goldfish (Bartheld et al., 1984).

Murakami et al. (1983) examined the intratelencephalic fibre connections in *Sebasticus marmoratus* using tract-tracing methods and showed that the cells in area dorsalis pars posterior (Dp), which is the main olfactory target (PTF) in this species, project to the ipsilateral ventral region of area dorsalis pars medialis (vDm). Fibres from vDm in turn project to the hypothalamus (inferior lobe, nucleus posterior tuberis). This indicates a close relationship between the olfactory sense and the hypothalamic functions. Connections from other olfactory target areas (MTF, LTF) remain largely unclear (Shiga et al., 1985).

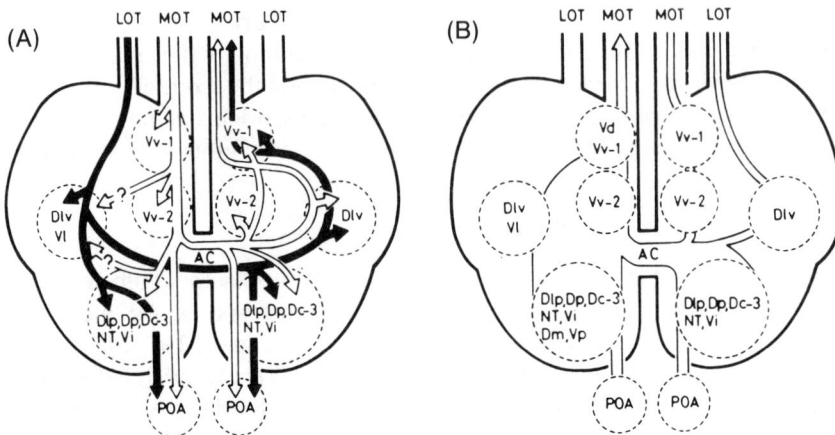

Fig. 3.6 Schematic drawing of the pathways of (A) mitral cell axons and (B) centrifugal fibres in the telencephalon of the carp. Dorsal view; rostral direction is towards top of figure. Filled and open arrows in (A) indicate projections of mitral cell axons in the LOT and those of mitral cell axons in the MOT, respectively. It remains uncertain whether projections from MOT to the precommissural ventrolateral region (Dlv, Vl) are via the precommissural ventromedial region (Vv-1, Vv-2) or via the postcommissural region (Dlp, Dp, Dc-3, NT, Vi) (the two possible pathways are illustrated with question marks). Abbreviations: AC, anterior commissure; D, area dorsalis telencephali; Dc, central zone of D; Dc-1,2,3, parts of Dc; Dd, dorsal zone of D; Dl, lateral zone of D; Dld, dorsal part of Dl; Dlp, posterior part of Dl; Dlv, ventral part of Dl; Dm, medial zone of D; Dp, posterior zone of D; LOT, lateral olfactory tract; MOT, medial olfactory tract; NT, nucleus taenia; POA, preoptic area; V, area ventralis telencephali; Vd, dorsal nucleus of V; Vi, intermediate nucleus of V; Vl, lateral nucleus of V; Vp, postcommissural nucleus of V; Vv, ventral nucleus of V; Vv-1,2, parts of Vv. (Reproduced with permission from Fujita et al., 1984.)

Electrophysiological studies of the higher olfactory areas are scarce (Döving et al., 1973; Satou, 1974). Specific areas in the telencephalic hemisphere (ventral and supracommissural parts of ventral telencephalon, Vv and Vs) were suggested to be involved in sexual behaviour of the goldfish (Kyle and Peter, 1982; Kyle et al., 1982; Koyama et al., 1984, 1985) and himé salmon, landlocked *Oncorhynchus nerka* (Satou et al., 1984; Satou, 1987). Electrolytic lesions in these areas severely impair male sexual behaviour, whereas electrical stimulation applied to these areas facilitates sexual behaviour both in the male and in the female. These areas overlap with the terminal fields of the medial olfactory system. However, the neural responses of these areas to sex pheromones are still to be investigated.

Origin of centrifugal fibres to olfactory bulb

The origin of the centrifugal fibres to the olfactory bulb has been studied using retrograde tracer techniques (Oka, 1980; Bass, 1981b; Ebbesson et al.,

1981; Murakami et al., 1983; Bartheld et al., 1984). In the channel catfish (Bass, 1981b), piranha (Ebbesson et al., 1981) and *Sebastiscus marmoratus* (Murakami et al., 1983), labelled cells are found in the ipsilateral telencephalon. In the goldfish they are found in the bilateral telencephalon, with fewer labelled cells in the contralateral side (von Bartheld et al., 1984). In the carp the regions where the electrical stimulation elicited the C_3 or C_4 waves (i.e. waves due to centrifugal fibre volleys) in the olfactory bulb are distributed bilaterally in the telencephalon and the preoptic area (Fig. 3.6(B)) (Fujita et al., 1984). This suggests that in the carp, either the cells of origin or their axon collaterals distribute bilaterally. The distribution of efferent cells to the olfactory bulb overlaps the projection areas of the secondary neurones, indicating that the olfactory bulb has reciprocal connections with the telencephalon and diencephalon.

Telencephalic areas receiving non-olfactory sensory inputs

Anatomical studies of the afferent and efferent connections of the teleost olfactory bulb indicate that relatively restricted regions of the telencephalon are directly related to olfactory function. The functions of other telencephalic regions are largely unknown, although lesion or electrical stimulation studies have assigned various functions concerning sexual and parental behaviour, aggressive behaviour, feeding behaviour, or learning processes (Segaar, 1961, 1965; Demski and Knigge, 1971; Demski et al., 1975; de Bruin, 1980; Savage, 1980; Aronson, 1981; Kyle and Peter, 1982; Kyle et al., 1982; Koyama et al., 1984, 1985; Satou et al., 1984; Satou, 1987).

Recently, non-olfactory sensory inputs have been reported to reach the telencephalon. Visual inputs reach the lateral part of the dorsal telencephalon (area dorsalis pars lateralis, Dl) (Ito et al., 1980; Ito and Vanegas, 1983; Murakami et al., 1983; Vanegas and Ito, 1983), while the acoustico-lateral line inputs reach the central and medial parts of the dorsal telencephalon (Dc and Dm) (Finger, 1980; Echteler, 1984, 1985; Murakami et al., 1986). These studies show that the teleost telencephalon is not merely a recipient area for olfactory information. It may process inputs from various senses and may function as a higher integrative centre, as in the land vertebrates. Whether higher-order olfactory information also reaches areas receiving other sensory inputs remains to be clarified.

ACKNOWLEDGEMENTS

This work is supported by grants from the Ministry of Education, Science and Culture of Japan, the Ministry of Agriculture, Forestry and Fisheries of Japan and the Mitsubishi Foundation.

ABBREVIATIONS

The following abbreviations are used in the text; those used in figures are explained in the legends.

Dp	Area dorsalis pars posterior
EPSP	Excitatory postsynaptic potential
FMRFamide	Molluscan cardioexcitatory peptide
GL	Glomerular layer (of olfactory bulb)
HRP	Horseradish peroxidase
ICL	Internal cell layer (of olfactory bulb)
IPSP	Inhibitory postsynaptic potential
LHRH	Luteinizing hormone-releasing hormone
LOT	Lateral olfactory tract
LTF	Lateral terminal field
MCL	Mitral cell layer (of olfactory bulb)
MOT	Medial olfactory tract
MTF	Medial terminal field
ONL	Olfactory nerve layer (of olfactory bulb)
PGF	F-series prostaglandin
POA	Preoptic area
PTF	Posterior terminal field
TN	Terminal nerve
vDm	Ventral region of area dorsalis pars medialis
Vs	Supracommissural part of ventral telencephalon
Vv	Ventral part of ventral telencephalon

REFERENCES

Adrian, E.D. (1953) Sensory messages and sensations: the response of olfactory organ to different smells. *Acta physiol. scand.*, **29**, 5–14.

Allison, A.C. (1953) The morphology of the olfactory system in the vertebrates. *Biol. Rev.*, **28**, 195–244.

Andres, K.H. (1970) Anatomy and ultrastructure of the olfactory bulb in fish, amphibia, reptiles, birds and mammals, in *Taste and Smell in Vertebrates* (Ciba Foundation Symp.) (eds G.E.W. Wolstenholm and J. Knight), Churchill, London, pp. 177–94.

Aronson, L.R. (1981) Evolution of telencephalic function in lower vertebrates, in *Brain Mechanisms of Behaviour in Lower Vertebrates* (ed. P.R. Laming), Cambridge University Press, Cambridge, pp. 33–58.

Bartheld, C.S. and Meyer, D.L. (1986) Tracing of single fibers of the nervus terminalis in the goldfish brain. *Cell Tissue Res.*, **245**, 143–58.

Bartheld, C.S., Meyer, D.L., Fiebig, E. and Ebbesson, S.O.E. (1984) Central connections of the olfactory bulb in the goldfish, *Carassius auratus. Cell Tissue Res.*, **238**, 475–87.

Bass, A.H. (1981a) Olfactory bulb efferents in the channel catfish, *Ictalurus punctatus. J. Morph.*, **169**, 91–111.

Bass, A.H. (1981b) Telencephalic efferents in the channel catfish, *Ictalurus punctatus*: projections to the olfactory bulb and optic tectum. *Brain Behav. Evol.*, **19**, 1–16.
Cajal, S. Ramon y (1911) *Histologie du Système Nerveux de l'Homme et des Vertébrés*, Vol. 2, Paris, Maloine, 993 pp.
Davis, R.E., Chase, R., Morris, J. and Kaufman, B. (1981) Telencephalon of the teleost *Macropodus*: experimental localization of secondary olfactory areas and of components of the lateral forebrain bundle. *Behav. neural Biol.*, **33**, 257–79.
de Bruin, J.P.C. (1980) Telencephalon and behavior in teleost fish. A neuroethological approach, in *Comparative Neurology of the Telencephalon* (ed. S.O.E. Ebbesson), Plenum Press, London, pp. 175–201.
Demski, L.S. and Knigge, K.M. (1971) The telencephalon and hypothalamus of the bluegill (*Lepomis macrochirus*): evoked feeding, aggressive and reproductive behavior with representative frontal sections. *J. comp. Neurol.*, **143**, 1–16.
Demski, L.S. and Northcutt, R.G. (1983) The terminal nerve: a new chemosensory system in vertebrates? *Science*, Wash., D.C., **220**, 435–7.
Demski, L.S. and Schwanzel-Fukuda, M. (eds) (1987) *The Terminal Nerve (Nervus Terminalis): Structure, Function, and Evolution*. Ann. N.Y. Acad. Sci., **519**, 1–469.
Demski, L.S., Bauer, D.H. and Gerald, J.W. (1975) Sperm release evoked by electrical stimulation of the fish brain: a functional-anatomical study. *J. exp. Zool.*, **191**, 215–32.
Döving, K.B. (1986) Functional properties of the fish olfactory system, in *Progress in Sensory Physiology*. Vol. 6 (ed. D. Ottoson), Springer-Verlag, Berlin, pp. 39–104.
Döving, K.B. and Selset, R. (1980) Behavior patterns in cod released by electrical stimulation of olfactory tract bundlets. *Science*, Wash., D.C., **207**, 559–60.
Döving, K.B., Enger, P.S. and Nordeng, H. (1973) Electrophysiological studies on the olfactory sense in char (*Salmo alpinus* L.). *Comp. Biochem. Physiol.*, **45A**, 21–4.
Döving, K.B., Selset, R. and Thommesen, G. (1980) Olfactory sensitivity to bile acids in salmonid fishes. *Acta physiol. Scand.*, **108**, 123–31.
Dubois-Dauphin, M., Döving, K.B. and Holley, A. (1980) Topographical relation between the olfactory bulb and the olfactory tract in tench (*Tinca tinca* L). *Chem. Senses*, **5**, 159–69.
Dulka, J.G., Stacey, N.E., Sorensen, P.W. and Van Der Kraak, G.J. (1987) Sex steroid pheromone synchronizes male–female spawning readiness in the goldfish. *Nature, Lond.*, **325**, 251–3.
Ebbesson, S.O.E., Meyer, D.L. and Scheich, H. (1981) Connections of the olfactory bulb in the piranha (*Serrasalmus nattereri*). *Cell Tissue Res.*, **216**, 167–80.
Echteler, S.M. (1984) Connections of the auditory midbrain in a teleost fish, *Cyprinus carpio*. *J. comp. Neurol.*, **230**, 536–51.
Echteler, S.M. (1985) Organization of central auditory pathways in a teleost fish, *Cyprinus carpio*. *J. comp. Physiol.*, **156A**, 267–80.
Echteler, S.M. and Saidel, W.M. (1981) Forebrain connections in goldfish support telencephalic homologies with land vertebrates. *Science*, Wash., D.C., **212**, 683–5.
Finger, T.E. (1975) The distribution of the olfactory tracts in the bullhead catfish, *Ictalurus nebulosus*. *J. comp. Neurol.*, **161**, 125–41.
Finger, T.E. (1980) Non-olfactory sensory pathway to the telencephalon in a teleost fish. *Science*, Wash., D.C., **210**, 671–3.
Fujita, I., Satou, M. and Ueda, K. (1984) A field-potential study of centripetal and centrifugal connections of the olfactory bulb in the carp, *Cyprinus carpio* (L.). *Brain Res. Amsterdam*, **321**, 33–44.
Fujita, I., Satou, M. and Ueda, K. (1985) Ganglion cells of the terminal nerve: morphology and electrophysiology. *Brain Res. Amsterdam*, **335**, 148–52.

Fujita, I., Satou, M. and Ueda, K. (1988) Morphology of physiologically identified mitral cells in the carp olfactory bulb: a light microscopic study after intracellular staining with horseradish peroxidase. *J. comp. Neurol.*, **267**, 253–68.

Fujita, I., Sorensen, P.W., Stacey, N.E. and Hara, T.J. The olfactory system, not the terminal nerve, functions as the primary chemosensory pathway mediating responses to sex pheromones in male goldfish. *Brain Behav. Evol.*, **38**, 313–21.

Grober, M.S., Baas, A.H., Burd, G., Marchaterre, M.A., Segil, N., Scholz, K. and Hodgson, T. (1987) The nervus terminalis ganglion in *Anguilla rostrata*: an immunocytochemical and HRP histochemical analysis. *Brain Res. Amsterdam*, **436**, 148–52.

Guselnikova, K.G. and Guselnikov, V.I. (1975) *Electrophysiology of Vertebrate Olfactory Analyzer*, Moscow University Press, Moscow, 256 pp. (In Russian)

Halpern-Sebold, L.R. and Schreibman, M.P. (1983) Ontogeny of centers containing luteinizing hormone-releasing hormone in the brain of platyfish (*Xiphophorus maculatus*) as determined by immunocytochemistry. *Cell Tissue Res.*, **229**, 75–84.

Hara, T.J. (1975) Olfaction in fish. *Progr. Neurobiol.*, **5**, 271–335.

Hara, T.J. (ed.) (1982) *Chemoreception in Fishes*, Elsevier, Amsterdam, 433 pp.

Hara, T.J. (1986) Role of olfaction in fish behaviour, in *The Behaviour of Teleost Fishes* (ed. T.J. Pitcher), Croom Helm, London, pp. 152–76.

Holmgren, N. (1920) Zur Anatomie und Histologie des Vorder- und Zwischenhirns der Knochenfische. *Acta zool., Stockh.*, **1**, 137–315.

Ichikawa, M. (1975) The central projections of the olfactory tract in the goldfish, *Carassius auratus*. *J. Fac. Sci. Tokyo Univ., Sec. IV*, **13**, 257–63.

Ichikawa, M. (1976) Fine structure of the olfactory bulb in the goldfish, *Carassius auratus*. *Brain Res. Amsterdam*, **115**, 43–56.

Ichikawa, M. and Ueda, K. (1977) Fine structure of the olfactory epithelium in the goldfish, *Carassius auratus*. A study of retrograde degeneration. *Cell Tissue Res.*, **183**, 445–55.

Ichikawa, M. and Ueda, K. (1978) Electron microscopic study of the terminal regions of the olfactory tract fibers in the goldfish telencephalic hemispheres. *J. Fac. Sci. Tokyo Univ., Sec. IV*, **14**, 127–34.

Ichikawa, M. and Ueda, K. (1979) Electron microscpic study of the termination of the centrifugal fibers in the goldfish olfactory bulb. *Cell Tissue Res.*, **197**, 257–62.

Ito, H. (1973) Normal and experimental studies on synaptic patterns in the carp telencephalon, with special reference to the secondary olfactory termination. *J. Hirnforsch.*, **14**, 237–53.

Ito, H. and Vanegas, H. (1983) Cytoarchitecture and ultrastructure of nucleus prethalamicus, with special reference to telencephalon, in a teleost (*Holocentrus ascensionis*). *J. comp. Neurol.*, **221**, 401–15.

Ito, H., Morita, Y., Sakamoto, N. and Ueda, S. (1980) Possibility of telencephalic visual projection in teleosts, Holocentridae. *Brain Res. Amsterdam*, **197**, 219–22.

Kah, O., Breton, B., Dulka, J.G., Nunez-Rodriguez, J., Peter, R.E., Corriga, A., Rivier, J.E. and Vale, W.W. (1986) A reinvestigation of the Gn–RH (gonadotrophin-releasing hormone) systems in the goldfish brain using antibodies of salmon Gn–RH. *Cell Tissue Res.*, **244**, 327–37.

Kleerekoper, H. (1969) *Olfaction in Fishes*, Indiana University Press, Bloomington, 222 pp.

Kosaka, T. (1980) Ruffed cell: a new type of neuron with a distinctive initial unmyelinated portion of the axon in the olfactory bulb of the goldfish (*Carassius auratus*). II. Fine structure of the ruffed cell. *J. comp. Neurol.*, **193**, 119–45.

Kosaka, T. and Hama, K. (1979) Ruffed cell: a new type of neuron with a distinctive initial unmyelinated portion of the axon in the olfactory bulb of the goldfish

(*Carassius auratus*). I. Golgi impregnation and serial thin sectioning studies. *J. comp. Neurol.*, **186**, 301–19.

Kosaka, T. and Hama, K. (1980) Presence of the ruffed cell in the olfactory bulb of the catfish, *Parasilurus asotus*, and the sea eel, *Conger myriaster. J. comp. Neurol.*, **193**, 103–17.

Kosaka, T. and Hama, K. (1981) Ruffed cell: a new type of neuron with a distinctive initial unmyelinated portion of the axon in the olfactory bulb of the goldfish (*Carassius auratus*). III. Three-dimensional structure of the ruffed cell dendrite. *J. comp. Neurol.*, **201**, 571–87.

Kosaka, T. and Hama, K. (1982) Structure of the mitral cell in the olfactory bulb of the goldfish (*Carassius auratus*). *J. comp. Neurol.*, **212**, 365–84.

Kosaka, T. and Hama, K. (1982–1983) Synaptic organization in the teleost olfactory bulb. *J. Physiol., Paris*, **78**, 707–19.

Koyama, Y., Satou, M., Oka, Y. and Ueda, K. (1984) Involvement of the telencephalic hemispheres and the preoptic area in sexual behavior of the male goldfish, *Carassius auratus*: a brain-lesion study. *Behav. neural Biol.*, **40**, 70–86.

Koyama, Y., Satou, M. and Ueda, K. (1985) Sexual behavior elicited by electrical stimulation of the telencephalic and preoptic areas in the goldfish, *Carassius auratus*. *Zool. Sci.*, **2**, 565–70.

Kyle, A.L. and Peter, R.E. (1982) Effects of brain lesions on spawning behaviour in the male goldfish. *Physiol. Behav.*, **28**, 1103–9.

Kyle, A.L., Stacey, N.E. and Peter, R.E. (1982) Ventral telencephalic lesions: effects on bisexual behavior, activity and olfaction in the male goldfish. *Behav. neural Biol.*, **36**, 229–41.

Levine, R.L. and Dethier, S. (1985) The connections between the olfactory bulb and the brain in the goldfish. *J. comp. Neurol.*, **237**, 427–44.

Liley, N.R. (1982) Chemical communication in fish. *Can. J. Fish. Aquat. Sci.*, **39**, 22–35.

MacLeod, N.K. (1976) Field potentials in the olfactory bulb of the codfish (*Gadus morhua*). *Comp. Biochem. Physiol.* **55A**, 297–9.

MacLeod, N.K. and Lowe, G.A. (1976) Field potentials in the olfactory bulb of the rainbow trout (*Salmo gairdneri*): evidence for a dendrodendritic inhibitory pathway. *Exp. Brain Res.*, **25**, 255–66.

Mori, K. (1987) Membrane and synaptic properties of identified neurons in the olfactory bulb. *Progr. Neurobiol.*, **29**, 275–320.

Mori, K., Kishi, K. and Ojima, H. (1983) Distribution of dendrites of mitral, displaced mitral, tufted, and granule cells in the rabbit olfactory bulb. *J. comp. Neurol.*, **219**, 339–55.

Moulton, D.G. (1967) Spatio-temporal patterning of response in the olfactory system. *Olfaction and Taste*, **2**, 109–16.

Moulton, D.G. (1976) Spatial patterning of response to odors in the peripheral olfactory system. *Physiol. Rev.*, **56**, 578–93.

Mozell, M.M. (1958) Electrophysiology of the olfactory bulb. *J. Neurophysiol.*, **21**, 183–96.

Münz, H. and Claas, B. (1987) The terminal nerve and its development in the teleost fishes, in *The Terminal Nerve (Nervus Terminalis). Structure, Function, and Evolution* (eds L.S. Demski and M. Schwanzel-Fukuda). Ann. NY Acad. Sci., **519**, 50–9.

Münz, H., Stumpf, W.E. and Jennes, L. (1981) LHRH-systems in the brain of platyfish. *Brain Res. Amsterdam*, **221**, 1–13.

Münz, H., Claas, B., Stumpf, W.E. and Jennes, L. (1982) Centrifugal innervation of the retina by luteinizing hormone releasing hormone (LHRH)-immunoreactive telencephalic neurons in teleostean fishes. *Cell Tissue Res.*, **222**, 313–23.

Murakami, T., Morita, Y. and Ito, H. (1983) Extrinsic and intrinsic fiber connections of the telencephalon in a teleost, *Sebastiscus marmoratus*. *J. comp. Neurol.*, **216**, 115-31.

Murakami, T., Fukuoka, T. and Ito, H. (1986) Telencephalic ascending acousticolateral system in a teleost (*Sebasticus marmoratus*), with special reference to the fiber connections of the nucleus preglomerulosus. *J. comp. Neurol.*, **247**, 383-97.

Nieuwenhuys, R. (1963) The comparative anatomy of the actinopterygian forebrain. *J. Hirnforsch.*, **6**, 171-92.

Nieuwenhuys, R. (1967) Comparative anatomy of olfactory centers and tracts, in *Progress in Brain Research*. Vol. 23, *Sensory Mechanisms* (ed. Y. Zotterman), Elsevier, Amsterdam, pp. 1-64.

Northcutt, R.G. and Braford, M.R., jun (1980) New observations on the organization and evolution of the telencephalon of actinopterygian fishes, in *Comparative Neurology of the Telencephalon* (ed. S.O.E. Ebbesson), Plenum, New York, pp. 41-98.

Northcutt, R.G. and Davis, R.E. (1983) Telencephalic organization in ray-finned fishes, in *Fish Neurobiology*. Vol. 2, *Higher Brain Areas and Functions* (eds R.E. Davis and R.G. Northcutt), University of Michigan Press, Ann Arbor, pp. 203-36.

Oka, Y. (1980) The origin of the centrifugal fibers to the olfactory bulb in the goldfish, *Carassius auratus*: an experimental study using the fluorescent dye primuline as a retrograde tracer. *Brain Res. Amsterdam*, **185**, 215-25.

Oka, Y. (1983) Golgi, electron-microscopic and combined Golgi–electron-microscopic studies of the mitral cells in the goldfish olfactory bulb. *Neuroscience.*, **8**, 723-42.

Oka, Y., Ichikawa, M. and Ueda, K. (1982) Synaptic organization of the olfactory bulb and central projection of the olfactory tract, in *Chemoreception in Fishes* (ed. T.J. Hara), Elsevier, Amsterdam, pp. 61-75.

Powers, J.B. and Winans, S.S. (1975) Vomeronasal organ: critical role in mediating sexual behavior of the male hamster. *Science*, Wash., D.C., **187**, 961.

Price, J.L. and Powell, T.P.S. (1970) The morphology of the granule cells of the olfactory bulb. *J. Cell Sci.* **7**, 91-123.

Raisman, G. (1972) An experimental study of the projection of the amygdala to the accessory olfactory bulb and its relationship to the concept of a dual olfactory system. *Exp. Brain Res.*, **14**, 395-408.

Satou, M. (1974) Electrical responses at various levels of the olfactory pathways in himé salmon, *Oncorhynchus nerka*. *Jap. J. Physiol.*, **24**, 389-402.

Satou, M. (1987) A neuroethological study of reproductive behavior in the salmon, in *Proc. 3rd Int. Symp. Reprod. Physiol. Fish* (eds D.R. Idler, L.W. Crim and J.M. Walsh), Memorial University Press, St. John's, Newfoundland, Canada, pp. 154-9.

Satou, M. (1990) Synaptic organization, local neuronal circuitry, and functional segregation of the teleost olfactory bulb. *Progr. Neurobiol.*, **34**, 115-42.

Satou, M., Ichikawa, M., Ueda, K. and Takagi, S.F. (1979) Topographical relation between olfactory bulb and olfactory tracts in the carp. *Brain Res. Amsterdam*, **173**, 142-6.

Satou, M., Fujita, I., Ichikawa, M., Yamaguchi, K. and Ueda, K. (1983) Field potential and intracellular potential studies of the olfactory bulb in the carp: evidence for a functional separation of the olfactory bulb into lateral and medial subdivisions. *J. comp. Physiol.*, **152A**, 319-33.

Satou, M., Oka, Y., Kusunoki, M., Matsushima, T., Kato, M., Fujita, I. and Ueda, K. (1984) Telencephalic and preoptic areas integrate sexual behavior in himé salmon (landlocked red salmon, *Oncorhynchus nerka*): results of electrical brain stimulation experiments. *Physiol. Behav.*, **33**, 441-7.

Savage, G.E. (1980) The fish telencephalon and its relation to learning, in *Comparative Neurology of the Telencephalon* (ed. S.O.E. Ebbesson), Plenum Press, London, pp. 129–74.
Scalia, F. and Ebbesson, S.O.E. (1971) The central projections of the olfactory bulb in a teleost (*Gymnothorax funebris*). *Brain Behav. Evol.*, **4**, 376–99.
Segaar, J. (1961) Telencephalon and behaviour in *Gasterosteus aculeatus*. *Behaviour*, **18**, 256–87.
Segaar, J. (1965) Behavioural aspects of degeneration and regeneration in fish brain: a comparison with higher vertebrates, in *Progr. Brain Res.* Vol. 14, *Degeneration in the Nervous System* (eds. M. Singer and J.P. Schade), Elsevier, Amsterdam, pp. 143–231.
Sheldon, R.E. (1909) The nervus terminalis in the carp. *J. comp. Neurol. Psychol.*, **19**, 191–201.
Sheldon, R.E. (1912) The olfactory tracts and centers in teleosts. *J. comp. Neurol.*, **22**, 177–339.
Shepherd, G.M. (1972) Synaptic organization of the mammalian olfactory bulb. *Physiol. Rev.*, **52**, 864–917.
Shepherd, G.M. (1977) The olfactory bulb: a simple system in the mammalian brain, in *Handbook of Physiology*, Section 1: *The Nervous System*. Vol. 1, Part 1, American Physiological Society, Bethesda, MD, pp. 945–68.
Shepherd, G.M. (1985) The olfactory system: the uses of neural space for a non-spatial modality. *Progr. clin. biol. Res.*, **176**, 99–114.
Shiga, T., Oka, Y., Satou, M., Okumoto, N. and Ueda, K. (1985) An HRP study of afferent connections of the supracommissural ventral telencephalon and the medial preoptic area in himé salmon (landlocked red salmon, *Oncorhynchus nerka*). *Brain Res.* Amsterdam, **361**, 162–77.
Sorensen, P.W., Hara, T.J. and Stacey, N.E. (1987) Extreme olfactory sensitivity of mature and gonadally-regressed goldfish to a potent steroidal pheromone, 17α, 20β-dihydroxy-4-pregnen-3-one. *J. comp. Physiol.*, **16A**, 305–13.
Sorensen, P.W., Hara, T.J., Stacey, N.E. and Goetz, F.Wm (1988) F prostaglandins function as potent olfactory stimulants that comprise the postovulatory female sex pheromone in goldfish. *Biol. Reprod.*, **39**, 1039–50.
Springer, A.D. (1983) Centrifugal innervation of goldfish retina from ganglion cells of the nervus terminalis. *J. comp. Neurol.*, **214**, 404–15.
Stacey, N.E. and Kyle, A.L. (1983) Effects of olfactory tract lesions on sexual and feeding behavior in the goldfish. *Physiol. Behav.*, **30**, 621–8.
Stacey, N.E., Sorensen, P.W., Dulka, J.G., Van Der Kraak, G.J. and Hara, T.J. (1987) Teleost sex pheromones: recent studies on identity and function, in *Proc. 3rd Int. Symp. Reprod. Physiol. Fish*, (eds D.R. Idler, L.W. Crim and J.M. Walsh), Memorial University Press, pp. 150–3.
Stell, W.K., Walker, S.E., Chohan, K.S. and Ball, A.K. (1984) The goldfish nervus terminalis: a luteinizing hormone-releasing hormone and molluscan cardioexcitatory peptide immunoreactive olfactoretinal pathway. *Proc. natn. Acad. Sci. U.S.A.*, **81**, 940–4.
Stell, W.K., Chohan, K.S. and Kyle, A.L. (1985) Substance P-immunoreactivity coexists with LHRH- and FMRFamide-immunoreactivity in nervus terminalis efferents to goldfish retina. *Invest. Ophthalmol. Vis. Sci.*, **26** (Supp.), 277.
Thommesen, G. (1978) The spatial distribution of odour induced potentials in the olfactory bulb of char and trout (Salmonidae). *Acta physiol. scand.*, **102**, 205–17.
Vanegas, H. and Ito, H. (1983) Morphological aspect of the teleostean visual system: a review. *Brain Res. Rev.*, **6**, 117–37.
Wirsig, C.R. and Leonard, C.M. (1987) Terminal nerve damage impairs the mating behavior of the male hamster. *Brain Res.* Amsterdam, **417**, 293–303.

Chapter four

Structure of the peripheral gustatory organ, represented by the siluroid fish Plotosus lineatus (Thunberg)

Klaus Reutter

4.1 INTRODUCTION

In recent years, our knowledge concerning the ultrastructure of fish taste buds (TBs) has been greatly enlarged through a series of studies done on different species of fish. Several reviews summarize the earlier light microscopical work and the more recent, especially the transmission and scanning electron microscopical work done on different vertebrates, and these reviews include the problems concerning fish TBs discussed here. The older work is extensively reviewed by Oppel (1900), Ebner (1902), Herrick (1902, 1903), Kolmer (1927), and Boeke (1934); the newer data and, in part, the earlier literature too, is collected and reported on by Cordier (1964), Bardach and Atema (1971), Murray (1971), Kapoor *et al.* (1975), Reutter (1978, 1982, 1986), Tucker (1983), Caprio (1988), Roper (1989) and Jakubowski and Whitear (1990). Besides this, there exists a bibliography on chemoreception in fishes which encompasses many articles concerning the morphology of TBs (Klaprat and Hara, 1984). All these sources together are a comprehensive compilation of fish TB literature from the early nineteenth century to now.

It seems to be generally accepted that the peripheral gustatory organ, the taste bud, of fishes consists of different types of cells, each of which possesses a different morphology, shape, and regional position within the sensory epithelium of the TB, and presumably different functions. Nevertheless, there is some confusion in the nomenclature of fish TB cells, particularly the functional interpretation of the elongated cells, which comprise the main part of the sensory epithelium. The so-called 'light sensory cells' and 'dark sensory cells' are synonymously named respectively 'sensory or gustatory cells' and 'sustentacular or supporting cells'. This controversy might partly be due to the extensive light microscopical work on fish taste buds in the decades before electron microscopy was introduced. As light microscopy involves the use of different stains, such as hematoxyline, azocarmine, methylene blue, trichrome- and silver stains, there are always some cells that stain more intensely than the others. I shall try to throw some light on these differences.

Another point of interest is that obviously the TBs of fishes belonging to different systematic groups are different. Most work in this field is focused on teleosts, especially on cyprinids and catfishes. The following describes the morphology of a teleostean TB, that of the plotosid *Plotosus lineatus* (synonym *P. anguillaris* Bloch). This species possesses very typical teleostean TBs which have not been described in detail up to now; so I feel free to describe the *Plotosus* TB without the need of repeating my previous findings, figures and interpretations. On the other hand, the description of the *Plotosus* TB might be of special interest to those who are working with these species to elucidate its physiological mechanisms of recognizing taste substances. The following description is restricted to the ultrastructure of the differentiated TB. Developmental aspects and histochemical work are mentioned only briefly and can be found in the above-listed reviews.

4.2 DISTRIBUTION AND GROSS ANATOMY OF TASTE BUDS

This description concentrates on the TBs occurring abundantly within the epidermis of all eight barbels of *Plotosus*. As in most other siluroids, many TBs also occur all over the body surface, especially that of the head. Inside the mouth cavity, the pharynx, the proximal oesophagus, the prebranchial region and on the gill arches, a large number of TBs are also found.

A barbel TB is of slender, ovoid shape and positioned upright within the stratified, squamous epithelium (epidermis) of the skin (Fig. 4.1). It is about 55 μm high and 35 μm wide. As a TB is regularly not as high as the epidermis itself, the organ is positioned on the top of an elongated papilla of the underlying connective tissue, the corium, which invades the basis of the

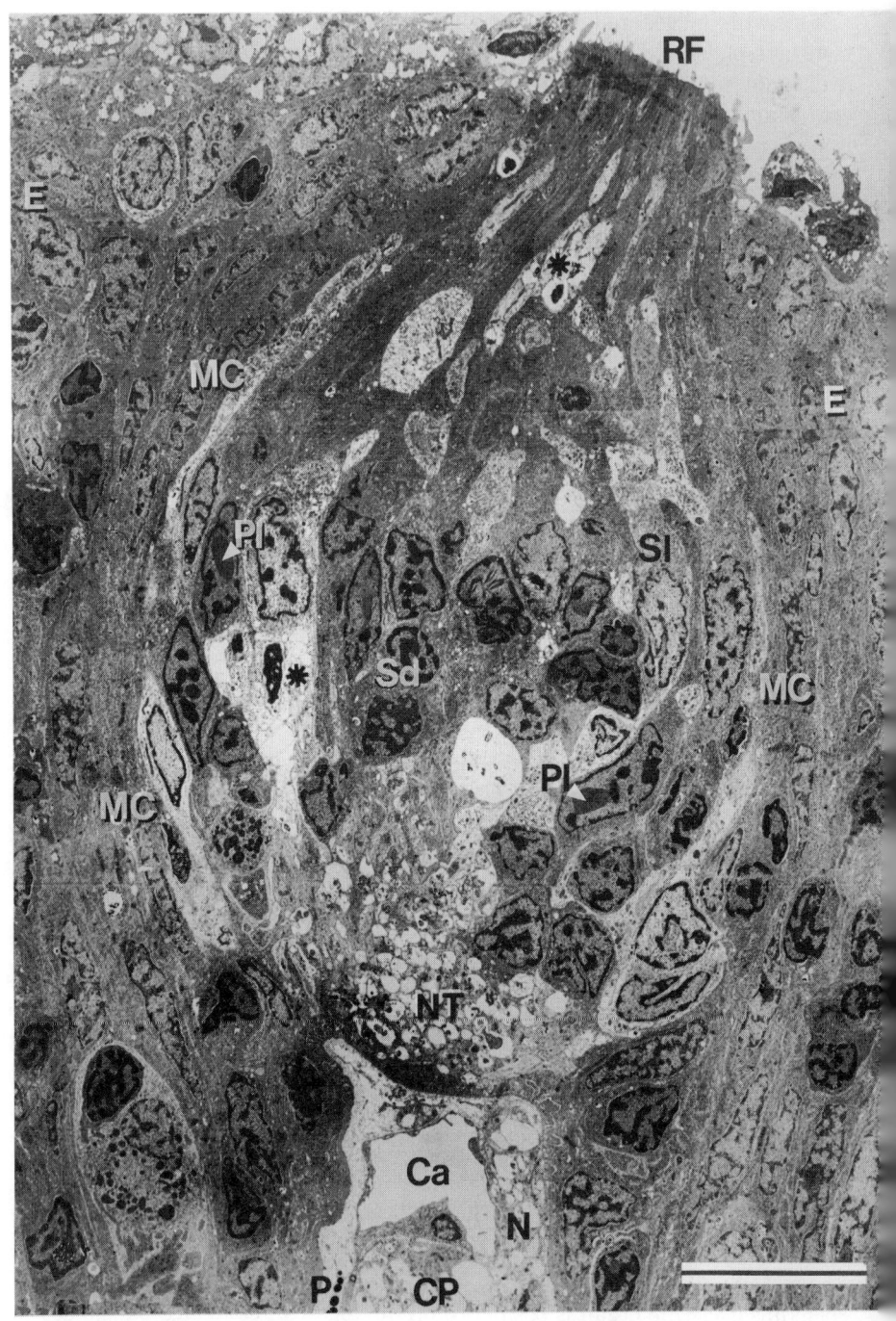

epidermis. The corium papilla is surrounded by a basal lamina and contains some fibrocytes, fibrillar collagenous structures, a reverse U-shaped capillary blood vessel and moderately myelinated TB nerves. Normally, the apex of the TB is somewhat elevated above the surface of the epidermis. The apical elongated part of the TB's sensory epithelium forms the TB's receptor field. It protrudes slightly from the epidermis so that the receptor field comes into a position above the epidermis. The sensory epithelium of a TB is surrounded by nonsensory slender and vertically flattened 'normal' epidermal cells, the marginal cells. Some of them surround the receptor field of the bud and help to form the apical TB hillock. The cellular constituents of the TB are shown in Fig. 4.1. The sensory epithelium consists of elongated cells which run more or less parallel and follow the longitudinal axis of the organ as seen in a TB sectioned longitudinally. The varying electron density within the sensory epithelium invites an interpretation of its cells as light and dark cells. The basal cells are situated at the basis of the TB, just on top of the basal lamina of the corium papilla. The disc-like cells are transversely arranged with respect to the longitudinal axis of the TB (Fig. 4.1). Each basal cell is reached by some of the slender and widely subdivided basal processes of the sensory cells. Between the basal cell(s) and the bases of the sensory cells, unmyelinated nerve fibres leading to the TB nerve are intermingled with these sensory cell processes (Fig. 4.1). All these structures together form the TB's nerve fibre plexus. This is the region where synaptic contacts occur. Within the main parts of the sensory epithelium no nerve fibres, and therefore no synapses, were found in this species.

4.3 FINE STRUCTURE OF TASTE BUDS

Sensory epithelium

The main part of a TB consists of the sensory epithelium. As mentioned above, the elongated cells of the sensory epithelium have been conventionally described as either light and dark cells, or as gustatory and supporting cells, respectively. Besides these extreme cell forms, intermediate and degenerating cells are also found.

Fig. 4.1 Taste bud situated in the epithelium of a maxillary barbel from *Plotosus lineatus* in longitudinal section (asterisk, degenerative cell). In this and the following electron micrographs the tissues were prepared as follows: barbel pieces taken from juvenile fish were fixed in 2% glutaraldehyde, osmicated in 1% osmium tetroxide, embedded into Araldite epoxy resin and cut with the ultramicrotome (Ultracut; Reichert, Wien, Austria). After staining with lead citrate and uranyl acetate, the ultrathin sections were viewed under an electron microscope (EM 300, Philips, The Netherlands) at 60 kV. Scale bar 10 μm. For abbreviations in this and other figures, see list on p. 75.

Light cells

The cytoplasm of light cells is relatively clear and electron-lucent, when the tissue is fixed and stained as described in Fig. 4.1. These cells are distinctly elongated and slender; they are wider only in the nuclear region (Figs 4.1 and 4.2). At their apical pole they terminate in only one club-shaped or conoid microvillar process, the large receptor villus. In cross-sections of TBs the light cells are roundish, especially at the level of their nuclei. Their basal cell processes are intensely divided.

The irregularly lobed and somewhat elongated nucleus is relatively light, and exhibits compartments or hetero- and euchromatin. It often contains a paracrystalline and relatively electron-dense structure. The nuclear envelope is rich in nuclear pores, and the perinuclear cisterna is well developed.

The cytoplasm comprises many organelles. In the perinuclear, especially the supranuclear, region it contains numerous Golgi systems, which are more and more elongated and vertically positioned to the middle part of the cell. The apical third of the cytoplasm contains no Golgi systems. In close association with Golgi systems are found many vesicular structures of different sizes. The rough endoplasmic reticulum is rare, but a number of tubular, curved and electron-dense smooth endoplasmic reticulum occur throughout the cell body. Free ribosomes, often organized as polyribosomes, are rich in the nuclear region. Elongated mitochondria are abundant in the apical region, running parallel to the cell. Intermediate filaments (tonofilaments) and microtubules occur in all parts of the cytoplasm. Lysosomes, multivesicular bodies, centrioles and glycogen granules are occasionally found. Within the basal processes of the cells, clear vesicles (30–60 nm in diameter) are numerous.

Dark cells

All in all, dark cells are relatively more electron-dense and rich in organelles compared with the light cells (Figs 4.1 and 4.2). Dark cells are slender and reach from the TB's base up to the receptor field, where they terminate in several small microvilli. Especially in the cross-sectioned TB, these cells are strongly divided into lobed processes which ensheath, as in the apical parts of the TB, the neighbouring light cells. In longitudinal sections the lobed processes of dark cells appear as thin sheet-like structures. In the perinuclear region the cell body is also divided into lobes which, in part, are wrapped

---

Fig. 4.2 Receptor field of a *Plotosus* TB in longitudinal section. Note the microfibrillar core of light cells and their large receptor villi. Between the sensory cells there are desmosomes and tight junctions (arrows). In this section only a few small receptor villi are cut. The inset is the continuation of the receptor field on the left. Scale bar, 1 μm.

around the light sensory cells. The basal parts of the cells are heavily divided into lobar processes which are intermingled with those of the light cells and fibres of the nerve plexus. They pass down to the basal cells and to the basal lamina that surrounds the corium papilla.

The nucleus is darker than in light cells, somewhat elongated, and runs parallel to the axis of the cells and that of the TB, respectively. Hetero- and euchromatin are distinct areas, and paracrystalline inclusions may also be found (Fig. 4.1). The nuclear envelope clearly shows nuclear pores and perinuclear cisterna. In the perinuclear region the cytoplasm contains some rough endoplasmic reticulum as well as elongated and tubular smooth endoplasmic reticulum. Numerous vesicles of different sizes are present in all parts of the cell. Golgi systems are well developed and mitochondria are numerous. In contrast to the light cells, the dark possess more intermediate filaments (tonofilaments), which are normally organized into bundles running longitudinally through the cell. In the apical region and within the basal processes of dark cells the filament bundles are found in large numbers. Microtubules are sparsely scattered in all parts of the cells. A centriole is occasionally found. Lysosomes, polyribosomes and glycogen granula occur. Clear vesicles (30–60 nm) and occasionally some dense-cored vesicles are also found.

Other cells

Intermediate cells

This group of cells comprises all cell types other than light or dark cells. They are situated in the nuclear region of the TB and do not reach either the receptor field or the basal structures of the TB. They contain organelles similar to those found in light and dark cells. It is thus assumed that they are at developmental stages leading to the main TB cell types, the light and dark cells.

Degenerate cells

Occasionally, degenerate cells are found within all parts of the sensory epithelium (Fig. 4.1). These cells are the final stages of light and dark ones. They have pycnotic nuclei and relatively light cytoplasm, with swollen mitochondria and irregularly lobed vesicles and membrane systems. Golgi systems are not usually found.

Receptor field

The receptor field consists of the apical endings of the sensory cells (Fig. 4.2). The large receptor villi (about 1.5 to 2 µm long and 0.2 µm wide), the end structures of the light cells, are cylindrical or conical. Their cytoplasm is filled with densely packed, parallel microfilaments, forming relatively electron-dense cores. The core passes down the apical part of its sensory cell for some distance. There are no other organelles within the villus, with the exception of some tubular profiles of the smooth endoplasmic reticulum.

The small villi of the receptor field are the terminal structures of the dark cells. Each cell may have several microvilli. Thus, more small villi than large ones are present within a receptor field. Small villi are cylindrical, about 1 µm long and 50–100 nm wide. Their cytoplasm contains filamentous structures which run parallel along the length of the villus. Histochemical ultrastructural preparations (for references see Discussion) show that the villi are covered with some mucous substances which are obviously the product of the dark cells, and the marginal cells. Occasionally a fuzzy coat of the glycocalyx is found. The large receptor villi penetrate the mucus of the receptor field's surface.

Basal cells

Basal cells are exclusively situated at the TB's base (Figs 4.1, 4.3 and 4.4). There are up to five basal cells lying directly on the basal lamina, which surrounds the corium papilla. Normally, they fill out distinct depressions of the basal lamina and are orientated transversely to the longitudinal axis of the TB. A basal cell is disc-like in shape, and its smooth basal plasmalemma runs parallel to the basal lamina and contacts it directly and without hemidesmosomes. The apical plasmalemma is adjacent to the basal side of the nerve fibre plexus. It is more irregular and impressed by nerve fibres and sensory cell processes. Sometimes the apical plasmalemma is extended into the nerve fibre plexus to form spine- or microvillus-like structures. These spines (about 0.5 µm long) occur only occasionally in the lateral parts of the cell and interdigitate with the nerve fibre plexus.

The nuclei of basal cells, elliptic and sometimes lobed, are orientated transversely. At the margin of the nucleus the karyoplasm consists of distinct areas of heterochromatin; euchromatin areas are located more centrally. The nuclear envelope shows perinuclear cisternae and some nuclear pores. The cytoplasm is rich in small roundish or elongated mitochondria and both rough and smooth endoplasmic reticulum. Golgi systems are common; some multivesicular bodies, small vesicles (30–60 nm in diameter), large, often dense-cored vesicles (60–80 nm in diameter), free ribosomes, and some glycogen granula are regularly found. Intermediate filaments and microtubules occur all over the cytoplasm, with no particular arrangement. The centrioles are occasionally found in the apical, perinuclear region of the cytoplasm.

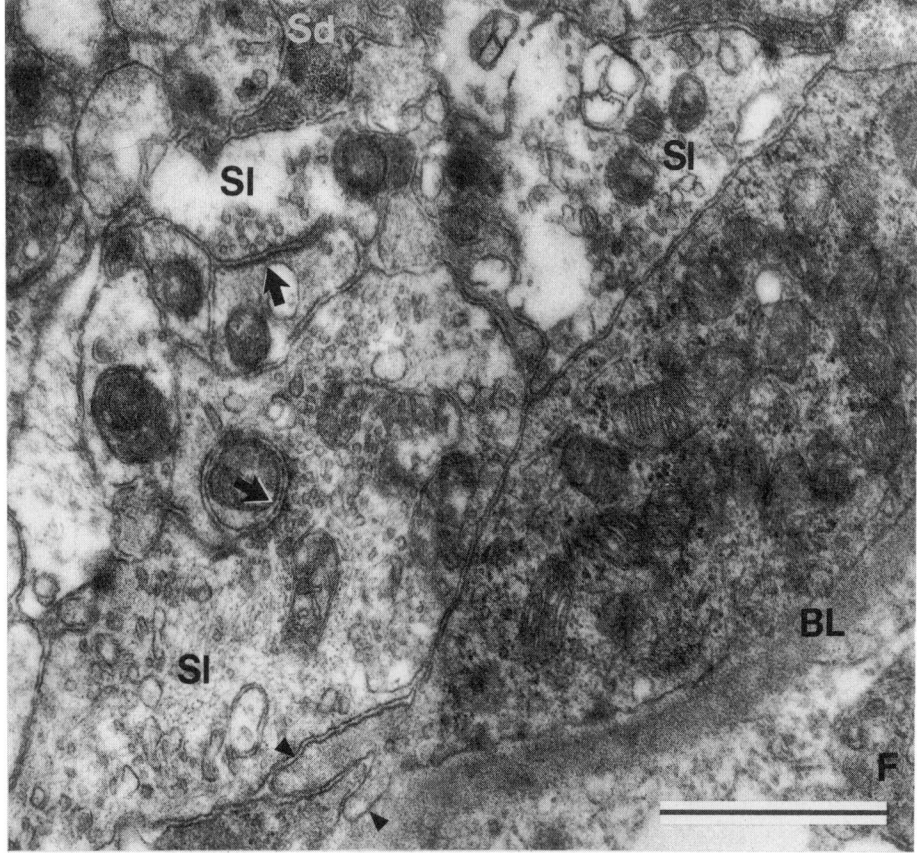

Fig. 4.3 Basal part of a *Plotosus* TB, longitudinal section. Small arrowheads (lower left) on the lateral side of the basal cell denote two spines; large arrows show two processes of light cells that synapse to nerve fibres. Scale bar, 1 μm.

Nerve fibre plexus

The nerve fibre plexus (Fig. 4.1) consists of at least three different structures which are mingled together. There are a number of roundish profiles of different sizes, more or less cross-cut nerve fibres. The number of nerve fibres is variable. They are myelin-free, and no signs of a Schwann or glial cell are found within the region of the plexus. The nerve fibres are rich in mitochondria, and contain some filamentous and tubular structures, sometimes some small vesicles as well. Degenerated nerve fibres with swollen mitochondria, which often contain myelin-like lamellae, are occasionally found.

Between the mass of nerve fibres the basal processes of both light and dark cells are intermingled. The course of the processes is irregular and, in most of

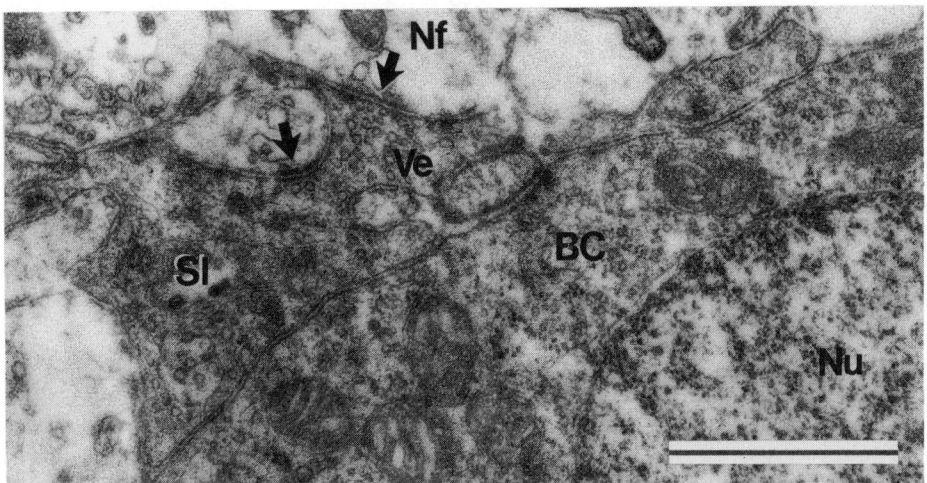

Fig. 4.4 Basal part of a *Plotosus* TB, longitudinal section. A light cell's process synapses to two nerve fibre profiles (arrows). Scale bar, 1 μm.

the ultrathin sections, they are cut only for a short distance. The processes are often deeply indented by the nerve fibres; in some cases, they fill up depressions of the nerve fibres. The cytoplasm of the processes of light cells is relatively clear and contains some mitochondria, intermediate filaments and small vesicles (30–60 nm in diameter) which are clustered together. The more electron-dense cytoplasm of the dark cell processes contains a few mitochondria, bundles of intermediate filaments and sometimes small vesicles.

Synaptic interconnections

The morphology of the synapses of *Plotosus* TBs, like those in other teleosts, is not clearly organized (Figs 4.3 to 4.6). The synaptic cleft is generally well organized, but the presynaptic membrane specializations, including the presynaptic web and the postsynaptic densities, are not always distinct. Synaptic vesicles are found on the presynaptic side. There occur small clear vesicles (30–60 nm in diameter) and, sporadically, some large dense-cored vesicles (60–100 nm in diameter).

Synapses are situated at the basal part of the perinuclear cell bodies of sensory cells, i.e. at their divided processes and at the basal cells. Light cells (presynaptic side) synapse to the fibres of the nerve fibre plexus. These synapses are relatively rich in small clear vesicles. Darker profiles (presynaptic side) seldom synapse to the nerve fibres of the plexus. These synapses contain clear vesicles and a few dense-cored ones. The basal cells (presynaptic side) form synapses to the fibres of the nerve fibre plexus. They possess small clear and some dense-cored vesicles. Synapses situated between the

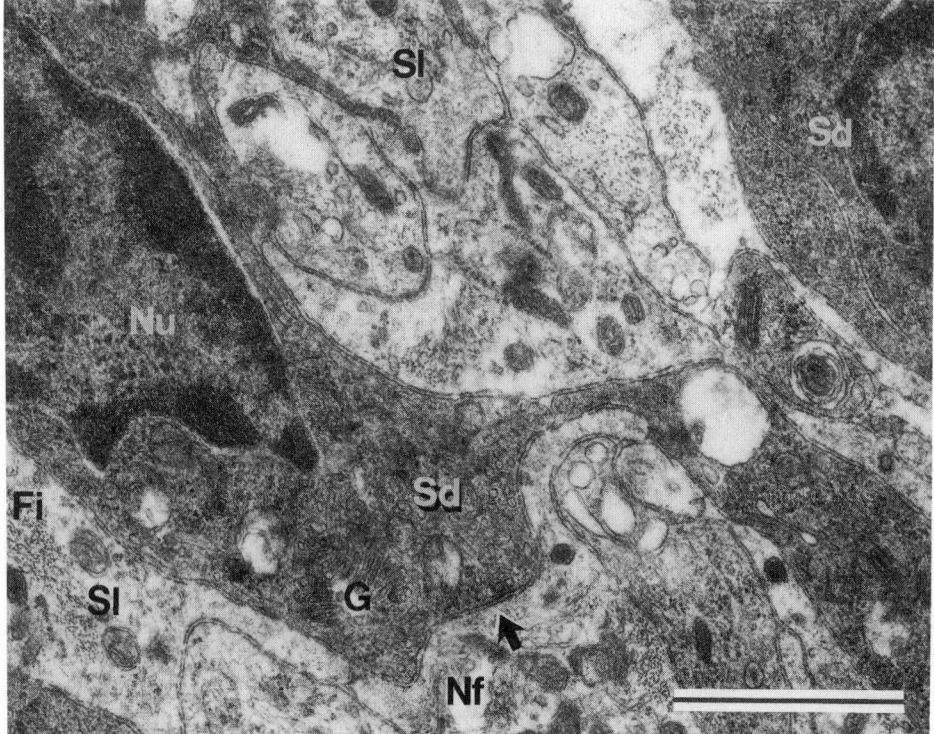

Fig. 4.5 Subnuclear region of a *Plotosus* TB in longitudinal section. A dark cell contains small clear vesicles and synapses to a nerve fibre (arrow). Scale bar, 1 μm.

nerve fibres of the plexus (presynaptic side) and the sensory or basal cells may exist. In rare cases, accumulations of some small vesicles are found within the nerve fibres, but there are no distinct signs of membrane specializations. Therefore the existence of such synapses remains uncertain.

Intercellular junctions

All the cellular components of the TB are densely packed, with an intercellular space of about 20–30 nm. As mentioned above, the dark cells possess lobe-like processes that ensheath the apical parts of the light cells. In the perinuclear region and in the upper part of a TB the plasmalemmata of the neighbouring dark cells' processes are interdigitated intensely, but interdigitations situated between dark and light cells are also found. Additionally, the cellular processes are attached to each other by small desmosomes. Desmosomes occur also between light and dark cells, and, in some cases, between neighbouring light cells. The basal processes of both types of cell contact each other by desmosomes. Furthermore, desmosomes are located

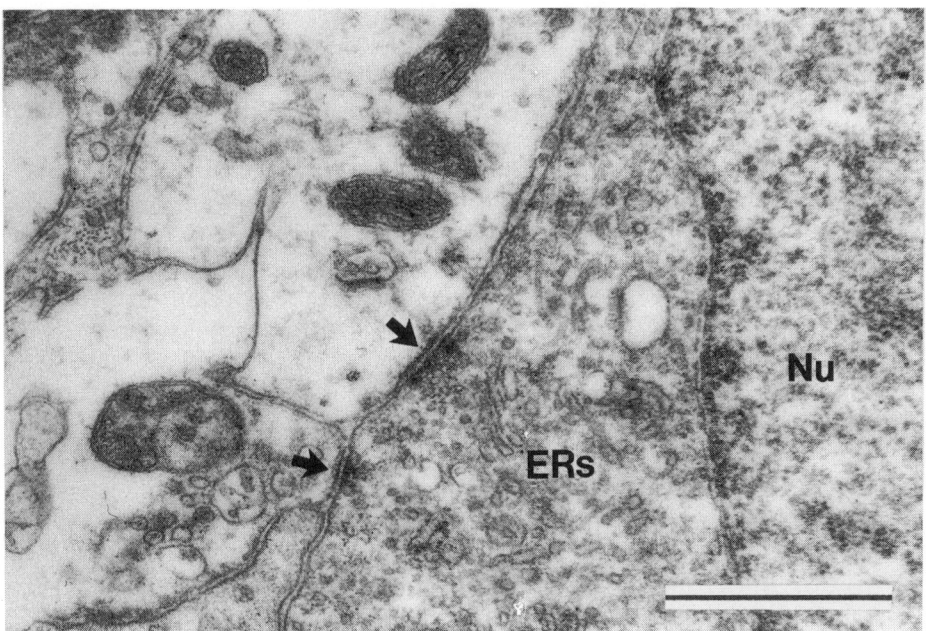

Fig. 4.6 Basal cell of a *Plotosus* TB in longitudinal section, showing two synapses (arrows) between the basal cell and fibres of the TB's nerve fibre plexus. Scale bar, 1 µm.

between these processes and the basal cells. Dark cell processes which reach down to the actual basis of the TB are rich in filamentous structures and contact the basal lamina by hemidesmosomes.

The apical endings of the light and dark cells are in close contact with each other through tight junctions (Fig. 4.2).

Marginal cells

Marginal cells do not count among the TB cells. They form, as normal epithelial cells, the borderline between a TB and the neighbouring stratified squamous epithelium (Fig. 4.1). They are so named because of their topical position. Normally, the marginal cells are elongated or flattened and in their arrangement follow the shape of the TB. Like other normal epithelial cells they form interdigitations and desmosomes. Marginal cells which are located adjacent to sensory cells are interdigitated and connected to each other by desmosomes. In the upper region of the epithelium the marginal cells are abundant, roundish, and somewhat elevated from the surface level of the skin, forming a small hillock with a TB in its centre. These apically situated marginal cells possess Golgi systems and vesicular structures of different sizes with secretory function.

Dermal (corium) papilla

Normally, a TB is situated directly atop a slender papilla, which consists of connective tissue and belongs to the dermis resp. corium (Fig. 4.1). The height of a papilla depends on the thickness of the epidermis, and varies between 20 and 100 µm. The papilla is marked off by the basal lamina. The apex of the papilla is flat, and the basal lamina shows some distinct depressions where the basal cells are located. The connective tissue of the papilla consists of fibrocytes, some of which are regularly positioned transversely to the end of the papilla. Some melanocytes lie within the papilla. The intercellular space of the papilla is rich in collagenous fibres. The loop-like capillary blood vessel and the TB nerve are the main structures of the papilla. Both structures run through the whole length of the papilla. The conventionally structured wall of the vessel includes endothelial cells rich in pinocytotic vesicles. Within the basal part of the corium papilla, the TB nerve consists of some slightly myelinated nerve fibres, whereas in its upper region they are reduced or totally lacking. These unmyelinated nerve fibres penetrate the basal lamina through a distinct hole and form the nerve fibre plexus. The TB nerve fibres are rich in mitochondria and contain some microtubules and intermediate filaments (neurofilaments), but vesicles are rare.

4.4 DISCUSSION

One of the most frequently discussed points in all the literature dealing with the fish gustatory system is the question of cell types present in the TBs. Recently, Roper (1989) published a table in which he summarizes 'a small sample of the extant literature'. It discloses that in different vertebrate groups, different nomenclatures are used for the same or corresponding cells or cell types. The table also shows the differences in TB synaptology found by different authors. The TB cell nomenclature commonly employed stems from early TB descriptions based on light microscopical studies. In these studies some of the TB cells clearly stain more intensely than the others. But in the transmission electron microscope (TEM) there are distinct differences caused by electron density, provided that fixation and staining of the tissues are done in the ordinary way, i.e. by the use of glutaraldehyde and osmium tetroxide fixation, and uranyl acetate and lead citrate staining. These differences in electron density originally led to the adoption of the light and dark cell nomenclature introduced by the light microscopists. As a further and essential point, the cells of both types possess synapses and therefore are involved in sensory and neuronal transmission. There is no doubt that the light cells are the ones that preferentially transmit chemosensory signals to the central nervous system. They possess ultrastructural equipment and numerous synapses which favour them for this task compared with the dark cells. Occasionally, synapses are also found in dark cells, and therefore they, too, may be classified as sensory. Judging from their structural organization, they also seem to fulfil other responsibilities, primarily of additional, such as supportive, nutritional and secretory, functions. This is in disagreement with Jakubowski and Whitear's view (1990); based on a large number of ultrastructural observations of fish TBs, they favour the 'gustatory cell'–'supporting cell' nomenclature. It is important that there also exist some species-specific differences. In my opinion, a highly developed teleostean TB as represented by siluroid fishes possesses a distinct sensory epithelium, nerve fibre plexus, and several basal cells. Possibly, fish TBs that are more simply organized have a reduced nerve fibre plexus and often no basal cells, as is the case for external TBs situated in the skin in the Gadidae and some other species (Jakubowski and Whitear, 1990). There is probably a correlation between the organizational state of a TB and the state of differentiation of its cells; well-developed TBs may have more specialized cells such as light and dark cells than TBs of lower organization.

The discussion about light and dark 'sensory' cells is still continuing. For this reason, in my laboratory we decided to do some work concerning TB synaptology using different fixatives and ultrastructural histochemical tests. This would be helpful, as even in glutaraldehyde/osmium-fixed and uranylacetate/lead citrate-stained tissues the light and dark cells may appear at nearly the same electron contrast (e.g. Grover-Johnson and Farbman, 1976). But at the moment there is no need to change the light–dark cell nomenclature. In previous papers on *Ameiurus* (= *Ictalurus*) TBs I described synapses which are situated between the basal processes (presynaptic side) of both sensory cells and basal cells (Reutter, 1971, 1978, 1982, 1986). In the present study I could not find any such contacts. Therefore, in respect of the *Plotosus* TB my schematical sketches on TB synapses (Reutter, 1971, 1986) must be revised.

At their apices the sensory cells of a TB end in microvillus-like structures, which together organize the receptor field. The term 'receptor field' may be more appropriate than 'taste pore' (Jakubowski and Whitear, 1990), because in fish TBs the receptor villi are normally arranged in a flat, disc-like order, as seen in the scanning electron microscope. Taste pores in the stricter sense are common in the TBs of mammals (for example Murray, 1973, 1978, 1986).

It seems to be generally accepted that the receptor villi are the loci where the initial events of chemoreception take place. In freeze-etch preparations obtained from the *Ameiurus* TB (Reutter and Bardele, 1983; Reutter, 1986), both the membranes of large and small receptor villi contain more intramembranous particles than the remaining sensory cell membranes. There seems to be a direct relationship between the intramembranous particles and the receptor sites or receptor proteins.

In conventional (see above) TEM preparations, the mucus covering the receptor field is hardly detected. But, by the use of scanning electron microscopy done on freeze-dried TBs (Reutter, 1980) or ruthenium-red-stained ultrathin TB-sections (Reutter, 1986, 1987), as well as light (Witt and Reutter, 1988) and electron microscopic lectin-binding studies (Witt and Reutter, 1990), this mucus layer can be identified. Lectin-binding studies also allow differentiation between the cells of the sensory epithelium. In general, the apical parts of dark cells contain more lectin-binding vesicles than light cells. Intermediate cells also have some lectin-binding abilities. In special lectin-binding tests, the apically situated marginal cells are also marked: beside the sensory epithelium, the marginal cells may provide the receptor field with mucous substances. For this reason the marginal cells are separately listed as special cells on the TB's periphery. They seem to function in a way similar to the 'perireceptor events' described in the olfactory organ (Getchell et al., 1988).

The basal cells of a teleostean TB are of special interest, too. It seems to be accepted that these cells are different from the basal cells of the mammalian TB, which are considered to be stem cells or regenerative cells (Murray, 1973). Especially in view of the villus-like processes or spines that they often show (Jakubowski, 1983; Reutter, 1986, 1987; Toyoshima, 1989; Jakubowski and Whitear, 1990), and as they are rich in a biogenic monoamine (serotonin; Reutter, 1971; Nada and Hirata, 1977; Toyoshima et al., 1984) and in neurone-specific enolase (NSE; Toyoshima, 1989), some discussion should be made in the context of Merkel cells, belonging to paraneurones. (Nevertheless, recently all the TB sensory cells have been considered to be paraneurones. Unfortunately, the basal cells of fish TBs were not mentioned in this context (Fujita et al., 1988).)

The function of the basal cells is still unclear. Judging from their synaptic connections they are thought to be interneurones (Reutter, 1971, 1986), and their resemblance to Merkel cells may suggest a mechanoreceptive function (Jakubowski, 1983; Reutter, 1986, 1987). However, Jakubowski and Whitear (1990) emphasized distinctions between basal cells and Merkel cells, especially in their synaptic morphology. As paraneurones in the stricter sense, they are possibly paracrine cells (Toyoshima et al., 1984). Since the basal cells of the Ameiurus TB possess at least two neurotransmitter systems, a cholinergic and serotoninergic (Reutter, 1971), they could possibly serve different functions.

Intermediate and degenerating cells are not normally considered to be distinct TB cell types. However, I should like to list them separately because cell renewal within the TB is not well understood. Cell renewal within the sensory epithelium starts from basally situated epithelial (marginal) cells; at 30 °C taste cell renewal takes 10–12 days (Raderman-Little, 1979). There is no information about cell renewal of basal cells in fish TBs.

All the synapses found in the *Plotosus* TB are afferent. The existence of efferent synapses is still an open question, although structures that have been interpreted as efferent synapses are discussed for vertebrate TBs in general by Roper (1989) and, in relation to fish TBs, by Jakubowski and Whitear (1990). Desgranges (1966) was the first to describe an efferent synapse in the *Ameiurus* TB. In gustatory (light) cells, Jakubowski and Whitear (1990) moderately often found flattened subsynaptic (postsynaptic) cisternae of the endoplasmic reticulum being in close relation to nerve fibres. Such cisternae are thought to be a distinct sign of efferent synapses, as they are typically found as the end structures of efferent nerve fibres in the cochlea (Emmerling *et al.*, 1990; review: Pappas and Waxman, 1972).

The afferent nerve fibres of a TB situated in the exterior skin, including barbel skin, run to the central nervous system by way of the facial nerve, cranial nerve VII (Herrick, 1902, 1903; Atema, 1971; Kapoor *et al.*, 1975; Reutter, 1986; Finger, 1988; Jakubowski and Whitear, 1990). (The TBs of the oropharyngeal and branchial region and of the oesophagus are innervated by cranial nerves IX (glossopharyngeal) and X (vagus; references as above)). It must be mentioned that within the barbel's nerve, and presumably within the TB nerve, fibres are also found that belong to cranial nerve V, the trigeminal nerve. Some of these trigeminal fibres belong to free nerve endings and perigemmal nerve fibres of the barbel skin, and are mechanoreceptive (tactile) in function (Herrick, 1902, 1903; Atema, 1971); the others serve the fish's common chemical sense (Parker, 1922).

ACKNOWLEDGEMENTS

Specimens of *Plotosus lineatus* were a gift from Dr T. Marui (Kagoshima, Japan). The tissues were fixed by Dr E. Zeiske (Hamburg, FRG) during a stay in the laboratory of Dr J. Caprio (Baton Rouge, LA). The help of these colleagues and friends is greatly acknowledged. Electron microscopical work was carried out by Mr Gerd Geiger, photographic work was done by Mr Manfred Mauz and the manuscript was typewritten by Ms Karin Tiedemann. Many thanks for all the skilful support I received! Many thanks also to the reviewers for their detailed and useful comments on the manuscript.

ABBREVIATIONS

BC	Basal cell
BL	Basal lamina
Ca	Capillary
CP	Corium papilla
D	Desmosome
E	Epidermis
ER	Endoplasmic reticulum

ERs	Smooth endoplasmic reticulum
F	Fibrocyte
Fi	Intermediate filament
Fm	Microfilament
G	Golgi system
I	Interdigitation
MC	Marginal cell
Mt	Microtubule
N	Taste bud nerve
Nf	Nerve fibre
NSE	Neurone-specific enolase
NT	Nerve fibre plexus of TB
Nu	Nucleus
P	Pigment cell
PI	Paracrystalline inclusion
RF	Receptor field
Sd	Dark cell
Sl	Light cell
TB	Taste bud
TEM	Transmission electron microscope
Ve	Synaptic vesicle
Vl	Large receptor villus
Vs	Small receptor villus

REFERENCES

Atema, J. (1971) Structures and functions of the sense of taste in the catfish (*Ictalurus natalis*). *Brain Behav. Evol.*, **4**, 273–94.

Bardach, J.E. and Atema, J. (1971) The senses of taste in fishes, in *Handbook of Sensory Physiology*. Vol. 4, *Chemical Senses*, Part 2, *Taste* (ed. L.M. Beidler), Springer, Berlin, pp. 293–336.

Boeke, J. (1934) Organe mit Endknospen und Endhügeln nebst eigesenkten Organen, in *Handbuch der vergleichenden Anatomie der Wirbeltiere* (eds L. Bolk, E. Göppert, E. Kallius and E. Lubosch), Urban and Schwarzenberg, Berlin, pp. 949–88.

Caprio, J. (1988) Peripheral filters and chemoreceptor cells in fishes, in *Sensory Biology of Aquatic Animals* (eds J. Atema, R.R. Fay, A.N. Popper and W.N. Tavolga), Springer-Verlag, New York, pp. 313–38.

Cordier, R. (1964) Sensory cells, in *The Cell: Biochemistry, Physiology, Morphology* (eds J. Brachet and A.E. Mirsky), Academic Press, New York, pp. 313–86.

Desgranges, J.C. (1966) Sur la double innervation des cellules sensorielles des bourgeons du goût des barbillons du poisson-chat. *C.r. Acad. Sci.*, [D] Paris, **263**, 1103–6.

Emmerling, M.R., Sobkowicz, H.M., Levenick, C.V., Scott, G.L., Slapnick, S.M. and Rose, J.E. (1990) Biochemical and morphological differentiation of acetylcholinesterase-positive efferent fibers in the mouse cochlea. *J. Electron Microsc. Tech.*, **15**, 123–43.

Ebner, V. von (1902) Verdauungsorgane – Von den Geschmacksknospen, in *A. Koelliker's Handbuch der Gewebelehre des Menschen*, Engelmann, Leipzig, pp. 18–31.

References

Finger, T.E. (1988) Organization of chemosensory systems within the brains of bony fishes, in *Sensory Biology of Aquatic Animals* (eds J. Atema, R.R. Fay, A.N. Popper and W.N. Tavolga), Springer-Verlag, New York, pp. 339–63.

Fujita, T., Kanno, T. and Kobayashi, S. (1988) *The Paraneuron*, Springer-Verlag, Tokyo, 367 pp.

Getchell, M.L., Zielinski, B. and Getchell, T.V. (1988) Odorant and autonomic regulation of secretion in the olfactory mucosa, in *Molecular Neurobiology of the Olfactory System* (eds F. Margolis and T.V. Getchell), Plenum, N.Y., pp. 71–98.

Grover-Johnson, N. and Farbman, A.J. (1976) Fine structure of taste buds in the barbel of the catfish, *Ictalurus punctatus*. *Cell Tissue Res.*, **169**, 395–403.

Herrick, C.J. (1902) The organ and sense of taste in fishes. *Bull. U.S. Fish. Comm.*, **1902**, 237–72.

Herrick, C.J. (1903) On the phylogeny and morphological position of the terminal buds of fishes. *J. comp. Neurol. Psychol.*, **13**, 121–38.

Jakubowski, M. (1983) New details of the ultrastructure (TEM, SEM) of taste buds in fishes. *Z. mikrosk.-anat. Forsch.*, **97**, 849–62.

Jakubowski, M. and Whitear, M. (1990) Comparative morphology and cytology of taste buds in teleosts. *Z. mikrosk.-anat. Forsch.*, **104**, 529–60.

Kapoor, B.G., Evans, H.E. and Pevzner, R.A. (1975) The gustatory system in fish. *Adv. mar. Biol.*, **13**, 53–108.

Klaprat, D.A. and Hara, T.J. (1984) A bibliography on chemoreception in fishes, 1807–1983. *Can. Tech. Rep. Fish. Aquatic Sci.* no. 1268, 47 pp.

Kolmer, W.C. (1927) Geschmacksorgan, in *Handbuch der mikroskopischen Anatomie des Menschen*. Bd. 3, Teil 1: *Haut, Milchdrüse, Geruchsorgan, Geschmacksorgan, Sehorgan* (ed. W. von Möllendorf), Springer-Verlag, Berlin, pp. 154–91.

Murray, R.G. (1971) Ultrastructure of taste receptors, in *Handbook of Sensory Physiology*. Vol. 4, *Chemical senses*, Part 2, *Taste* (ed. L.M. Beidler), Springer-Verlag, Berlin, pp. 31–50.

Murray, R.G. (1973) The ultrastructure of taste buds, in *The Ultrastructure of Sensory Organs* (ed. J. Friedemann), North Holland Publishing Company, Amsterdam, pp. 1–81.

Murray, R.G. (1978) Gustatory receptor cells, in *Functional Morphology of Receptor Cells* (ed. J.W. Rohen), Akademie der Wissenschaften und der Literatur Mainz, Steiner Verlag, Wiesbaden, pp. 88–118.

Murray, R.G. (1986) The mammalian taste bud type III cell: a critical analysis. *J. Ultrastruct. molec. Struct. Res.*, **95**, 175–88.

Nada, O. and Hirata, K. (1977) The monoamine-containing cell in the gustatory epithelium of some vertebrates. *Arch. Histol. Jpn., Niigata, Jpn*, **40**, 197–206.

Oppel, A. (1900) *Lehrbuch der vergleichenden mikroskopischen Anatomie der Wirbeltiere*. III. *Mundhöhle, Bauchspeicheldrüse und Leber–Nerven und Sinnesorgane der Zunge und Mundhöhle*, Fischer, Jena, pp. 439–86.

Pappas, G.D. and Waxman, S.G. (1972) Synaptic fine structure–morphological correlates of chemical and electronic transmission, in *Structure and Function of Synapses* (eds G.D. Pappas and D.P. Purpura), North Holland Publishing Company, Amsterdam, pp. 1–43.

Parker, G.H. (1922) *Smell, Taste and Allied Senses in the Vertebrates*. Lippincott, Philadelphia, 192 pp.

Raderman-Little, R. (1979) The effect of temperature on the turnover of taste bud cells in catfish. *Cell Tissue Kinet.*, **12**, 269–80.

Reutter, K. (1971) Die Geschmacksknospen des Zwergwelses *Amiurus nebulosus* (Lesueur). Morphologische und histochemische Untersuchungen. *Z. Zellforsch.*, **120**, 280–308.

Reutter, K. (1978) Taste organ in the bullhead (Teleostei). *Adv. Anat. Embryol. Cell Biol.*, **55**, 1–98.

Reutter, K. (1980) SEM-study of the mucus layer on the receptor field of fish taste buds, in *Olfaction and Taste* VII (ed. H. van der Starre), IRL Press, London, p. 107.

Reutter, K. (1982) Taste organ in the barbel of the bullhead, in *Chemoreception in Fishes* (ed. T.J. Hara), Elsevier, Amsterdam, pp. 77–91.

Reutter, K. (1986) Chemoreceptors, in *Biology of the Integument*. Vol. II (eds J. Bereiter-Hahn, A.G. Matoltsy and K.S. Richards), Springer, Berlin, pp. 586–604.

Reutter, K. (1987) Specialized receptor villi and basal cells within the taste bud of the European silurid fish, *Silurus glanis* (Teleostei), in *Olfaction and Taste* IX (eds D. Roper and J. Atema), *Annals N.Y. Acad. Sci.*, **510**, 570–3.

Reutter, K. and Bardele, C. (1983) Ultrastrukturelle Untersuchung der Geschmacksorgane des Zwergwelses anhand von Gefrierätz-Präparaten. *Verh. anat. Ges. Jena.*, **77**, 747–9.

Roper, S.D. (1989) The cell biology of vertebrate taste receptors. *A. Rev. Neurosci.*, **12**, 329–53.

Toyoshima, K. (1989) Ultrastructure and immunohistochemistry of the basal cells in the taste buds of the loach, *Misgurnus anguillicaudatus*. *J. Submicrosc. Cytol. Pathol.*, **21**, 771–4.

Toyoshima, K., Nada, O. and Shimamura, A. (1984) Fine structure of monoamine-containing basal cells in the taste buds on the barbels of three species of teleosts. *Cell Tissue Res.*, **235**, 479–84.

Tucker, D. (1983) Fish chemoreception: peripheral anatomy and physiology, in *Fish Neurobiology: Brainstem and Sensory Organs*, Vol. I (eds R.G. Northcutt and R.E. Davis), University of Michigan Press, Ann Arbor, pp. 311–49.

Witt, M. and Reutter, K. (1988) Lectin histochemistry on mucous substances of the taste buds and adjacent epithelia of different vertebrates. *Histochemistry*, **88**, 453–61.

Witt, M. and Reutter, K. (1990) Electron microscopic demonstration of lectin binding sites in the taste buds of the European catfish *Silurus glanis* (Teleostei). *Histochemistry*, **94**, 617–28.

Chapter five

Central representation and projections of gustatory systems

Jagmeet S. Kanwal and Thomas E. Finger

5.1 INTRODUCTION

Fish have historically served as a vertebrate model for the study of the gustatory (taste) system. Taste buds in fish were first described as terminal buds in the early 1800s (Weber, 1827; Leydig, 1851) before their function as organs of taste became firmly established (Merkel, 1880). However, the peripheral and central organization of the gustatory pathways in fishes was elucidated (Herrick, 1901, 1904, 1905) long before similar studies were pursued in mammals (Norgren and Leonard, 1973; Norgren, 1974, 1976). The taste system of fishes is interesting because, unlike tetrapods, many species of fish can use the taste system as a distance sense to localize a chemical source in the external environment, i.e. for food search (Bardach *et al.*, 1967; Johnsen and Teeter, 1980). Thus, taste receptors in fishes, like their olfactory receptors, can respond to chemical stimuli originating from a distant source. In some species of fish, gustatory and olfactory receptors may even show equally high sensitivity to the same type of chemicals such as amino acids (Caprio, 1982). The classical distinctions between olfaction and gustation in tetrapods, based upon the medium of stimulus delivery (air vs water) and the proximity (distant vs contact) of the stimulus source (Parker, 1910), therefore, do not exist for aquatic vertebrates. Why then did multiple chemosensory systems evolve in the aquatic ancestors of tetrapods? We can partly resolve this question by examining the behavioral role and the neural organization of the gustatory system in the present-day fishes.

To understand the evolution and the behavioural roles of olfactory, gustatory and other chemosensory systems, one must look beyond the physical and chemical parameters of stimulus delivery and transduction. While all chemosensory systems in teleosts (including olfactory, trigeminal and spinal) collect information from chemicals delivered to the receptor through the aqueous medium, each system is in fact anatomically distinct with respect to its (1) end-organ morphology and position, and (2) innervation and central representation. These differences, especially regarding innervation and central pathways, lead to the modulation of separate behaviours (action patterns) by separate chemosensory systems in any one species. This makes it especially interesting to examine the central taste pathways in teleosts, where in several species gustation can be used for orientation and/or navigation towards a food source in addition to selective ingestion (Herrick, 1904; Bardach et al., 1967; Atema, 1971; Holland, 1978).

In this chapter, the gustatory system is examined at the cellular and connectional levels of organization within the medulla. In addition, the connectivity between the medulla and the forebrain is also described. The contents of this chapter are primarily based on studies on the organization of the gustatory system within the brains of two groups of teleosts, the ictalurids and the cyprinids, in which the gustatory system is highly developed (Fig. 5.1). The intramedullary connections and the ascending projections from the primary gustatory centres (facial and vagal lobes) have been extensively studied in the bullhead catfish, *Ictalurus nebulosus* and the channel catfish, *Ictalurus punctatus* (ictalurids) and in the crucian carp, *Carassius carassius* and the goldfish, *Carassius auratus* (cyprinids). Representatives of these two groups as well as other species such as a goatfish (Mullidae) and a rockling (Gadidae) provide unique models to study the central organization of gustatory pathways and together illustrate the phylogenetic diversity in the organization of this highly developed sensory system (Fig. 5.2). In fact, the specialized viscerosensory (gustatory) brainstem structures in many groups of teleosts show more interspecific variability than those structures associated with any other sense (Kotrschal and Junger, 1988).

In ictalurids the extraoral gustatory system is highly developed, whereas in cyprinids the intraoral gustatory system is highly developed. A comparison of the extraoral and intraoral gustatory systems within teleosts illustrates well the close relationship between the behavioural goal and the anatomical organization of a chemosensory system. Thus, the extraoral, facial gustatory centre (facial lobe) in ictalurid catfish (Ictaluridae) is disproportionately enlarged and organized into lobules for representation of taste and tactile inputs from the barbels, lips, snout and flank. Similarly, the dorsal portion of the facial lobe in a goatfish (Mullidae) is also exceptionally large and consists of lobules with taste inputs from a pair of

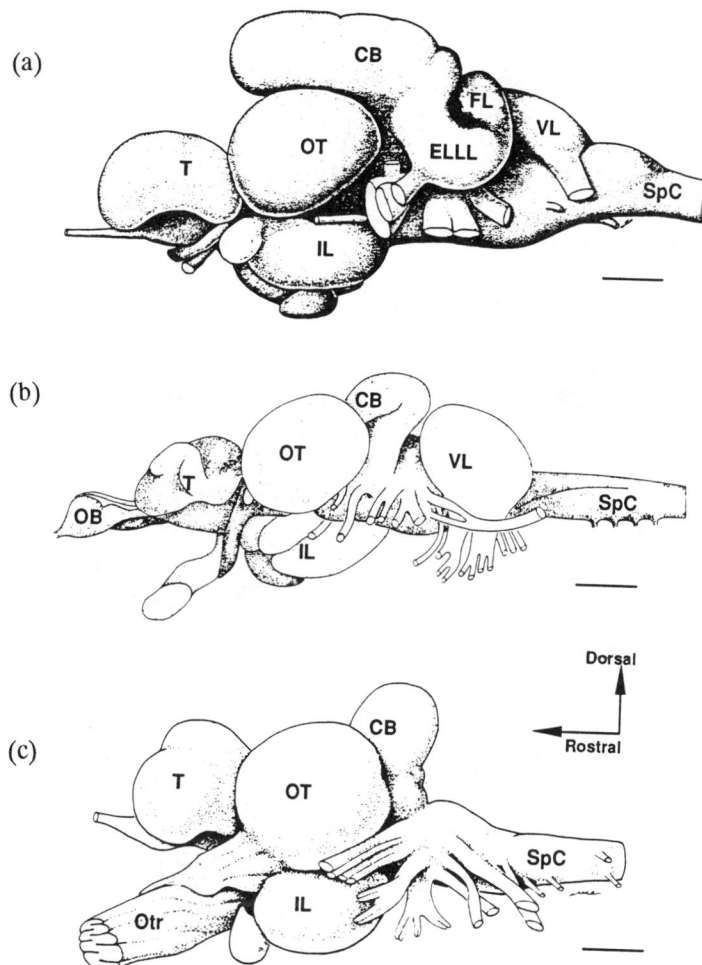

Fig. 5.1 Lateral views of the brains of (a) a catfish, *Ictalurus punctatus* (after Striedter, 1990), (b) the common goldfish, *Carassius auratus* and (c) a sunfish, *Lepomis cyanellus*. (b and c after Wulliman and Northcutt, 1988). Note the enlarged gustatory structures (facial and/or vagal lobes) in the catfish and the goldfish. Other species such as the sunfish have relatively few taste buds and do not have a facial or a vagal lobe. Scale bar represents approximately 1 mm. (For abbreviations see list p. 21).

mandibular (chin) barbels (Barry and Norton, 1989). In both a catfish and a goatfish, the extraoral gustatory system coordinates equivalent behaviours of stimulus localization and food acquisition. In cyprinids such as a goldfish, the medullary visceral centre for intraoral taste (vagal lobe) is enlarged and is organized anatomically into layers. This structure is also involved in stimulus localization and contains a representation of

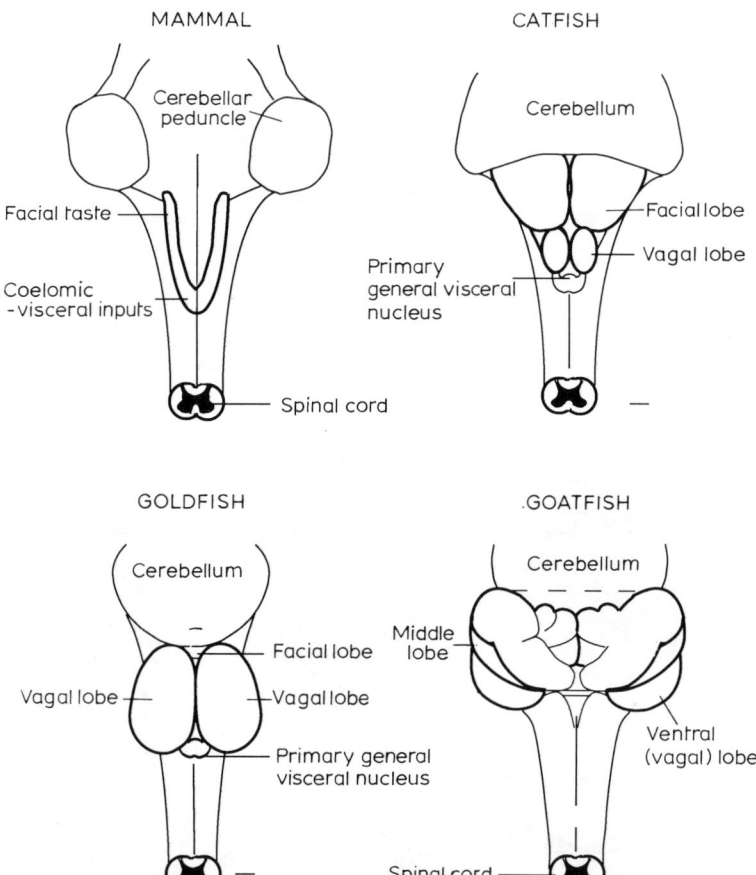

Fig. 5.2 Diagram of gustatory representation in the medulla (dorsal views) of a generalized mammal and three species of teleosts. In mammals, the gustatory inputs are restricted to the rostral two-thirds of the nucleus of the solitary tract, which is a narrow zone in the medulla. In some fishes, which utilize gustatory information for stimulus localization (catfishes, codfishes and cyprinids), either the rostral portion (facial lobe) or the caudal portion (vagal lobe) of the medulla is enlarged for processing gustatory inputs. In the goatfish, both the facial (dorsal) and the vagal (ventral) lobes are large. Scale bar represents approximately 1 mm.

the taste and tactile inputs from the palatal organ in the oral cavity (Morita and Finger, 1985b). The palatal organ accomplishes selective ingestion of food by contraction of specific sets of muscles, which can selectively hold on to 'tasty' food particles (Sibbing, 1982; Sibbing *et al.*, 1986).

5.2 PRIMARY GUSTATORY CENTRES

Gross morphology and subdivisions of the gustatory system

For the olfactory (cranial nerve I), visual (cranial nerve II) and auditory (cranial nerve VIII) systems in vertebrates, a single cranial nerve carries all the peripheral sensory information for that particular sense (Angevine and Cotman, 1981). For the gustatory system of teleosts and most other vertebrates, however, peripheral gustatory inputs reach the brain via three separate cranial nerves, i.e. facial (VII), glossopharyngeal (IX) and vagal (X). The facial nerve transmits gustatory information from the extraoral surface, the glossopharyngeal from the anterior part of the oral cavity, and the vagal nerve from the oropharynx. The facial nerve is well developed in ictalurid catfish and the goatfish, innervating taste buds on all extraoral structures such as the lips, barbels, snout region, flank and fins. In cyprinids the facial nerve is large, but the distribution of extraoral innervation varies with the species in question. Thus, in the carp, the facial nerve innervates the one or more pairs of barbels, while in goldfish, which do not have any barbels, it innervates predominantly the upper and lower lips, while the recurrent branch of the facial nerve in all species innervates the flank. The vagal nerve is developed best in the goldfish and carp, where it innervates the enlarged, muscular palatal organ.

Centrally, the facial nerve terminates in a rostral medullary structure known as the facial lobe. This is a paired, lobulated structure in both catfish and the goatfish, but it is restricted to a single midline lobe in the goldfish. The vagal nerve terminates in the vagal lobes located in the caudal medulla. These are relatively small in catfish, but are enlarged into a pair of layered structures in the goldfish. The glossopharyngeal nerve terminates in a dorsal medullary region between the facial and vagal lobes. This region is not distinguishable into a separate glossopharyngeal lobe in either the goldfish or catfish. In the goatfish, the medullary gustatory centres are arranged in a largely dorsoventral fashion, in contrast to the anteroposterior arrangement in catfish. Moreover, in a goatfish, the dorsal (facial) and ventral (glossopharyngeal and vagal) lobes are relatively large (Barry, M., 1990 pers. comm.)

The taste system is anatomically divisible into two distinct, though interrelated subsystems. Each of these subsystems is specialized to coordinate a separate phase of feeding. The extraoral part of the facial taste subsystem can detect chemical stimuli at a distance and can therefore subserve a food-localization function during food search, while the oral or vagal taste subsystem performs a discriminative sensorimotor function leading to the selective ingestion of food.

The role of the facial taste system in stimulus localization has been elucidated also on the basis of neuroethological experiments in the bullhead

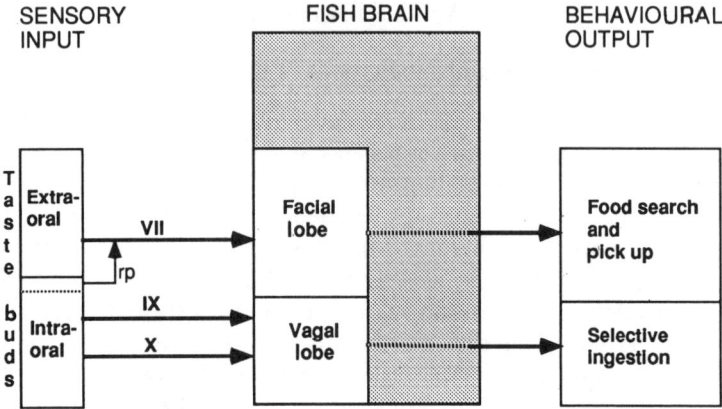

Fig. 5.3 Anatomical and behavioural segregation of gustatory function in the bullhead catfish, *Ictalurus nebulosus*. The extraoral taste subsystem is anatomically distinct from the oropharyngeal subsystem in its innervation and central representation. The two subsystems are also behaviourally distinct in that they coordinate different phases of feeding. rp, ramus palatinus—a branch of the facial nerve which innervates the anterior palate.

catfish, *I. nebulosus* (Atema, 1971). In these experiments, animals in which the vagal lobe was ablated managed to find food and pick it up in their mouths but were unable to swallow it, while those without facial lobes were unable to search for food but were able to swallow the food introduced into the oral cavity (Fig. 5.3). The gross anatomical seperation of the two gustatory lobes is thus reflected in the different functions and the intrinsic organization of the neural substrate in these two regions.

In ictalurid catfish, 'food search' and 'pick up' are the critical phases for determining what the animal finds to feed on, whereas in cyprinids 'selective ingestion' is the more decisive phase. Accordingly, differences in the gross morphology of the two taste subsystems described above are accompanied by differences in the neural circuitry within the medulla. Thus, the facial subsystem is well developed in a catfish, while the vagal subsystem is well developed in the goldfish. To incorporate the diversity of gustatory representations in the brains of teleosts, it is necessary to describe the neural representation of facial taste in ictalurids and of vagal taste in cyprinids. Most other teleostean species have facial and vagal subsystems that are intermediate in their relative development compared with these two species.

Central representation of gustatory inputs

Facial and vagal taste inputs terminate in the facial and vagal lobes, respectively, in the medulla of teleosts. In the majority of teleostean species, neither the facial lobe nor the vagal lobe is well developed, and the two may even be continuous and grossly indistinguishable from each other. This general pattern of organization of primary gustatory nuclei applies to elasmobranchs as well (Barry, 1987). The primary gustatory nucleus (usually called vagal lobe by previous authors) is contained in a continuous visceral sensory column protruding slightly from the floor of the fourth ventricle. The rostral end of this column receives facial nerve (gustatory) input, a relatively small central section receives glossopharyngeal nerve input, and the caudal half receives vagus nerve input. The vagus nerve territory is, however, not purely gustatory, but contains, at its caudal end, the general visceral afferentation, e.g. from swim bladder, heart and gut (Kanwal and Caprio, 1987; Morita and Finger, 1987). The gustatory and general visceral territories appear to be separable, based on histochemistry and morphology as well as on position within the visceral sensory column (Finger, 1981; Kanwal and Caprio, 1987; Kanwal and Finger, unpublished).

The facial lobe of the channel catfish is an elaboration of the general teleostean pattern. The facial lobe is large and organized into six longitudinal columns or lobules extending rostrocaudally and arranged alongside each other (Hayama and Caprio, 1989). Each lobule receives segregated input from discrete portions of the external body surface (Fig. 5.4(a)). This arrangement is similar to the one reported for the facial lobe in *Plotosus lineatus* (Marui *et al.*, 1988). Nerve-tracing studies utilizing horseradish peroxidase (HRP) as an anterograde tracer show that facial inputs terminate

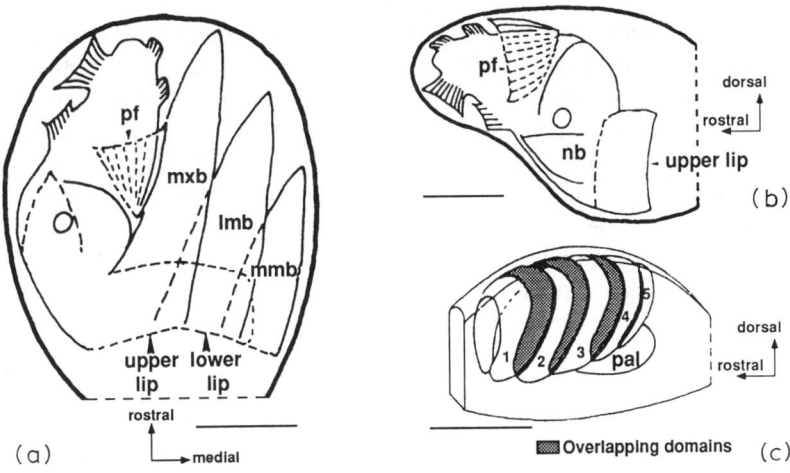

Fig. 5.4 (a and b) Extraoral inputs are mapped sharply along well-defined boundaries in the facial lobe of the channel catfish, *Ictalurus punctatus* (after Hayama and Caprio, 1989); (c) Oropharyngeal inputs in the same species are represented in a diffuse manner, with some regions in the vagal lobe receiving overlapping inputs from adjacent gill arches. Scale bars represent approximately 1 mm. Abbreviations: lmb, lateral mandibular barbel; mmb, medial mandibular barbel; mxb, maxillary barbel; nb, nasal barbel; pf, pectoral fin.

within restricted regions of the facial lobe (Finger, 1976). The three medial lobules in the facial lobe receive input (from medial to lateral) from the medial and lateral mandibular barbels and the maxillary barbel, respectively (Hayama and Caprio, 1989). The two ventral lobules receive inputs from the nasal barbel and pectoral fin. The face and flank are represented in a large dorsolateral lobule. In short, these and previous anatomical studies indicate that the entire extraoral surface of a catfish is mapped onto the facial lobe in a well-defined manner. Additional electrophysiological studies have confirmed the functional integrity of this map (Marui and Caprio, 1982). Similarly, anatomical and electrophysiological studies of the vagal lobe in catfish (Kanwal and Caprio, 1987, 1988) indicate that the oral cavity is mapped in the vagal lobe (Fig. 5.4(b)). A mapping of the peripheral gustatory apparatus onto the gustatory sensory column may be a feature common to all teleosts.

There is a distinct difference, however, in the nature of these mappings in the facial and vagal gustatory centres in catfish. Although the boundaries of extraoral structures mapped in the facial lobe are quite sharply defined, as indicated by electrophysiological and neuroanatomical studies, the oropharyngeal map in the vagal lobe is relatively 'diffuse' so that some neurones in the area of the recording site are responsive to stimulation of more than one oropharyngeal structure (Kanwal and Caprio, 1987). The vagal lobe also lacks a subnuclear organization into either lobules or layers. The only anatomically distinct regions within the vagal lobe are the dorsal cap (Finger, 1981; Kanwal and Caprio, 1987) and the nucleus intermedius of the vagal lobe (Herrick, 1905; Kanwal and Caprio, 1987). Both these regions have been shown to receive primary sensory inputs from topographically separated regions of the oropharyngeal cavity and are suggested as being correlation centres (Herrick, 1905; Kanwal and Caprio, 1987).

The well-defined, non-overlapping nature of the sensory map within the facial lobe coincides with its role in stimulus localization, where precise representation of spatial information may be essential. Such representations are known to be common characteristics of any sensory system (auditory, visual or somatosensory) across unrelated species where stimulus localization is of importance. In the peripheral auditory system, a linear array of receptors by itself is unable to encode three-dimensional information regarding the location of a stimulus source in auditory space. Since auditory space cannot be topographically represented at the sensory level, this information is generated from binaural inputs and represented in the form of computational maps in higher-order centres within the brain (Konishi and Knudsen, 1982). Similarly, pulse-echo time differences are computed by target range maps in bats (Suga, 1990), where the auditory system is critical for the localization of a target (food source). In short, the presence of a precise map in the facial lobe of catfish may be taken as indirect evidence that these species use facial gustatory information for spatial localization of a chemical stimulus.

In a goatfish, the dorsally placed facial lobe consists of many more lobules than does the facial lobe of a catfish (Fig. 5.2). This lobe receives topographically mapped terminations of afferent fibers innervating the chin barbels (Barry and Norton, 1989). Furthermore, these lobules contain clusters of cells surrounded by neuropil. Each cluster has a laminar organization with at least three types of cells. Densely packed round neurones with small dendritic fields are located most superficially, followed by a layer of more heterogeneous small and medium-sized neurones. Fusiform neurones constitute another cell type and are located at the border between the neuropils associated with adjacent clusters. Large, multipolar cell bodies are scattered at the base of the clusters and may be the primary projection neurones (Barry and Norton, 1989). Detailed mapping studies of the other gustatory centres (middle and ventral lobes) in this species will surely provide an interesting comparison for correlating structure and function, as described for the catfish.

In cyprinids (such as the common carp and goldfish), gustatory detection and localization follow oral acquisition of a potential food source. Thus, the food-search phase is mostly non-gustatory and may be mediated primarily by visual and lateral line systems. In the mirror carp, *Cyprinus carpio* L., acquisition of potential food involves relatively non-discriminatory mechanisms of particulate feeding and gulping (Sibbing, 1982; Sibbing et al., 1986). In this type of feeding behaviour, repositioning of food particles, spitting and recollection of food retained in the branchial sieve are the important steps which permit post-capture gustatory selection of food (Sibbing et al., 1986). The postponement of gustatory discrimination subsequent to food search and acquisition in the feeding behaviour of cyprinids allows them to exploit an ecological niche different from that of the silurids. Concomitantly, the neural circuitry for gustatory localization is 'shifted' posteriorly from the facial lobe to the vagal lobes in the caudal medulla. This shift in function in the cyprinids is associated with the vagal lobes becoming enlarged, highly organized structures, while the facial lobes form a single and relatively small midline lobe (Figs 5.1 and 5.2).

Selective injections of HRP into either palatal or branchial nerve roots in a goldfish indicate a different laminar pattern of termination of each nerve within the vagal lobe (Morita and Finger, 1985b). The branchial nerves terminate in layers 2, 4 and 9, while the palatal nerves terminate in layers 6 and 9. Moreover, for the branchial nerves, a local injection of peroxidase results in labelling across a dorsoventrally orientated slab of the vagal lobe. In contrast, local injections into the palatal organ result in punctate labelling of layer 6 in the vagal lobe. These studies also indicate that the oral cavity is mapped onto the vagal lobe so that the anterior part of the mouth is represented anteriorly in the lobe and the posterior oropharynx posteriorly (Morita and Finger, 1985b). The lateral portions of the mouth are represented dorsally in the lobe, and the median part of the palatal organ is represented ventrally. Interestingly, the palatal organ, which is active in localization and selection of 'tasty' particles, appears to be more precisely mapped than the less active branchial structures. This difference in the precision of mapping of oropharyngeal structures within a single lobe is comparable to the above-mentioned differences between maps within the two (facial and vagal) lobes in catfish. The important point which is evident from a comparative description of non-homologous medullary regions in two different species is that the function of a neural structure, rather than its embryological origin, is the primary determinant of its substructural organization and intrinsic neuronal connectivity.

5.3 MEDULLARY NETWORKS AND DESCENDING GUSTATORY PROJECTIONS

To elucidate the behavioural role(s) of a specific sense, it is necessary to examine both the sensory representation within relevant nuclear regions and the long ascending and descending pathways within the whole brain. Within the medulla, local gusto-motor networks and descending pathways form the neural substrate for reflexive movements such as orientation, food pick up or biting and swallowing or disgorging. To delineate the functional organization of these networks, both neuroanatomical and electrophysiological studies are necessary. At present, most of the anatomical information about the sensorimotor circuits within the gustatory lobes of teleosts comes from tract-tracing studies utilizing HRP as an anterograde and a retrograde tracer. The results of these studies are described briefly for the facial lobe in the channel catfish and for the vagal lobe in the goldfish.

Local networks in the facial lobe of the channel catfish

Cell bodies of neurones in the facial lobe of ictalurid catfish are organized into small clusters or glomeruli. According to Herrick (1905), the incoming primary gustatory fibres terminate in these glomeruli. Moreover, on the basis of cytoarchitectural characteristics and position, at least three types of neurones can be described in the facial lobe of both the carp and the channel catfish. The smallest of these cell types is considered to be an intrinsic neurone. Each of the 'small intrinsic neurones' receives input from a number of primary gustatory fibres. These intrinsic neurones have a local dendritic field and project to the 'large neurones' (second type) generally distributed around the peripheral margin of the lobules (Fig. 5.5(a)), and possibly to other facial lobe cells. Recent immunohistochemical studies in the channel catfish also indicate that many facial lobe neurones with small, round cell bodies contain the inhibitory neurotransmitter, GABA (gamma amino butyric acid) (Kanwal and Finger, unpublished obs.). These neurones are uniformly scattered throughout the facial lobe and may form another type of intrinsic facial lobe neurone. Golgi studies in both carp and catfishes also show a third type of neurone, the 'medium-sized cells', which have a relatively large dendritic spread and constitute the ascending output neurones of the facial lobe. In the channel catfish, a fourth neurone type, totalling only 15 to 20 neurones, was identified after HRP injections into the dorsolateral region of the spinal cord (Kanwal and Finger, 1989; Kanwal, unpublished observation). These 'giant output neurones' are present only in the lateral lobule of the facial lobe and seem to have non-overlapping dendritic receptive fields.

Vagal sensorimotor circuits in the goldfish

The relatively direct projection from the vagal gustatory lobe to motoneurones driving the oropharynx exists in goldfish (Finger, 1988) as well as catfish (Morita and Finger, 1985a). That is, primary vagal gustatory afferent fibres terminate within the vagal lobe and second-order neurones form an obligatory link in the reflex pathway connecting to the vicinity of the orobranchial motor neurones.

The principal difference between the catfish and goldfish systems lies in the morphological arrangement of these circuit elements. In catfish, the vagal lobe is purely sensory, containing the primary gustatory nerve terminals and the second- and higher-order interneurones. The axons of these interneurones leave the lobe to reach the vicinity of the nucleus ambiguus dendrites in the ventrolateral medulla. In goldfish, the vagal lobe is a rigidly laminated structure containing the orobranchial motoneurones (=nucleus ambiguus) as well as the gustatory afferent fibres and interneurones (Herrick, 1905; Morita and Finger, 1985b). The elaborate vagal lobe in goldfish appears organized to facilitate point-to-point reflex connections between the affector and effector systems of the orobranchial apparatus (Fig. 5.5(b)). The palatal and branchial surfaces of the mouth are mapped in a systematic fashion onto the vagal lobe. Furthermore, the palatal musculature is innervated in an organized, mapped fashion so that motoneurones innervating a particular part of the palatal organ lie radially inward from that portion of the sensory layers that carry the sensory representation from that same point in the oral cavity. A simple surface-to-deep-layer reflex system connects each point of the sensory layers of the vagal lobe to its corresponding motor pool (Finger, 1988). This system of connectivity appears to be involved in the intraoral food-sorting mechanism employed by this species (Sibbing, 1982; Sibbing et al., 1986).

Descending facial lobe projections in the channel catfish

As explained earlier, the facial lobe in ictalurid catfish is quite large, corresponding to the extensive peripheral distribution of taste buds over the entire

Lateral lobule of the facial lobe
Large, dorsal secondary neurone (descending funicular and ascending lemniscal projections)
Giant neurone (spinal projection)
Intrinsic neurone
Intrinsic neurones
Medium-sized isthmic projection neurone
Afferent fibre
Cell body in facial ganglion
50 µm
Rostral
Lateral

(a)

Vagal lobe
Sensory layers
Motor layer
Fibre layer
mlf
Vagus nerve
BRAIN
MOUTH
Palatal muscles
Palatal organ
Food particle
Oral cavity
Taste buds
Gill Arch

(b)

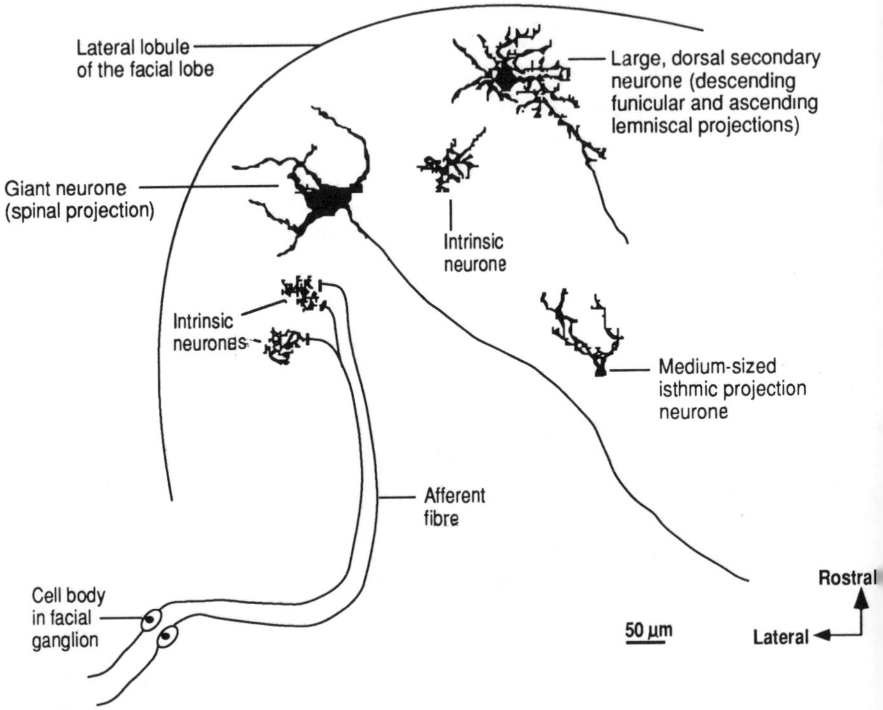

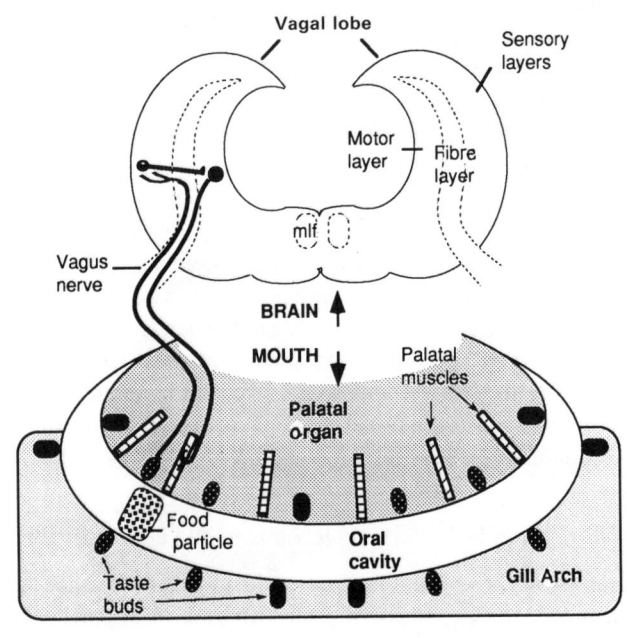

extraoral surface. Ictalurid catfish use their vagal taste system for selective food ingestion and the facial taste system for food search (Atema, 1971). Stimulation of the vagal taste system may result in swallowing, regurgitation or gaping, i.e. movements of the orobranchial and pharyngeal musculature. Stimulation of the facial taste system may produce swimming, orientation or turning, i.e. movements of the body and fins. This behavioural dichotomy is proposed on the basis of differences in the local projections from the two lobes. Thus cells in the vagal lobe project to the nucleus ambiguus and not to the medial reticular formation, whereas cells in the facial lobe project to the medial reticular formation but none to the nucleus ambiguus (Finger and Morita, 1985). A subset of these reticular cells wthin the medial reticular formation (the reticulospinal cells) project downstream through the ventral horn of the spinal cord and terminate at various anterior–posterior levels (Kanwal, 1988; Kanwal, unpublished obs.). Since the spinal cord has the circuitry to coordinate swimming movements, the descending connections of the facial taste system probably coordinate a component of the food-search behaviour via the reticulospinal system. Although gustatory representations in each facial lobe are restricted to the ipsilateral half, reticular neurones exhibit unilateral as well as bilateral receptive fields, indicating convergence of inputs from both facial lobes (Kanwal and Finger, 1989). Other studies also indicate that the lateral lobule of the facial lobe projects to the spinal cord via a second, direct pathway (Herrick, 1906; Finger, 1978). This pathway originates from the giant (50–70 µm) cells scattered within the lateral lobule which project downstream through the dorsal horn of the spinal cord (Kanwal and Finger, 1989). The exact function of this projection is not yet known.

5.4. SUPRAMEDULLARY GUSTATORY PATHWAYS

In addition to the local medullary networks and descending pathways, the primary gustatory centres also project to nuclei within the diencephalon and thence to the telencephalon in teleosts. These gustatory projections ascend along multiple routes within the brain. Much less is known of their functional significance as compared with the medullary and spinal connections. Furthermore, it is difficult to homologize any of these pathways with

Fig. 5.5 (a) Semischematic drawing of a horizontal section of the facial lobe showing the four neurone types, their connectivity patterns in the facial lobe and their projection targets within the brain of a catfish; (b) Diagram of the basic neuronal circuitry for the palatal sorting system in the common goldfish. The neural elements are diagrammed on a transverse section through the head of a goldfish, at the level of the vagal lobe. The essential components of this network include (1) gustatory primary afferents, (2) one or more neurones in the sensory layer of the vagal lobe, and (3) α-motor neurones in the motor layer that innervate palatal musculature.

those of the gustatory system in the mammalian forebrain. One reason for this difficulty is that the telencephalon of teleosts develops by the process of eversion, in contrast to inversion, the latter being the normal process for the reorganization of telencephalic nuclei during embryonic development in tetrapods (Gage, 1893). Therefore, comparison of telencephalic connections of the teleostean gustatory and other lemniscal (labelled-line sensory pathways to the pallial part of the telencephalon) systems with those of amniote vertebrates is relatively difficult. Nevertheless a general scheme of the central organization of gustatory pathways among teleosts can be constructed, based upon studies in the channel catfish, the crucian carp and a few other species. In the case of some diencephalic and telencephalic nuclei, however, homologies even among the different species of teleosts are unclear at present. Therefore, the forebrain connections of the gustatory system will be described primarily for the channel catfish, in which these are best known (Kanwal et al., 1988).

As a working hypothesis, the ascending gustatory system of teleosts can be considered to consist of at least two types of pathways. One route ascends from the medulla to the telencephalon after making one synaptic connection in the ventroposterior diencephalic region. This route may be comparable to the lemniscal gustatory pathways leading to pallial (gustatory cortex) areas in mammals. Gustatory afferents to the medulla reach the ventral diencephalic (hypothalamic) regions of the brain by yet another route. This pathway exhibits a multiplicity of connections between diencephalic nuclei and may be equivalent or even homologous to the hypothalamo–limbic connections of the gustatory system in mammals (Finger, 1987; Kanwal et al., 1988).

Connections with the isthmus and diencephalon (non-telencephalic)

The majority of secondary ascending projections from facial and vagal lobes in all teleosts studied so far terminate in a large isthmic nucleus known as the superior secondary gustatory nucleus (nGS; abbreviations are listed on p. 98). This nucleus has several sets of tertiary projections to the diencephalic region (Fig. 5.6). A few fibres from this nucleus ascend to the mid-diencephalic region and terminate in the vicinity of the nucleus lateralis thalami (nLT). This is a small nucleus at the level of the lateral sulcus formed by the junction of the inferior lobe with the torus semicircularis and consists of a mixed population of large and small cells. Another set of projections arise from the large neurones of the secondary gustatory nucleus and reach the nucleus diffusus in the posteroventral region of the inferior lobes. Finally, the majority of fibres from cells within the secondary gustatory nucleus ascend to the level of the nLT (thalamic taste nucleus) and proceed ventrally into the inferior lobe via the tractus thalamolobaris, where they turn posteriorly to terminate along longitudinally orientated columns (longitudinal terminal fields) within the ipsilateral inferior lobe.

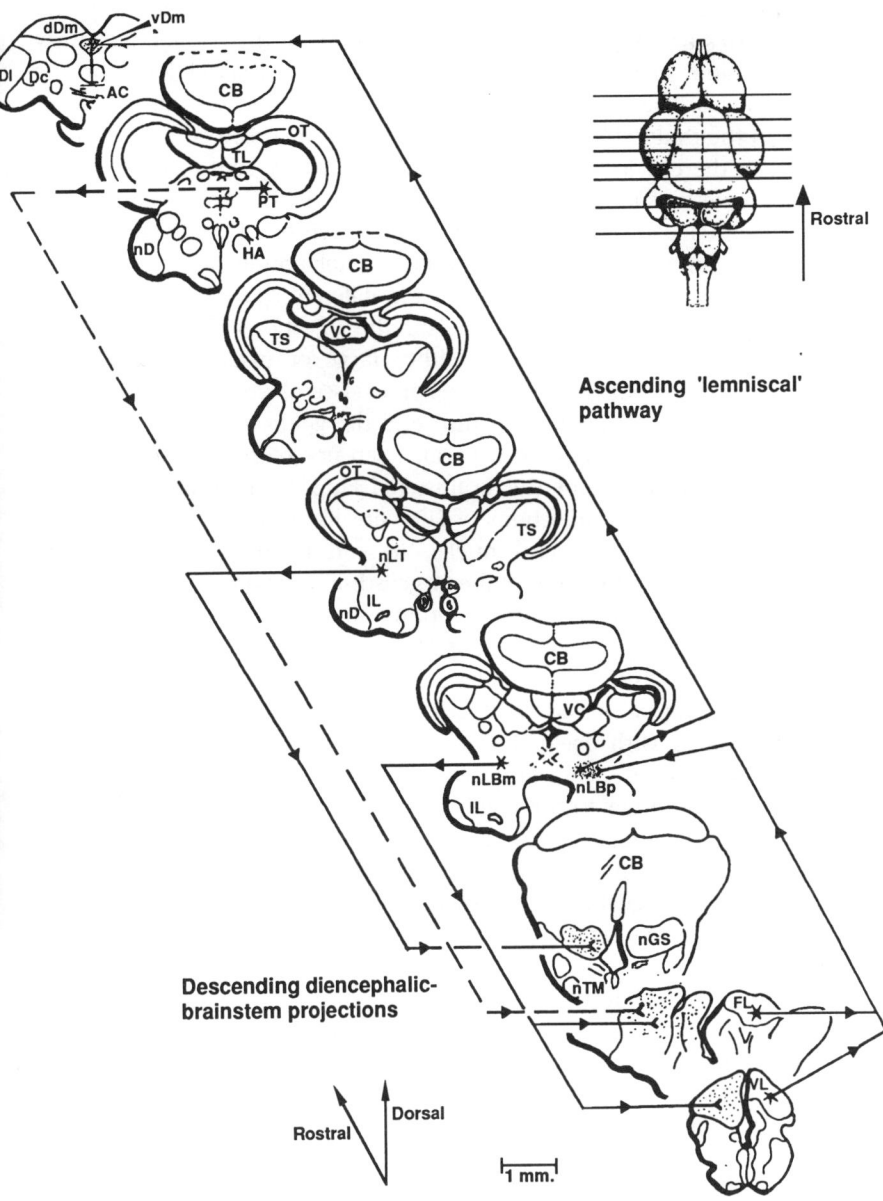

Fig. 5.6 Schematic representation of the isthmic–diencephalic projections originating from the secondary gustatory nucleus. Stars in the transverse sections indicate cell bodies; stippled regions indicate areas of termination. Insert at top right shows the caudal to rostral levels of the sections on a dorsal view of the catfish brain (For abbreviations see list p. 98.)

An important point to note is that in catfish, where the medullary and isthmic centres are large, the thalamic projections of the isthmic gustatory centre are minor except for those terminating in the inferior lobe and the nucleus diffusus. This is in contrast to the major 'gustatory' termination sites reported in the thalamus of the crucian carp and the green sunfish, *Lepomis cyanellus*. In the crucian carp, this thalamic gustatory centre is named the nucleus 'glomerulosus' (Morita et al., 1980), but in the green sunfish it is termed the 'nucleus gustatorius tertius' – a subdivision of the preglomerular complex (Wullimann, 1988). An alternative explanation not listed for this discrepancy between the results obtained in different species may be that in both the crucian carp and the sunfish, the injections of HRP into the secondary gustatory nucleus were either large or extended far enough laterally to include non-gustatory, coelomic–visceral areas of the isthmus. From recent studies in ictalurid catfish and the goldfish, coelomic–visceral nuclei are known to constitute a parallel sensory pathway with separate nuclei in the medulla and the brainstem (Finger and Kanwall, 1992). Therefore, large isthmic injections of HRP are likely to include portions of the adjacent coelomic–visceral nuclei. In catfish, HRP injections were confined within the boundaries of the large isthmic gustatory nucleus by use of combined electrophysiological recording and iontophoretic techniques (Kanwal et al., 1988). The tertiary gustatory centre in both the crucian carp and the green sunfish may be a largely non-gustatory, coelomic–visceral centre with perhaps a minor vagal–gustatory component, since the latter is closely associated with coelomic–visceral projections. As in catfishes, the gustatory component probably accounts for most of the non-thalamic diencephalic projections to the inferior lobes and to the nucleus diffusus. In light of this information, the true pattern of tertiary gustatory projections is probably consistent in silurids, cyprinids and percomorphs. The variable extent of thalamic 'gustatory' representations reported for different species could be a direct consequence of the inclusion of non-gustatory visceral areas at the injection site.

The superior secondary gustatory nucleus also receives descending projections from the nLT (Fig. 5.7). Descending projections from magnocellular neurones of the nucleus lobobulbaris (nLB) in the ventral diencephalon and from the pretectum bypass the secondary gustatory nucleus and terminate in the facial lobes within the medulla. Magnocellular neurones in the nLB also project to the vagal lobe and constitute the only known descending output from a diencephalic gustatory nucleus. These projections are similar to those reported for the nPT in the crucian carp (Morita et al., 1983).

Telencephalic connections

The gustatory system is represented in a small area in the medial telencephalon. This area includes the ventral portion of the area dorsalis pars

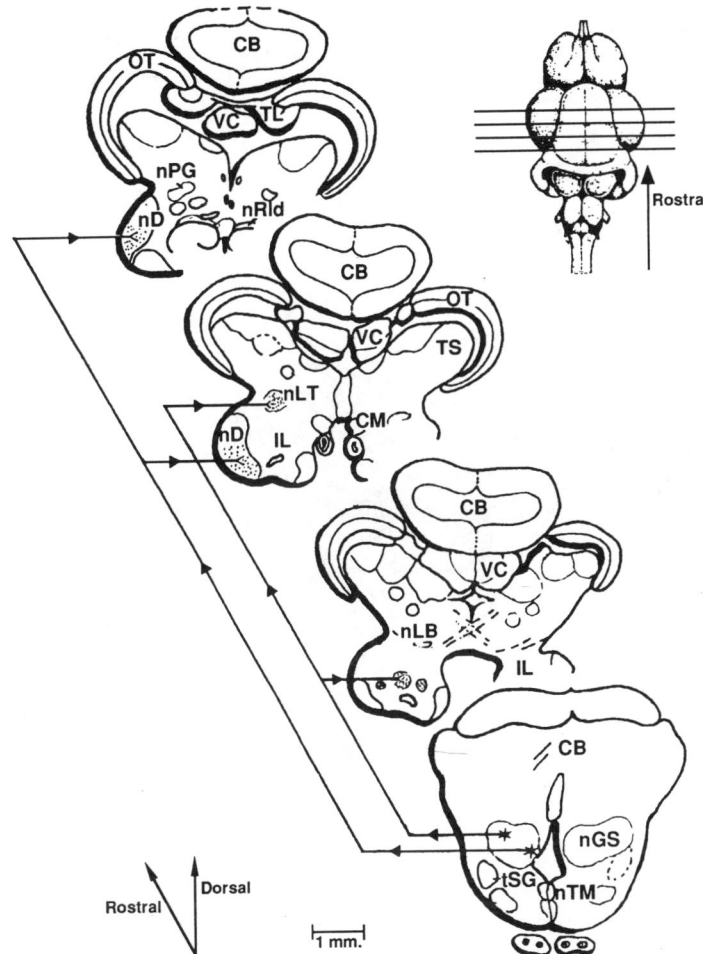

Fig. 5.7 Schematic representation of the descending projections from the diencephalon to the secondary gustatory nucleus (left) and the ascending lemniscal pathway from the medulla to the telencephalon (right). Stars in the transverse sections indicate cell bodies; stippled regions indicate areas of termination. Insert at top right shows the caudal to rostral levels of the sections on a dorsal view of the catfish brain. (For abbreviations see list p. 98.)

medialis (vDm) and the medial region of the area dorsalis pars centralis (Dc). Ascending gustatory inputs to this region seem to originate from a group of small, fusiform cells in the ventroposterior diencephalon which constitute the nucleus lobobulbaris parvicellularis (nLBp) (Kanwal et al., 1988). This diencephalic nucleus receives gustatory inputs from the primary sensory nuclei in the medulla. Thus, the nLBp forms the relay in a monosynaptic gustatory pathway from the medulla to the telencephalon (Fig. 5.7).

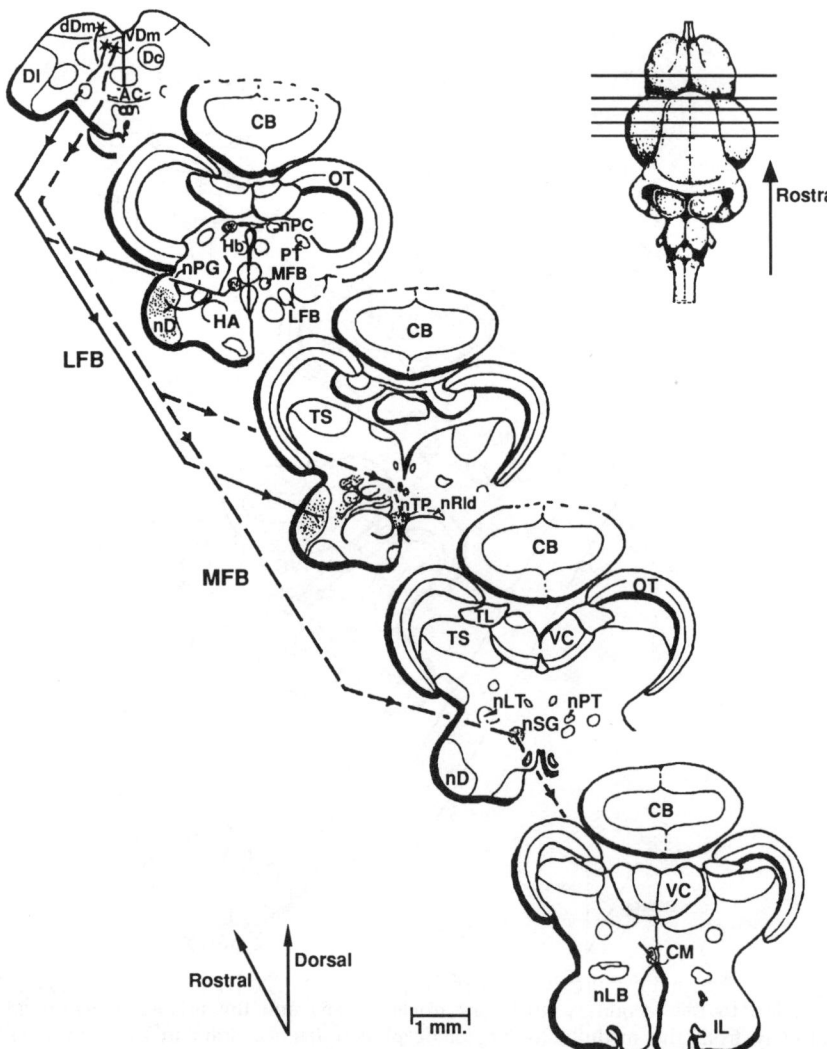

Fig. 5.8 Schematic representation of descending diencephalic projections from the telencephalic taste area in ictalurid catfish. Stars in the transverse section indicate cell bodies; stippled regions indicate areas of termination. Insert at top right shows the caudal to rostral levels of the sections on a dorsal view of the catfish brain. (For abbreviations see list p. 98.)

The telencephalic gustatory region, including the area dorsalis pars medialis (Dm) and the medial portion of the area dorsalis pars centralis (Dc), has several descending projection targets in the diencephalon. These include the nucleus preglomerulosus (nPG), the anterior ventromedial nucleus of the thalamus (nAVm), the nucleus of the lateral recess (nRl), the nucleus paracommissuralis (nPC) and the nucleus of the posterior tubercle (nTP) in the anterior diencephalon. The nucleus diffusus (nD), the nucleus of the posterior thalamus (nPT), the nucleus subglomerulosus (nSG) and the corpus mammillare (CM) in the posterior diencephalon also receive inputs from the gustatory portion of the telencephalon (Fig. 5.8). Because the forebrain gustatory connections described here have been studied up to the level of the telencephalon in only one species (*I. punctatus*), the general organizational plan of the gustatory system within the forebrain (unlike the medullary region) cannot be analysed at present.

The functional significance of forebrain connections of the gustatory system in teleosts is also unknown. It is interesting, however, to note that gustatory neurones in the ictalurid telencephalon have complex response patterns (mixed excitatory and inhibitory) to simple stimuli (amino acids) and have multiple receptive fields which converge from segregated regions of the body (Kanwal *et al.*, 1988) such as those seen in the gustatory cortex of mammals (Yamamoto *et al.*, 1980; Yamamoto, 1984; Azuma *et al.*, 1984). In addition, some neurones respond best to complex food tastes such as liver extract instead of the elementary tastants such as amino acids (Kanwal et al., 1988). These responses are reminiscent of the tuning of some of the neurones in the primary insular taste cortex of primates, which respond best to blackcurrant juice (Yaxley *et al.*, 1990). Therefore, the level of information processing at the telencephalic level in teleosts and certain cortical areas in mammals is comparable in at least some respects. Extensive neuroanatomical, immunocytochemical, electrophysiological and behavioural studies in several teleostean species are necessary before any conclusive statements can be made about the neurobiological significance of gustatory representations within the forebrain.

5.5 SUMMARY

In several species of teleosts the gustatory system has evolved to facilitate feeding by discriminating and localizing food stimuli either in the environment (via the facial subsystem) or in the oropharynx (via the vagal subsystem). In general, the primary gustatory centres exhibit similar types of anatomical modifications, such as lobular or layered morphology (exhibiting highly organized units of structure) and topographic patterns of peripheral input for the representation of sensory information. In any one species, the facial, glossopharyngeal, and vagal chemosensory inputs are encoded and distributed in parallel, at least up to the level of the primary sensory centre.

The gustatory tracts from the facial and vagal lobes ascend and descend within the brain. Fibres within the descending tract terminate in the funicular nuclei. In addition, direct (faciospinal) and indirect (reticulospinal) pathways transmit facial taste information to the spinal cord, while vagal taste projects directly to the nucleus ambiguus. The supramedullary

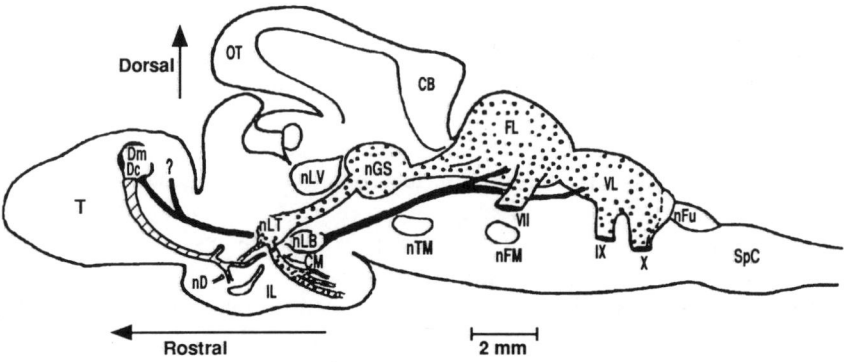

Fig. 5.9 Schematic representation of supramedullary gustatory pathways on a sagittal view of the brain of ictalurid catfish. The facial and vagal lobes in the medulla project directly to the nGS and have reciprocal connections with the nLB (shown as a filled tract). The majority of fibres from the nGS terminate either in the nD or in longitudinally orientated terminal fields in the vicinity of the nucleus centralis in the inferior lobe (stippled tracts). The nLT also receives a minor input from the nGS and projects back to the nGS (stippled tracts). Small neurones in nLB constitute an ascending 'lemniscal channel' which projects to the taste area (vDm and Dc) in the telencephalon (filled tract). This region (Dm and Dc) of the telencephalon sends axons to several targets within the diencephalon (hatched tract). (For abbreviations see list p. 98.)

gustatory pathways (Fig. 5.9) may belong to either of two categories. One category of reciprocal connections constitutes a loop between the diencephalic and the brainstem nuclei. The other is a 'lemniscal-type' labelled-line sensory pathway which ascends from the medulla to the telencephalon. However, these ascending pathways are not necessarily homologous to similar pathways (limbic and lemniscal systems) described in mammals. Finally, several descending projections also originate from the telencephalic taste centre. These descend ipsilaterally via the medial and the lateral forebrain bundles and terminate in the vicinity of the preglomerular complex, the nucleus paracommissuralis in the habenula and at several other sites within the ventral diencephalon.

ACKNOWLEDGEMENT

We wish to thank Dr Michael Barry for his comments and useful suggestions regarding this manuscript.

ABBREVIATIONS

AC	Anterior commissure
CB	Cerebellum

Abbreviations

CM	Corpus mammillare
Dc	Area dorsalis pars centralis
dDm	Dorsal portion of area dorsalis pars medialis
Dl	Area dorsalis pars lateralis
Dm	Area dorsalis pars medialis
ELLL	Electrosensory lateral line lobe
FL	Facial lobe
GABA	Gamma amino butyric acid
HA	Anterior hypothalamic nucleus
Hb	Habenula
HRP	Horseradish peroxidase
IL	Inferior lobe
LFB	Lateral forebrain bundle
MFB	Medial forebrain bundle
mlf	Medial longitudinal fasciculus
nD	Nucleus diffusus
nAVm	Anterior ventromedial nucleus of the thalamus
nFM	Facial motor nucleus
nFu	Funicular nuclei
nGS	Secondary gustatory nucleus
nLBm	Magnocellular division of nucleus lobobulbaris
nLBp	Parvicellular division of nucleus lobobulbaris
nLT	Nucleus lateralis thalami
nLV	Nucleus of the lateral valvula
nPG	Nucleus preglomerulosus
nPC	Paracommissural nucleus
nPT	Nucleus of the posterior thalamus
nRl	Nucleus recessus lateralis
nRld	Nucleus recessus lateralis pars dorsalis
nSG	Nucleus subglomerulosus
nTM	Trigeminal motor nucleus
nTP	Nucleus of the posterior tubercle
OB	Olfactory bulb
OT	Optic tectum
Otr	Optic tract
PT	Pretectum
SpC	Spinal cord
T	Telencephalon
TL	Torus longitudinalis
TS	Torus semicircularis
tSG	Secondary gustatory tract
VC	Valvula cerebelli
vDm	Ventral portion of area dorsalis pars medialis
VL	Vagal lobe

REFERENCES

Angevine, J.B. and Cotman, C.W. (1981) *Principles of Neuroanatomy*, Oxford University Press, New York, 22 pp.

Atema, J. (1971) Structures and functions of the sense of taste in the catfish (*Ictalurus natalis*). *Brain Behav. Evol.*, **4**, 273–94.

Azuma, S., Yamamoto, T. and Kawamura, Y., (1984) Studies on gustatory responses of amygdaloid neurons in rats. *Exp. Brain Res.*, **56**, 12–22.

Bardach, J.E., Todd, J.H. and Crickmer, R. (1967) Orientation by taste in fish of the genus *Ictalurus*. *Science, Wash. D.C.*, **155**, 1276–8.

Barry, M.A. (1987) Central connections of the IX and X cranial nerves in the clearnose skate, *Raja eglentaria*. *Brain Res. Amst.*, **425**, 159–66.

Barry, M.A. and Norton, L.E. (1989) Organization of primary gustatory nuclei in a goatfish, *Parupenus multifaciatus*. *Am. Zool.*, **29**, 13A.

Caprio, J. (1982) High sensitivity and specificity of olfactory and gustatory receptors of catfish to amino acids, in *Chemoreception in Fishes* (ed. T.J. Hara), Elsevier, Amsterdam, pp. 109–34.

Finger, T.E. (1976) Gustatory pathways in the bullhead catfish. I. Connections of the anterior ganglion. *J. comp. Neurol.*, **165**, 513–26.

Finger, T.E. (1978) Gustatory pathways in the bullhead catfish. II. Facial lobe connections. *J. comp. Neurol.*, **180**, 691–706.

Finger, T.E. (1981) Enkephalin-like immunoreactivity in the gustatory lobes and visceral nuclei in the brains of goldfish and catfish. *Neuroscience*, **6**, 2747–58.

Finger, T.E. (1987) Gustatory nuclei and pathways in the central nervous system, in *Neurobiology of Taste and Smell* (ed. T.E. Finger and W.L. Silver), John Wiley & Sons, New York. pp. 285–310.

Finger, T.E. (1988) Sensorimotor mapping and oropharyngeal reflexes in goldfish, *Carassius auratus*. *Brain Behav. Evol.*, **31**, 17–24.

Finger, T.E. and Kanwal, J.S. (1992) Ascending general visceral pathways within the brainstems of two teleost fishes: *Ictalurus punctatus* and *Carassius auratus*. *J. Comp. Neurol.*, in press.

Finger, T.E. and Morita, Y. (1985) Two gustatory systems: facial and vagal gustatory nuclei have different brainstem connections. *Science, Wash., D.C.*, **227**, 776–8.

Gage, S.P. (1893) The brain of *Diemyctylus viridescence* from larval to adult life and comparison with the brain of *Amia* and *Petromyzon*, in *Wilder Quarter Century Book*, Ithaca, NY, pp. 259–314.

Hayama, T. and Caprio, J. (1989) Lobule structure and somatotopic organization of the medullary facial lobe in the channel catfish *Ictalurus punctatus*. *J. comp. Neurol.*, **285**, 9–17.

Herrick, C.J. (1901) The cranial nerves and cutaneous sense organs of the North American silurid fishes. *J. comp. Neurol. Psychol.*, **11**, 177–249.

Herrick, C.J. (1904) The organ and sense of taste in fishes. *Bull. U.S. Fish. Comm.*, **22**, 237–72.

Herrick, C.J. (1905) The central gustatory paths in the brains of bony fishes. *J. comp. Neurol. Psychol.*, **15**, 375–456.

Herrick, C.J. (1906) On the centers of taste and touch in the medulla oblongata of fishes. *J. comp. Neurol. Psychol.*, **16**, 403–39.

Holland, K. (1978) Chemosensory orientation to food by a Hawaiian goatfish (*Parupeneus porphyreus*, Mullidae). *J. Chem. Ecol.*, **4**, 173–86.

Johnsen, P.B. and Teeter, J.H. (1980) Spatial gradient detection of chemical cues by catfish. *J. comp. Physiol.*, **140A**, 95–9.

Kanwal, J.S. (1988) How the catfish tracks its prey: an interactive 'pipelined' processing system may direct foraging via reticulospinal neurons, in *Neural Information Processing Systems* (ed. D.Z. Anderson), Am. Inst. Physics, NY, pp. 402–11.

Kanwal, J.S. and Caprio, J. (1987) Central projections of the glossopharyngeal and vagal nerves in the channel catfish, *Ictalurus punctatus*: clues to differential processing of visceral inputs. *J. comp. Neurol.*, **264**, 216–30.

Kanwal, J.S. and Caprio, J. (1988) Overlapping taste and tactile maps of the oropharynx in the vagal lobe of the channel catfish, *Ictalurus punctatus*. *J. Neurobiol.*, **19**, 211–22.

Kanwal, J.S. and Finger, T.E. (1989) Functional organization of faciospinal pathways in a catfish: neural correlates of gustatory food search, in *Neural Mechanisms of Behavior* (eds J. Erber, R. Menzel, H.-J. Pflüger and D. Todt), Georg Thieme Verlag, Stuttgart, pp. 107.

Kanwal, J.S., Finger, T.E. and Caprio, J. (1988) Forebrain connections of the gustatory system in Ictalurid catfishes. *J. comp. Neurol.*, **278**, 353–76.

Konishi, M. and Knudsen, E.I. (1982) A theory of neural auditory space: auditory representation in the owl and its significance, in *Cortical Sensory Organization* Vol. 3, (ed. C.N. Woolsey), Humana Press, Clifton, N.J., pp. 219–29.

Kotrschal, K. and Junger, H. (1988) Patterns of brain morphology in mid-European cyprinids (Cyprinidae, Teleostei): a quantitative histological study. *J. Hirnforsch.*, **29**, 341–53.

Leydig, F. (1851) Ueber die aussere Haut einiger Susswasserfische. *Z. wiss. Zool.*, **3**, 1–12.

Marui, T. and Caprio, J. (1982) Electrophysiological evidence for the topographical arrangement of taste and tactile neurons in the facial lobe of the channel catfish. *Brain Res. Amst.*, **231**, 185–90.

Marui, T., Caprio, J., Kiyohara, S. and Kasahara, Y. (1988) Topographical organization of taste and tactile neurons in the facial lobe of the sea catfish, *Plotosus lineatus*. *Brain Res. Amst.*, **446**, 178–82.

Merkel, F. (1880) *Ueber die Endigungen der sensiblen Nerven in der Haut der Wirbeltiere*. Rostock.

Morita, Y. and Finger, T.E. (1985a) Reflex connections of the facial and vagal gustatory systems in the brainstem of bullhead catfish, *Ictalurus punctatus*. *J. comp. Neurol.*, **231**, 547–58.

Morita, Y. and Finger, T.E. (1985b) Topographic and laminar organization of the vagal gustatory system in the goldfish, *Carassius auratus*. *J. comp. Neurol.*, **238**, 187–201.

Morita, Y. and Finger, T.E. (1987) Topographic representation of the sensory and motor roots of the vagus nerve in the medulla of the goldfish, *Carassius auratus*. *J. comp. Neurol.*, **264**, 231–49.

Morita, Y., Ito, H. and Masai, H. (1980) Central gustatory paths in the crucian carp, *Carassius carassius*. *J. comp. Neurol.*, **191**, 119–32.

Morita, Y., Murakami, T. and Ito, H. (1983) Cytoarchitecture and topographic projections of the gustatory centers in a teleost, *Carrassius carrassius*. *J. comp. Neurol.*, **218**, 378–91.

Norgren, R. (1974) Gustatory afferents to ventral forebrain. *Brain Res. Amst.*, **81**, 285–95.

Norgren, R. (1976) Taste pathways to hypothalamus and amygdala. *J. comp. Neurol.*, **166**, 17–30.

Norgren, R. and Leonard, C.M. (1973) Ascending central gustatory pathways. *J. comp. Neurol.*, **150**, 217–38.

Parker, G.H. (1910) Olfactory reactions in fishes. *J. exp. Zool.*, **8**, 535–42.

Sibbing, F.A. (1982) Pharyngeal mastication and food transport in the carp (*Cyprinus carpio*): a cinematographic and electromyographic study. *J. Morph.*, **172**, 223–58.

Sibbing, F.A., Osse, J.W. and Terlouw, A. (1986) Food handling in the carp (*Cyprinus carpio*): its movement patterns, mechanisms and limitations. *J. Zool., Lond.*, **210**, 161–203.

Striedter, G. (1990) The diencephalon of the channel catfish, *Ictalurus punctatus:* I. Nuclear organization. *Brain Behav. Evol.*, **36**, 329–54.

Suga, N. (1990) Cortical computational maps for auditory imaging. *Neural networks*, **3**, 3–21.

Weber, E.H. (1827) Meckel's Arch. Anat. Physiol. pp. 309–15.

Wullimann, M.F. (1988) The tertiary gustatory center in sunfishes is not nucleus glomerulosus. *Neurosci. Lett.*, **86**, 6–10.

Wullimann, M.F. and Northcutt, R.G. (1988) Connections of the corpus cerebelli in the green sunfish and the common goldfish: a comparison of perciform and cypriniform teleosts. *Brain Behav. Evol.*, **32**, 293–316.

Yamamoto, T. (1984) Taste responses of cortical neurons. *Progr. Neurobiol.*, **23**, 273–315.

Yamamoto, T., Matsuo, R. and Kawamura, Y. (1980) Localization of cortical gustatory area in rats and its role in taste discrimination. *J. Neurophysiol.*, **44**, 440–55.

Yaxley, S., Rolls, E.T. and Sienkiew, Z.J. (1990) Gustatory responses of single neurons in the insula of the Macaque monkey. *J. Neurophysiol.*, **63**, 689–700.

Chapter six

Solitary chemosensory cells

Mary Whitear

6.1 INTRODUCTION

In tetrapod vertebrates, taste buds differentiate in oral epithelium under the influence of the appropriate innervation. In the gnathostome fishes, taste buds are found in the oropharyngeal epithelium, and may also occur in the external skin of the actinopterygian and dipnoan fishes. Taste bud-like organs are also found on the oral tentacles of hagfish, and in the pharynx of larval and adult lampreys. In addition to these organs, there exists a system of differentiated epithelial sensory cells, which closely resemble gustatory receptor cells but are not organized into discrete end organs. In certain examples, such cells have been shown to be chemosensory, but the physiology, the neural connections, and the distribution of the system are poorly known. Cells of comparable histological character have been recorded from a number of bony fishes, including dipnoans, from hagfish, lampreys, and from some anuran larvae. In selachian fishes, similar cells may occur in the oropharyngeal epithelium.

Some record of the existence of the system can be found in the nineteenth-century literature, although it is not always easy to reconcile early reports with modern observations. For instance, it was known since 1872 that the free pectoral fin rays of triglids did not bear taste buds; there were a few reports of the presence of sensory cells, of which that of Morrill (1895) appears to be the most accurate (Whitear, 1971b). In 1876, Foettinger and Langerhans independently described bipolar cells in the epidermis of lampreys, *Lampetra fluviatilis* and *L. planeri* (Fahrenholz, 1936), some of which correspond to the chemo- sensory cells investigated more recently (Whitear and Lane, 1983a). In ranid tadpoles, Kölliker (1886) found Stiftchenzellen, which have been the subject of later histological and fine-structural studies (Meyer, 1962; Whitear, 1976). Knowledge of epidermal sensory cells in hagfish epidermis still depends on Schreiner (1918). Bipolar cells in teleost epidermis were stained intra-vitam with methylene blue by Whitear (1952), but it was not until the cells could be identified by electron microscopy that their resemblance to the gustatory cells of the corresponding species could be appreciated and their innervation confirmed (Whitear, 1965, 1971b). Physiological evidence is even more recent, and still scanty. Available data sustain the hypothesis that the skin and oropharyngeal surfaces of the primarily aquatic vertebrates are provided with a diffuse system of chemoreceptors, related to, but distinct from, the gustatory system. The distribution and morphology of these solitary sensory cells is reviewed herein, and their possible functions are discussed.

6.2 DISTRIBUTION

Figure 6.1 is a summary diagram of the appearance of the solitary chemosensory cells in diverse species where they have been studied by transmission electron miscroscopy (TEM). The cells are bipolar, with an apical process, or processes, at the surface of the epidermis, and the proximal region associated with a neurite profile. They may be found in the external skin and oropharyngeal epithelium, including the gills, in various families of teleost fishes. Whitear (1971b) illustrated the appearance by TEM of cells from *Phoxinus*, *Pomatoschistus*, *Gasterosteus* and *Trigla*, comparing them with the gustatory receptors in each species. Lane (1977) confirmed their existence in *Ictalurus melas*, and they have since been seen in various other catfishes, several species of cyprinids, and members of other families including the gadids.

The distribution of the sensory cells can be studied in scanning EM (SEM) preparations that show the apical processes protruding between epithelial cells at the surface of the skin (Lane and Whitear, 1982). In lampreys, the corresponding cells were called 'oligovillous' to distinguish them from other bipolar cells present in the epidermis (Whitear and Lane, 1983a). The cytology is peculiar but sufficiently similar to that in the teleosts to warrant comparison (Figs 6.1(a) and 6.8). Oligovillous cells are especially numerous on the papillae bordering the gill vents and dorsal fin of the adult brook lamprey, *Lampetra planeri* (Fig. 6.2), also on the genital papilla, and on the branching cirrhi under the oral hood of the ammocoete. In *Acipenser ruthenus*, Kotrschal (1984, pers. comm.) found typical apical processes by SEM (Fig. 6.3). Here, and in the cyprinids, the apical processes often end in finger-like extensions (Figs 6.4 and 6.5). The sand goby, *Pomatoschistus minutus*, an acanthopterygian, has blunt-ended apical processes (Fig. 6.6). In this species the solitary cells tend to occur in groups; some apices are close to taste bud pores, as in Fig. 6.7, where it is not possible to know whether the two outlying apices belong to the solitary cell system, or are gustatory cells. A similar situation has been reported from sea bass, *Dicentrarchus labrax* (Connes et al., 1988). Outlying sensory cells are found near taste bud pores also in *Trigla* (Whitear, 1971b) and occasionally in the loach, *Cobitis* (Jakubowski, 1983). In external taste buds of the rocklings

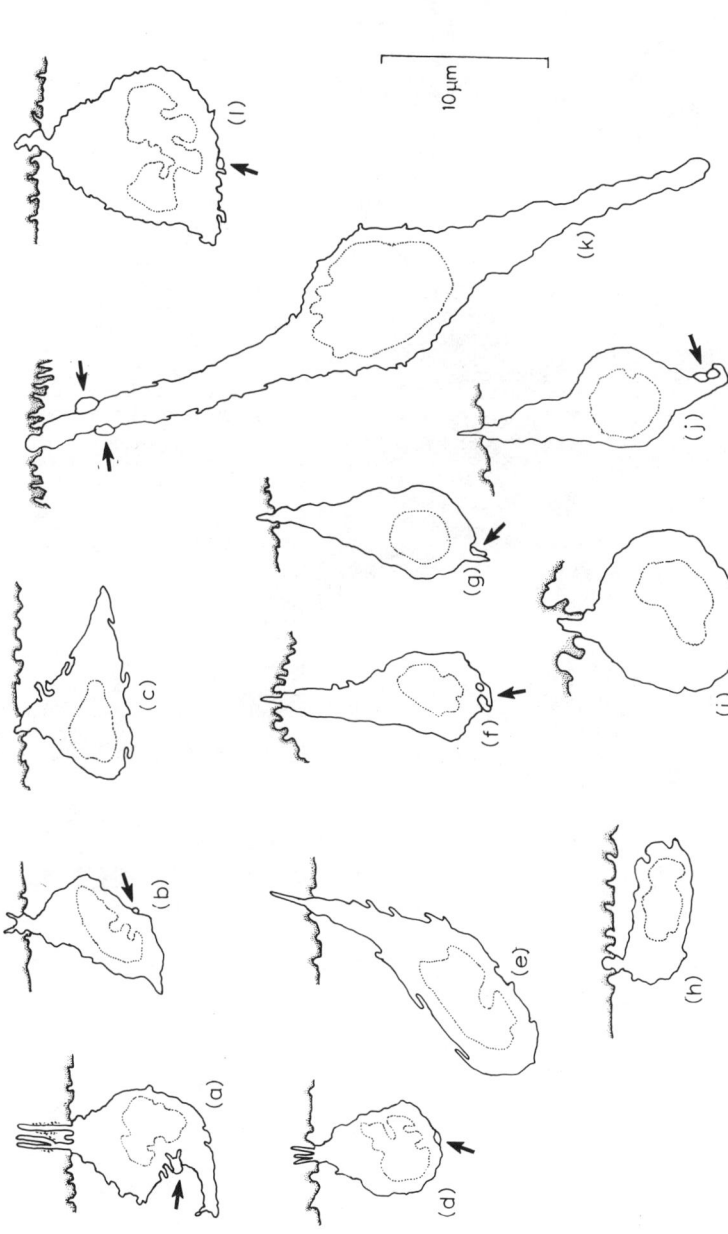

Fig. 6.1 Outline drawings of profiles of representative solitary chemosensory cells, as seen by TEM in sections vertical to the skin surface. The relationship to the surface mucus when in the living state is problematical, and neurite profiles (arrows) will not necessarily be seen in particular sections. (a) *Lampetra planeri*, brook lamprey, ammocoete, from the oral hood; (b) *Clupea harengus*, herring, from skin of flank; (c) *Phoxinus phoxinus*, minnow, from underside of opercular valve; (d) *Cobitis taenia*, spined loach, oral epithelium; (f) *Ictalurus melas*, black bullhead, skin of flank; (f) *Pomatoschistus minutus*, sand goby, oral epithelium; (g) *Gasterosteus aculeatus*, three-spined stickleback, flank; (h) *Poecilia reticulata*, guppy, skin near mouth; (i) *Trigla lucerna*, sapphirine gurnard, skin over sternohyoid muscles; (j) *Ciliata mustela*, five-bearded rockling, vibratile ray of dorsal fin; (k) *Protopterus aethiopicus*, a dipnoan, pelvic fin; (l) *Rana temporaria*, frog tadpole, dorsal skin at three-toed stage. (Originals of (a) and (e) by E.B. Lane, of (d) by M. Jakubowski.)

106   Solitary chemosensory cells

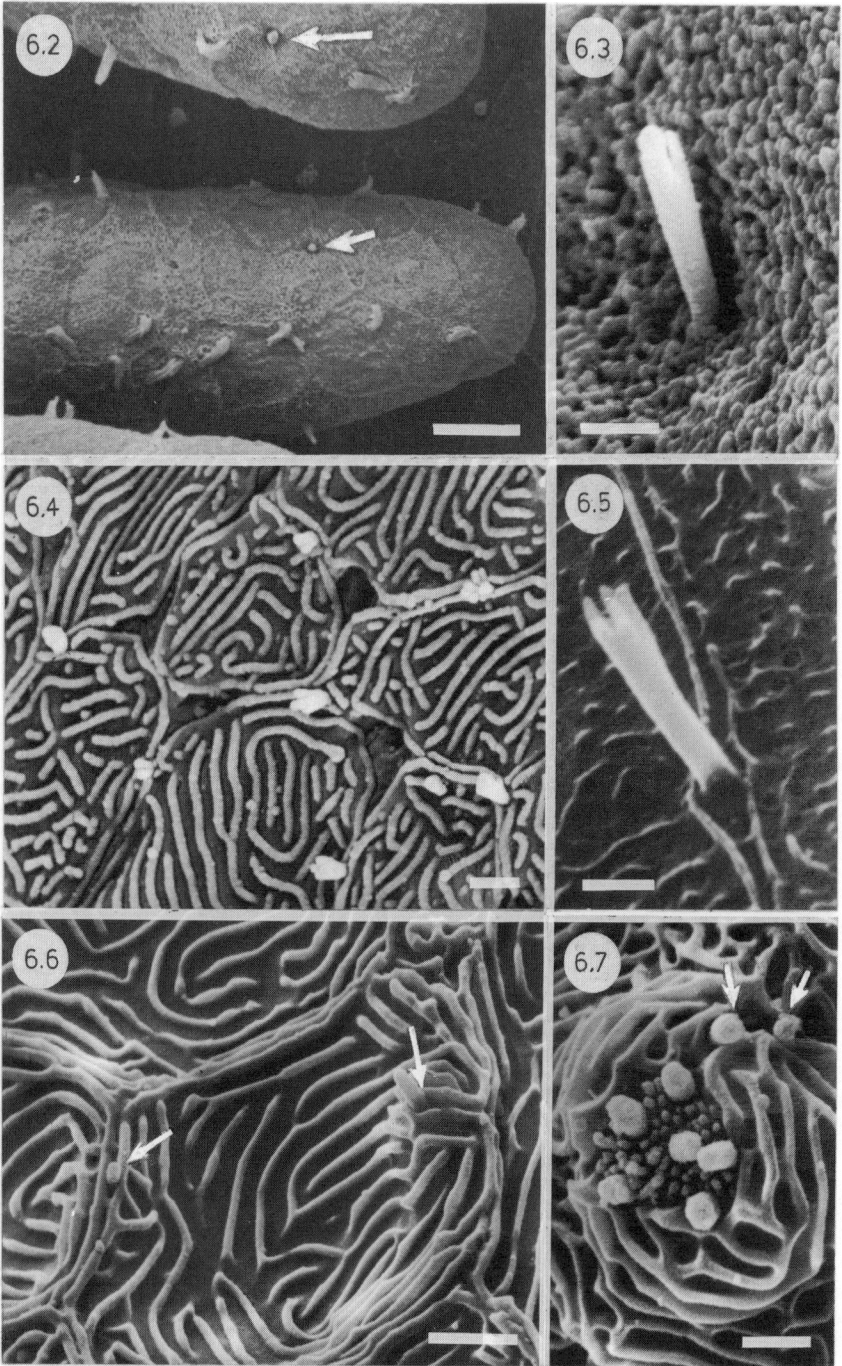

(Gadidae), the pore is surrounded by numerous sensory apices of an uncertain nature (Whitear and Kotrschal, 1988; Jakubowski and Whitear, 1990).

Although solitary chemosensory cells are widespread among teleost fishes, they are not equally common in all situations, nor in all species. The greatest concentrations recorded are on the fin rays of the anterior dorsal fin of rocklings, which are gadids that have modified this fin by reduction of the web and multiplication of the rays; in life the fin displays a rapid flickering motion. It has been estimated that the anterior dorsal fin of *Gaidropsarus mediterraneus*, a shore rockling which grows to about 200 mm in length, carries between three million and six million sensory cells; only the first ray of this fin bears taste buds. Elsewhere on the skin of the rockling, the solitary chemosensory cells occur at a much lower density, of the order of hundreds rather than in excess of 100 000 per square millimetre. On the head of *Phoxinus*, Kotrschal *et al.* (1984) gave a figure of over 1000 mm^{-2}, but some recent counts in cyprinid species are higher (Kotrschal, 1992). The number of solitary sensory cells greatly exceeded that estimated for the gustatory cells within taste buds in the same areas of skin. In six species of cyprinid, the numbers of solitary sensory cells in external skin ranged about 1000 mm^{-2}, fairly evenly distributed on head and trunk although with some local variation. The highest count recorded, on the head of *Rutilus*, approached 4000 mm^{-2}. In the catfishes *Ictalurus nebulosus* and *Clarias batrachus*, and in the cichlid *Haplochromis burtoni*, Kotrschal's figures range from 600 to 1200 mm^{-2}; in a percid, *Stizostedion lucioperca*, from 400 to 800, and in a characid, *Hyphessobrycon*

Figs 6.2 to 6.7 Apical processes of chemosensory cells as seen by SEM. (Originals of Figs 6.3, 6.4, 6.6 and 6.7 by K. Kotrschal.)

Fig. 6.2 *Lampetra planeri*, brook lamprey, papillae on hind border of a gill aperture. Two apices (arrows) belong to multivillous sensory cells; the other projections are the clumped microvilli of oligovillous cells. Scale bar, 10 μm.

Fig. 6.3 *Acipenser ruthenus*, sterlet, apex of a solitary sensory cell near the mouth. Scale bar, 1 μm.

Fig. 6.4 *Rutilus rutilus*, roach, behind eye. Apices of seven solitary sensory cells among epithelial cells with microridges; the triangular apices are undischarged goblet cells. Scale bar, 2 μm.

Fig. 6.5 *Phoxinus phoxinus*, minnow, operculum. Process of a solitary chemosensory cell projecting between two epithelial cells. Scale bar, 1 μm.

Fig. 6.6 *Pomatoschistus minutus*, sand goby, flank. Processes of two solitary chemosensory cells (arrows), one wrapped in the rim of an epithelial cell. Scale bar, 2 μm.

Fig. 6.7 *P. minutus*, near nostril. A taste bud pore with six gustatory cell processes, and two sensory processes (arrows) that are outside the area marked by the small microvilli of supporting cells. Scale bar, 1 μm.

innesi, from 200 to 300 mm^{-2}. In the goby, *Pomatoschistus minutus*, where apices were counted in scanning electron micrographs of comparable areas around the nares, on two individuals, the mean of the two sides was 103 mm^{-2} in one fish, and 181 mm^{-2} in the other. In the second fish, the number counted in an area of 3 mm^2 between the eyes was 295, but the distribution was not entirely even; there were greater numbers of the apices at the anterior end of the field, in the vicinity of groups of free neuromasts (Whitear, pers. obs., 1988). The numbers matched the estimates from histological preparations (Whitear, 1971b). Concentration of sensory apices in the vicinity of free neuromasts was also noticed in cyprinids (Kotrschal, 1992). If the solitary sensory cells are scattered at a density of even 100 mm^{-2}, they will be encountered in sections only rarely, and the records from a number of species of teleost, for instance the knifefish, *Notopterus notopterus*, the goldfish, *Carassius auratus*, the saithe, *Pollachius virens*, the guppy, *Poecilia reticulata*, and swamp eel, *Monopterus cuchia*, depend on limited observations. In several species examined extensively by SEM and TEM, such as the pogge, *Agonus cataphractus*, the European eel, *Anguilla anguilla*, and the fifteen-spined stickleback, *Spinachia spinachia*, no solitary sensory cells have been recorded, nor were they seen by TEM in the mud skipper, *Periophthalmus koelreuteri*; in these cases it can be assumed that the system is either absent, or unusually poorly developed. In the shanny, *Blennius pholis*, the cells have not been seen by electron microscopy, in spite of extensive examination of the skin, although they were stained with methylene blue in that species (Whitear, 1971b) and were recorded by TEM in *B. tentacularis* by Schulte and Holl (1972). On the other hand, the catfishes are well supplied with solitary sensory cells in addition to taste buds (Lane and Whitear, 1982, and see above). The skin and oral epithelium of two clupeids, the herring, *Clupea harengus*, and the sprat, *Sprattus sprattus*, also contains numerous solitary sensory cells. Other authors have illustrated them in the climbing perch, *Anabas testudineus* (Hughes and Munshi, 1973), the sword-tail, *Xiphophorus helleri* (Reutter *et al.*, 1974), the brown trout, *Salmo trutta* (Harris and Hunt, 1975), the cichlid, *Oreochromis aureus* (Fishelson, 1984), the loach, *Cobitis taenia* (Jakubowski, pers. comm., 1984, and Fig. 6.1(d)), and possibly in the armoured catfish, *Ancistrus* (Ono, 1980).

Whitear (1971b) examined material of the bowfin, *Amia calva* and the reedfish, *Erpetoichthys calabaricus*, by TEM and found cells that were probably of the chemosensory type; for technical reasons the identification could not be certain. In the dipnoans *Protopterus aethiopicus* and *P. amphibius*, Lane and Whitear (1978 unpubl. obs.; Fox *et al.*, 1980) found bipolar cells in the epidermis of the fins, which had many of the expected characteristics of chemosensory cells, but cannot as yet be confirmed as such. The only nerves seen were adjacent to the side of the cell rather than at the base, as expected (Fig. 6.1(k)), although there was some indication of synaptic contact (Fig. 6.13). The investigation was preliminary only, and hampered by the large size of dipnoan cells. Reutter and Vogel (1989) found that the nerves in external taste buds of dipnoans were also associated with the sides rather than the bases of sensory cells. However, the common association of nerve fibres with the flask cells of amphibian skin, which are a type of ionocyte (see below p. 117), warrants caution in the interpretation, although the cytological characters in *Protopterus* were more consistent with a sensory than with an ionocytic function. Neurites were frequently adjacent to the sides, as well as the base, of the Stiftchenzellen of ranid tadpoles (Figs 6.1(l) and 6.11), where the cytology allows the cells to be classed as chemoreceptors even in the absence of direct evidence (Whitear, 1976). They are present at all stages of larval life, on the body, the tail, and even on the developing limbs. They disappear at metamorphosis, and have not been reported from other types of amphibian larvae.

6.3 CYTOLOGY

Differences in the morphology of the sensory cells often correspond to the variation in gustatory cell cytology between particular species. The apical process may be a single stout projection 1–2 μm in length, occasionally 4 μm, or it may have a terminal division into several extensions (Figs 6.4 and 6.5; Lane and Whitear, 1982). In the clupeids there is usually a cluster of microvilli set on the apex of the cell (Fig. 6.1(b); Fox et al., 1980). Similar clustered, slender microvilli have been recorded in *Cobitis* (Fig. 6.1(d); Jakubowski, pers. comm., 1984) and are a rare variation in the apical processes of gustatory cells in taste buds (Jakubowski and Whitear, 1990). In the oropharyngeal epithelium of the selachian *Raja clavata*, apices belonging to probable solitary chemosensory cells bear 4 or 5 microvilli (Whitear and Moate, unpubl. obs., 1991), similar to those reported on the gustatory cells of selachian taste buds (Pevsner, 1976). The apical processes of the supposed chemosensory cells in *Protopterus* were swollen (Lane and Whitear, 1982). Swollen processes were also recorded on dipnoan electroreceptors (Roth and Tscharntke, 1976). The lamprey oligovillous cells have from 2 to 30 prominent microvilli with a conspicuous filamentous core (Whitear and Lane, 1983a).

The form of the cell as a whole depends on its situation; in a thin epidermis the cell body may be inclined to one side, as in Fig. 6.1(c) and (h). Where the epidermis is thick, the nucleus of the sensory cell often lies at the level of the second tier of epithelial cells from the surface, but in other situations the cell may be elongated, with its deep pole immediately above the basal layer of the epidermis. Usually the apical process is of sufficient length to raise the presumed receptive membrane above the mucus covering the surface of the epithelium (as seen when fixed). To reveal the apices for SEM, it is necessary to remove the mucous layer. In some examples the glycocalyx on the membrane of the process is clearly differentiated from the mucous cover on the adjacent epithelial cells, for instance on the vibratile fin in rocklings (Whitear and Kotrschal, 1988). In the triglids, however, which have a particularly thick

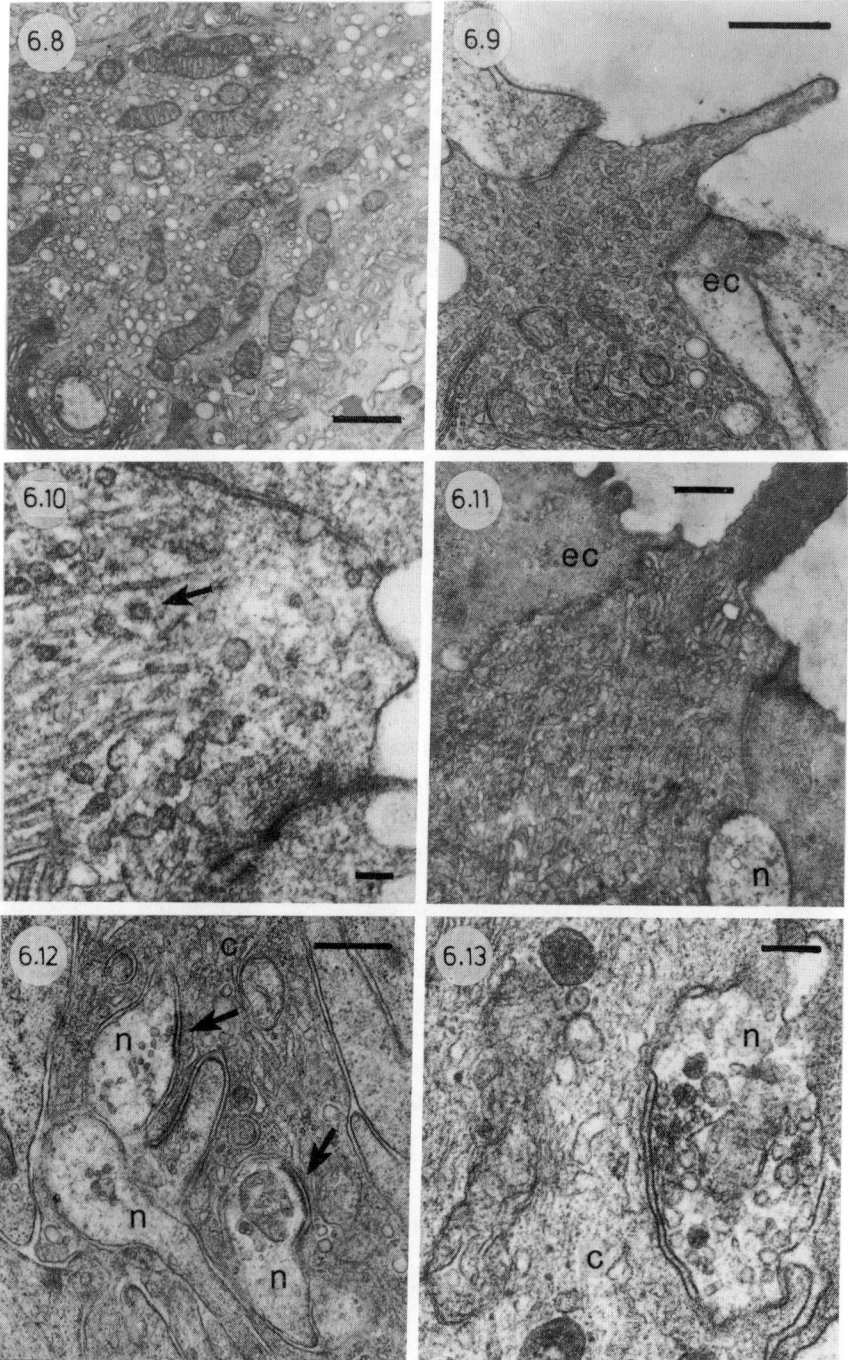

mucoid cuticle on the skin, the chemosensory cell processes appear not to penetrate the covering layer, and on the modified part of the pectoral fin rays often have no integral projection (Whitear, 1971b). This probably accounts for the failure of Silver and Finger (1984) to identify the sensory cells on the free fin rays of the related species *Prionotus carolinus* by SEM; the apices would be hidden among the tall microridges.

The filamentous core of an apical process extends some way into the distal cytoplasm, in *Gasterosteus* even as far as, or beyond, the nucleus. Lamprey oligovillous cells also have long microvillar cores which may extend throughout the neck of the cell. There is cytological evidence of activity in the distal half of the cell, which contains numerous mitochondria, Golgi systems giving rise to characteristic vesicles, endoplasmic reticulum, coated vesicles, and a few multivesicular bodies and lysosomes (Fig. 6.8). There are also well-developed, vertically set, microtubules. Below the apex of the cell these microtubules characteristically surround the core of actin filaments subtending the apical process, or processes; this description applies to the solitary chemosensory cells of other groups besides lampreys (Figs 6.9, 6.10, and 6.11). The distal vesicles are arranged around, and in some examples appear to be aligned along, the microtubules. Vesicles are occasionally seen within the apical projection and may open to the surface, but the microtubules do not extend so far.

It can be speculated that the microtubules in the neck of the sensory cell are involved in transporting the vesicles formed in the Golgi system to the apex of

---

Figs 6.8 to 6.13 Parts of solitary sensory cells as seen as TEM.

Fig. 6.8 *Lampetra planeri*, brook lamprey, site as in Fig. 6.2. Distal cytoplasm of an oligovillous cell, with mitochondria, vesicles, Golgi system and a lysosome. Scale bar, 1 µm.

Fig. 6.9 *Sprattus sprattus*, sprat, oral epithelium. Apex of a solitary chemosensory cell with mitochondria and vesicles, part of a Golgi system, and the apical process; ec epithelial cell. Scale bar, 0.5 µm.

Fig. 6.10 *Pomatoschistus minutus*, sand goby, ventral skin of body. Apex of a solitary chemosensory cell with microtubules and vesicles; arrow points to a coated vesicle and there is a coated pit at the lateral membrane. Scale bar, 0.1 µm.

Fig. 6.11 *Rana temporaria*, frog tadpole, skin of foot shortly before metamorphosis. Apex of a Stiftchenzelle with a nerve profile (n) beside it; ec, epithelial cell. Scale bar, 0.5 µm.

Fig. 6.12 *Pomatoschistus minutus*, sand goby, oral epithelium. Base of a solitary chemosensory cell (c) associated with neurite profiles (n). Arrows point to two of several synaptic densities. Scale bar, 0.5 µm.

Fig. 6.13 *Protopterus aethiopicus*, a dipnoan, pelvic fin. Neurite profile (n) beside the neck of a cell (c) similar to that in Fig. 6.1(k). Scale bar, 0.2 µm. (Original of Fig. 6.13 by E.B. Lane.)

the cell, for the purpose of renewing the receptive membrane covering the apical process or processes. In the cyprinids and silurids especially, these vesicles, 50–70 nm in diameter, have a characteristic appearance due to relatively electron-dense contents. In some other cases the vesicular contents are more lucent. These vesicles correspond to the distal vesicles of gustatory cells, which sometimes take the form of 70 nm tubules, although rows of vesicles are more usual (Jakubowski and Whitear, 1990). In the solitary sensory cell shown in Fig. 6.10, some vesicles are fused together. In the sensory cells of the vibratile fin of the rocklings, the subapical vesicles are particularly well developed, large, and lined with material resembling the glycocalyx covering the apical process (Whitear and Kotrschal, 1988). Vesicles are smaller and denser in solitary sensory cells elsewhere in the rockling skin. These observations suggest that there may be a specialized secretion on the sensory cells of the vibratile fin. In some examples of sensory cells, for instance in *Pomatoschistus*, there may be whorls and channels of smooth endoplasmic reticulum (SER) in the distal process of the cell, in addition to the vesicles. This can be paralleled in the gustatory cells of taste buds (Jakubowski and Whitear, 1990). In lamprey oligovillous cells, more than one type of vesicle is present in the subapical region (Whitear and Lane, 1983a); possibly this is a general character. On the free pectoral fin rays of the triglids, instead of an apical process there is often a cluster of vesicular profiles in the mucous cuticle (Whitear, 1971b). This might have been caused by delayed fixation due to the presence of the cuticle, or may represent membranous material that had been shed from the apex of the sensory cell and trapped in the mucus.

Vesicles and microtubules are likewise characteristic features of the cytoplasm in the oligovillous cells of lampreys, the Stiftchenzellen of ranid tadpoles, and the bipolar cells of *Protopterus*. The oligovillous cells provide evidence of cyclical activity in the system. Some represent juvenile, and some senile, stages, but the majority show features that could be interpreted as due to the periodic renewal of the membrane covering the processes and to the eventual withering of individual microvilli (Whitear and Lane, 1983a). Although membranous components are shed from the microvillar surface, it seems unlikely that apical membrane is recycled, because coated invaginations are not a feature of the solitary sensory cell apices, in lampreys or in other fish. One is present in Fig. 6.10, but it is on the lateral face of the cell.

The nucleus is usually embayed; in the tadpole Stiftchenzellen it may be lobulated (Meyer, 1962; Whitear, 1976). At the nuclear level and below, the characteristic organelles are cisternae of ER, often ribosomal. There may be vesicular profiles, but these are not numerous. In occasional sensory cells of *Pomatoschistus* and *Trigla*, the base of the cell contains a stack of ER cisternae (Whitear, 1971b). A large stack of ER, distinct from the Golgi apparatus, is a regular feature of lamprey oligovillous cells (Whitear and Lane, 1983a).

6.4 INNERVATION

In teleosts, slender nerve fibres make passing contact with the solitary sensory cells, usually near the base and sometimes indenting the membrane. Neurite profiles do not occur in every section (Fig. 6.1), and synaptic specializations are rarely seen, except in favourable situations like the rockling vibratile fin, or the head and oral epithelium of *Pomatoschistus* (Fig. 6.12). In some cases, a nerve fibre is seen passing on beyond the sensory cell; this may imply that it also innervates other sensory cells, or is one of the free-ending nerve fibres (Whitear, 1971a) which happened to lie adjacent to the sensory cell. When synapses could be identified, they appear like those on the gustatory cells, consisting of membrane densities, with a postsynaptic density under the neurite membrane (Fig. 6.12); occasionally the base of the cell was full of vesicles. In the dipnoan *Protopterus*, not enough is known of the detail of taste-bud innervation to make a comparison, but there appears to be some increase of membrane density in Fig. 6.13 where the nerve is indented into the side of the supposed solitary chemosensory cell.

In lampreys the nerve fibre is relatively thick and bears spur-like processes, which indent the base of the oligovillous cell (Figure 6.1(a)). Narrow channels, apparently of ER, converge on the neurite spurs. There is increased density of the cell and neurite membrane, and the cleft material has the form of rows of prickles on the cell membrane (Whitear and Lane, 1983a).

The transmitter involved in the solitary sensory cells is unknown. There is some evidence of the presence of acetylcholinesterase in nerves to taste buds, and possibly to solitary cells, in the barbels of *Ictalurus* (Reutter, 1974), but the histochemical activity extended along the nerve fibres, and the subsequent discovery of Merkel cells, which are associated with tactile nerve fibres, in the tissue (Lane and Whitear, 1977) introduces uncertainty. The frequent absence of vesicles near synaptic densities is of interest because Roper (1989) suggested that if the transmitters in taste cells were amino acids, vesicular storage would not be necessary.

Solitary sensory cells on the head and oropharyngeal epithelium could share innervation with the taste buds in the same area, from the facial or vagal special visceral component. In species with a recurrent facial nerve passing to taste buds on the body, this can also carry nerve fibres to solitary sensory cells, as in the case of the anterior dorsal fin of rocklings, where denervation eliminated the facial component and left nearly all the sensory cells without a nerve supply. However, a small number of sensory cells were then associated with nerve fibres that were evidently of spinal origin (Whitear and Kotrschal, 1988); whether this was a new or formerly established innervation is not known. In the triglids, the sensory innervation of the modified pectoral fin rays is undoubtedly from spinal nerves (Finger, 1982, 1983, 1988). The nerve endings concerned lie in small projections of the epidermis, the claviform bodies, each of which contains a sensory cell and a loop of free nerve ending that is presumably tactile in function (Bardach et al., 1967; Whitear, 1971a, b). In *Gasterosteus aculeatus*, plentiful solitary sensory cells are present in the epidermis of the body, although a recurrent facial nerve is thought not to exist in this species, implying that they are innervated by spinal nerves (Whitear, 1971b). In tadpoles, the sensory cells on the developing limbs are associated with nerve fibres that must be spinal in origin, although it could be argued that on the body an alternative pathway is available along the lateralis nerves.

In most vertebrates, taste buds are maintained by a trophic action of the nerves supplying them, and degenerate if the nerve is cut, with a possible exception in anurans (Toyoshima et al., 1984). It is not known whether the solitary sensory cells are similarly dependent on an intact innervation for their survival. Lane (1977) cut the recurrent facial nerve in *Ictalurus melas*, and monitored the disappearance of taste buds on the flanks; most taste buds had degenerated within a week, but solitary sensory cells could still be found after 120 days. However, this does not rule out the possibility that the nerve supply to the solitary cells was initially spinal, so that, if they require neural maintenance, it could have come from spinal nerves. In the rocklings, *Ciliata* and *Gaidropsarus*, the great majority of the sensory cells of the vibratile dorsal fin rays were deprived of innervation by cutting the recurrent facial nerve, and persisted for as long as 3 or 4 weeks (Kotrschal et al., 1985; Whitear and Kotrschal, 1988). This persistence might reflect a slow turnover rate, as denervated taste buds were still recognizable after 3 weeks, or it might be that the cells were maintained by the intact spinal sensory innervation of the fin, even though only a few sensory cells were in synaptic contact with these nerve fibres.

In the rocklings, the fine facial nerve fibres supplying the vibratile fin originate from a small population of geniculate ganglion sensory neurones. These terminate exclusively within a distinct, dorsal, part of the medullary facial lobe (Kotrschal and Whitear, 1988). However, this localization might be topological and due to the derivation of the sensory input from the particular fin; nothing is known as yet of the central connections of solitary chemosensory cells elsewhere. The relative abundance of solitary chemosensory cells in the skin of cyprinid species does not correlate with the numbers of external taste buds, nor with the relative development of the facial and vagal medullary lobes, nor with the feeding habits of the fish (Kotrschal, 1991b).

6.5 PHYSIOLOGY

Direct investigation of stimulation of solitary chemosensory cells is complicated by the frequent presence of taste buds in the same region. There are a few situations in which it has been possible to exploit more discrete systems. The free pectoral fin rays of the Triglidae bear no taste buds and have sensory nerves passing to spinal centres. Bardach and Case (1965) showed that chemical stimulation of these fin rays by appropriate substances induced searching behaviour, the first stage of a feeding reaction. Silver and Finger

(1984) analysed the electrophysiological response in the nerves leading from the fin rays: ten compounds tested were stimulatory, five evoked no response. The general sensitivity to amino acids was similar to that found in the fish gustatory system (Caprio, 1975), although not identical in detail.

In the rocklings, the anterior dorsal fin has taste-buds on the first ray only, which can easily be eliminated. External taste-buds in *Ciliata* respond to mechanical stimuli and to amino acids, but not to fish mucus, which is a potent stimulus to the sensory cells on the vibratile fin. The nerve from the vibratile rays does not respond to amino acids or to touch (Peters *et al.*, 1987, 1989, 1991).

Mucoid substances are also stimulants for lamprey skin chemoreceptors. Baatrup and Døving (1985), using suction electrodes, recorded responses to chemical and tactile stimuli from many areas of lamprey epidermis. Only chemical stimuli were effective on the papillae by the gill vents. The epidermis in this region contains numerous oligovillous cells, but no Merkel cells (assumed to be associated with tactile nerve fibres) have been found (Whitear and Lane, 1983a). There is a strong presumption that the oligovillous cells are involved in chemoreception, although the papillae by the gill vents also contain sensory cells of another type (Fig. 6.2), the so-called multivillous cells that are a part of the lateralis system (Whitear and Lane, 1983b). These are thought to correspond to certain receptors that Parker (1912) considered distinct from those for the chemosensitivity of the ammocoete skin. The pharyngeal terminal buds of the ammocoete responded to classical taste stimuli and to amino acids (Baatrup, 1985), while the receptors of lamprey skin responded to acetic acid, strong solutions of NaCl, sialic acid, mucus from trout, and thaw-water from frozen trout (Baatrup and Døving, 1985).

It is clear that in the rocklings, the chemosensory vibratile fin has a function distinct from that of the external taste-buds, which responded to amino acids and to extracts from invertebrates but not to mucus (Peters *et al.*, 1989, 1991). The sensitivity of the vibratile fin receptors to mucoid substances provoked speculation about possible interspecific interactions (Peters *et al.*, 1987; Whitear and Kotrschal, 1988). It would be expected that smell would be involved in, for instance, sibling recognition (Olsén, 1989 and Chapter 11 herein), and it has now been shown that behavioural responses of *Ciliata* to mucus from other species depend on integration of the vibratile fin receptors with olfaction (Kotrschal *et al.*, 1989); this can be compared with the interaction of nasal trigeminal receptors with the olfactory system (p. 118). The undulation of the vibratile fin creates a current that serves to smooth out peaks of chemical stimuli in the ambient water and allows sampling at very low stimulus concentration (Kotrschal, 1990, pers. comm.). Possibly such a device is advantageous to nocturnal and cryptic species, to test the environs for predators before venturing from the lair. The responses of typical scattered chemosensory cells remain to be determined. Herrings, *Clupea harengus*, have plentiful solitary sensory cells, but predator avoidance in their young appears to involve mechanical stimuli conveyed to the otic bullae (Blaxter and Fuiman, 1990). It is not known when the solitary chemosensory cells first differentiate in this species, nor in any other fish.

Human saliva, besides mucus, is a stimulus to the rockling and lamprey skin chemosenses. Tucker (1983), discussing fish chemoreception in general, remarked that the taste buds of Swedish and Japanese carp react differently to saliva. He also pondered how the mucous coat covering the skin could act simultaneously to protect the receptors and to admit stimuli, apparently assuming that this mucus comes from goblet cells and is continuous over taste bud pores. In fact the secretion at a taste bud pore is derived from the supporting cells, and is histochemically different from the mucoid coat on the surrounding epithelial cells (Kiyohara et al., 1984; Witt and Reutter, 1988, 1990; Jakubowski and Whitear, 1990). The processes of gustatory cells and of solitary chemosensory cells normally project beyond the mucous layer, as it is seen in fixed material (except in the triglids, where the cuticular mucus is particularly thick). During life, the surface mucus of any fish epidermis presumably exists as a thicker layer than is seen after fixation, and becomes diluted in contact with the water. This mucus might well affect events at the receptor membrane, perhaps by influencing ion transfer. In an experimental situation, both the epithelial mucus and the glycocalyx on the sensory apical process could be affected by applied mucoid substances, although specific components need to be identified. Caprio (1988) concludes that there must be binding to a membrane if chemoreception is to occur; Chapters 8 and 9 herein deal with transduction processes in olfactory and gustatory receptors; there is no information on similar aspects of the solitary chemosensory cells.

6.6 THE COMMON CHEMICAL SENSE

Parker (1912) distinguished a chemical sensitivity in the exposed mucous surfaces of Man and in the skin of lower vertebrates, from that involving taste-buds, and called it the 'common chemical sense'. By differential nerve section and observation of behaviour in *Ictalurus*, he demonstrated that spinal nerves to the flank skin were capable of responding to acid, alkaline and salt solutions, which in appropriate concentrations initiated avoidance reactions. Ammocoete larvae also swam away from sufficiently concentrated solutions of salts or acids pipetted against the body. Similar observations had been made on a selachian by Sheldon (1909), who concluded that the reactions to chemical stimuli were mediated by trigeminal and spinal nerves. Parker thought that the nerves concerned with the common chemical sense ended freely in the epidermis; so did Beidler (1965), when discussing chemosensitivity of free nerve endings. The demonstration that solitary sensory cells are present (Whitear, 1965, 1971b; Lane, 1977) implicated them as possible receptors for the common chemical sense of fish epithelia.

However, it should be stated at once that solitary sensory cells are not a prerequisite for chemical sensitivity in an epithelium.

Tactile and chemical sensitivities in frog skin were first separated by von Anrep (1880), using cocaine. The Stiftchenzellen of ranid tadpoles do not survive metamorphosis, but, in adult frogs, methylene blue will stain differentiated cells, the flask cells, in a manner similar to that by which it picks out the chemosensory cells in fish skin. There were cytological reasons for considering the mitochondria-rich flask cells to be ionocytes rather than sensory cells (Whitear, 1975), and it is now demonstrated that the flask cells are responsible for chloride transport across the epidermis (Katz, 1986). Although not themselves sensory, it is possible that the flask cells are associated with the detection of chemical stimuli by frog skin, because there are commonly nerves passing up the sides of the flask cells, often partially embracing the neck of the cell. These nerve endings approach close to the skin surface near a site of ionic movement (through the flask cell apex), where the overlying keratinized layer may also be particularly permeable because the junctions over the flask cells form only at the previous moult. One can therefore speculate that these nerve fibres may be preferentially exposed to chemical excitation. In amphibian larvae, the mitochondria-rich cells that presumably correspond to the flask cells are not specially associated with nerve fibres, and the association in the adult may depend on mechanical factors connected with the pattern of moulting (Whitear, 1974, 1975, 1983). Spinal reflex reactions of frogs to salt solutions applied to the skin are slow, but can be differentiated from tactile and painful stimuli (Cole, 1910; Crozier, 1916).

It is reasonably certain from numerous investigations that the corneal and conjunctival epithelium of tetrapods does not contain bipolar sensory cells, although the skin covering the cornea in fish can do so (pers. obs.). Beidler (1965) cited experiments showing response in corneal nerves to certain odorous substances. Silver (1987), summarizing his work on chemical sensitivity mediated by the trigeminal nerves in mammals, birds and amphibians, suggested that if the non-gustatory receptors for chemical stimuli in fish skin are sensory cells, then the term 'common chemical sense' should be reserved for systems such as the trigeminal, or other, free nerve endings. Parker (1912) used the term in respect of fish as well as applying it to mucous surfaces of tetrapods, but it certainly cannot be assumed that his common chemical sense is an entity. At this stage, the nomenclature of a system so little understood is not a thing to worry about. Studies on the oral and nasal trigeminal senses in tetrapods (e.g. Silver et al., 1988) shed no light on the function of the solitary chemosensory cells of fish. This problem is also discussed by Kotrschal (1991).

In teleosts, bipolar cells differing from olfactory cells have been identified in the nasal mucosa (Whitear, 1971b; Yamamoto, 1982), but the rod-like apical process of these cells may be an artefact derived from degenerate cilia (Caprio, 1988). Belousova et al. (1983) recorded from branches of the trigeminal nerve supplying the olfactory sacs of cod, Gadus morhua, and carp, Cyprinus carpio.

There were responses in different fibres to mechanical stimulation and to a variety of chemical stimuli. Some fibres had specific sensitivities; the threshold concentrations were similar to those for taste. The authors concluded that the system resembled the trigeminal nasal innervation of tetrapods and probably interacted with the olfactory system. There is, however, no information on the receptors concerned. There are chemosensory cells in the skin very near the nares, at least in cyprinids (Kotrschal and Whitear, unpubl., 1988), but the epidermis of teleosts is also permeated with free nerve endings which may lie under or even between the most superficial epithelial cells (Whitear, 1971a). In rainbow trout, *Oncorhynchus mykiss*, Yamashita *et al.* (1987) studied responses in the palatal nerve and thought that different receptors were reacting to carbon dioxide and to hydrochloric acid; again, there is no morphological information. Recent studies on the central connections of the trigeminal and facial nerves of fishes have not considered the peripheral receptors (Bartheld and Meyer, 1985; Kiyohara *et al.*, 1986; Puzdrowski, 1988).

It has been asserted that the free neuromasts are concerned in reception of chemical stimuli in fish (Katsuki *et al.*, 1970; Katsuki *et al.*, 1971; Katsuki and Yanagisawa, 1982) and in the clawed toad *Xenopus laevis* (Onoda and Katsuki, 1972; Blanchi *et al.*, 1976). Onoda and Katsuki (1972) state: 'It is now certain that the function of the chemoreception of the lateral line organ of *Xenopus* is to provide the common chemical sense...'. There are, however, serious flaws in the experimental evidence reported by these various authors. The electrophysiological recordings were made with the animals out of water and partly covered with gauze, circumstances in which it is almost impossible that the cupulae of the superficial neuromasts could retain their normal relation to the sensory surface, which ensures a high potassium ion concentration at the receptive apices (Russell and Sellick, 1976). Significantly, potassium was the cation most effective in enhancing the responses in the lateralis nerves, but was not effective on the canal neuromasts of fish where the cupulae would be protected from damage in the experimental situation. Blanchi and Guardabassi (1977) found that the response to KCl solutions in lateralis nerves of *Xenopus* was reduced when animals had been kept out of water, but did not state what was the condition of the neuromasts and their cupulae after such treatment. These are among the few authors who have accepted the hypothesis of dual mechano- and chemoreceptive function of the free neuromasts; Russell (1976) rejects it.

Theoretically, the electroreceptors (ampullary organs) would be capable of detecting changes in salinity of the ambient water. Parker (1912) was not aware of the existence of electroreceptors, but his experiments had eliminated the ampullary organs of *Ictalurus*, along with the rest of the lateralis system, from participating in his common chemical sensitivity.

The diffuse nature of the solitary chemosensory cell system in fishes makes it difficult to devise experiments that could separate it from the olfactory and gustatory senses, without so crippling the fish's sensory abilities as to make behavioural observations unreliable. The sensitivity of rockling vibratile fin and lamprey skin to mucus has been mentioned; mucous secretions from other animals must be a significant part of the content of natural waters. Water of low pH can be detected and avoided by salmonids (Bishai, 1962); the detection of such background information would be admirably suited to a diffuse sensory system, but again experimental evidence about the receptors concerned is lacking. The lamprey oligovillous cells responded to acid stimuli at pH below 5 (Baatrup and Døving, 1985). Acidly reacting chemicals also caused strong responses in the recurrent facial nerves of rocklings, recorded both from the vibratile fin (with no taste buds) and from the flanks, which have taste buds and solitary sensory cells in the skin (Peters et al., 1987). In *Prionotus*, Silver and Finger (1984) did not specifically consider pH effects, nor the responses from the skin other than on the modified fin rays, although triglids possess solitary chemosensory cells in the epidermis of the body as well as on the fin rays (Fig. 6.1(I); Whitear, 1971b). A demonstration that the transduction mechanism for bitter sensation in gustatory cells differs from that for other stimulants (Akabas et al., 1988) should induce caution in arguing from one set of chemoreceptors to another, even in a single species. Furthermore, one cannot be confident that all the responses so far studied in situations where solitary chemosensory cells are present, are necessarily passing through that channel and not affecting the unmodified elements of the epithelium.

6.7 PARANEURONES

Specialized neuroendocrine cells in the internal epithelia of vertebrates can be compared with the chemoreceptors of fish skin, for both sets can be classified as paraneurones (Fujita et al., 1988) and most, although not all, paraneurones are innervated. The brush cells of the bile duct of the rat have a fine structure strikingly similar to that of the chemosensory cells in fish, especially lampreys, with apical microvilli and microtubules and vesicles in the subapex (Luciano et al., 1981). Similar cells occur in the tracheal epithelium, where they are considered to be chemosensitive (Luciano et al., 1968). Bipolar cells reacting for serotonin extend through the oesophageal epithelium of frog (Nada et al., 1984). The apical cell of neurepithelial bodies of toad lung bears a cilium (Rogers and Haller, 1980), but so do the receptor cells of the sensory corpuscles of the lamprey pharynx (Baatrup, 1983). The bronchopulmonary paraneurones (review: Fujita et al., 1988) include single and multicelled types, analogous to the systems of solitary chemosensory cells and taste buds in the skin and oral epithelium of fishes. This analogy prompts a speculation that the chemosensory cells may have functions that bypass their neural connections. The chemosensory cells of fish do not, it is true, contain the dense-cored vesicles that are found in many neuroendocrine cells, and also in the gustatory and basal cells of amphibian taste buds, but that does not exclude a local endocrine function. If such a function exists, it may well affect the mucous secretion of the epidermis, via both goblet cells and superficial epithelial cells. Mucification responds to systemic endocrine stimulation and also to the ionic composition of the environment (Wendelaar Bonga and Meis, 1981; other references cited by Whitear, 1986). Iger *et al.* (1988) reported increased mucous secretion from the skin of carp kept in dirty water. Since the goblet cells of fish epidermis are not directly innervated, a local, rather than a central, endocrine circuit, from sensory cells to mucus-producing cells, is at least feasible.

The existence of solitary chemosensory cells so similar to the gustatory cells should draw attention to the role of the supporting cells in taste bud function. Morphological differences between the solitary chemosensory cells and the gustatory cells within taste buds concern the relationship to surrounding cells and the secretions potentially affecting the receptor membrane. In the neck of a taste bud, the gustatory cells are wrapped by the supporting cells, which not only insulate the gustatory cells one from another, but also may play a part in sheathing the distal part of the cells, in a manner analogous to the role of the sheath cell on a nerve fibre. This arrangement could control the immediate ionic environment and perhaps facilitate the passage of an electrotonic signal down the gustatory cell. In a solitary chemosensory cell, the membrane of the neck of the cell is adjacent to several epithelial cells, rather than to a single specialized element. In a taste bud, the protuberant gustatory cell processes are at least partially bathed in the special secretion of the supporting cell apices, whereas a solitary cell microvillus is surrounded by the general mucoid secretion of the surface epithelial cells. Such factors should be taken into account when analysing the physiological responses of the chemosensory receptors, especially if gustatory responses are studied on dissociated cells. Some internal integration of the taste bud is probably also due to another type of paraneurone, the basal cell of fish taste buds, where it occurs (Whitear, 1989; Jakubowski and Whitear, 1990). It is assumed that the supporting cells are not themselves primarily sensory, despite differing opinions of other authorities (Reutter, Chapter 4 herein; Roper, 1989), because the postulated synaptic association with nerves has not been confirmed (Jakubowski and Whitear, 1990), but this interpretation does not deny the supporting cells functional importance.

6.8 CONCLUSION

It can be stated that secondary sensory cells capable of responding to chemical stimuli differentiate in the epithelia of primarily aquatic vertebrates, and that certain of these chemosensory cells are incorporated in discrete taste buds while others, the solitary chemosensory cells, do not form a special association with surrounding epithelial cells. It is, however, not necessary that all chemical stimuli to an epithelium involve specialized receptor cells, although such cells would be capable of conveying stimuli to nerve endings within the epithelium, that otherwise might not be accessible to non-noxious substances

(Whitear, 1971b, 1983; Whitear and Kotrschal, 1988). The solitary sensory cells certainly exist, and the elucidation of their function presents a considerable challenge.

ACKNOWLEDGEMENTS

Professor M. Jakubowski, Dr K. Kotrschal and Dr E.B. Lane kindly gave permission to use their original micrographs, and Dr Kotrschal made constructive suggestions on reading a draft manuscript. Thanks are also due to P.M. Lees for technical assistance at University College London.

REFERENCES

Akabas, M.H., Dodd, J. and Al-Awqati, Q. (1988) A bitter substance induces a rise in intracellular calcium in a subpopulation of rat taste cells. *Science*, **242**, 1047–50.
Anrep, B. von, (1880) Ueber die physiologische Wirkung des Cocain. *Pflugers Arch. ges. Physiol.*, **21**, 38–77.
Baatrup, E. (1983) Terminal buds in the branchial tube of the brook lamprey (*Lampetra planeri* (Bloch))–putative respiratory monitors. *Acta zool. Stockh.*, **64**, 139–47.
Baatrup, E. (1985) Physiological studies on the pharyngeal terminal buds in the larval brook lamprey, *Lampetra planeri* (Bloch). *Chem. Senses*, **10**, 549–58.
Baatrup, E. and Døving, K.B. (1985) Physiological studies on solitary receptors of the oral disc papillae in the adult brook lamprey, *Lampetra planeri* (Bloch). *Chem. Senses*, **10**, 559–66.
Bardach, J., Fujiya, M. and Holl, A. (1967) Investigations of external chemoreceptors of fishes, in *Olfaction and Taste* II (ed. T. Hayashi), Pergamon Press, Oxford, pp. 647–65.
Bardach, J.E. and Case, J. (1965) Sensory capabilities of the modified fins of squirrel hake (*Urophycis chuss*) and searobins (*Prionotus carolinus* and *P. evolans*). *Copeia* **1965**, 194–206.
Bartheld, C.S. von and Meyer, D.L. (1985) Trigeminal and facial innervation of cirri in three teleost species. *Cell Tissue Res.*, **241**, 615–22.
Beidler, L.M. (1965) Comparison of gustatory receptors, olfactory receptors, and free nerve endings. *Cold Spring Harb. Symp. quant. Biol.*, **30**, 191–200.
Belousova, T.A., Devitsina, G.V. and Malyukina, G.A. (1983) Functional peculiarities of fish trigeminal system. *Chem. Senses*, **8**, 121–30.
Bishai, H.M. (1962) Reactions of larval and young salmonids to different hydrogen ion concentrations. *J. Cons. perm. int. Explor. Mer*, **27**, 181–91.
Blanchi, D. and Guardabassi, A. (1977) Changes in neuromast chemosensitivity in *Xenopus laevis* kept under various environmental conditions. *J. Endocrinol.*, **74**, 157–8.
Blanchi, D., Camino, E. and Guardabassi, A. (1976) Chemoreception of the lateral-line organs in intact, hypophysectomized, and prolactin-treated hypophysectomized *Xenopus laevis* specimens. *Comp. Biochem. Physiol.*, **55A**, 301–7.
Blaxter, J.H.S. and Fuiman, L.A. (1990) The role of the sensory systems of herring larvae in avoiding predatory fishes. *J. mar. biol. Ass. U.K.*, **70**, 413–27.
Caprio, J. (1975) High sensitivity of catfish taste receptors to amino acids. *Comp. Biochem. Physiol.*, **52A**, 247–51.
Caprio, J. (1988) Peripheral filters and chemoreceptor cells in fishes, in *Sensory Biology of Aquatic Animals* (eds J. Atema, R.R. Fay, A.N. Popper and W.N. Tavolga), Springer-Verlag, New York, pp. 313–38.

Cole, L.W. (1910) Reactions of frogs to chlorides of ammonium, potassium, sodium and lithium. *J. comp. Neurol.*, **20**, 601–14.

Connes, R., Granie-Prie, M., Diaz, J.P. and Paris, J. (1988) Ultrastructure des bourgeons de goût du téléostéen marin *Dicentrarchus labrax* L. *Can. J. Zool.*, **66**, 2135–42.

Crozier, W.J. (1916) Regarding the existence of the 'common chemical sense' in vertebrates. *J. comp. Neurol.*, **26**, 1–8.

Fahrenholz, C. (1936) Die sensiblen Einrichtungen der Neunaugenhaut. *Z. mikrosk. -anat. Forsch.*, **40**, 323–80.

Finger, T.E. (1982) Somatotopy in the representation of the pectoral fin and free fin rays in the spinal cord of the sea robin, *Prionotus carolinus*. *Biol. Bull. mar. biol. Lab., Woods Hole.*, **163**, 154–61.

Finger, T.E. (1983) The gustatory system in teleost fish, in *Fish Neurology and Behaviour*. Vol. 1 (eds R.G. Northcutt and R.E. Davis), Univ. Michigan Press, Ann. Arbor, pp. 285–309.

Finger, T.E. (1988) Organization of chemosensory systems within the brains of bony fishes, in *Sensory Biology of Aquatic Animals* (eds J. Atema, R.R. Fay, A.N. Popper and W.N. Tavolga), Springer-Verlag, New York, pp. 339–63.

Fishelsen, L. (1984) A comparative study of ridge mazes on surface epithelial cell-membranes of fish scales (Pisces, Teleostei). *Zoomorphology*, **104**, 231–8.

Fox, H., Lane, E.B. and Whitear, M. (1980) Sensory nerve endings and receptors in fish and amphibians, in *The Skin of Vertebrates* (Linn. Soc. Symp. Ser. No. 9) (eds R.I.C. Spearman and P.A. Riley), Acad. Press, London, pp. 271–81.

Fujita, T., Kanno, T. and Kobayashi, S. (1988) *The Paraneuron*, Springer-Verlag, Tokyo, 367 pp.

Harris J.E. and Hunt, S. (1975) The fine structure of the epidermis of two species of salmonid fish, the Atlantic salmon (*Salmo salar* L.) and the brown trout (*Salmo trutta* L.). I. General organization and filament-containing cells. *Cell Tissue Res.*, **157**, 553–65.

Hughes, G.M. and Munshi, J.S. Datta (1973) Fine structure of the respiratory organs of the climbing perch, *Anabas testudineus* (Pisces: Anabantidae). *J. Zool. Lond.*, **170**, 201–25.

Iger, Y., Abraham, M., Dotan, A., Fattal, B. and Rahamin, E. (1988) Cellular responses in the skin of carp maintained in organically fertilized water. *J. Fish Biol.*, **33**, 711–20.

Jakubowski, M. (1983) New details of the ultrastructure (TEM, SEM) of taste buds in fishes. *Z. mikrosk. -anat. Forsch.*, **97**, 849–62.

Jakubowski, M. and Whitear, M. (1990) Comparative morphology and cytology of taste buds in teleosts. *Z. mikrosk. -anat. Forsch.*, **104**, 529–60.

Katsuki, Y. and Yanagisawa, K. (1982) Chemoreception in the lateral line organ, in *Chemoreception in Fishes* (ed. T.J. Hara), Dev. in Aquaculture and Fisheries Sci., **8**, Elsevier, Amsterdam, pp. 227–42.

Katsuki, Y., Hashimoto, T. and Yanagisawa, K. (1970) The lateral-line organ of shark as a chemoreceptor. *Adv. Biophys.*, **1**, 1–51.

Katsuki, Y., Hashimoto, T. and Kendall, J.I. (1971) The chemoreception in the lateral-line organs of teleosts. *Jap. J. Physiol.*, **21**, 99–118.

Katz, U. (1986) The role of amphibian epidermis in osmoregulation and its adaptive response to changing environment, in *Biology of the Integument*. Vol. 2. *Vertebrates* (eds J. Bereiter-Hahn, A.G. Matoltsy and K.S. Richards), Springer-Verlag, Berlin, pp. 473–98.

Kiyohara, S., Yamashita, S. and Kitoh, J. (1984) Rapid location of fish taste buds by a selective staining method. *Bull. Jap. Soc. scient. Fish.*, **50**, 1293–7.

Kiyohara, S., Houman, H., Yamashita, S., Caprio, J. and Marui, T. (1986) Morphological evidence for a direct projection of trigeminal nerve fibres to the primary gustatory center in the sea catfish *Plotosus anguillaris. Brain Res. Amsterdam*, **379**, 353–7.

Kölliker, A. (1886) Histologische Studien an Batrachierlarven. *Z. wiss. Zool.*, **43**, 1–40.

Kotrschal, K., (1991) Solitary chemosensory cells: taste, common chemical sense or what? *Rev. Fish Biol. Fish.*, **1**, 3–22.

Kotrschal, K. (1992) Quantitative electron microscopy of solitary chemoreceptor cells in cyprinids and other teleosts. *Env. Biol. Fishes.*, [in press].

Kotrschal, K. and Whitear, M. (1988) Chemosensory anterior dorsal fin in rocklings (*Gaidropsarus* and *Ciliata*, Teleostei, Gadidae): somatotopic representation of the ramus recurrens facialis as revealed by transganglionic transport of HRP. *J. comp. Neurol.*, **268**, 109–20.

Kotrschal, K., Whitear, M. and Adam, H. (1984) Morphology and histology of the anterior dorsal fin of *Gaidropsarus mediterraneus* (Pisces Teleostei), a specialized sensory organ. *Zoomorphology*, **104**, 365–72.

Kotrschal, K., Goldschmid, A., Adam, H. and Whitear, M. (1985) The first dorsal fin of *Gaidropsarus mediterraneus* (Teleostei), a specialized chemosensory organ. *Fortschr. Zool.*, **30**, 727–30.

Kotrschal, K., Peters, R.C. and Atema, J. (1989) A novel chemosensory system in fish: do rocklings (*Ciliata mustela*, Gadidae) use their solitary chemosensory receptor cells as fish detectors? *Biol. Bull. mar. biol. Lab., Woods Hole*, **177**, 328–9.

Lane, E.B. (1977) Structural aspects of skin sensitivity in the catfish *Ictalurus*. PhD Thesis, University of London, 143 pp.

Lane, E.B. and Whitear, M. (1977) On the occurrence of Merkel cells in the epidermis of teleost fishes. *Cell Tissue Res.*, **182**, 235–46.

Lane, E.B. and Whitear, M. (1982) Sensory structures at the surface of fish skin. I. Putative chemoreceptors. *Zool. J. Linn. Soc.*, **75**, 141–51.

Luciano, L., Reale, E. and Ruska, H. (1968) Über eine 'chemorezeptive' Sinneszelle in der Trachea der Ratte. *Z. Zellforsch.*, **85**, 350–75.

Luciano, L., Castellucci, M. and Reale, E. (1981) The brush cells of the common bile duct of the rat. *Cell Tissue Res.*, **218**, 403–20.

Meyer, M. (1962) Kegel- und andere Sonderzellen der larvalen Epidermis von Froschlurchen. *Z. mikrosk. -anat. Forsch.*, **68**, 79–131.

Morrill, A.D. (1895) The pectoral appendages of *Prionotus* and their innervation. *J. Morph.*, **11**, 177–92.

Nada, O., Hiratsuka, T. and Komatsu, K. (1984) The occurrence of serotonin-containing cells in the esophageal epithelium of the bullfrog *Rana catesbiana*: a fluorescence histochemical and immunohistochemical study. *Histochemistry*, **81**, 115–18.

Olsén, K.H. (1989) Sibling recognition in juvenile Arctic charr, *Salvelinus alpinus* (L.). *J. Fish Biol.*, **34**, 571–81.

Ono, R.D. (1980) Fine structure and distribution of epidermal projections associated with taste buds on the oral papillae in some loricariid catfishes (Siluroidei: Loricariidae). *J. Morph.*, **164**, 139–59.

Onoda, N. and Katsuki, Y. (1972) Chemoreception of the lateral line organ of an aquatic amphibian, *Xenopus laevis. Jap. J. Physiol.*, **22**, 87–102.

Parker, G.H. (1912) The relations of smell, taste, and the common chemical sense in vertebrates. *J. Acad. nat. Sci. Philad.*, Ser. 2, **15**, 221–34.

Peters, R.C., Steenderen, G.W. van and Kotrschal, K. (1987) A chemoreceptive function for the anterior dorsal fin in rocklings (*Gaidropsarus* and *Ciliata*: Teleostei: Gadidae): electrophysiological evidence. *J. mar. biol. Ass. U.K.*, **67**, 819–23.

Peters, R.C., Kotrschal, K., Krautgartner, W.-D. and Atema, J. (1989) A novel chemosensory system in fish: electrophysiological evidence for mucus detection by solitary chemoreceptor cells in rocklings (*Ciliata mustela*, Gadidae). *Biol. Bull. mar. biol. Lab., Woods Hole.*, **177**, 329.

Peters, R.C., Kotrschal, K. and Krautgartner, W.-D. (1991) Solitary chemosensory cells of *Ciliata mustela* (Gadidae, Teleostei) are tuned to mucoid stimuli. *Chem. Senses*, **16**, 31–42.

Pevsner, R. (1976) Electron microscope study of the taste buds of elasmobranchs, *Trigon pastinaca* and *Raja clavata. Tsitologiya*, **18**, 561–6. [In Russian with English summary].

Puzdrowski, R.L. (1988) Afferent projections of the trigeminal nerve in the goldfish, *Carassius auratus. J. Morph.*, **198**, 131–47.

Reutter, K. (1974) Cholinergic innervation of scattered sensory cells in fish epidermis. *Cell Tissue Res.*, **149**, 143–6.

Reutter, K. and Vogel, W.O.P. (1989) Ultrastructure of taste buds in the lungfishes *Protopterus* and *Lepidosiren. Chem. Senses*, **14**, 192.

Reutter, K., Breipohl, W. and Bijvank, G.J. (1974) Taste bud types in fishes. II. Scanning electronmicroscopical investigations on *Xiphophorus helleri* Heckel (Poeciliidae, Cyprinodontiformes, Teleostei). *Cell Tissue Res.*, **153**, 151–65.

Rogers, D.C. and Haller, C.J. (1980) The ultrastructural characteristics of the apical cell in the neurepithelial bodies of the toad lung (*Bufo marinus*). *Cell Tissue Res.*, **209**, 485–98.

Roper, S.D. (1989) The cell biology of vertebrate taste receptors. *A. Rev. Neurosci.*, **12**, 329–53.

Roth, A. and Tscharntke, H. (1976) Ultrastructure of the ampullary receptors in lungfish and Brachiopterygii. *Cell Tissue Res.*, **173**, 95–108.

Russell, I.J. (1976) Amphibian lateral line receptors, in *Frog Neurobiology* (eds R. Llinas and W. Precht), Springer-Verlag, Berlin, pp. 513–50.

Russell, I.J. and Sellick, P.M. (1976) Measurement of potassium and chloride ion concentrations in the cupulae of the lateral lines of *Xenopus laevis. J. Physiol., Lond.*, **257**, 245–55.

Schreiner, K.E. (1918) Zur Kenntnis der Zellgranula. Untersuchungen über die feineren Bau der Haut von *Myxine glutinosa*. I. Teil. Zweite Hälfte. *Arch. Mikrosk. Anat. Entwkges.*, **92**, 1–63.

Schulte, E. and Holl, A. (1972) Feinbau der Kopftentakel und ihrer Sinnesorgane bei *Blennius tentacularis* (Pisces, Blenniiformes). *Mar. Biol.*, **12**, 67–80.

Sheldon, R.E. (1909) The reactions of the dogfish to chemical stimuli. *J. comp. Neurol. Psychol.*, **19**, 273–311.

Silver, W.L. (1987) The common chemical sense, in *Neurobiology of Taste and Smell* (eds T.L. Finger and W.L. Silver), Wiley, New York, pp. 65–87.

Silver, W.L. and Finger, T.E. (1984) Electrophysiological examination of a non-olfactory, non-gustatory chemosense in the searobin, *Prionotus carolinus. J. comp. Physiol.*, **154A**, 167–74.

Silver, W.L., Arzt, A.H. and Mason, J.R. (1988) A comparison of the discriminatory ability and sensitivity of the trigeminal and olfactory systems to chemical stimuli in the tiger salamander. *J. comp. Physiol.*, **164A**, 55–66.

Toyoshima, K., Honda, E., Nakahara, S. and Shimamura, S. (1984) Ultrastructural and histochemical changes in the frog taste organ following denervation. *Arch. Histol. Jpn. (Niigata, Jpn)*, **47**, 31–42.

Tucker, D. (1983) Fish chemoreception: peripheral anatomy and physiology, in *Fish Neurobiology*. Vol. 1 (eds R.G. Northcutt and R.E. Davis), Univ. Michigan Press, Ann Arbor, pp. 311–49.

Wendelaar Bonga, S.E. and Meis, S. (1981) Effects of external osmolality, calcium and prolactin on growth and differentiation of the epidermal cells of the cichlid teleost *Sarotherodon mossambicus*. *Cell Tissue Res.*, **221**, 109–23.

Whitear, M. (1952) The innervation of the skin of teleost fishes. *Q. J. Microsc. Sci.*, **93**, 289–305.

Whitear, M. (1965) Presumed sensory cells in fish epidermis. *Nature. Lond.*, **208**, 703–4.

Whitear, M. (1971a) The free nerve endings in fish epidermis. *J. Zool., Lond.*, **163**, 231–6.

Whitear, M. (1971b) Cell specialization and sensory function in fish epidermis. *J. Zool., Lond.*, **163**, 237–64.

Whitear, M. (1974) The nerves in frog skin. *J. Zool., Lond.*, **172**, 503–29.

Whitear, M. (1975) Flask cells and epidermal dynamics in frog skin. *J. Zool., Lond.*, **175**, 107–49.

Whitear, M. (1976) Identification of the epidermal 'Stiftchenzellen' of frog tadpoles by electron microscopy. *Cell Tissue Res.*, **175**, 391–402.

Whitear, M. (1983) The question of free nerve endings in the epidermis of lower vertebrates. *Acta Biol. Hung.*, **34**, 303–19.

Whitear, M. (1986) The skin of fishes including cyclostomes; epidermis, dermis, in *Biology of the Integument*. Vol. 2. *Vertebrates* (eds J. Bereiter-Hahn, A.G. Matoltsy and K.S. Richards), Springer-Verlag, Berlin, pp. 8–64.

Whitear, M. (1989) Merkel cells in lower vertebrates. *Arch. Histol. Cytol.*, **52** (Supp.), 415–22.

Whitear, M. and Kotrschal, K. (1988) The chemosensory anterior dorsal fin in rocklings (*Gaidropsarus* and *Ciliata*, Teleostei, Gadidae): activity, fine structure and innervation. *J. Zool., Lond.*, **216**, 339–66.

Whitear, M. and Lane, E.B. (1983a) Oligovillous cells of the epidermis: sensory elements of lamprey skin. *J. Zool., Lond.*, **199**, 359–84.

Whitear, M. and Lane, E.B. (1983b) Multivillous cells: epidermal sensory cells of unknown function in lamprey skin. *J. Zool., Lond.*, **201**, 259–72.

Witt, M. and Reutter, K. (1988) Lectin histochemistry on mucous substances of the taste buds and adjacent epithelia of different vertebrates. *Histochemistry*, **88**, 453–61.

Witt, M. and Reutter, K. (1990) Electron microscopic demonstration of lectin binding sites in the taste buds of the European catfish *Silurus glanis* (Teleostei). *Histochemistry*, **94**, 617–28.

Yamamoto, M. (1982) Comparative morphology of the peripheral olfactory organ in teleosts, in *Chemoreception in Fishes* (ed. T.J. Hara), Elsevier, Amsterdam, pp. 39–59.

Yamashita, S., Evans, R.E. and Hara, T.J. (1987) Responses of the palatine nerve of the rainbow trout (*Salmo gairdneri*) to carbon dioxide and to hydrochloric acid. *Chem. Senses*, **12**, 513.

Chapter seven

Molecular mechanisms of chemosensory transduction: gustation and olfaction

J.G. Brand and R.C. Bruch

7.1 INTRODUCTION

Understanding of vertebrate chemoreception has been greatly aided by several aquatic models. The fishes in particular are well suited as biological models because they often possess specific, sensitive and readily accessible chemosensory organs. The ability to define receptor specificity and sensitivity – using classical receptor binding techniques – has permitted a critical evaluation of the several transduction sequences that follow the initial receptor binding step. It is now clear that the peripheral specificity hypothesized with neurophysiological techniques is present at the receptor membrane level and that these receptor events are coupled to changes in intracellular ionic activity via either GTP-binding regulatory protein stimulation (inhibition) of second messengers or direct translocation of ions via stimulus-gated ion channel receptors. The recent progress in characterizing these sequences in fish chemosensory systems has stimulated the search for similar mechanisms in mammalian chemoreception.

While the fishes have long been a valuable model for chemoreception, aquatic invertebrates also possess strikingly unique and specific chemoreceptive transduction sequences. While these will not be reviewed here, it is appropriate to point to some of these well-defined systems. In crayfish, Hatt and colleagues (Hatt, 1989) have for several years been documenting a nicotinamide-specific taste response characterized by very rapid onset and desensitization of a receptor–ion channel complex. Hatt (1989) has also recently reported a nucleotide-specific receptor channel using a patch-clamp recording technique of outside-out soma from taste cells of crayfish (1989). Trapido-Rosenthal and colleagues (1989) have characterized ectoenzymatic-receptor-uptake systems in the olfactory organs of the Florida spiny lobster. These enzyme-receptor-uptake systems function through dephosphorylation of AMP, ADP and ATP (adenosine 5'-monophosphate, -diphosphate, and -triphosphate respectively) with subsequent internalization of adenosine.

This chapter will review evidence for receptor-mediated transduction in gustation and olfaction in fishes. The primary emphasis will be on amino acid receptors and their associated transduction sequences in ictalurids. For other recent general reviews on the peripheral mechanisms of taste and olfaction, the reader is referred to recent articles (Bruch et al., 1988; Lancet, 1988; Roper, 1989) and books (Margolis and Getchell, 1988; Brand et al., 1989b; Cagan, 1989).

7.2 RECEPTOR EVENTS IN CHEMORECEPTION

Gustation

Biochemical studies of the initial event in gustation – binding of the stimulus to a presumed receptor protein – have only been carried out using the channel catfish, *Ictalurus punctatus*, and the brown bullhead catfish, *Ictalurus nebulosus*. For both of these species, amino acids act as potent taste stimuli. In *I. punctatus*, electrophysiological evidence is consistent with the hypothesis that there are at least two major receptor classes for the cutaneous taste system: one which responds to short-chain neutral amino acids such as alanine, glycine, serine and threonine, the other of which responds to basic amino acids, primary L-arginine (Caprio, 1982; Marui and Caprio, Chapter 9 herein). Recent studies characterizing the specificity and kinetics of these two sites and the differential ability of lectins to inhibit the binding of amino acids to these sites supports the hypothesis that there are, at least, two distinct taste receptor populations in *I. punctatus*.

Binding sites for neutral amino acids

The neutral amino acid, L-alanine, is a potent stimulus for the cutaneous taste system of the catfish, *I. punctatus*. Electrophysiological thresholds have been estimated at 1–10 nM (Caprio, 1982; Brand et al., 1987). In the mid 1970s, biochemical binding experiments began characterizing the presumed receptor site for this amino acid. A partial membrane preparation (Fraction P2) of taste-bud-rich epithelium from the taste organs (barbels) of *I. punctatus* was developed and used as the receptor population for equilibrium binding and kinetic studies (Krueger and Cagan, 1976). Results indicated an apparent dissociation constant, K_{Dapp}, for L-[^{3}H]alanine of 1.5–6.0 μM. Heterogeneity in the saturation binding profile was observed, and incubation of the tissue with high concentrations of L-alanine unmasked additional (or spare) receptor sites for L-alanine and for other short-chain neutral amino acids including L-serine, glycine and D-alanine (Cagan, 1979, 1986). On the basis of competitive inhibition studies and the specificity of these 'spare' receptor sites, Cagan hypothesized two subsets of the neutral amino acid site, one of which was specific for L-alanine, L-serine, glycine and D-alanine, the other of which was specific for L-threonine, L-serine, L-alanine and somewhat for D-alanine and β-alanine (Cagan, 1986).

Neurophysiological evidence indicated that L-alanine and D-alanine were both agonists to the taste system of the catfish, while cross-adaptation and mixture experiments suggested some independent sites for D-alanine (Brand *et al.*, 1987). This enantiomeric specificity was corroborated by binding experiments in which additional sites for D-alanine at high D-alanine concentrations were demonstrated (Brand *et al.*, 1987).

The binding of L-[^{3}H]leucine to a membrane fraction from the barbels of *I. nebulosus* has been reported by Sheiko *et al.* (1983). The reported K_{Dapp} values in the range of 1 nM are unusually low given the fact that L-leucine is a relatively poor stimulus for the catfish and, in receptor assays, is unable to compete well with L-[^{3}H]alanine for binding to neutral amino acid sites (Bryant, pers. comm., 1985).

Binding sites for basic amino acids

The other hypothesized major receptor type of the catfish cutaneous taste system is activated primarily by L-arginine. Rate constants for association and dissociation of L-[^{3}H]arginine have indicated heterogeneous binding with two K_{Dapp} values of 18 nM and 1.3 μM (Kalinoski *et al.*, 1989a).

Using competitive binding techniques, Kalinoski *et al.* (1989a) reaffirmed the relative independence of these sites from the short-chain neutral site(s) because both L-alanine and glycine competed poorly for bound L-[^{3}H]arginine (Fig. 7.1). Two analogues of L-arginine, L-arginine methyl ester and L-α-amino-β-guanidino propionic acid (L-AGPA), both potent stimuli to the taste system, competed well with L-[^{3}H]arginine. In contrast, the poor stimulus D-arginine competed as well as L-AGPA for the L-arginine site. D-Arginine was, however, a good cross-adaptor of L-arginine stimulation, suggesting that this enantiomer may be an inhibitor to activation of the basic amino acid receptor by L-arginine. Results of reconstitution studies of the L-arginine receptor are consistent with this hypothesis (see Stimulus-activated ion channels, p. 142).

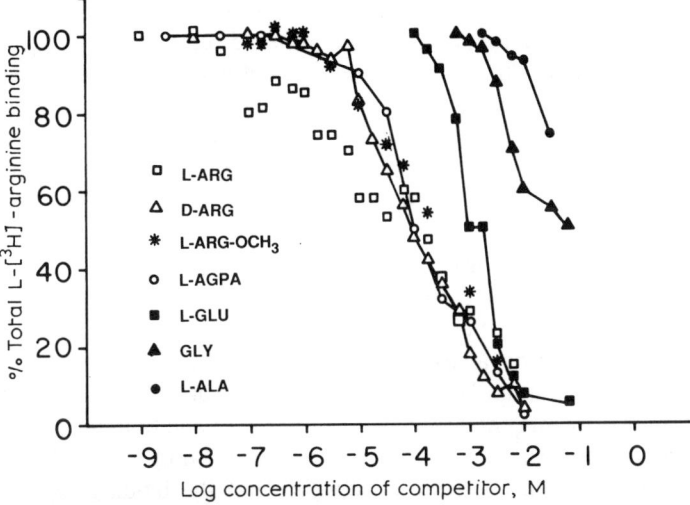

Fig. 7.1 Percent total binding of L-[^{3}H]arginine (1 μM) versus log concentration of competitor. Binding of L-[^{3}H]arginine was to a sedimentable partial membrane preparation from taste epithelium of *Ictalurus punctatus*. Abbreviations in key: L-ARG, L-arginine; D-ARG, D-arginine; L-ARG-OCH$_3$, L-arginine methyl ester; L-AGPA, L-α-amino-β-guanidino propionic acid; L-GLU, L-glutamate, GLY, glycine; L-ALA, L-alanine. (Reproduced with permission from Kalinoski *et al.*, 1989a.)

Olfaction

Owing to their proximity to the external environment and the evidence implicating them in olfactory reception, the apical dendritic cilia of peripheral olfactory neurones are generally regarded as the site of the initial interaction of stimuli with the chemosensory membrane (Rhein and Cagan, 1981; Getchell *et al.*, 1985; Getchell, 1986; Lancet, 1988). Although most olfactory neurones in vertebrates are ciliated, microvillar receptor cells are also prominent in fish (Rhein *et al.*, 1981; Yamomoto, 1982; Cancalon, 1983; Muller and Marc, 1984; Zielinski and Hara, 1988; Zeiske *et al.*, Chapter 2 herein). In addition to their morphological differences, the two receptor cell types also respond differentially in electrophysiological assays to amino acid and bile salt stimuli in some (Thommesen, 1983), but not all, species (Erickson and Caprio, 1984). It is generally accepted that detection and recognition of olfactory stimuli are mediated by macromolecular receptors localized in the ciliary and microvillar membranes of the receptor cells (Rhein and Cagan, 1981; Getchell, 1986; Lancet, 1988; Bruch, 1989). Although the molecular nature, neuronal distribution, and cellular expression of the receptors remain fundamentally unresolved problems, the presumed localization of the receptors in the apical dendritic membrane supports the use of isolated cilia as appropriate membrane preparations for characterizing the nature of stimulus interaction with the putative receptors.

The interaction of stimulus amino acids with membranes derived from olfactory epithelium was initially investigated by Cagan and co-workers (Rhein and Cagan, 1981). In their early experiments, these investigators showed that binding sites for amino acids were present in a particulate fraction derived from homogenates of rainbow trout, *Oncorhynchus mykiss* (formerly *Salmo gairdneri*), olfactory epithelium (Cagan and Zeiger, 1978). These investigators also first described the isolation of olfactory cilia and showed that the amino acid binding sites detected in crude membranes were retained in the isolated cilia (Rhein and Cagan, 1980, 1983). The interaction of stimulus amino acids with the binding sites in isolated cilia was consistent with the minimal criteria of specificity, saturability, and reversibility expected for membrane-associated receptors. In addition, good agreement was obtained between the rank order of binding and electrophysiological potency, further supporting the conclusion that the binding data reflected the interaction of stimulus amino acids with physiologically relevant receptors.

Competition studies indicated that the selectivity of stimulus binding was consistent with the existence of a limited number of receptors that recognized structurally similar ligands. Neutral L-amino acids, such as alanine, serine, and threonine, but not basic amino acids, competed for a common binding site, suggesting the existence of a discrete site (site TSA) for these stimuli. Similarly, basic amino acids, such as L-arginine and L-lysine, that interacted with a common binding site (site L), did not interact with site TSA. Although similar results were also obtained in another study using membranes derived from trout olfactory epithelium, it was concluded that the binding data represented amino acid transport rather than ligand–receptor interaction (Brown and Hara, 1981). However, the good correlation between the rank order of binding of a stimulus and its neurophysiological potency, and the reversibility of amino acid binding to olfactory cilia suggest that transport contributes minimally, if at all, to the observed binding data.

The channel catfish has been shown to respond to amino acids as both taste and olfactory stimuli. The relative potencies and the molecular groupings of these amino acids are, however, distinct for both sensory systems. Therefore, it cannot be assumed that the receptor that binds L-alanine in the taste system is the same molecular entity as the one that binds L-alanine in the olfactory system. Bryant *et al.* (1987) reported that a monoclonal antibody, raised against the taste receptor plasma membrane, and which recognized a glycoprotein after polyacrylamide gel electrophoresis in sodium dodecyl sulphate (SDS-PAGE) of taste epithelial membranes did not recognize any protein bands on an SDS-PAGE from the olfactory cilia. In addition, the specificities of lectin inhibition of amino acid binding to taste and olfactory tissues are distinct (Kalinoski et al., 1987).

Using a suspension of olfactory receptor cells from *I. punctatus*, Cancalon (1978) reported binding of stimulus amino acids to this fraction at levels of hundreds of pmol mg^{-1} protein. Most importantly, binding was readily and almost completely reversible, supporting the hypothesis that stimulus amino acids are not taken up by the receptor cells for metabolism or incorporation into protein. The relative binding of 12 amino acids to receptor cells and their relative biological potency, as determined by electrophysiological effectiveness, were well correlated ($r = 0.88$, $P < 0.01$). Correlations were poor when electrophysiological effectiveness was compared with binding of these amino acids to respiratory cells.

Ligand binding studies in isolated cilia from *I. punctatus* showed that stimulus amino acids interacted with binding sites with affinity and selectivity comparable to the sites described in trout (Bruch and Teeter, 1989). Competitive binding experiments indicated that the selectivity of amino acid binding paralleled the specificity of *in vivo* neural responses (Kalinoski et al., 1987; Bruch and Rulli, 1988). These results suggested that the specificity of receptor cell responses to acidic, basic, long-chain neutral, and short-chain neutral amino acids (Caprio and Byrd, 1984) was accounted for by the binding of stimuli to discrete macromolecular receptors of similar specificity. The binding of amino acids was also stereoselective, because D-enantiomers exhibited lower affinity for the receptors than the corresponding L-isomers, paralleling their lower potency in neural assays (Caprio, 1978; Bruch and Rulli, 1988). Competitive binding studies also indicated that derivatives of L-alanine with substituted alpha-amino and/or carboxyl groups interacted with the receptor less effectively than L-alanine itself (Bruch and Rulli, 1988). The binding data paralleled the specificity, stereoselectivity, and structure–activity relationship expected for the receptors from neurophysiological recordings.

Consistent with the expected glycoprotein nature of the receptors, lectins differentially inhibited the binding of L-alanine and L-arginine to their receptors by a mechanism that decreased the number of binding sites with no change in affinity. The binding inhibition in the presence of lectins was completely reversed by the appropriate monosaccharides, indicating that the inhibition was a consequence of specific carbohydrate recognition (Kalinoski et al., 1987). Major characteristics of two of the more potent stimuli are presented in Table 7.1.

Neutral amino acid binding sites have also been reported in isolated membrane preparations from the olfactory epithelium of salmon (Rehnberg and Schreck, 1986), carp (Parfenova and Etingof, 1988), and skate (Novoselov et al., 1980). In coho salmon, *Oncorhynchus kisutch*, competitive binding experiments indicated that neutral amino acids interacted with a binding site in isolated olfactory membranes with specificity similar to site

Table 7.1 Properties of the L-alanine and L-arginine olfactory receptors

Property	L-Ala	L-Arg
K_{Dapp} (μM)*	2	5
B_{max} (pmol mg^{-1} protein)	38	78
Lectin inhibition (%)†		
Con A	60	72
WGA	54	73
PNA	64	19
IC_{50} (μM)‡		
−GTP	1.8	2.8
+GTP	4.0	10

*Apparent dissociation constants (K_{Dapp}) and binding capacities (B_{max}) were obtained from homologous competitive binding curves (Kalinoski et al., 1987).
†The percent lectin inhibition values are the amount of ligand-binding inhibition following preincubation of isolated cilia with each lectin, expressed relative to parallel control samples preincubated in the absence of lectins (Kalinoski et al., 1987). Abbreviations: Con A, concanavalin A; WGA, wheat germ agglutinin; PNA, peanut agglutinin.
‡IC_{50} is the concentration of unlabelled ligand required to inhibit 50% of radioligand binding in homologous competition experiments in the absence (−GTP) and presence (+GTP) of guanine nucleotide (Bruch and Kalinoski, 1987). (Table reproduced from Bruch and Teeter, 1989.)

TSA in trout (Rehnberg and Schreck, 1986). However, in behavioural cross-adaptation assays, threonine was recognized as a stimulus qualitatively distinct from serine and alanine, which were perceived as identical stimuli. The binding and behavioural results in combination therefore suggested that a functional olfactory receptor for serine and alanine (site SA), existed in the peripheral olfactory system of salmon. In carp, *Cyprinus carpio*, binding sites for L-alanine and L-leucine were also detected in isolated olfactory membranes (Parfenova and Etingof, 1988). These binding sites were also sensitive to guanine nucleotides, suggesting functional coupling of these sites to G-proteins (p. 133, GTP-binding regulatory proteins). In skate, *Dasyatis pastinaca*, binding sites with unusually high affinity for L-alanine were also reported (Novoselov et al., 1980). A 98 kDa glycoprotein (gp98) was subsequently isolated from detergent-solubilized olfactory membranes from skate by elution from amino acid affinity columns (Novoselov et al., 1988a). It was therefore proposed that gp98 represented a putative olfactory receptor protein, although confirmatory evidence that the isolated protein exhibits the appropriate biological activity expected of a receptor has not been reported.

Experimental procedures to characterize olfactory processes in fishes have taken advantage of the sensitivity and selectivity for amino acids in certain fishes. Binding sites have been delineated, and binding data have been correlated to known functional parameters such as neurophysiological and/or behavioural responses. A variety of tissue preparative procedures and receptor binding techniques have been used. Taken together, the data favour the hypothesis that specific classes of olfactory receptor binding sites exist for these stimuli in salmonids, skate and catfish.

7.3 RECEPTOR-MEDIATED CHEMOSENSORY TRANSDUCTION

A variety of receptor-mediated transduction events are known. Among these are transduction processes that produce second messengers, others that act via internalization of the stimulus, others that result from stimulus-gated ion channels, and others that enhance coupled enzyme processes. Chemoreception in fishes uses at least two of these general processes: one that alters the concentration of second messengers, mediated most probably by intervening GTP-binding regulatory proteins (G-proteins), and another that affects intracellular ionic activity via stimulus-gated ion channels.

Transduction processes generating second messengers

GTP-binding regulatory proteins

Signal-transducing GTP-binding regulatory proteins (G-proteins) have been identified in chemosensory membranes from catfish by bacterial toxin-catalysed ADP-ribosylation and by immunoblotting with subunit-specific antisera (Bruch and Kalinoski, 1987). In membranes derived from taste and olfactory epithelium, a 45 kDa cholera toxin substrate was identified, corresponding to the alpha-subunit of G_s, the G-protein generally associated with stimulation of adenylyl cyclase (Gilman, 1987). A 41 kDa pertussis toxin substrate was also identified in taste membranes, corresponding to the alpha-subunit of G_i, the G-protein mediating inhibition of adenylyl cyclase (Gilman, 1987). In olfactory cilia, a 40 kDa pertussis toxin substrate was identified that cross-reacted with antisera to a common amino acid sequence of G-protein alpha-subunits. In contrast to other vertebrates (Bruch, 1990), immunoblotting studies of fish further showed the 40 kDa pertussis toxin substrate to be distinct from G_i (41 kDa) and G_o (39 kDa). Thus, the 40 kDa pertussis toxin substrate may be homologous to the variant of G_i, G_i2 (Gilman, 1987), recently identified by molecular cloning in rat olfactory epithelium (Jones and Reed, 1987).

The presence of G-proteins in chemosensory membranes suggested the possibility that these proteins interacted with gustatory and olfactory amino acid receptors. This hypothesis was tested by radioligand binding experiments in the absence and presence of guanine nucleotides (Bruch and

Kalinoski, 1987). In olfactory cilia, the affinities of the L-alanine and L-arginine receptors for their ligands were decreased in the presence of GTP (guanosine-5′-triphosphate) or a hydrolysis-resistant analogue (Table 7.1). These results satisfied a well-established pharmacological criterion expected for G-protein-linked receptors (Gilman, 1987). Similar results were also reported for L-alanine and L-leucine binding sites in olfactory membranes from carp (Parfenova and Etingof, 1988). In contrast, guanine nucleotides did not affect the binding of L-alanine or L-arginine in taste membranes from catfish (Bruch and Kalinoski, 1987). However, coupling of the gustatory L-alanine receptor to G-proteins was subsequently demonstrated by stimulus enhancement of guanine nucleotide-dependent second messenger formation.

The ability of the alpha-subunit to bind and hydrolyse GTP is a common property shared by all members of the family of heterotrimeric G-proteins (Gilman, 1987). High-affinity GTPase activity was detected in membranes from the olfactory epithelium of carp (Parfenova and Etingof, 1988). The membrane-associated GTPase activity was partially inhibited by the hydrolysis-resistant GTP analogue Gpp(NH)p, an expected property of G-protein GTPase activity. It was therefore proposed that Gpp(NH)p-inhibitable GTPase activity could serve as a specific marker for G-proteins in olfactory membranes. A 56 kDa glycoprotein, gp56, was also isolated from detergent-solubilized olfactory membranes from skate (Novoselov et al., 1988a,b); it coeluted as a complex with the putative olfactory receptor protein gp98 and was dissociated from gp98 in the presence of Mg^{2+} and GTPγS. In addition, gp56 exhibited GTPase activity that was enhanced by L-alanine, but only in the presence of gp98 (Novoselov et al., 1988b). Although these results were consistent with established characteristics of signal-transducing G-proteins (Gilman, 1987), the molecular size and glycoprotein nature of gp56 are unexpected properties for G-protein alpha-subunits. Thus, like gp98, the molecular identity and functional role of gp56 in olfactory signal transduction remain to be established.

Metabolism of second messengers in taste

Several studies in the early 1970s attempted to obtain evidence for taste-stimulus-generated cyclic AMP (Kurihara, 1972; Price, 1973; Cagan, 1976). For the most part, these studies were equivocal, with the exception that xanthines and a few other bitter-tasting compounds, acting presumably as phosphodiesterase inhibitors, caused changes in cAMP concentration. In the mid 1980s, interest in the possibility that second messengers may mediate taste transduction was renewed. The wealth of data on the taste receptor specificity in *I. punctatus* made it a model of choice for these studies.

Metabolism of cyclic AMP by amino acid taste stimuli

Fractions of the taste receptor epithelium of *I. punctatus* contain modest activity of adenylyl cyclase. The average basal level of adenylyl cyclase activity, 1.7 pmol of cyclic AMP per minute per milligram of protein, was approximately 100-fold lower than that of the olfactory system of *I. punctatus* (Kalinoski et al., 1989b). Adenylyl cyclase of the taste epithelium of *I. punctatus* displayed several properties of the classical hormone-regulated enzyme (Gilman, 1987). Forskolin directly activated the enzyme, resulting in a 10-fold increase in catalytic activity. Sodium fluoride increased cyclase activity in a dose-dependent manner, with 20 mM NaF yielding maximal stimulation of cyclase activity. G-Protein effectors, including GTP and the nonhydrolysable GTP analogues Gpp(NH)p and GTPγS, stimulated cyclic AMP formation. The nonhydrolysable ATP analogue, App(NH)p, did not increase cyclase catalytic activity, indicating a guanine nucleotide specificity for activation of adenylyl cyclase. Calcium ion concentrations in the range of 1 μM to 1 mM inhibited the drug forskolin's ability to stimulate cyclase activity in a dose-related manner. Levels of calcium ion in the range of 10 nM to 1 μM enhanced adenylyl cyclase activity (Kalinoski et al., 1989b).

In the presence of the taste stimulus L-alanine (0.5–30 μM), adenylyl cyclase activity was rapidly (<1 min) stimulated nearly twofold. This range of concentration of L-alanine is within the electrophysiological and biochemically active range. Basal adenylyl cyclase activities of barbel (taste) epithelium and skin from the flank of the fish (a control tissue) were essentially identical. However, in the presence of 1 μM L-alanine, there was no stimulation of cyclase activity from the flank skin fraction, while a 2.5-fold stimulation was observed from the barbel fraction. When L-alanine stimulation of cyclase was measured in the presence of GTP (1 nM–100 μM), cyclic AMP accumulation was approximately additive. The potent taste stimulus L-arginine, which binds to a different class of receptor sites (Cagan, 1986; Kalinoski et al., 1989a), had little effect on adenylyl cyclase catalytic activity, even at concentrations of 10 mM (Kalinoski et al., 1989b). These results suggest that the different classes of taste receptors may act through separate transduction sequences.

Metabolism of polyphosphoinositides by amino acid taste stimuli

The lipid composition of the barbel taste epithelium of *I. punctatus* has been reported (Brand et al., 1989a). No unusual types or quantities of neutral or polar lipids were observed. Radiolabelled acetate was readily incorporated into all of the major lipid species. Phosphatidylinositols accounted for about 5% of the total lipid (Brand et al., 1989a). Turnover studies with ^{32}P suggested rapid metabolism of phosphatidylcholines, lysophosphatidylcholines and phosphatidylinositols (Rabinowitz et al., 1990). Phosphatidylcholine was rapidly metabolized by a phospholipase A1 and a phospholipase D, enzymes which may be involved in alternate pathways of second messenger production (Exton, 1988).

The potential second messengers, inositol trisphosphate (IP3) and diacylglycerol (DAG), are formed from phosphatidylinositol bisphosphate (PIP2) via a specific phospholipase C (Berridge, 1984). It is generally held that IP3 acts to release calcium ion from intracellular stores and that DAG functions within the plane of the membrane to modulate protein kinase C activity (Abdel-Latif, 1986). The first indications that these polyphosphoinositide metabolites may be involved in transduction in taste and olfaction were obtained in experiments on agonist and G-protein effector studies in *I. punctatus* (Huque and Bruch, 1986; Huque et al., 1987; Kalinoski et al., 1989b). In a standard assay at pH 7.1, phospholipase C of the taste system of *I. punctatus* exhibited similar activity against PIP2 (18.4 nmol min^{-1} per mg protein) and phosphatidylinositol-4-phosphate (22.7 nmol min^{-1} per mg protein), but had weak activity against phosphatidylinositol (2.4 nmol min^{-1} per mg protein). In the presence of Ca^{2+}/EGTA buffers alone, and with free Ca^{2+} concentrations in the range of 14 nM and 12 μM, the enzyme readily degraded PIP2 in a calcium-dependent manner (Kalinoski et al., 1989b).

When homogenates of taste tissue were incubated in the presence of various concentrations of the potent taste stimulus L-alanine, phospholipase C activity was enhanced nearly twofold at amino acid concentrations in the range of 1–100 μM. While basal phospholipase C activities of barbel (taste) epithelium and ventral skin (a control tissue) were virtually identical, the ventral skin enzyme exhibited no enhancement of activity to 10 μM L-alanine, whereas the barbel enzyme was stimulated 1.6-fold. Since the density of taste buds and L-alanine receptors is higher in the barbel epithelium, these data support the hypothesis that phospholipase C activity is regulated by taste-stimulus-specific receptors (Huque et al., 1987; Kalinoski et al., 1989b).

The role of G-proteins in the regulation of enzyme activity was evaluated by examining the ability of NaF to stimulate phospholipase C. When NaF, at levels in the range of 1–50 mM, was incubated with taste tissue homogenates, maximal activity was observed at 10 mM. With L-alanine at 5 μM and NaF at 10 mM, an additive effect of the two agonists on phospholipase C activity was noted (Huque et al., 1987; Kalinoski et al., 1989b).

In the presence of 10 μM L-alanine and 10 mM NaF, IP3 production was maximally stimulated at the earliest time point sampled (10 s), and by 30 s had returned to basal levels. The rapid production and equally rapid degradation of IP3 are consistent with a second-messenger role for this molecule (Huque et al., 1987; Kalinoski et al., 1989b).

It has been demonstrated that an increase in IP3 levels of only 10–20% above basal leads to maximal release of Ca^{2+} in many cells (Lynch et al.,

1985; Williamson, 1986). Assuming that IP3 production in taste tissue leads to the release of calcium from the endoplasmic reticulum, as it does in many cell types (Abdel-Latif, 1986; Berridge and Irvine, 1984), it seems likely that this level of IP3 production in taste epithelium is sufficient to support a role for inositol phospholipids as second messengers in the taste system.

Metabolism of second messengers in olfaction

Metabolism of cyclic AMP by amino acid olfactory stimuli

A role for adenylyl cyclase in mediating olfactory signal transduction in aerosmic vertebrates has been indicated by a variety of biochemical and neurophysiological evidence (Lancet, 1988; Snyder et al., 1988; Bruch, 1989; Bruch and Gold, 1990). Thus, it is now generally accepted that stimulus binding to olfactory receptors activates a G_s-linked adenylyl cyclase in the cilia. However, olfactory neurones in rat also express a unique G-protein, G_{olf}, that shares extensive sequence homology to G_s and also mediates stimulation of adenylyl cyclase (Jones and Reed, 1989). G_{olf} may therefore represent an olfactory-specific variant of G_s, analogous to transducin, the G-protein mediating visual transduction. The possibility that adenylyl cyclase also participates in olfaction in aquatic organisms was investigated in isolated cilia from the channel catfish (Bruch and Teeter, 1989, 1990). Ten stimuli, representative of each receptor subtype and covering a wide range of electrophysiological potency, were tested for the ability to affect guanine nucleotide-dependent cAMP formation. Dose–response studies indicated that significant elevation of cAMP levels was obtained only when receptor occupancy approached or exceeded 50% (Fig. 7.2). When tested at the same (100 μM) concentration, all stimuli elicited similar 2–3-fold increases in cAMP levels. These responses were poorly correlated ($r = -0.42$, $P > 0.05$) with phasic neural responses evoked by the same stimulus concentrations. In addition, the stereoselectivity exhibited by the enantiomers of alanine in ligand binding (Bruch and Rulli, 1988) and neurophysiological (Caprio, 1978) assays was not reflected in their ability to stimulate adenylyl cyclase. Taken together, these results suggested that cAMP mediated olfactory responses only under appropriate conditions of receptor occupancy. The high levels of receptor occupancy required to stimulate adenylyl cyclase suggested further that cAMP may be involved in mediating tonic responses resulting from high stimulus concentration or prolonged exposure to stimulus.

Metabolism of polyphosphoinositides by amino acid olfactory stimuli

The possibility that inositol phospholipid hydrolysis was also involved in olfaction was investigated in isolated cilia from the channel catfish. Subcellular fractionation of the olfactory epithelium indicated that the majority of

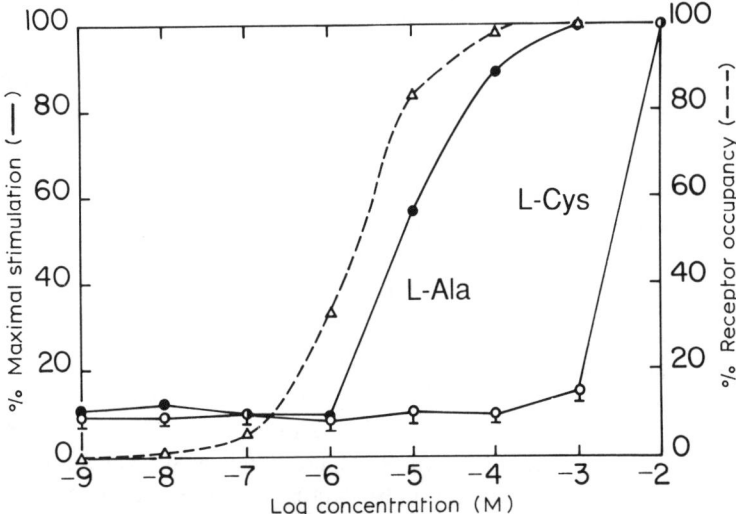

Fig. 7.2 Effect of L-amino acid concentration on adenylyl cyclase activity. Adenylyl cyclase activity was measured after 10 min in the presence of 10 μM Gpp(NH)p and the indicated concentrations of L-alanine (closed circles) and L-cysteine (open circles). The data are expressed as the percentage of maximal stimulation over basal levels (312 ± 36 pmol/min/mg protein) for each amino acid, and are the mean $\pm$ S.D. of three experiments for each stimulus assayed in duplicate. Errors in the curve for L-alanine were no larger than the data points on the figure. The maximal-fold stimulation over basal at 10^{-2} M amino acid was 7.25 ± 0.13 for L-alanine and 11.8 ± 1.0 for L-cysteine. The percentage receptor occupancy (broken line) for L-[^{3}H]alanine binding measured under the conditions of the adenylyl cyclase assay was calculated from triplicate competition curves (Bruch and Rulli, 1988). Maximal L-[^{3}H]alanine binding was 50 ± 6 pmol/mg protein with apparent K_D of 2.4 ± 0.5 μM. (Reproduced with permission from Bruch and Teeter, 1990.)

the phospholipase C activity was retained in the soluble fraction and about 5% of the total activity was reproducibly associated with the isolated cilia (Boyle *et al.*, 1987). L-Alanine and L-arginine stimulated guanine nucleotide-dependent IP3 formation at concentrations an order of magnitude lower than those required to stimulate adenylyl cyclase (Bruch *et al.*, 1987). In the absence of stimulus, GTP and its hydrolysis-resistant analogues also stimulated IP3 formation, suggesting the likely involvement of a G-protein in mediating phosphoinositide hydrolysis in the cilia (Huque and Bruch, 1986; Bruch *et al.*, 1987). In contrast to the slow rate of stimulus-dependent cAMP accumulation, significant stimulation of IP3 formation was observed within several seconds of exposure to stimulus. The rapid rise in IP3 levels was transient and gradually declined within 2 min to basal levels. During this period, no elevation in cAMP levels could be detected. The rapid rate and transient nature of stimulus-dependent IP3 formation in the cilia was consistent with the characteristically rapid kinetics of agonist-stimulated phosphoinositide hydrolysis. The rapid rate of IP3 generation at much lower stimulus concentrations than were required to activate adenylyl cyclase suggested that IP3 may mediate olfactory responses under conditions of low receptor occupancy or following brief exposure to stimulus.

Cyclic GMP and olfactory transduction

Guanylyl cyclase activity was also identified in the olfactory epithelium of the channel catfish (Bruch and Teeter, 1989) and in olfactory cilia from rat and pig (Steinlen et al., 1990). The enzyme was distributed in catfish in both particulate and soluble fractions, with about two-thirds of the activity associated with the particulate fraction. Although the enzyme was also detected in isolated cilia, the activity retained in these preparations represented only about 2% of the total activity of the olfactory epithelium. In all fractions, Mn^{2+} was markedly more effective than Mg^{2+} as a cofactor for the enzyme. Mn^{2+}-dependent cGMP formation was also stimulated by exogenous Ca^{2+}, but only at high (mM) concentrations; this effect was also markedly potentiated by the calcium ionophore A23187. These results suggested a role for Ca^{2+} regulation of cGMP formation within olfactory cilia. Elevation of intracellular Ca^{2+}, possibly within localized subcellular domains (Alkon and Rasmussen, 1988), as a result of Ca^{2+} release from intracellular stores and/or influx of extracellular Ca^{2+}, may act to increase cGMP levels (Ui, 1986).

Fate of second messengers

IP3-mediated Ca^{2+} release

In olfactory cilia from the channel catfish, fractionation of inositol phosphate products by anion-exchange chromatography showed that IP3 (inositol triphosphate) was the major product and accounted for about 90% of total inositol phosphates formed under basal and stimulated conditions (Boyle et al., 1987). The ability of IP3 to release Ca^{2+} from isolated microsomes from the olfactory epithelium was therefore investigated (Bruch, 1990). Isolated olfactory microsomes sequestered $^{45}Ca^{2+}$ by a Mg^{2+}/ATP-dependent mechanism and about 10% of this uptake was inhibited by oligomycin, an inhibitor of mitochondrial Ca^{2+} transport. In the presence of oligomycin, the calcium ionophore A23187 rapidly released about 90% of the sequestered Ca^{2+}. IP3 released about 30% of the sequestered Ca^{2+} in the same preparations. Maximal IP3-mediated Ca^{2+} release was obtained by 2 min and was followed by re-uptake. These results therefore confirmed the well-established second messenger role of IP3 in mediating release of Ca^{2+} from intracellular stores (Putney, 1987) and suggested a role for intracellular Ca^{2+} in olfactory signal transduction.

Inositol trisphosphate has also been reported directly to gate a calcium channel in olfactory ciliary membranes incorporated into phospholipid

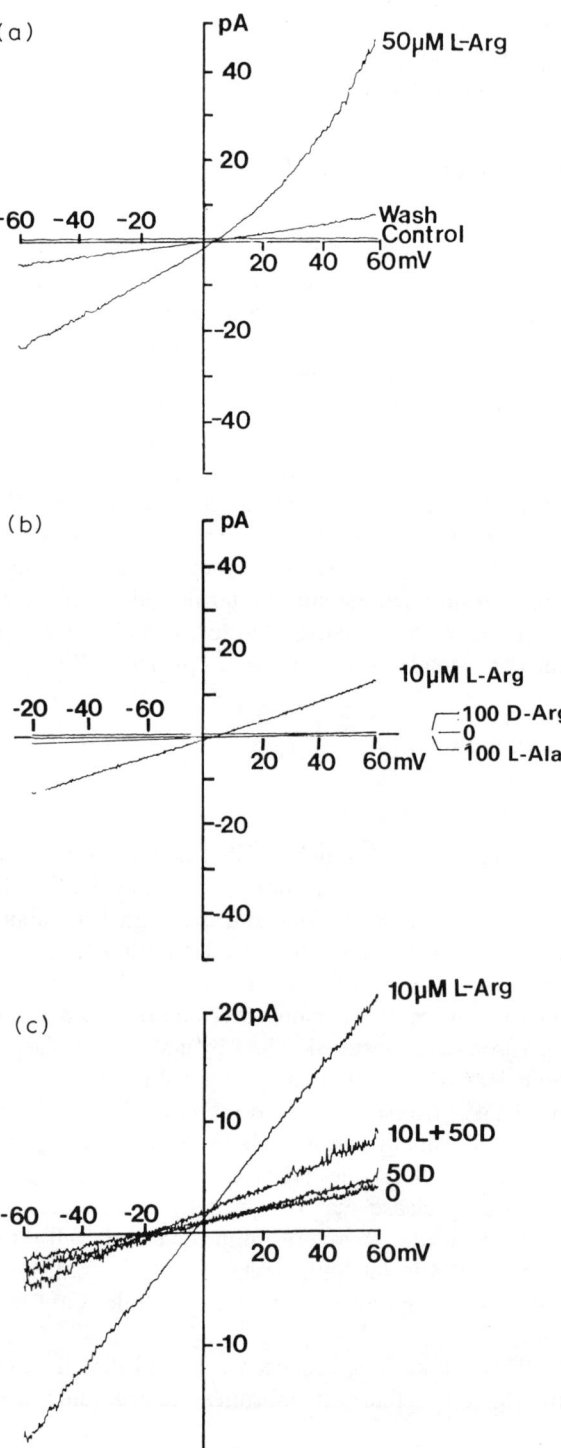

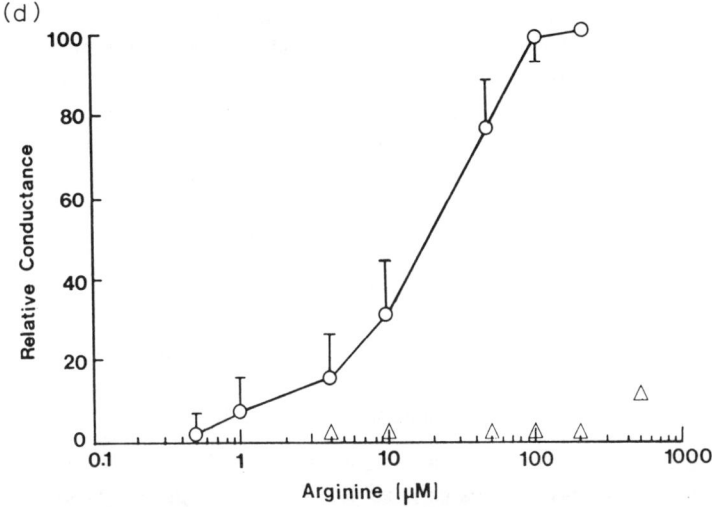

Fig. 7.3 Conductance changes evoked by amino acid stimuli in azolectin bilayers into which fragments of plasma membranes from catfish taste epithelium were incorporated. (a) Current-voltage curves in a bilayer before addition (control), after addition, and after removal (wash) of 50 μM L-arginine. (b) Effects of 10 μM L-arginine, 100 μM D-arginine, and 100 μM L-alanine on the conductance of a bilayer containing taste epithelial membranes. (c) Suppression of 10 μM L-arginine-activated conductance by 50 μM D-arginine (10 L + 50 D). D-Arginine alone at 50 μM (50 D) had little effect on the conductance of the bilayer. Control conductance noted as '0'. Note changed vertical scale. (d) Concentration-response curves for L-arginine (○) and D-arginine (Δ); points are mean ±SD for three bilayers. (Reproduced with permission from Teeter et al., 1989.)

bilayers (Bruch et al., 1989; Restrepo et al., 1990). Isolated olfactory neurones displayed transient increases in intracellular calcium ion activities through at least two pathways when monitored using the intracellular calcium ion fluorescent indicator, fura-2 (Bruch et al., 1989; Restrepo and Boyle, 1991).

Cyclic-nucleotide-gated channels in olfactory cilia

A cyclic-nucleotide-gated conductance has been described in excised membrane patches from olfactory cilia from toad (Nakamura and Gold, 1987), and in homogenates of rat olfactory epithelium reconstituted in artificial phospholipid bilayers (Vodyanoy, 1989). The same or similar cyclic-nucleotide-gated conductance was also detected in isolated cilia from the channel catfish incorporated into azolectin phospholipid bilayers (Bruch and Teeter, 1989, 1990). Both cAMP and cGMP were equally effective in modulating membrane conductance, with 2–3 μM of either nucleotide eliciting half-maximal increases in conductance. The cyclic-nucleotide-dependent current reversed near 0 mV, indicating that the underlying ion channels were nonselective between Na^+ and K^+. The slope conductance of the cyclic-nucleotide-gated channels averaged 45 pS. As in previous reports (Nakamura and Gold, 1987; Vodyanoy, 1989), exogenous nucleotide triphosphates were not required for the reversible modulation of membrane conductance elicited by cyclic nucleotides to be observed, suggesting that cyclic nucleotides directly gated the channels mediating the increase in conductance without intermediary protein phosphorylation. In addition, the observation of a cyclic-nucleotide-gated conductance in diverse species suggests that it may represent a common, highly conserved, transduction event in vertebrate olfaction.

Stimulus-activated ion channels

Intracellular recordings from taste cells *in situ* in *I. punctatus* suggest that different transduction processes exist for different populations of receptors (Teeter et al., 1989). Most catfish taste cells responded to L-alanine with small graded depolarizations and little or no clear change in membrane resistance. L-Arginine also depolarized taste cells, but with a small decrease in membrane resistance. Cells with low resting potentials usually showed no change in membrane resistance. Some cells displayed hyperpolarizing decreases in membrane resistance in response to L-arginine, indicating a fundamentally different mechanism, perhaps an increase in K^+ (or Cl^-) conductance.

Catfish taste cells are small, making stable intracellular recordings difficult. Whole-cell patch-clamp recordings should obviate many of the difficulties associated with recording from these small cells. However, so far attempts to dissociate and record from cutaneous taste cells from the catfish have been only partially successful. Consequently, the conductance properties imparted to planar bilayers after incorporation of purified membrane fragments derived from catfish taste epithelia have been examined.

Fragments of purified membranes (Bo) derived from Fraction P2 used in binding studies (Cagan and Boyle, 1984) were incorporated into azolectin bilayers at the tip of patch pipettes, and the effects of voltage ramps and pulses and taste stimuli were examined. Although a variety of spontaneous and voltage-dependent currents were often observed, about 40% of the bilayers displayed no voltage-dependent conductances in response to voltage ramps between -80 and $+80$ mV. In about half of these 'silent' bilayers, micromolar concentrations of L-arginine elicited a rapid, reversible increase in conductance. The L-arginine-elicited conductance was concentration dependent, usually becoming activated at about 0.5 μM and saturated between 100 and 200 μM. The conductance was specific for L-arginine, with neither L-alanine nor D-arginine having an effect at concentrations up to 200 μM. At 50 μM, D-arginine, a partial antagonist of L-arginine binding

(Fig. 7.1), markedly suppressed the increase in conductance elicited by 10 μM L-arginine (Fig. 7.3). D-Arginine at 50 μM had no effect on conductance. L-Arginine-activated single-channel currents had slope conductances of about 40 pS and reversed near 0 mV in normal internal and external solutions, indicating that they were not selective between Na^+ and K^+ (Teeter et al., 1989, 1990).

These results are consistent with available binding data and suggest that the L-arginine taste receptor in the catfish may be part of, or closely associated with, a cation channel. This hypothesis is also consistent with the apparent lack of effect of L-arginine on potential second-messenger pathways

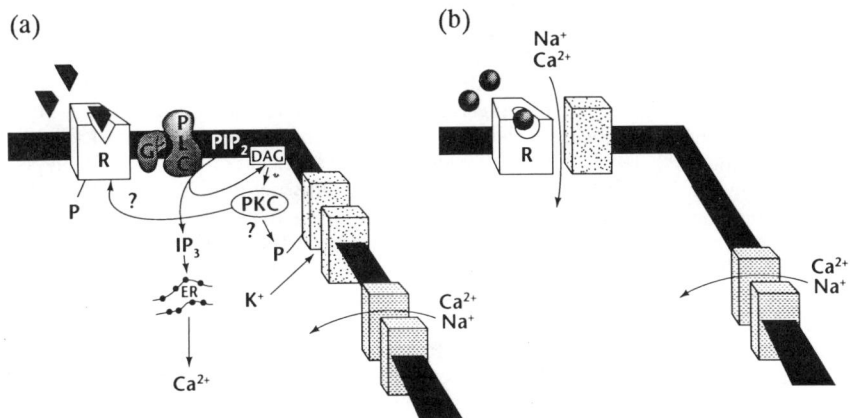

Fig. 7.4 Two possible transduction sequences for taste. (a) A transduction scheme involving receptor-mediated production of second messenger. The binding of the stimulus to receptor (R) activates GTP-binding regulatory proteins (G) which, in turn, activate second-messenger producing enzymes – in this case, phospholipase C (PLC). PLC metabolizes phosphatidylinositol bisphosphate (PIP2) into two second messengers: inositol trisphosphate (IP3) and diacylglycerol (DAG). IP3 may release calcium from internal stores, such as endoplasmic reticulum (ER), while DAG may activate protein kinase C (PKC). PKC may phosphorylate susceptible proteins including ion channels and the receptors. Phosphorylation of potassium channels may close them leading to a build-up of intracellular K^+. Depolarization brought about by build-up of K^+ may activate voltage-dependent Na^+ and Ca^{2+} channels leading to further depolarization and to influx of calcium ion. An increase in calcium ion activity would trigger neurotransmitter release. This transduction scheme is postulated to be active for reception of L-alanine and other neutral amino acids in the catfish cutaneous taste system (Brand et al., 1991; Kalinoski et al., 1989b). (b) A transduction scheme involving receptor mediated modification of a closely applied or contiguous ion channel. Stimulus binding to receptor (R) directly opens an ion channel which allows influx of sodium and calcium ions. As with scheme a, influx of Na^+ and Ca^{2+} leads to membrane depolarization and release of neurotransmitter. This transduction scheme is postulated to be active for reception of L-arginine and L-proline in the catfish cutaneous taste system (Brand et al., 1991; Teeter et al., 1990).

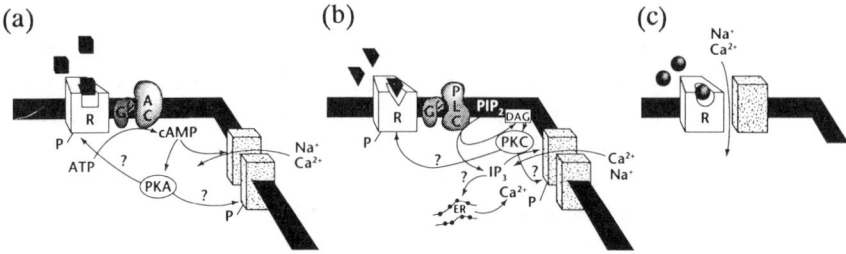

Fig. 7.5 Three possible transduction sequences in olfaction. (a) Stimulus binding to receptor (R) in the ciliary membrane activates GTP-binding regulatory proteins (G) which stimulate the activity of adenylyl cyclase (AC). AC generates cyclic AMP (cAMP) from ATP. In the olfactory system, it is probable that cAMP directly gates an ion channel in the ciliary plasma membrane, allowing influx of sodium and calcium ion. In the olfactory receptor neurone, a generator potential then ensues. cAMP may also activate a cyclic AMP dependent protein kinase A which may phosphorylate susceptible proteins, including ion channels and receptors, to modulate their activity. This transduction scheme producing cyclic AMP is active in amphibians (Lancet et al., 1988; Nakamura and Gold, 1987) and rats (Snyder et al., 1988; Boekhoff et al., 1990), and is also activated in cilia from catfish olfactory neurones (Bruch and Teeter, 1990). (b) Stimulus binding to receptor (R) in the ciliary membrane activates (G) which stimulates activity of a phospholipase C (PLC). PLC metabolizes phosphatidyl inositol bisphosphate (PIP2) into two second messengers, inositol trisphosphate (IP3) and diacylglycerol (DAG). In the olfactory system, IP3 may directly gate an IP3 sensitive ion channel in the ciliary plasma membrane allowing influx of calcium and sodium. In the olfactory receptor neurone a generator potential ensues. IP3 may also release calcium from intracellular stores, and Ca ions may regulate additional Ca release or influx, as well as Ca-dependent ion channels. DAG may activate protein kinase C (PKC) which can phosphorylate proteins, including ion channels and receptors, to modulate their activity. This transduction scheme producing IP3 is known to be active in the olfactory system of catfish (Bruch and Kalinoski, 1987; Bruch and Teeter, 1989; Bruch et al., 1989; Huque and Bruch, 1986; Restrepo and Boyle, 1991; Restrepo et al., 1990) and rats (Boekhoff et al., 1990). (c) Stimulus binding to receptor (R) directly opens an ion channel which allows influx of sodium and calcium ions. In the olfactory receptor neurone, a generator potential ensues. This scheme is at present only postulated for olfaction in teleosts. There is evidence for it in amphibians (Anholt, 1989; Labarca et al., 1988).

that are stimulated by L-alanine, a representative stimulus for the other major class of amino acid receptors.

A recent study has reported conductance changes in *Xenopus* oocytes to L-arginine after the oocytes had been injected with mRNA from taste epithelium of *I. punctatus* (Getchell et al., 1989). The responses were rapid and of short duration. Messenger RNA isolated from catfish brain or liver did not lead to the expression of these L-arginine-stimulated currents in oocytes.

7.4 CONCLUSION

The receptor/transduction sequences for gustation and olfaction in fishes are receiving increased attention, particularly as regards amino-acid-sensitive transductory pathways in certain specialized models. The sensitivity of the receptor pathways has permitted the use of classical receptor binding techniques to define the specificity of these receptors. Subsequent studies of the probable events in transduction, such as second-messenger accumulation and ion-channel activation, have permitted comprehensive hypotheses to be drawn concerning the likely pathways of these sensory systems at the molecular level. It appears as though multiple mechanisms exist for the transduction of chemosensory information in teleosts, and that in some cases multiple transduction pathways can be stimulated by a single receptor class.

Figures 7.4 and 7.5 show possible receptor and transduction pathways for taste and olfaction, respectively. While the molecular nature of the receptors in taste and olfaction in teleosts remains unknown, it is likely that the receptors will soon be identified by biochemical isolation and cloning techniques. Putative receptors from the olfactory system of rat, for example, have recently been characterized using this technique (Buck and Axel, 1991). Isolation of the receptors for taste and olfactory stimuli, and positive characterization of the remainder of the transduction pathways, will lead to a more complete understanding of chemosensory transduction.

ACKNOWLEDGEMENTS

This work was supported by awards from the National Institutes of Health (DC-00356 and DC-00327), by the Veterans Affairs Department and by the Office of Naval Research (N00014-90-J-1503). We thank Mrs Janice Blescia for processing the manuscript.

REFERENCES

Abdel-Latif, A.A. (1986) Calcium-mobilizing receptors, polyphosphoinositides and the generation of second messengers. *Pharmacol. Rev.*, **38**, 227–72.

Alkon, D.L. and Rasmussen, H. (1988) A spatial–temporal model of cell activation. *Science, Wash., D.C.*, **239**, 998–1005.

Anholt, R.R.H. (1989) Molecular physiology of olfaction. *Am. J. Physiol.*, **257**, C1043–54.

Berridge, M.J. (1984) Inositol trisphosphate and diacylglycerol as second messengers. *Biochem. J.*, **220**, 345–60.

Berridge, M.J. and Irvine, R. (1984) Inositol trisphosphate, a novel second messenger in cellular signal transduction. *Nature*, **312**, 315–21.

Boekhoff, I., Tareilus, E., Strotmann, J. and Breer, H. (1990) Rapid activation of alternative second messenger pathways in olfactory cilia from rats by different odorants. *EMBO J.*, **9**, 2453–8.

Boyle, A.G., Park, Y.S., Huque, T. and Bruch, R.C. (1987) Properties of phospholipase C in isolated olfactory cilia from the channel catfish (*Ictalurus punctatus*). *Comp. Biochem. Physiol.*, **88B**, 767–75.

Brand, J.G., Bryant, B.P., Cagan, R.H. and Kalinoski, D.L. (1987) Biochemical studies of taste sensation. XIII. Enantiomeric specificity of alanine taste receptor sites in catfish, *Ictalurus punctatus*. *Brain Res. Amst.*, **416**, 119–28.

Brand, J.G., Huque, T., Rabinowitz, J.L. and Bayley, D.L. (1989a) Lipid characterization and ^{14}C-acetate metabolism in catfish taste epithelium. *Experientia*, **45**, 77–81.

Brand, J.G., Teeter, J.H., Cagan, R.H. and Kare, M.R. (eds) (1989b) *Chemical Senses 1. Receptor Events and Transduction in Taste and Olfaction*, Marcel Dekker, New York, 529 pp.

Brand, J.G., Teeter, J.H., Kumazawa, T., Huque, T. and Bayley, D.L. (1991) Transduction mechanisms for the taste of amino acids. *Physiol. Behav.*, **49**, 899–904.

Brown, S.B. and Hara, T.J. (1981) Accumulation of chemostimulatory amino acids by a sedimentable fraction isolated from olfactory rosettes of rainbow trout (*Salmo gairdneri*). *Biochim. biophys. Acta*, **675**, 149–62.

Bruch, R.C. (1989) Signal transduction in olfaction and taste, in *G-Proteins* (eds R. Iyengar, and L. Birnbaumer), Academic Press, New York, pp. 411–27.

Bruch, R.C. (1990) G proteins in olfactory neurons, in *G-Proteins and Calcium Signalling* (ed. P.H. Naccache), CRC Press, Boca Raton, FL, pp. 123–34.

Bruch, R.C. and Gold, G.H. (1990) G-protein-mediated signalling in olfaction, in *G-Proteins as Mediators of Cellular Signalling Processes* (eds M.D. Houslay and G. Milligan), Wiley, London, pp. 113–24.

Bruch, R.C. and Kalinoski, D.L. (1987) Interaction of GTP-binding regulatory proteins with chemosensory receptors. *J. biol. Chem.*, **262**, 2401–4.

Bruch, R.C. and Rulli, R.D. (1988) Ligand binding specificity of a neutral L-amino acid olfactory receptor. *Comp. Biochem. Physiol.*, **91B**, 535–40.

Bruch, R.C. and Teeter, J.H. (1989) Second messenger signalling mechanisms in olfaction, in *Chemical Senses 1: Receptor Events and Transduction in Taste and Olfaction* (eds J.G. Brand, J.H. Teeter, M.R. Kare and R.H. Cagan), Marcel Dekker, New York, pp. 283–98.

Bruch, R.C. and Teeter, J.H. (1990) Cyclic AMP links amino acid chemoreceptors to ion channels in olfactory cilia. *Chem. Senses*, **15**, 419–30.

Bruch, R.C., Kalinoski, D.L. and Kare, M.R. (1988) Biochemistry of vertebrate olfaction and taste. *A. Rev. Nutr.*, **8**, 21–42.

Bruch, R.C., Rulli, R.D. and Boyle, A.G. (1987) Olfactory L-amino acid receptor specificity and stimulation of potential second messengers. *Chem. Senses*, **12**, 642–3.

Bruch, R.C., Restrepo, D. and Teeter, J.H. (1989) Cyclic AMP and IP3 link G protein-coupled chemoreceptors to ion channels in olfactory cilia. *J. gen. Physiol.*, **94**, 4a.

Bryant, P.B., Brand, J.G., Kalinoski, D.L., Bruch, R.C. and Cagan, R.H. (1987) Use of monoclonal antibodies to characterize amino acid taste receptors in catfish: effects on binding and neural responses, in *Olfaction and Taste IX* (eds S.D. Roper and J. Atema), New York Acad. Sci., New York, pp. 208–9.

Buck, L. and Axel, R. (1991) A novel multi-gene family may encode odorant receptors: a molecular basis for odor recognition. *Cell*, **65**, 175–87.

Cagan, R.H. (1976) Biochemical studies of taste sensation. II. Labeling of cyclic AMP of bovine taste papillae in response to sweet and bitter stimuli. *J. Neurosci. Res.*, **2**, 363–71.

Cagan, R.H. (1979) Biochemical studies of taste sensation. VII. Enhancement of taste stimulus binding to a catfish taste receptor preparation by prior exposure to the stimulus. *J. Neurobiol.*, **10**, 207–20.

Cagan, R.H. (1986) Biochemical studies of taste sensation. XII. Specificity of binding of taste ligands to a sedimentable fraction from catfish taste tissue. *Comp. Biochem. Physiol.*, **85A**, 355–8.

Cagan, R.H. (ed.) (1989) *Neural Mechanisms in Taste*, CRC Press, Boca Raton, FL 232 pp.

Cagan, R.H. and Boyle, A.G. (1984) Biochemical studies of taste sensation. XI. Isolation, characterization and taste ligand binding activity of plasma membranes from catfish taste tissue. *Biochim. biophys. Acta*, **799**, 230–7.

Cagan, R.H. and Zeiger, W.N. (1978) Biochemical studies of olfaction: binding specificity of radioactively labeled stimuli to an isolated olfactory preparation from rainbow trout (*Salmo gairdneri*). *Proc. natn. Acad. Sci. U.S.A.*, **75**, 4679–83.

Cancalon, P. (1978) Isolation and characterization of the olfactory epithelial cells of the catfish. *Chem. Senses Flavour*, **3**, 381–96.

Cancalon, P. (1983) Receptor cells of the catfish olfactory mucosa. *Chem. Senses*, **8**, 203–9.

Caprio, J. (1978) Olfaction and taste in the channel catfish: an electrophysiological study of the responses to amino acids and derivatives. *J. comp. Physiol.*, **123**, 357–71.

Caprio, J. (1982) High sensitivity and specificity of olfactory and gustatory receptors of catfish to amino acids, in *Chemoreception in Fishes* (ed. T.J. Hara), Elsevier, Amsterdam, pp. 109–34.

Caprio, J. and Byrd, R.P., jun. (1984) Electrophysiological evidence for acidic, basic and neutral amino acid olfactory receptor sites in the catfish. *J. Gen. Physiol.*, **84**, 403–22.

Erickson, J.R. and Caprio, J. (1984) The spatial distribution of ciliated and microvillous olfactory receptor neurons in the channel catfish is not matched by a differential specificity to amino acid and bile salt stimuli. *Chem. Senses*, **9**, 127–41.

Exton, J.H. (1988) Mechanisms of action of calcium-mobilizing agonists: some variations on a young theme. *FASEB J.*, **2**, 2670–6.

Getchell, T.V. (1986) Functional properties of vertebrate olfactory receptor neurons. *Physiol. Rev.*, **66**, 772–818.

Getchell, T.V., Margolis, F.L. and Getchell, M.L. (1985) Perireceptor and receptor events in vertebrate olfaction. *Progr. Neurobiol.*, **23**, 317–45.

Getchell, T.V., Grillo, M., Tate, S., Urade, R., Teeter, J.H. and Margolis, F.L. (1989) Expression of amino acid taste receptors in *Xenopus* oocytes. *Neurochem. Res.*, **15**, 449–56.

Gilman, A.G. (1987) G-Proteins: transducers of receptor-generated signals. *A. Rev. Biochem.*, **56**, 615–49.

Hatt, H. (1989) Stimulus-driven chemosensory membrane channels on crayfish sensory cells, in *Chemical Senses 1. Receptor Events and Transduction in Taste and Olfaction* (eds. J.G. Brand, J.H. Teeter, R.H. Cagan and M.R. Kare), Marcel Dekker, New York, pp. 363–87.

Huque, T. and Bruch, R.C. (1986) Odorant- and guanine nucleotide-stimulated phosphoinositide turnover in olfactory cilia. *Biochem. Biophys. Res. Commun.*, **137**, 36–42.

Huque, T., Brand, J.G., Rabinowitz, J.L. and Bayley, D.L. (1987) Phospholipid turnover in catfish barbel (taste) epithelium with special reference to phosphatidyl-4,5-bisphosphate. *Chem. Senses*, **12**, 666–7.

Jones, D.T. and Reed, R.R. (1987) Molecular cloning of five GTP-binding protein cDNA species from rat olfactory neuroepithelium. *J. biol. Chem.*, **262**, 14241–9.

Jones, D.T. and Reed, R.R. (1989) G_{olf}: an olfactory neuron specific G-protein involved in odorant signal transduction. *Science*, Wash., D.C., **244**, 790–5.

Kalinoski, D.L., Bruch, R.C. and Brand, J.G. (1987) Differential interaction of lectins with chemosensory receptors. *Brain Res. Amst.*, **418**, 34–40.

Kalinoski, D.L., Bryant, B.P., Shaulsky, G., Brand, J.G. and Harpaz, S. (1989a) Specific L-arginine taste receptor sites in the catfish, *Ictalurus punctatus*: biochemical and neurophysiological characterization. *Brain Res. Amst.*, **488**, 163–73.

Kalinoski, D.L., Huque, T., La Morte, V.J. and Brand, J.G. (1989b) Second messenger events in taste, in *Chemical Senses 1. Receptor Events and Transduction in Taste and Olfaction* (eds J.G. Brand, J.H. Teeter, R.H. Cagan and M.R. Kare), Marcel Dekker, New York, pp. 85–101.

Krueger, J.M. and Cagan, R.H. (1976) Biochemical studies of taste sensation. IV. Binding of L-[^{3}H]alanine to a sedimentable fraction from catfish barbel epithelium. *J. biol. Chem.*, **259**, 88–97.

Kurihara, K. (1972) Inhibition of cyclic 3',5'-nucleotide phosphodiesterase in bovine taste papillae by bitter taste stimuli. *FEBS Lett.*, **27**, 279–81.

Lancet, D. (1988) Molecular components of olfactory reception and transduction, in *Molecular Neurobiology of the Olfactory System* (eds F.L. Margolis and T.V. Getchell), Plenum Press, New York, pp. 25–50.

Labaraca, P., Simon, S.A. and Anholt, R.R.H. (1988) Activation by odorants of a multistate cation channel from olfactory cilia. *Proc. Natl. Sci. U.S.A.*, **85**, 944–7.

Lynch, C.J., Blackmore, P.F., Charest, R. and Exton, J.H. (1985) The relationships between receptor binding capacity for norepinephrine, angiotensin II and vasopressin and release of inositol trisphosphate, Ca^{2+} mobilization and phosphorylase activation in rat liver. *Mol. Pharmacol.*, **28**, 93–9.

Margolis, F.L. and Getchell, T.V. (eds) (1988) *Molecular Neurobiology of the Olfactory System*, Plenum Press, New York, 379 pp.

Muller, J.F. and Marc, R.E. (1984) Three distinct morphological classes of receptors in fish olfactory organs. *J. comp. Neurol.*, **222**, 482–95.

Nakamura, T. and Gold, G.H. (1987) A cyclic nucleotide-gated conductance in olfactory receptor cilia. *Nature*, **325**, 442–4.

Novoselov, V.I., Krapivinskaya, L.D. and Fesenko, E.E. (1980) Molecular mechanisms of odor sensing. V. Some biochemical characteristics of the alanineous receptor from the olfactory epithelium of the skate, *Dasyatis pastinaca*. *Chem. Senses Flavour*, **5**, 195–203.

Novoselov, V.I., Krapivinskaya, L.D. and Fesenko, E.E. (1988a) Amino acid binding glycoproteins from the olfactory epithelium of skate (*Dasyatis pastinaca*). *Chem. Senses*, **13**, 267–78.

Novoselov, V.I., Krapivinskaya, L.D., Krapivinsky, G.B. and Fesenko, E.E. (1988b) GTP-binding protein associated with amino acid binding proteins from olfactory epithelium of skate, *Dasyatis pastinaca*. *FEBS Lett.*, **234**, 471–4.

Parfenova, E.V. and Etingof, R.N. (1988) The participation of GTP-binding proteins in olfactory receptors in vertebrates. *Biokhimiya*, **53**, 435–43.

Price, S. (1973) Phosphodiesterase in tongue epithelium: activation by bitter taste stimuli. *Nature*, **241**, 54–5.

Putney, J.W., jun. (1987) Formation and actions of calcium-mobilizing messenger, inositol 1,4,5-trisphosphate. *Am. J. Physiol.*, **252**, G149–57.

Rabinowitz, J.L., Huque, T., Brand, J.G. and Bayley, D.L. (1990) Lipid metabolic interrelationships and phospholipase activity in gustatory epithelium of *Ictalurus punctatus in vitro*. *Lipids*, **25**, 181–6.

Rehnberg, B.G. and Schreck, C.B. (1986) The olfactory L-serine receptor in coho salmon: biochemical specificity and behavioral response. *J. comp. Physiol.*, **159**, 61–7.

Restrepo, D. and Boyle, A.G. (1991) Stimulation of olfactory receptors alters regulation of $[Ca_i]$ in olfactory neurons of the catfish (*Ictalurus punctatus*). *J. Membrane Biol.*, **120**, 223–32.

Restrepo, D., Miyamoto, T., Bryant, B.P. and Teeter, J.H. (1990) Odor stimuli trigger influx of calcium into olfactory neurons of the channel catfish. *Science, Wash., D.C.*, **249**, 1166–8.

Rhein, L.D. and Cagan, R.H. (1980) Biochemical studies of olfaction: isolation, characterization and odorant binding activity of cilia from rainbow trout olfactory rosettes. *Proc. natn. Acad. Sci. U.S.A.*, **77**, 4412–16.

Rhein, L.D. and Cagan, R.H. (1981) Role of cilia in olfactory recognition, in *Biochemistry of Taste and Olfaction* (eds R.H. Cagan and M.R. Kare), Academic Press, New York, pp. 47–68.

Rhein, L.D. and Cagan, R.H. (1983) Biochemical studies of olfaction: binding specificity of odorants to a cilia preparation from rainbow trout olfactory rosettes. *J. Neurochem.*, **41**, 569–77.

Rhein, L.D., Cagan, R.H., Orkand, P.M. and Dolack, M.K. (1981) Surface specializations of the olfactory epithelium of rainbow trout, *Salmo gairdneri*. *Tissue and Cell*, **13**, 577–87.

Roper, S.D. (1989) The cell biology of vertebrate taste receptors. *A. Rev. Neurosci.*, **12**, 329–53.

Sheiko, L.M., Ostretsova, I.B., Leuko, A.V., Volotvskii, I.D., Etingof, R.N. and Konev, S.V. (1983) Interaction of leucine with plasma membranes of catfish taste buds. *Molek. Biol.*, **17**, 1220–6.

Snyder, S.H., Sklar, P.B. and Pevsner, J. (1988) Olfactory receptor mechanisms: odorant-binding protein and adenylate cyclase, in *Molecular Neurobiology of the Olfactory System* (eds F.L. Margolis and T.V. Getchell), Plenum Press, New York, pp. 3–24.

Steinlen, S., Klumpp, S. and Schultz, E. (1990) Guanylate cyclase in olfactory cilia from rat and pig. *Biochim. biophys. Acta*, **1054**, 69–72.

Teeter, J.H., Brand, J.G. and Kumazawa, T. (1990) A stimulus-activated conductance in isolated taste epithelial membranes. *Biophys. J.*, **58**, 253–9.

Teeter, J.H., Sugimoto, K. and Brand, J.G. (1989) Ionic currents in taste cells and reconstituted taste epithelial membranes, in *Chemical Senses 1. Receptor Events and Transduction in Taste and Olfaction* (eds J.G. Brand, J.H. Teeter, R.H. Cagan and M.R. Kare), Marcel Dekker, New York, pp. 151–70.

Thommesen, G. (1983) Morphology, distribution, and specificity of olfactory receptor cells in salmonid fishes. *Acta physiol. scand.*, **117**, 241–9.

Trapido-Rosenthal, H.G., Carr, W.E.S. and Gleeson, R.A. (1989) Biochemistry of purinergic olfaction. The importance of nucleotide dephosphorylation, in *Chemical Senses 1. Receptor Events and Transduction in Taste and Olfaction* (eds J.G. Brand, J.H. Teeter, R.H. Cagan and M.R. Kare), Marcel Dekker, New York, pp. 243–81.

Ui, M. (1986) Pertussis toxin as a probe of receptor coupling to inositol lipid metabolism, in *Phosphoinositides and Receptor Mechanisms* (ed. J.W. Putney, jun.), Liss, New York, pp. 163–95.

Vodyanoy, V. (1989) Cyclic nucleotide-gated electrical activity in olfactory receptor cells, in *Chemical Senses 1: Receptor Events and Transduction in Taste and Olfaction* (eds J.G. Brand, J.H. Teeter, R.H. Cagan and M.R. Kare), Marcel Dekker, New York, pp. 319–46.

Williamson, J.R. (1986) Role of inositol lipid breakdown in the generation of intracellular signals. *Hypertension*, **8** (Supp. II), 140–56.

Yamomoto, M. (1982) Comparative morphology of the peripheral olfactory organs in telosts, in *Chemoreception in Fishes* (ed. T.J. Hara), Elsevier, Amsterdam, pp. 39–59.

Zielinski, B. and Hara, T.J. (1988) Morphological and physiological development of olfactory receptor cells in rainbow trout (*Salmo gairdneri*) embryos. *J. comp. Neurol.* **271**, 300–11.

Chapter eight

Mechanisms of olfaction

Toshiaki J. Hara

8.1 INTRODUCTION

The interaction of an odorant molecule with the membrane of an olfactory neurone initiates a series of membrane phenomena associated with sensory transduction and subsequent electrical events. The binding of odorants to receptor molecules, believed to be a reversible reaction (Brand and Bruch, (Chapter 7 herein), in turn causes conformational changes of the receptor membrane leading to spike generation in the sensory cell. The electrical signal is subsequently transmitted to the higher-order nervous system, where the sensory perception of odours is produced.

The purpose of this chapter is to describe the sequence of electrical events in the peripheral olfactory system in response to chemical stimulation. The chapter focuses on the sensitivity and specificity of olfactory responses to representative chemicals known to be potent odorants for fish. Multiplicity of olfactory receptors as well as possible functional separation of the central olfactory system will be discussed. Because electrophysiological investigations have provided basic information on functioning of olfactory neurones, the major focus is a detailed discussion of the physiological activity of olfactory neurones as shown primarily by electrophysiological techniques. In the following discussion, the term 'olfactory neurone' is primarily used for the widespread use of 'receptor cell' to avoid confusion with 'molecular receptor' involving a ligand-binding molecular structure. The term 'receptor site' is loosely used here to accommodate the stimulus-binding site and all subsequent events (transduction processes) leading to the production of generator potentials within the olfactory neurone.

8.2 SENSORY TRANSDUCTION

Electrical characteristics of the olfactory neurone

Primarily because olfactory neurones are small, studies of membrane characteristics have been hampered. Inevitably, the following description is based on a single report available on intracellular recordings from fish olfactory neurones (Fig. 8.1). The resting membrane potentials of the olfactory neurones of lamprey, *Entosphenus japonicus*, range from -36.7 to -60.0 mV with an input resistance of 74.6–356.9 MΩ (Suzuki, 1977). The membrane time constants average 3.46 and 0.34 ms. These data are in agreement with those obtained from the bullfrog and mudpuppy, the only other vertebrates in which membrane characteristics of the olfactory neurones have been studied. The resting membrane potentials in other excitable cells are primarily dependent on the external and internal concentrations of K^+. However, the ion concentration over the olfactory epithelium seems unlikely to be a determining factor of the resting potential of the lamprey olfactory neurone, because alteration of K^+, Na^+, and Ca^{2+} concentrations bathing the olfactory mucosa has no significant influence on the resting potential (Suzuki, 1977).

Generation of action potentials

The injection of depolarizing currents into an olfactory neurone elicits spike responses. The thresholds for spike activation range from 4×10^{-11} to

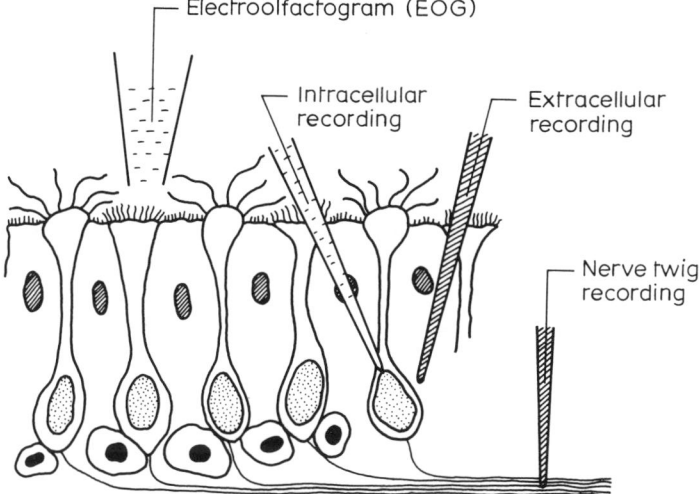

Fig. 8.1 Schematic representation of electrophysiological recordings from the peripheral olfactory system of fish.

1.9×10^{-10} A. The neurone exhibits repetitive firing when prolonged depolarizing current is applied. The firing frequency increases linearly with an increase in the current intensity. However, the spike height, normally smaller than the resting potential, decreases with an increase in current strength. In contrast, the neurone responds to hyperpolarizing currents by exhibiting a transient hyperpolarizing response, sometimes followed by a single spike activity upon the cessation of current injection.

Application of L-arginine to the olfactory epithelium produces a depolarizing receptor (generator) potential. The magnitude of the generator potential increases (up to 25 mV) with an increase in stimulus concentration, on which spikes are normally superimposed. The input resistance of the receptor membrane decreases during the depolarizing receptor potential, suggesting involvement of ionic conductance changes in the transduction process. However, application of high concentrations of L-glutamic acid often causes hyperpolarization (hyperpolarizing receptor potential).

The neural response to L-arginine is blocked by bathing the receptor surface in a Ca-free solution containing a Ca-chelating agent EGTA (ethylene glycol-bis (β-amino-ethyl ether) N,N,N',N'-tetra-acetic acid). Based on these findings, combined with the fact that specific binding of radiolabelled L-arginine to lamprey olfactory tissue increases with an increase in Ca concentration, Suzuki (1978, 1982) proposed a hypothetical model for the olfactory transduction. The model assumes that the receptor neuronal membrane is heterogeneous and comprises two zones, receptive and electrogenic. The receptive zone, assumed to be located in the olfactory knob and/or cilia, initiates the interaction between stimulus molecules and specific receptors, which leads to the changes in permeability to ions in the electrogenic zones. The coupling between the binding of stimulus molecules and changes in permeability is believed to be enzymatic in nature. Recent studies using the patch-clamp technique indicate that excised patches of ciliary plasma membrane obtained from dissociated olfactory neurones of the toad *Bufo marinus* contain a conductance which is gated directly by cyclic AMP (Nakamura and Gold, 1987; Brand and Bruch, herein). In patch-clamp method, a glass micropipette (tip diameter of 2–3 μm), filled with electrically conducting solution, is used to seal very tightly to the membrane of excitable cells. The membrane encircled by the tip is called patch; analyses can be carried out with the electrical potential across the membrane clamped at different values in order to obtain the current that flows through the patch.

Electroolfactogram (EOG)

Summated stimulus-evoked activity of olfactory neurones can be studied by electroolfactogram (EOG) recordings from the epithelial surface using a relatively large electrode (Fig 8.1). The EOG primarily represents a summation of generator potentials of the olfactory neurones, similar to those found in other sensory systems (Ottoson, 1956, 1971; Getchell, 1974). The EOG, sometimes known as the underwater EOG for fish (Silver *et al.*, 1976), is a large (of the order of millivolts) negative voltage transient consisting of two major components, phasic and tonic (Evans and Hara, 1985). The phasic response completes within a few seconds of stimulation, but the tonic response is normally maintained throughout the stimulus duration. The EOG differs considerably in size and shape, depending upon the type of stimulants. However, to date no systematic study has been carried out to relate the temporal characteristics of EOG to the physico-chemical properties of stimuli. EOG recording is widely used as an experimental tool in analysis of olfactory mechanisms.

The mucosal neural recording (MNR) is also used to record summated multiunit activity from the epithelium (Suzuki and Tucker, 1971; Silver, 1982). The method, like EOG, has the advantage of obtaining olfactory activity with a minimum of surgery. However, movement of the olfactory lamellae caused by continuous flow of water causes instability in recordings.

Extracellular recordings of action potentials

The action potential can be recorded extracellularly using microelectrodes from a receptor neurone intraepithelially, or from olfactory nerve twigs at some distance from the nares (Fig. 8.1). Unitary action potential recordings are instrumental in that they provide the quantitative and qualitative properties of the sensory neurones. Olfactory neurones normally exhibit spontaneous activity, i.e. impulse discharges without chemical stimuli, and respond to stimuli by (in most cases) increasing firing frequency.

The integrated neural activity from small bundles of primary olfactory nerve is also used to obtain olfactory neuronal activity (Sveinsson and Hara, 1990a,b). This olfactory nerve twig recording (NTR) is more stable than the MNR recording. The integrated NTR response to a stimulus normally exhibits a rapid phasic response followed by a tonic response that is maintained throughout the stimulus period (Sveinsson and Hara, 1990a). The magnitude of both components increases with an increase of the stimulus intensity, but their ratio remains relatively constant throughout for a given stimulus.

Site of olfactory transduction

It is generally believed that the receptor sites lies on the membranes of olfactory cilia, since these are the first points of contact with odorous molecules. Particularly in fishes, the olfactory knob does not protrude as much beyond the limiting surface of the epithelium as in other vertebrates. This view is further supported by recent biochemical findings that odorants activate adenylate cyclase in frog olfactory cilia, and application of cyclic nucleotides to a patch of ciliary membrane increases the membrane conductance (Chapter 7). Removal of cilia from the frog olfactory epithelium with a detergent, Triton X-100, decreases EOG responses (Adamek et al., 1984). However, in a recent study with the carp, *Cyprinus carpio*, nearly normal olfactory bulbar responses to amino acids were maintained after deciliation by the use of an 'ethanol–calcium shock' method (Kashiwayanagi et al., 1988). This result suggests that the cilia do not play an essential part in olfactory transduction in carp. Although ciliated receptor cells do respond to amino acids in rainbow trout, *Oncorhynclus mykiss*, embryos, microvillar sensory cells may also contribute to amino acid reception (Zielinski and Hara, 1988; Hara and Zielinski, 1989). It has been hypothesized that the microvillar form contains high-affinity amino acid receptor sites, and the ciliated form, the low-affinity sites. The ability of microvillar sensory cells to respond to amino acid stimulation has been suggested in salmonid species (Thommesen, 1983). The data presented by Kashiwayanagi et al. (1988) showed that the lower-affinity portion of the concentration–response curve for L-alanine was significantly decreased, but the higher-affinity portion remained almost intact after deciliation.

8.3 SIGNAL TRANSMISSION

Electrical activities of the olfactory bulb

Electrical signals evoked in the olfactory neurones in response to chemical stimuli are transmitted through the olfactory nerve fibres, the centrally directed axons of the neurones, to the first relay station, the olfactory bulb. The olfactory nerve synapses with second-order bulbar neurones, mitral cells, as glomeruli. The axon of the olfactory neurone does not terminate in more than one glomerulus, and each glomerulus receives impulses only from a limited group of olfactory sensory cells. Since the total numbers of the sensory neurones and mitral cells in average teleosts are estimated as 10^7 and 10^4, respectively (Döving and Gemne, 1965; Yamamoto, 1982), a convergence ratio approximates 1000:1. This large convergence ratio is implicated in higher response probability (van Drongelen et al., 1978). A single mitral cell of fishes has from one to five dendrites, each of which ends in one or more glomeruli (Allison, 1953; Nieuwenhuys, 1967). This is remarkably different from the mammalian mitral cells, in which only a single main dendrite ends in each glomerulus.

Rhythmic oscillatory responses are recorded from the surface of the olfactory bulb (induced waves) when the nares are infused with water containing odorous chemicals. The induced waves, also termed electroencephalographic (EEG) responses, are interpreted as being produced in the dendritic network within the glomeruli as the result of synchronous activity of the secondary bulbar neurones (Ottoson, 1959a,b; Hara, 1975;

Thommesen, 1978). The induced waves mainly reflect the extracellular current caused by the synaptic depolarization of granule cell dendrites, and the rhythm of the induced wave is due primarily to the dendrodendritic synaptic interaction between the mitral and granule cells (Yamaguchi et al., 1988). The oscillatory response is superimposed upon the slow DC potential shift, when recorded through a DC-coupled preamplifier (Ottoson, 1959b; Thommesen, 1978). The responses induced by a given stimulus differ in magnitude at different regions of the bulb (Thommesen, 1978; Hara, 1982b), suggesting that specific populations of receptor neurones in the epithelium project to specific regions of the bulb (see also Chapter 3). A high correlation exists among the olfactory epithelial (EOG), extracellular single unit (olfactory nerve twig), and olfactory bulbar (EEG) responses, when examined in the same species (Silver, 1982; Caprio et al., 1989).

Mitral cells discharge spontaneously, when single unit recordings are obtained. An average discharge rate is about 2–6 impulses per second. A majority of cells respond to chemical stimulation by increasing firing rate, which results in a rhythmic discharge when the firing rate becomes greater than 5 impulses per second (Yamaguchi and Ueda, 1984). This rhythmic mitral cell discharge is often synchronized with the rhythmic induced waves, suggesting a common underlying neural mechanism. The synchronized rhythmic discharge of the mitral cell populations will provide a strong input to the higher olfactory centre and may play an important role in the intensity coding for olfactory processing.

Electrical activities of the olfactory tract

More than 50 years ago, Adrian and Ludwig (1938) recorded electrical activities of the olfactory tract in response to natural chemical stimulation of the olfactory organ in catfish, carp, and tench, *Tinca tinca*. This marked the very first study of the vertebrate olfactory system using an electrophysiological technique. They found that mechanical stimulation is an effective stimulus for the olfactory system, and emphasized the importance of solid particles in chemical stimulants. Boudreau (1962) also reported that mechanical stimulation of the olfactory organ consistently produces large spike discharges in the olfactory tract of catfish, *Ictarulus catus* and *I. melas*. Some chemicals (morpholine, acetic acid, and butanol) were effective at extremely low concentrations (10^{-13}–10^{-15}M). However, instability of many of his recordings makes their evaluation difficult.

Four functionally distinct olfactory tract fibre types are identified electrophysiologically in the gold crucian carp (*Carassius carassius grandoculis*): chemosensitive, thermosensitive, chemothermosensitive, and mechanosensitive (Sato and Suzuki, 1969). Two types of chemosensitive fibres further exist: phasic and tonic. One type of thermosensitive fibre is excited by warming and inhibited by cooling of the olfactory organ, and the other is excited by cooling and inhibited by warming. None of these thermosensitive fibres respond to chemical stimuli. Chemothermosensitive fibres respond to chemical stimuli only when accompanied by thermal changes (warming or cooling).

Mechanosensitive fibres, centrifugal in origin, have two types, one which is inhibited by tactile stimulation of the skin, or by photic stimulation, and the other which is stimulated by the tactile and chemical stimulation of the skin. These units are similar to the terminal nerve found in goldfish, *Carassius auratus* (Fujita *et al.*, 1991). Terminal ganglion cells, having extremely regular spontaneous firing rates, respond to tactile stimuli applied to the body surface by reducing their firing rate. No chemical stimulus is effective. Sensitivity of terminal nerve to tactile stimulation is implicated in their modulatory role associated with the physical interactions characterizing spawning (Fujita *et al.*, 1991). Involvement of both vibrational and visual cues in eliciting spawning behaviour in land-locked male sockeye salmon, *Oncorhynchus nerka*, has recently been reported (Satou *et al.*, 1987; Satou *et al.*, 1991).

8.4 SENSITIVITY AND SPECIFICITY

Aquatic odorants for fish

The aquatic environment surrounding fishes makes their olfaction unique. The entire process of olfaction takes place in water, therefore volatility of odorants is less relevant than it is for animals in air. Earlier studies on fish olfaction almost exclusively employed chemicals primarily odorous to humans. However, since Sutterlin and Sutterlin's (1971) and Suzuki and Tucker's (1971) pioneering works on amino acid stimulation of the olfactory receptors in Atlantic salmon, *Salmo salar*, and white catfish, *Ictalurus catus*, respectively, research on fish olfaction has centred on response characteristics to amino acids in a variety of species (reviews: Hara, 1982, 1986; Caprio, 1984, 1988). New groups of chemicals have recently been identified and their olfactory stimulatory characteristics for fish investigated. These include steroid hormones, bile salts, and prostaglandins (Fig. 8.2). In the following discussion, olfactory sensitivity and specificity of aquatic odorants for fishes will be described in terms of concentration–response and structure–activity relationships.

Thresholds: concentration–response relationships

Amino acids

The olfactory system of fish is capable of responding to low concentrations of amino acids. Thresholds obtained electrophysiologically from over 20 species, both fresh and sea water, lie between 10^{-7} and 10^{-9} M (Fig. 8.3; Table 8.1). Considering variabilities in experimental conditions, these figures

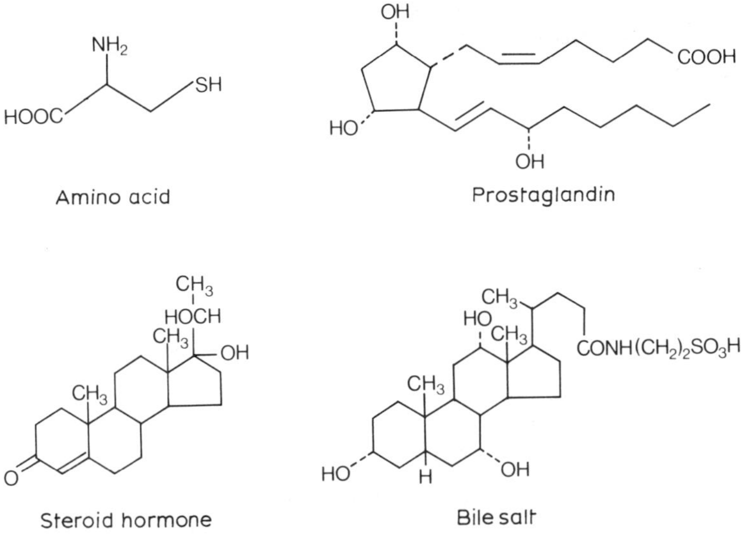

Fig. 8.2 General chemical structures of four major aquatic odorants for fish.

are extremely consistent throughout species and probably closely approximate levels of amino acids found in natural waters (Gardner and Lee, 1975). The threshold varies, depending upon the background level of amino acids; the more amino acids, the higher the apparent threshold. Olfactory responses to amino acids, whether recorded from the epithelium or from the nerve, increase in magnitude with higher concentrations. The concentration–response relationship has been variously described as exponential, sigmoidal, or as a power function (cf. Suzuki and Tucker, 1971; Caprio, 1978; Hara, 1982). However, over limited ranges of magnitude, all are reasonable approximations and the curves can be very similar. What is more important is their broad dynamic ranges of sensitivity, covering over 6–7 log units (Fig. 8.2). This represents a 1-to-10-million-fold change in amino acid concentration. The broad dynamic range of the concentration–response curve is most evident in EOG recordings from the olfactory epithelium. The wide dynamic range exhibited by the fish olfactory receptors may be the result of binding of amino acid molecules to several different populations of receptor sites, each with different binding affinities, or the result of negative cooperativity among receptor sites (Boeynaems and Dumont, 1980; Hara, 1982).

Steroids

Bile acids, especially those of taurine and sulphate conjugates, are potent olfactory stimuli for some fish species (Döving et al., 1980; Thommesen, 1983; Hara et al., 1984; Sorensen et al., 1987). Detection thresholds range between 10^{-10} and 10^{-8} M in salmonids (Fig. 8.3; Table 8.1). The most

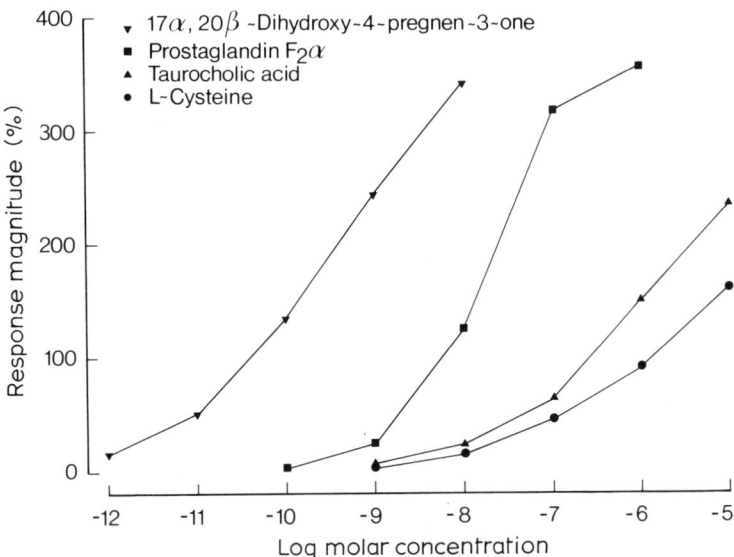

Fig. 8.3 Comparison of electroolfactogram (EOG) responses to 17α,20β-dihydroxy-4-pregnen-3-one, prostaglandin $F_{2\alpha}$, bile salt, and amino acid (L-cysteine). All responses are obtained from mature male goldfish. Response magnitude is represented as a percentage of that induced by standard amino acid L-serine at 10^{-5} M. (Adapted from Sorensen et al., 1987, 1988).

stimulatory bile salts tested are sulphotaurolithocholate, taurolithocholate, taurodeoxycholate, and taurochenodeoxycholate. Cholate, taurocholate and taurolithocholate are stimulatory for the channel catfish olfactory receptors, but only at high concentrations, 10^{-4} M and higher (Erickson and Caprio, 1984). EOG responses to bile salts are normally not affected by adaptation to high concentrations of amino acids, indicating the existence of separate receptor mechanisms for the two groups of chemicals (Hara et al., 1984; Sorensen et al., 1987).

Preovulatory goldfish release the oocyte maturational steroid hormone, 17α,20β-dihydroxy-4-pregnen-3-one (17,20P), to the surrounding water, where it functions as a pheromone. Water-borne 17,20P stimulates the endocrine system of mature males via their olfactory system to increase milt production by the time of spawning (Sorensen, Chapter 10 herein). Among 24 related steroid hormones tested for their EOG responses, 17,20P has the lowest detection threshold of 10^{-13}–10^{-12} M (Fig. 8.3; Table 8.1; Sorensen et al., 1987, 1990). This is one of the lowest olfactory threshold ever recorded for fish using a modern electrophysiological technique. Other stimulatory steroid hormones include 17α,20β,21-triol-4-pregnen-3-one, progesterone, and 17α-hydroxy-4-pregnen-3-one. EOG response magnitude for 17,20P increases sharply with an increase in concentration, and at 10^{-8} M responses reach 2–3 times that evoked by a 10^{-5} M L-serine standard. The relationship

Table 8.1 Olfactory thresholds for four major groups of aquatic odorants (amino acids, bile salts, steroid hormones, and prostaglandins) for fish, determined electrophysiologically

Species	Threshold (M)			Source
	Amino acids	Steroids	Prostaglandins	
Myxine glutinosa	10^{-6}–10^{-5}	10^{-6}–10^{-5}		Døving and Holmberg (1974)
Negaprion brevirostris	10^{-8}–10^{-7}			Zeiske et al. (1986)
Dasyatis sabina	10^{-8}–10^{-6}			Silver (1979)
Salmo salar	10^{-9}–10^{-5}			Sutterlin and Sutterlin (1971)
Salvelinus fontinalis	10^{-8}–10^{-7}			Hara et al. (1973)
Salvelinus alpinus	10^{-9}–10^{-7}	10^{-9}–10^{-8}	10^{-11}–10^{-9}	Belghaug and Døving (1977), Døving et al. (1980), Sveinsson (unpublished)
Salvelinus namaycush	10^{-8}–10^{-7}	10^{-9}–10^{-8}		Hara et al. (1989), Hara (unpublished)
Oncorhynchus mykiss (*Salmo gairdneri*)	10^{-8}–10^{-7}	10^{-10}–10^{-9}		Hara (1973), Hara et al. (1984)
Oncorhynchus nerka	10^{-7}–10^{-6}			Hara (1972)
Oncorhynchus kisutch	10^{-7}–10^{-6}			Hara (1972)
Coregonus clupeaformis	10^{-7}–10^{-6}			Hara et al. (1973)
Ictalurus catus	10^{-9}–10^{-7}			Suzuki and Tucker (1971), Caprio (1980)
Ictalurus punctatus	10^{-9}–10^{-7}	10^{-5}–10^{-4}		Caprio (1978), Erickson and Caprio (1984)
Ictalurus serracanthus	10^{-9}–10^{-7}			Caprio (1980)
Clarias gariepinus		10^{-11}–10^{-10}		Resink et al. (1989)
Cyprinus carpio	10^{-9}–10^{-7}			Goh and Tamura (1978), Ohno et al. (1984)
Carassius auratus	10^{-9}–10^{-8}	10^{-13}–10^{-11}	10^{-12}–10^{-10}	Sorensen et al. (1987, 1988)
Anguilla rostrata	10^{-9}–10^{-7}			Silver (1982)
Misgurnus anguillicaudatus			10^{-13}–10^{-10}	Kitamura and Ogata (1990)
Chrysophyrys major	10^{-7}			Goh et al. (1979), Goh and Tamura (1980)
Mugil cephalus	10^{-7}			Goh and Tamura (1980)
Seriola quinqueradiata	10^{-7}			Kobayashi and Fujiwara (1987)
Pagrus major	10^{-7}–10^{-6}			Kobayashi and Goh (1985)
Conger myriaster	10^{-9}–10^{-8}			Goh et al. (1979)

between 17,20P concentration and response magnitude is generally sigmoidal, approaching saturation at 10^{-6} M and higher. The olfactory system of female African catfish, *Clarias gariepinus*, is sensitive to steroid glucuronides including 5β-pregnen-3α,17β-diol-20-one, 5β-androstane-3α,17β-diol, etiocholanolone, 5β-dihydrotestosterone, testosterone, and 5β-androstane-3β-ol-17-one. The threshold concentration for 5β-pregnen-3α,17α-diol-20-one-3α-glucuronide, the most stimulatory steroid glucuronide tested, is approximately 10^{-11} M when recorded by EOG (Resink *et al.*, 1989). 17,20P does not elicit EOG responses in African catfish. Steroid glucuronides, on the other hand, are not effective olfactory stimuli for goldfish (Sorensen *et al.*, 1987). None of the sex steroids tested for the goldfish is stimulatory for the olfactory system of rainbow trout, *O. mykiss*, and Arctic charr, *Salvelinus alpinus* (Sveinsson and Hara, unpublished), suggesting high species specificity for olfactory detection of steroid hormones by fishes.

Prostaglandins

The olfactory epithelium of male goldfish is acutely sensitive to F-type prostaglandins (PGFs), postovulatory pheromones that stimulate male sexual arousal (Chapter 10). $PGF_{2\alpha}$ and its metabolite 15-keto-$PGF_{2\alpha}$ are detected at a threshold of approximately 10^{-10} M (Fig. 8.3; and Table 8.1; Sorensen *et al.*, 1988). At 10^{-6} M, where response magnitude starts to saturate, $PGF_{2\alpha}$ elicits responses three times those elicited by 10^{-5} M L-serine (Fig. 8.2). $PGF_{1\alpha}$ (having an extra double bond), $PGF_{3\alpha}$ (lacking a double bond), and PGE_2 (having a ketone group, instead of $-OH$) show curves similar in shape to that for $PGF_{2\alpha}$ but shifted to the right by one log unit. Responses to L-serine, taurocholate, and 17,20P are not reduced by adaptation to either $PGF_{2\alpha}$ or 15-keto-$PGF_{2\alpha}$, or vice versa, suggesting a separate receptor mechanism for PGs. High sensitivity of olfactory receptors to PGs is also demonstrated in Arctic charr and lake charr, *Salvelinus mamaycush* (Sveinsson and Hara, unpublished).

Specificity: structure–activity relationships

Amino acids

Olfactory responses to amino acids are not species specific. Generally, L-α-amino acids containing unbranched and uncharged side chains are the most effective olfactory stimuli for fish (Hara, 1982; Caprio, 1984). Ionically charged α-amino and α-carboxyl groups appear essential for effective stimulation in the majority of cases; acetylation of the former or esterification of the latter eliminates or significantly reduces activity. The free α-hydrogen atom is essential. Peptide bonds eliminate the stimulatory effectiveness of amino acids. This high stereospecificity suggests the existence of an amino acid receptor site with a rigid molecular framework involving two charged subsites, one anionic and one cationic, capable of interacting with ionized amino and carboxyl groups of the stimulant amino acid molecules (Hara, 1982). For the basic amino acids, a basic terminal, instead of α-carboxyl, is required for effective stimulation. L-Proline is the only natural amino acid that has no or little olfactory stimulatory effectiveness, but may have extreme importance in gustation in most fish species examined. L-Proline is one of the most stimulatory amino acids for gustation in all species examined and some salmonids seem to have gustatory amino acid receptors only for L-proline and its derivatives.

Steroids

Unlike amino acid receptors, olfactory receptors which detect the sex pheromone 17,20P in goldfish seem extremely specific. Of the 24 closely related steroid hormones (C_{21} steroids) tested, only 17,20P and 17α,20β,21P evoke EOG responses at 10^{-8} M. Although the receptor mechanisms responding to 17,20P recognize most aspects of the steroid molecule, an OH group at 17α or 20β position and a double bond between carbons 4 and 5 seem very important for effective stimulation. Addition of extra hydroxyl groups to 17,20P reduces its effectiveness dramatically. Of the nearly two dozen variations on 17,20P tested, only one, the addition of an OH to carbon 21 (17α,20β,21P), does not diminish olfactory potency. Receptor site specificity is further indicated by cross-adaptation experiments in which exposure to 10^{-8} M 17,20P has little effect on the magnitude of EOG responses elicited by L-serine and taurocholate. Thus, 17,20P and related hormones may be detected by receptors differing from those for amino acids and bile salts.

So far, taurolithocholic sulphate is the most stimulatory bile salt for the fish olfactory system. All the evidence from structure–activity relationship studies in rainbow trout and Arctic charr indicates that there are two types of receptor mechanisms for bile salts: one specific for bile salts with sulphate group at position 3 and taurine at position 23, and another specific for those with hydroxyl groups at positions 3 and 7 or 12, and taurine or glycine at position 23 (Sveinsson and Hara, unpublished).

Prostaglandins

Minor structural changes in $PGF_{2\alpha}$ molecules alter their stimulatory effectiveness drastically (Sorensen et al., 1988). Addition ($PGF_{1\alpha}$) or lack ($PGF_{3\alpha}$) of one double bond and substitution of hydroxyl group with ketone group (PGE_2) reduce effectiveness significantly. PGF_1, differing from $PGF_{2\alpha}$ at two positions, is far less stimulatory. A structural isomer of $PGF_{2\alpha}$, 11β-$PGF_{2\alpha}$, is also considerably less stimulatory than $PGF_{2\alpha}$. Again, responses to L-serine, taurocholate, and 17,20P are not reduced by adaptation to either $PGF_{2\alpha}$ or 15-keto-$PGF_{2\alpha}$, or vice versa, suggesting a separate receptor mechanism for PGs.

Multiplicity of receptor types

Amino acids

All the experimental evidence suggests that multiple receptor mechanisms exist for amino acid detection in fish olfaction. Under the circumstances where single unit recordings from olfactory neurones are difficult owing to the small size of the neurones, cross-adaptation experiments using multiunit receptor recordings provide an alternative method for isolating or eliminating receptor types specific for a given stimulus (Sutterlin and Sutterlin, 1971; Hara, 1982; Thommesen, 1982; Caprio and Byrd, 1984; Ohno et al., 1984). In cross-adaptation, chemoreceptors are exposed to one chemical stimulus (adapting stimulus) for a prolonged period of time. The receptors for that particular stimulus become adapted and less responsive, while other types of receptors remain unaffected. The degree of cross-adaptation, or inhibition, between stimuli indicates the extent to which they compete for the same receptor mechanisms. However, cross-adaptation utilizing multi-unit olfactory receptor recordings can only provide evidence for the possible different types of 'receptors' and/or 'transduction mechanisms' present within the epithelium. Whether they exist in different neurones (i.e. individual neuronal chemospecificity) still requires testing by unit recording.

Four relatively independent L-α-amino acid receptor sites have been identified in the channel catfish olfactory epithelium (Caprio and Byrd, 1984), specific for: (1) acidic amino acids (aspartate, glutamate), (2) basic amino acids (arginine, lysine), (3) short-chained neutral amino acids (glycine, alanine, serine), and (4) long-chained neutral amino acids (methionine, valine, leucine). This is consistent in principle with the earlier hypothesis, based on olfactory bulbar recordings, that the rainbow trout olfactory system contains at least three receptor types (Hara, 1982). All amino acid receptors, except for those involved in detection of basic amino acids, seem largely non-exclusive. Therefore, in cross-adaptation, all adapting amino acids affect, to varying degrees, the olfactory responses to the test amino acids. In addition, even when two amino acids are matched in intensity, their reciprocal cross-adapting effects may often be asymmetrical. For this reason, it is important in cross-adaptation to examine responses to the test stimulus over a wide concentration range at various concentrations of adapting stimulus.

Multiplicity of receptors for amino acids has been further characterized in Arctic charr, using kinetic analysis and non-linear regression model fitting of concentration–response curves of olfactory nerve-twig recording (NTR) responses (Sveinsson and Hara, 1990a). In these analyses, the interaction between the stimulus molecule, SM, and the receptor molecule, RM, is assumed to obey the principle of mass action (Beidler, 1954). The complex initiates a series of events leading eventually to the observed biological response, Q:

$$RM + SM \underset{k_{-1}}{\overset{k_1}{\rightleftharpoons}} RM\text{-}SM \text{ complex} \rightarrow Q$$

If all the receptor sites are involved in the initiation of a response, and the response measured is linearly related to the number of stimulus molecules bound, then:

$$Q = Q_{max} S/K_d + S \tag{8.1}$$

where K_d is the dissociation constant and Q_{max} is maximum response when stimulus concentration, S, is very high. When receptors are cross-adapted, the concentration–response curves are normally shifted in non-uniform fashion. These results are only predictable if the multiple receptor type model is considered. Analyses indicate that each of the more stimulatory amino acids interacts with two or more types of receptors with different affinities. A concentration–response curve forms a compound sigmoidal curve consisting of a series of component sigmoidal curves added on top of one another. Under the multiple receptor situation, the response of each type to a given stimulus is assumed to be independent of the others, thus:

$$Q = Q_1 + Q_2 + Q_3 + \cdots + Q_n$$
$$= (Q_{max1} S/K_{d1} + S) + (Q_{max2} S/K_{d2} + S) + \cdots + (Q_{maxn} S/K_{dn} + S) \tag{8.2}$$

If the relative independence of the respective binding sites of the component stimuli are known from cross-adaptation experiments, responses to amino acid mixtures may be predicted (Caprio et al., 1989). When binary or trinary mixtures of amino acids are tested, the magnitude of the response to a mixture whose component amino acids show significant cross-reactivity is equivalent to the response to any single component used to form that mixture. A mixture whose component amino acids show minimum cross-adaptation produces a significantly larger response than a mixture whose components exhibit considerable cross-reactivity.

Steroid hormones

Electrophysiological recordings (EOG) suggest that the olfactory system of the goldfish detects the preovulatory female sex pheromone 17,20P in a more specific manner than amino acids (Sorensen et al., 1987, 1990). The concentration–response curves for C_{21} steroids appear to approach the same asymptote and parallel each other, with the response of those least effective compounds being shifted furthest to the right. Adaptation to 10^{-8}M 17,20P completely inhibits responses to 17P and $17\alpha,20\beta,21$P. This level of receptor specificity has not previously been reported for a fish olfactory system. Because responses to androstendione, a C_{19} steroid, are not inhibited by adaptation to 17,20P, there may be another class of olfactory receptor sites which detect this steroid.

Prostaglandins

Goldfish appear to possess at least two classes of olfactory receptors for PGFs, one specific for $PGF_{2\alpha}$ and the other highly specific for 15-keto-$PGF_{2\alpha}$ (Sorensen et al., 1988). EOG responses to $PGF_{1\alpha}$, $PGF_{3\alpha}$, and PGE_2 are abolished during adaptation to equal concentration of either $PGF_{2\alpha}$ or 15-keto-$PGF_{2\alpha}$. EOG responses to 15-keto-$PGF_{2\alpha}$ approach saturation at a concentration of 10^{-8} M, where they approximate those evoked by $PGF_{2\alpha}$. At this equi-concentration, adaptation never eliminates responses to the other PGFs, suggesting the existence of separate olfactory receptors for $PGF_{2\alpha}$ and 15-keto-$PGF_{2\alpha}$.

8.5 NEURAL CODING

Existence of olfactory neurone types

Based primarily on cross-adaptation experiments, four groups of olfactory receptor populations, or transduction mechanisms, have been identified in fishes, each interacting exclusively with amino acids, bile salts, steroid hormones, and prostaglandins. Adapting the receptors to any one of the groups does not have a significant neural effect on the others, implying independent coding mechanisms. Within each group, a distinct molecular structure and binding mechanism seems responsible for transduction of each chemical. Although the existence of several receptor subtypes within the groups seems also evident, further work will be needed before those receptor subtypes are unequivocally assigned to be true receptor sites (Chapter 7).

Chemospecificity of individual olfactory neurones, i.e. distribution of receptor sites per olfactory neurone is not at all clear at present. It is generally agreed that the teleost olfactory epithelium contains two types of sensory cell, ciliated and microvillar (Zeiske et al., Chapter 2 herein). Nevertheless, their functional differentiation is not fully understood. In their electron microscopical and electrophysiological studies, Zielinski and Hara (1988) found that in rainbow trout, ciliated and microvillar sensory cells differentiate distinctly, and the former respond to amino acids at very early stages of their development. However, microvillar receptor cells may also contribute to amino acid reception, because they differentiate during the period of increased sensitivity of the olfactory mucosal neural responses. In fact, the ability of microvillar sensory neurones to respond to amino acid stimulation has been suggested (Thommesen, 1983).

In analogy with the vomeronasal sensory receptors in other vertebrates, microvillar sensory cells in fish olfactory organs may be responsible for pheromonal detection. Anatomically, vomeronasal sensory cells are located in the rostral end of the nasal region and possess microvilli (Wysocki and Meredith, 1987). During ontogeny, the ciliated sensory cells predominate in early stages of teleost development and occupy the rostral part of the olfactory epithelium. The origin of the vomeronasal organ remains obscure, but the segregated development of ciliated and microvillar sensory regions observed in early teleost development may reflect their prototypical origin. In adult carp, the medial bundle of the olfactory nerve is derived from the more rostral lamellae, while the lateral bundle arises from the more caudal lamellae. A diameter histogram of the olfactory nerve axon of burbot, Lota lota, clearly shows a bimodal distribution, indicating two distinct size groups (Gemne and Döving, 1969). However, the fibres of the two bundles often cross before they reach the bulb, so fibres from each reach all parts of the bulb (Sheldon, 1912).

Functional separation of the olfactory system

The fish olfactory system is not a uniform whole. Anatomical studies indicate that a clear topographical relationship exists throughout the olfactory epithelium, olfactory nerve, and olfactory bulb (Satou, Chapter 3 herein). Differential functioning of these anatomically distinct subdivisions of the olfactory system has been demonstrated in several fish species. Selective stimulation of the lateral part of the olfactory tract of cod, *Gadus morhua*, induces a behaviour pattern associated with feeding, and stimulation of the medial tract induces behaviour patterns involved in spawning (Döving and Selset, 1980). In goldfish, bilateral section of the olfactory tract reduces both sexual behaviour and feeding response. However, lateral olfactory section does not affect courtship behaviour, whereas medial olfactory section reduces courtship to the low level seen in goldfish with the whole tract sectioned (Stacey and Kyle, 1983). Furthermore, selective lesions of the area ventralis telencephali pars supracommissuralis and the posterior area ventralis telencephali pars ventralis severely impair male courtship and spawning behaviour (Kyle et al., 1982). Feeding behaviour evoked by a food odour remains unaffected by these lesions. Recent studies in goldfish further demonstrated that stimulation by hormonal pheromones 17,20P and prostaglandin $F_{2\alpha}$ is primarily mediated by the medial olfactory tract (Sorensen et al., 1991). The lateral olfactory tract appears to be largely responsible for conveying responses to amino acids. The specific role of amino acids has yet to be defined. The terminal nerve, running within the medial olfactory tract, may be responsible for pheromonal stimulation (Demski and Northcutt, 1983; Kyle et al., 1987); however, the firing rate of terminal nerve cell bodies in goldfish does not change in response to 17,20P and prostaglandin $F_{2\alpha}$ (Fujita et al., 1991).

8.6 SUMMARY AND CONCLUSION

The olfactory neurone of fish exhibits membrane characteristics similar to those determined in other vertebrate species. Unlike other excitable cells, the ionic concentrations across the olfactory epithelium are not determining factors of the resting membrane potential of the fish olfactory neurones. The injection of depolarizing electric current or the application of odorants to the olfactory neurones leads to firing of spike potentials. The firing frequency increases linearly with an increase in the current intensity or stimulant concentration. A hypothetical model for the olfactory transduction proposes that specific receptors in the receptive membrane zone, assumed to be located in the olfactory knob and/or cilia, bind to stimulus molecules. This results in the conductance change in the electrogenic zone, leading to production of a generator potential.

The entire process of fish olfaction takes place in water, therefore olfactory receptors are normally stimulated by water-soluble compounds. Four major chemical groups have been identified as potent odorants for fishes: amino acids, bile salts, sex hormones, and prostaglandins. Amino acids appear to be detected by at least three receptor sites, or transduction mechanisms, with varying affinity and specificity (generalist type), while the three other chemical groups are detected by respective receptors with high specificity (specialist type). Furthermore, multiple receptor subtypes with distinct binding characteristics exist within each chemical group. Structural and functional separation of the olfactory system exists in fishes. Information associated with reproductive behaviour is mediated through the medial olfactory tract, and stimulation by amino acids is mediated through the lateral olfactory tract.

ACKNOWLEDGEMENTS

I thank Scott Brown, Robert Evans, Dorthy Klaprat, Torarinn Sveinsson, and Chunbo Zhang for technical and Carol Catt for secretarial assistance. Supported by a Natural Science and Engineering Research Council of Canada grant (OGP0007576).

REFERENCES

Adamek, G.D., Gesteland, R.C., Mair, R.G. and Oakley, B. (1984) Transduction physiology of olfactory receptor cilia. *Brain Res. Amst.*, **310**, 87–97.

Adrian, E.D. and Ludwig, C. (1938) Nervous discharges from the olfactory organs of fish. *J. Physiol., Lond.*, **94**, 441–60.

Allison, A.C. (1953) The morphology of the olfactory system in the vertebrates. *Biol. Rev.*, **28**, 195–244.

Beidler, L.M. (1954) A theory of taste stimulation. *J. gen. Physiol.*, **38**, 133–9.

Belghaug, R. and Döving, K.B. (1977) Odour threshold determined by studies of the induced waves in the olfactory bulb of the char (*Salmo alpinus*). *Comp. Biochem. Physiol.*, **57A**, 327–30.

References

Boeynaems, J.M. and Dumont, J.E. (1980) *Outlines of Receptor Theory*, Elsevier/North-Holland, Amsterdam, 226 pp.

Boudreau, J.C. (1962) Electrical activity in the olfactory tract of the catfish. *Jap. J. Physiol.*, **12**, 272–8.

Caprio, J. (1978) Olfaction and taste in the channel catfish: an electrophysiological study of the responses to amino acids and derivatives. *J. comp. Physiol.*, **123A**, 357–71.

Caprio, J. (1980) Similarity of olfactory receptor responses (EOG) of freshwater and marine catfish to amino acids. *Can. J. Zool.*, **58**, 1778–84.

Caprio, J. (1984) Olfaction and taste in fish, in *Comparative Physiology of Sensory Systems* (eds L. Bolis, R.D. Keynes and S.H.P. Madrell), Cambridge University Press, Cambridge, pp. 257–83.

Caprio, J. (1988) Peripheral filters and chemoreceptor cells in fishes, in *Sensory Biology of Aquatic Animals* (eds J. Atema, R.R. Fay, A.N. Popper and W.N. Tavolga), Springer-Verlag, New York, pp. 313–38.

Caprio, J. and Byrd, R.P., jun. (1984) Electrophysiological evidence for acidic, basic, and neutral amino acid olfactory receptor sites in the catfish. *J. gen. Physiol.*, **84**, 403–22.

Caprio, J., Dudek, J. and Robinson, J.J. II (1989) Electro-olfactogram and multiunit olfactory receptor responses to binary and trinary mixtures of amino acids in the channel catfish, *Ictalurus punctatus*. *J. gen. Physiol.*, **93**, 245–62.

Demski, L.S. and Northcutt, R.G. (1983) The terminal nerve: a new chemosensory system in vertebrates? *Science, Wash., D.C.*, **202**, 435–7.

Döving, K.B. and Gemne, G. (1965) Electrophysiological and histological properties of the olfactory tract of the burbot (*Lota lota* L.). *J. Neurophysiol.*, **28**, 139–53.

Döving, K.B. and Holmberg, K. (1974) A note on the function of the olfactory organ of the hagfish *Myxine glutinosa*. *Acta physiol. scand.*, **91**, 430–2.

Döving, K.B. and Selset, R. (1980) Behavior patterns in cod released by electrical stimulation of olfactory tract bundlets. *Science, Wash., D.C.*, **207**, 559–60.

Döving, K.B., Selset, R., and Thommesen, G. (1980) Olfactory sensitivity to bile acids in salmonid fishes. *Acta physiol. scand.*, **108**, 123–31.

Drongelen, W. van, Holley, A. and Döving, K.B. (1978) Convergence in the olfactory system: quantitative aspects of odour sensitivity. *J. theor. Biol.* **71**, 39–48.

Erickson, J.R. and Caprio, J. (1984) The spatial distribution of ciliated and microvillous olfactory receptor neurons in the channel catfish is not matched by a differential specificity to amino acid and bile salt stimuli. *Chem. Senses*, **9**, 127–41.

Evans, R.E. and Hara, T.J. (1985) The characteristics of the electro-olfactogram (EOG): its recovery following olfactory nerve section in rainbow trout (*Salmo gairdneri*). *Brain Res. Amst.*, **330**, 65–75.

Fujita, I., Sorensen, P.W., Stacey, N.E. and Hara, T.J. (1991) The olfactory system not the terminal nerve, functions as the primary chemosensory pathway mediating responses to sex pheromones in male goldfish. *Brain Behav. Evol.*, **38**, 313–21.

Gardner, W.S. and Lee, G.F. (1975) The role of amino acids in the nitrogen cycle of Lake Mendota. *Limnol. Oceanogr.*, **20**, 379–88.

Gemne, G. and Döving, K.B. (1969) Ultrastructural properties of primary olfactory neurons in fish (*Lota lota* L.). *Am. J. Anat.*, **126**, 457–76.

Getchell, T.V. (1974) Electrogenic sources of slow voltage transients recorded from frog olfactory epithelium. *J. Neurophysiol.*, **37**, 1115–30.

Goh, Y. and Tamura, T. (1978) The electrical responses of the olfactory tract to amino acids in carp. *Bull. Jap. Soc. scient. Fish.*, **44**, 341–4.

Goh, Y. and Tamura, T. (1980) Olfactory and gustatory responses to amino acids in two marine teleosts–red sea bream and mullet. *Comp. Biochem. Physiol.*, **66C**, 217–24.

Goh, Y., Tamura, T. and Kobayashi, H. (1979) Olfactory responses to amino acids in marine teleosts. *Comp. Biochem. Physiol.*, **62A**, 863–8.

Hara, T.J. (1972) Electrical responses of the olfactory bulb of Pacific salmon *Oncorhynchus nerka* and *Oncorhynchus kisutch*. *J. Fish. Res. Bd Can.*, **29**, 1351–5.

Hara, T.J. (1973) Olfactory responses to amino acids in rainbow trout, *Salmo gairdneri*. *Comp. Biochem. Physiol.*, **44A**, 407–16.

Hara, T.J. (1975) Olfaction in fish, in *Progress in Neurobiology* (eds G.A. Kerkut and J.W. Phillis), Pergamon, Oxford, pp. 271–335.

Hara, T.J. (ed.) (1982a) *Chemoreception in Fishes*, Elsevier, Amsterdam, 334 pp.

Hara, T.J. (1982b) Structure–activity relationships of amino acids as olfactory stimuli, in *Chemoreception in Fishes* (ed. T.J. Hara), Elsevier, Amsterdam, pp. 135–58.

Hara, T.J. (1986) Role of olfaction in fish behaviour, in *The Behaviour of Teleost Fishes* (ed. T.J. Pitcher), Croom Helm, London, pp. 152–76.

Hara, T.J. and Zielinski, B. (1989) Structural and functional development of the olfactory organ in teleosts. *Trans. Am. Fish. Soc.*, **118**, 183–94.

Hara, T.J., Law, Y.M.C. and Hobden, B.R. (1973) Comparison of the olfactory response to amino acids in rainbow trout, brook trout, and whitefish. *Comp. Biochem. Physiol.*, **45A**, 969–77.

Hara, T.J., Macdonald, S., Evans, R.E., Marui, T. and Arai, S. (1984) Morpholine, bile acids and skin mucus as possible chemical cues in salmonid homing: electrophysiological re-evaluation, in *Mechanisms of Migration in Fishes* (eds J.D. McCleave, G.P. Arnold, J.J. Dodson and W.H. Neill), Plenum, New York, pp. 363–78.

Hara, T.J., Sveinsson, T., Evans, R.E. and Klaprat, D.A. (1989) Morphological and functional characteristics of the chemosensory organs of Canadian charr species. *Physiol. Ecol. Jpn*, Spec. Vol. **1**, 506.

Kashiwayanagi, M., Shoji, T. and Kurihara, K. (1988) Large olfactory responses of the carp after complete removal of olfactory cilia. *Biochem. biophys. Res. Commun.*, **154**, 437–42.

Kitamura, S. and Ogata, H. (1990) Olfactory responses of male loach, *Misgurnus anguillicadatus*, to F-type prostaglandins. *Taste and Smell*, **24**, 163–6.

Kobayashi, H. and Fujiwara, K. (1987) Olfactory response in the yellowtail *Seriola quinquerradiata*. *Nippon Suisan Gakkaishi*, **53**, 1717–25.

Kobayashi, H. and Goh, Y. (1985) Comparison of the olfactory responses to amino acids obtained from receptor and bulbar levels in a marine teleost. *Exp. Biol.*, **44**, 199–210.

Kyle, A.L., Stacey, N.E. and Peter, R.E. (1982) Ventral telencephalic lesions: effects on bisexual behaviour, activity, and olfaction in the male goldfish. *Behav. neural Biol.*, **36**, 229–41.

Kyle, A.L., Sorensen, P.W., Stacey, N.E. and Dulka, J.G. (1987) Medial olfactory tract pathways controlling sexual reflexes and behavior in teleosts, in *The Terminal Nerve* (nervus terminalis). *Structure, Function and Evolution* (eds L.S. Demski and M. Schwanzel-Fukuda), New York Academy of Science, New York, pp. 97–107.

Nakamura, T. and Gold, G.H. (1987) A cyclic nucleotide-gated conductance in olfactory receptor cilia. *Nature, Lond.*, **325**, 442–4.

Nieuwenhuys, R. (1967) Comparative anatomy of olfactory centres and tracts, in *Progress in Brain Research*. Vol. 23 (ed. Y. Zotterman), Elsevier, Amsterdam, pp. 1–64.

Ohno, T., Yoshii, K. and Kurihara, K. (1984) Multiple receptor type for amino acids in the carp olfactory cells revealed by quantitative cross-adaptation method. *Brain Res. Amst.*, **310**, 13–21.

Ottoson, D. (1956) Analysis of the electrical activity of the olfactory epithelium. *Acta physiol. scand.*, **35** (Supp. 122), 1–83.

Ottoson, D. (1959a) Studies on slow potentials in the rabbit's olfactory bulb and nasal mucosa. *Acta physiol. scand.*, **47**, 136–48.

Ottoson, D. (1959b) Comparison of slow potentials evoked in the frog's nasal mucosa and olfactory bulb by natural stimulation. *Acta physiol. scand.*, **47**, 149–59.

Ottoson, D. (1971) The electro-olfactogram, in *Handbook of Sensory Physiology*. Vol. 4, Part 1 (ed. L.M. Beidler), Springer-Verlag, Berlin, pp. 95–131.

Resink, J.W., Voorthuis, P.K., Hurk, R. van den, Peters, R.C. and Oordt, P.W.J. van (1989) Steroid glucuronides of the seminal vesicle as olfactory stimuli in African catfish, *Clarias gariepinus*. *Aquaculture*, **83**, 153–66.

Sato, Y. and Suzuki, N. (1969) Single unit analysis of the olfactory tract of the crucian carp. *J. Fac. Sci. Hokkaido Univ.*, Ser. 6, **17**, 208–23.

Satou, M., Shiraishi, A., Matsushima, T. and Okumoto, N. (1991) Vibrational communication during spawning behaviour in the himé salmon (landlocked red salmon, *Oncorhynchus nerka*). *J. comp. Physiol. A*, **168**, 417–28.

Satou, M., Takeuchi, H., Takei, K., Hasegawa, T., Okumoto, N. and Ueda, K. (1987) Involvement of vibrational and visual cues in eliciting spawning behaviour in male hime salmon (landlocked red salmon, *Oncorhynchus nerka*). *Anim. Behav.*, **35**, 1556–8.

Sheldon, R.E. (1912) The olfactory tracts and centers in teleosts. *J. comp. Neurol.*, **22**, 177–339.

Silver, W.L. (1979) Olfactory responses from a marine elasmobranch, the Atlantic stingray, *Dasyatis sabina*. *Mar. Behav. Physiol.*, **6**, 297–305.

Silver, W.L. (1982) Electrophysiological responses from the peripheral olfactory system of the American eel, *Anguilla rostrata*. *J. comp. Physiol.*, **148A**, 379–88.

Silver, W.L., Caprio, J., Blackwell, J.F. and Tucker, D. (1976) The underwater electro-olfactogram: a tool for the study of the sense of smell of marine fishes. *Experientia*, **32**, 1216–17.

Sorensen, P.W., Hara, T.J. and Stacey, N.E. (1987) Extreme olfactory sensitivity of mature and gonadally-regressed goldfish to a potent steroidal pheromone, $17\alpha,20\beta$-dihydroxy-4-pregnen-3-one. *J. comp. Physiol.*, **160A**, 305–13.

Sorensen, P.W., Hara, T.J. and Stacey, N.E. (1991) Sex pheromones selectively stimulate the medial olfactory tracts of male goldfish. *Brain Res. Amst.*, **558**, 343–7.

Sorensen, P.W., Hara, T.J., Stacey, N.E. and Dulka, J.G. (1990) Extreme olfactory specificity of male goldfish to the preovulatory steroidal pheromone $17\alpha,20\beta$-dihydroxy-4-pregnen-3-one. *J. comp. Physiol.*, **166A**, 373–83.

Sorensen, P.W., Hara, T.J., Stacey, N.E. and Goetz, F.W. (1988) F prostaglandins function as potent olfactory stimulants that comprise the postovulatory female sex pheromone in goldfish. *Biol. Reprod.*, **39**, 1039–50.

Stacey, N.E. and Kyle, A.L. (1983) Effects of olfactory tract lesions on sexual and feeding behavior in the goldfish. *Physiol. Behav.*, **30**, 621–8.

Sutterlin, A.M. and Sutterlin, N. (1971) Electrical responses of the olfactory epithelium of Atlantic salmon (*Salmo salar*). *J. Fish. Res. Bd Can.*, **28**, 565–72.

Suzuki, N. (1977) Intracellular responses of lamprey olfactory receptors to current and chemical stimulation, in *Food Intake and Chemical Senses* (eds Y. Katsuki, M. Sato, S.F. Takagi and Y. Oomura), University of Tokyo Press, Tokyo, pp. 13–22.

Suzuki, N. (1978) Effects of different ionic environments on the responses of single olfactory receptors in the lamprey. *Comp. Biochem. Physiol.*, **61A**, 461–7.

Suzuki, N. (1982) Responses of olfactory receptor cells to electrical and chemical stimulation, in *Chemoreception in Fishes* (ed. T.J. Hara), Elsevier, Amsterdam, pp. 93–108.

Suzuki, N. and Tucker, D. (1971) Amino acids as olfactory stimuli in freshwater catfish, *Ictalurus catus* (Linn.). *Comp. Biochem. Physiol.*, **40A**, 399–404.

Sveinsson, T. and Hara, T.J. (1990a) Analysis of olfactory responses to amino acids in Arctic char (*Salvelinus alpinus*) using a linear multiple-receptor model. *Comp. Biochem. Physiol.*, **97A**, 279–87.

Sveinsson, T. and Hara, T.J. (1990b) Multiple olfactory receptors for amino acids in Arctic char (*Salvelinus alpinus*) evidenced by cross-adaptation experiments. *Comp. Biochem. Physiol.*, **97A**, 289–93.

Thommesen, G. (1978) The spatial distribution of odour induced potentials in the olfactory bulb and char and trout (Salmonidae). *Acta physiol. scand.*, **102**, 205–17.

Thommesen, G. (1982) Specificity and distribution of receptor cells in the olfactory mucosa of char (*Salmo alpinus* L.). *Acta physiol. scand.*, **115**, 47–56.

Thommesen, G. (1983) Morphology, distribution, and specificity of olfactory receptor cells in salmonid fishes. *Acta physiol. scand.*, **117**, 241–9.

Wysocki, C.J. and Meredith, M. (1987) The vomeronasal system, in *Neurobiology of Taste and Smell* (eds T.E. Finger and W.L. Silver), Wiley-Interscience, New York, pp. 125–50.

Yamaguchi, K., Satou, M. and Ueda, K. (1988) Induced wave and its generation mechanism in the carp olfactory bulb. *Comp. Biochem. Physiol.*, **89A**, 605–8.

Yamaguchi, K. and Ueda, K. (1984) Rhythmic discharge of mitral cells in the carp olfactory bulb. *Brain Res. Amst.*, **322**, 378–81.

Yamamoto, M. (1982) Comparative morphology of the peripheral olfactory organ in teleosts, in *Chemoreception in Fishes* (ed. T.J. Hara), Elsevier, Amsterdam, pp. 39–60.

Zeiske, E., Caprio, J. and Gruber, S.H. (1986) Morphological and electrophysiological studies on the olfactory organ of the lemon shark, *Negaprion brevirostris* (Poey), in *Indo-Pacific Fish Biology: Proc. Second Int. Conf. Indo-Pacific Fishes* (eds T. Uyeno, R. Arai, T. Taniuchi and K. Mutsuura), Ichthyological Society of Japan, Tokyo, pp. 381–91.

Zielinski, B. and Hara, T.J. (1988) Morphological and physiological development of olfactory receptor cells in rainbow trout (*Salmo gairdneri*) embryos. *J. comp. Neurol.*, **271**, 300–11.

Chapter nine

Teleost gustation

Takayuki Marui and John Caprio

9.1 INTRODUCTION

Vertebrates, including fishes, possess two principal chemoreceptive systems, termed olfaction and gustation (taste), which are adapted to respond to specific chemical substances in the environment. In all vertebrates, chemical information that is detected and transmitted directly to the central nervous system (CNS) by bipolar neurones of cranial nerve I is termed olfaction, whereas chemical information detected by specialized epithelial cells (i.e. taste cells) and transmitted to the CNS by neurones of cranial nerve VII (facial), IX (glossopharyngeal) or X (vagus) is termed gustation. Receptor cells of both systems are required to discriminate relevant chemical stimuli from background chemical 'noise' existing in the environment of all organisms. Receptor molecules (most probably glycoproteins) that detect and preferentially pass biologically important information have evolved and have been positioned in the membranes of the receptor cells. These receptor molecules, upon being activated by their specific stimulus or stimuli, initiate a series of cellular molecular events that can result eventually in behavioural responses, such as food search and ingestion. Because both olfactory and gustatory systems in fishes are activated by water-soluble substances, it is often difficult to determine the specific role that each system plays in a particular behaviour; this difficulty adds to the confusion of taste/smell distinctions in fish. The present interpretation is that gustation is involved in the detection, selection and ingestion of food and protection against noxious (i.e. generally bitter) substances, and olfaction is involved in a generalized alerting response and is involved in possible specific pheromonal responses associated with fright, mating, spawning, territorial and homing behaviours.

Of the relatively few species of fishes studied with respect to their chemical senses, the Cyprinidae and Siluridae (including Ictaluridae) evolved some of the more highly developed taste systems of all vertebrates. Over the past 20

years, electrophysiological, biochemical and behavioural experiments clearly indicated that rather simple chemical compounds (e.g. amino acids and nucleotides) are particularly stimulatory to the chemosensory systems and appear to play important roles in inducing feeding in a number of aquatic organisms.

This chapter will focus on the physiology of the peripheral taste neurones in teleosts. For recent information on the molecular mechanisms of taste receptor transduction in teleosts see Brand and Bruch, Chapter 7 herein. The organization of gustatory neural pathways within the central nervous system of teleosts is reviewed by Kanwal and Finger, Chapter 5 herein.

9.2 HISTORICAL BACKGROUND OF GUSTATORY PERIPHERAL NERVE PHYSIOLOGY

The first electrophysiological recordings of taste activity in a teleost was obtained in an isolated head preparation from the facial/trigeminal nerve complex innervating the barbel of the North American bullhead catfish, *Ictalurus nebulosus* (formerly *Amiurus nebulosus*) (Hoagland, 1933). In this study, acetic acid, sodium chloride and meat juice evoked small potentials in the barbel nerve, whereas mechanoreceptive activity was appreciable. Since this pioneer work, a number of reports concerning the physiology of the taste system of teleosts have appeared (reviews: Bardach and Atema, 1971; Tucker, 1983; Caprio, 1982, 1984, 1988). For studies prior to the middle 1970s, most investigations of fish taste physiology utilized as test stimuli the four classical taste stimuli (sodium chloride, quinine hydrochloride, sucrose and a mineral or organic acid) used in tetrapods. The first investigation to suggest that the fish taste system is highly responsive to particular organic chemicals was performed by Konishi *et al.* (1966). They analysed single fibre activity of taste neurones in the sea catfish, *Plotosus lineatus* (formerly *P. anguillaris*) and found that some of the fibres that were not very responsive to the classical taste stimuli responded well to nereid worm extract, blood sera, human saliva, lecithin and betaine. This result suggested the existence of different fibre types based on chemical specificity. Bardach *et al.* (1967) examined the facial taste system of bullhead catfish (the yellow bullhead, *Ictalurus natalis*, and the brown bullhead, *I. nebulosus*) and tomcod, *Microgadus tomcod*, and indicated that amino acids were effective gustatory stimuli. A critical finding that olfactory receptors of Atlantic salmon, *Salmo salar* (Sutterlin and Sutterlin, 1971) and the white catfish, *I. catus* (Suzuki and Tucker, 1971) were highly sensitive to specific amino acids (estimated thresholds for the more stimulatory compounds ranged between 10^{-7} M and 10^{-9} M) was the additional impetus that resulted in the finding that facially innervated taste receptors on the maxillary barbel of the channel catfish, *I. punctatus*, were also highly sensitive (estimated thresholds $\leq 10^{-9}$ M) to specific amino acid stimuli

(Caprio, 1975, 1978). Subsequently, amino acids were shown to be potent stimuli to the facial taste systems of other species of teleosts (Fig. 9.1), although the taste specificities across the species could be very different (Table 9.1).

Although taste buds innervated by the facial nerve in catfish are broadly distributed, such that they are found within the rostral oral cavity, on the lips, barbels and flank, the sensitivities and specificities of these taste buds for amino acids are similar (Caprio, 1975, 1978; Davenport and Caprio, 1982; Kanwal et al., 1987). The oropharyngeal taste buds innervated by the glossopharyngeal and vagal nerves, although shown to have similar amino acid specificities, were generally less sensitive to specific amino acids by at least 1.5 to greater than 2 orders of magnitude than were facially innervated taste buds (Kanwal and Caprio, 1983). Thus, irrespective of the location of amino-acid-sensitive taste buds, the L-isomers of alanine and arginine are

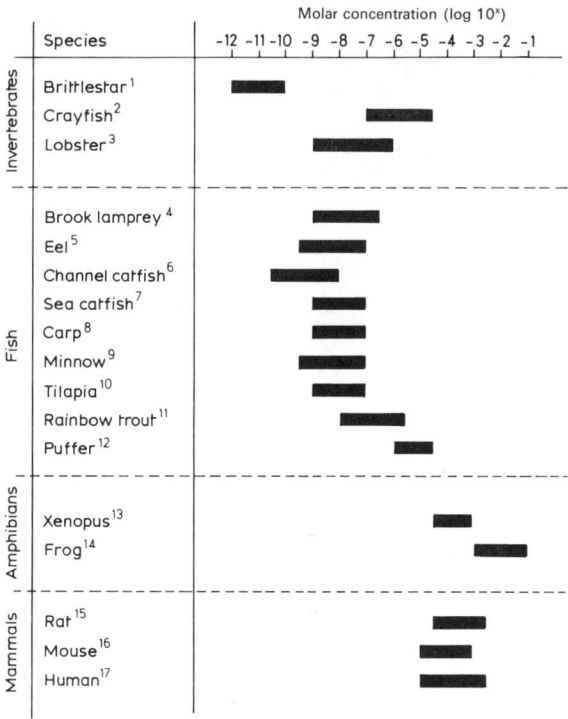

Fig. 9.1 Ranges of taste thresholds for the more potent amino acids, obtained electrophysiologically. Sources: 1, Moore and Cobb (1985); 2, Hatt (1984); 3, Derby and Atema (1982a); 4, Baatrup (1985); 5, Yoshii et al. (1979); 6, Caprio (1975); 7, Kohbara (unpublished); 8, Marui et al. (1983a); 9, Kiyohara et al. (1981); 10, Marui (unpublished); 11, Marui et al. (1983b); 12, Kiyohara et al. (1975); 13, Yoshii et al. (1982); 14, Gordon and Caprio (1985); 15, Harada et al. (1983); 16, Harada et al. (1982, 1983); 17, Schiffman et al. (1981).

Table 9.1 Grouping of fish according to the specificity of gustatory responses to amino acids*

Species	Stimulatory amino acids†	Concentration tested (mM)	Source
	Wide response range		
Channel catfish, Ictalurus punctatus	Neutral, basic, and acidic	0.1	Caprio (1975, 1978), Davenport and Caprio (1982), Kanwal and Caprio (1983)
Topmouth minnow, Pseudorasbora parva	Neutral, basic, and acidic	1.0	Kiyohara et al. (1981)
Tiger fish, Therapon oxyrhynchus	Neutral, basic, and acidic	0.1	Hidaka and Ishida (1985)
Rabbitfish, Siganus fuscescens	Neutral, basic, and acidic	0.1	Ishida and Hidaka (1987)
African cichlids, Tilapia nilotica	Neutral, basic, and acidic	1.0	Marui (unpublished)
Red sea bream, Chrysophrys major	Neutral and basic	10.0	Goh and Tamura (1980a)
Mullet, Mugil cephalus	Neutral and basic	10.0	Goh and Tamura (1980a)
Isaki grunt, Parapristipoma trilineatum	Neutral and basic	0.1	Ishida and Hidaka (1987)
Jack mackerel, Trachurus japonicus	Neutral and basic	10.0	Ishida and Hidaka (1987)
Chub mackerel, Scomber japonicus	Neutral and basic	10.0	Ishida and Hidaka (1987)

Species	Limited response range		Reference
Japanese eel, *Anguilla japonica*	Arg, Gly, Ala. Pro, Lys Ser	0.1	Yoshii et al. (1979)
Rainbow trout, *Oncorhynchus mykiss*‡	AGPA, Pro. Hpr, Bet, Leu, Ala, Phe	1.0	Marui et al. (1983b)
Common carp, *Cyprinus carpio*	Pro, Ala, CysH, Glu, Bet, Gly	1.0	Marui et al., (1983a)
Arctic charr, *Salvelinus alpinus*	Pro, Hpr, Ala, AGPA	1.0	Hara (pers. comm.)
Brook trout, *Salvelinus fontinalis*	Pro, Hpr, Ala, Phe, AGPA	1.0	Hara (pers. comm.)
Lake trout, *Salvelinus namaycush*	Pro, Hpr, Ala, Arg	1.0	Hara (pers. comm.)
Puffer, *Fugu pardalis*	Pro, Ala, Gly, Bet	10.0	Kiyohara et al. (1975)
Yellowtail, *Seriola quinqueradiata*	Pro, Bet, Try, Val, Ala	10.0	Hidaka et al. (1985)
Amberjack, *Seriola dumerili*	Pro, Try, Bet, Ala	10.0	Ishida and Hidaka (1987)

*Adapted from Hara and Zielinski (1989).
†Abbreviations: AGPA, L-α-amino-β-guanidino propionic acid; L-Ala, L-alanine; Arg, arginine; Asn, L-asparagine; Asp, L-aspartic acid; Bet, betaine; CysH, L-cysteine; Glu, L-glutamic acid; Gly, glycine; Hpr, hydroxy-L-proline; Leu, L-leucine; Met, methionine; Phe, L-phenylalanine; Pro, L-proline; Ser, L-serine; Try, L-tryptophan; Val, valine.
‡Formerly *Salmo gairdneri*.
Note: amino acids are listed in the order of relative effectiveness based solely on the mean magnitude of the integrated taste response to each stimulus.

highly effective stimuli, which suggests that similar amino acids may provide important cues for both food searching and ingestive behaviour in the channel catfish. The difference in sensitivity of oral and extra-oral taste buds appears to reflect the difference in the functions of the two (cranial nerves VII v. IX, X) taste systems in catfishes. The extraoral, facial taste system is important in food-searching behaviour and in the picking up of potential food material, while the glossopharyngeal/vagal, oral taste system is primarily involved in the final acceptance or rejection of these substances. Thus, in food search, high sensitivity of facially innervated taste buds is more beneficial to the organism than in the final decision over whether to accept or reject material already present in relatively high concentration within the oropharyngeal cavity. Accordingly, recent investigations have indicated different patterns of central neuronal connectivity for each system, suggesting differential processing of the gustatory inputs to the central nervous system (Finger and Morita, 1985; Morita and Finger, 1985; Kanwal and Finger, Chapter 5 herein). The only other work to have contrasted the sensitivities and specificities of taste buds innervated by different cranial nerves was performed by Sutterlin and Sutterlin (1970), who showed some differences in taste sensitivity to salts, acid and sugars from recordings from the facial and glossopharyngeal nerves in the Atlantic salmon, *Salmo salar*. Interestingly, there was an inverse relationship in the rank order of stimulatory effectiveness for the salts tested between the facial and glossopharyngeal nerves. No other reports exist for comparisons of taste responses across the different cranial nerves.

Most electrophysiological studies of fish taste responses to amino acids are based on whole nerve or nerve twig recordings, where the peak height of the integrated response of the multiunit activity is the measured response parameter. These electrophysiological recordings suggest that the gustatory response to amino acids is highly species specific (Table 9.1); in contrast, the olfactory response spectra to amino acids are relatively similar among the fishes tested (Hara, 1975; Goh et al., 1979; Caprio, 1984; Hara, Chapter 8 herein). Although different species of teleosts may have very different rank orders of effective taste stimuli, certain amino acids, such as L-alanine and L-proline, appear to be effective across the species studied (Table 9.1).

The dose–response (D–R) relationships for amino acids are described in the literature as power functions (channel catfish, Caprio, 1975; topmouth minnow, *Pseudorasbora parva*, Kiyohara et al., 1981), logarithmic functions (Japanese eel, *Anguilla japonica*, Yoshii et al., 1979; red sea bream, *Chrysophrys major*, Goh and Tamura, 1980a) or sigmoidal functions (common carp, *Cyprinus carpio* L., and rainbow trout, *Oncorhynchus mykiss*, Marui et al., 1983a,b). With increasing suprathreshold concentrations of a stimulus, the magnitude of the response will generally increase until saturation of the receptors occurs or the solubility limit of the stimulus is achieved. In some species the D–R functions for different amino acids are different (Marui et al., 1983b). Also, the choice of a particular mathematical function is due primarily to the range of stimulus concentrations the experimenter chooses for the function to best fit the experimental data (Caprio, 1984). For example, power and logarithmic functions do not truncate, but extend infinitely, which is certainly different from biological processes, such as the binding of chemical stimuli to chemoreceptor molecules. Thus, some D-R relations appear to fit a power function relationship over the range of the dynamic response, but because of response saturation occurring at high stimulus concentrations, a sigmoidal relationship is presented (Marui et al., 1983a,b). Also critical in the determination of the best fit function to represent the D-R relation is the duration of the interstimulus interval (ISI) between successive increases in stimulus concentration. As the concentration is raised, longer ISIs are required to allow for the maximal response of the system, owing to increasing effects of sensory adaptation. Thus, different mathematical relationships between dose and response may be obtained dependent solely upon the experimental ISI. Therefore, because of these variables, it is often difficult directly to compare stated D-R relations across species in an attempt to glean possible differences in mechanisms.

Electrophysiological thresholds determined from facial nerve recordings for the more stimulatory amino acids in the majority of the species of teleosts studied generally ranged between 10^{-6} M and 10^{-9} M (Caprio, 1984). Although there appear to be major sensitivity differences for amino acids between certain teleost species, such as the channel catfish (Caprio, 1978) and the puffer (Hidaka et al., 1976), some of the minor differences reported in taste thresholds may in part be due to the solvent and perfusate used by the different laboratories, e.g. distilled water, artificial pond or natural (tap) water, or artificial or natural seawater. Differences in background levels of amino acids in these waters would have profound effects on the experimentally determined thresholds.

9.3 RESPONSE FEATURES FOR SIMPLE CHEMICAL COMPOUNDS

Amino acids

The results of investigations concerning amino acid taste in teleosts indicate that taste receptor molecules have rather rigid requirements for the molecular structure of the stimulus molecules (Bryant et al., 1989; Bryant and Leftheris, 1991; Marui et al., 1983a,b; Wegert and Caprio, 1991). This indicates that taste receptors require specific conformations of stimuli to bind and activate the transduction process(es) within the taste receptor cell. For example, gustatory responses to amino acids in teleosts are highly stereospecific (Caprio, 1978; Yoshii et al., 1979; Kiyohara et al., 1981; Marui et al., 1983a,b, 1986; Marui and Kiyohara, 1987). The L-isomer of an amino acid has for most species of teleosts studied been shown to be more stimulatory than its D-enantiomer (Caprio, 1978; Yoshii et al., 1979; Kiyohara et al., 1981; Marui et al., 1983a,b, 1986; Marui and Kiyohara, 1987). However, recently reported data on the facial taste system of the sea catfish, *Arius felis*, revealed a group of taste fibres that were more responsive to D-alanine (Michel and Caprio, 1991; Section 9.5). It remains to be determined whether D>L amino acid sensitivity is more commonplace among marine than freshwater teleosts, or whether the taste system of the D-alanine fibre type of *Arius* is just peculiar to that species. Nevertheless, whether L- or D-amino acids are more stimulatory than their enantiomers, stereospecificity of the taste responses to amino acids is the general rule.

Unsubstituted, L-alpha amino acids are primarily the more stimulatory compounds to the taste systems of teleosts. The primary exceptions are: (1) betaine (N-trimethyl-glycine) is highly stimulatory to a number of marine (Ishida and Hidaka, 1987), but not freshwater, teleosts; (2) both alpha- and beta-alanine are equally stimulatory to palatal taste buds in the Japanese eel, *Anguilla japonica* (Yoshii et al., 1979), and (3) primary carboxyl esters of some amino acids may be (a) similarly stimulatory (Caprio, 1978), or (b) more stimulatory (Kiyohara et al., 1981), or (c) less stimulatory (Yoshii et al., 1979; Marui et al., 1983a; Bryant et al., 1989; Bryant and Leftheris, 1991), or (d) either more or less stimulatory, depending upon the specific ester (Marui et al., 1983b), than the parent free amino acid. However, a recent report (Bryant and Leftheris, 1991) suggested that the high level of gustatory neural activity in response to esters of L-alanine in the channel catfish (Caprio, 1978) was likely due to stimulation by free L-alanine which was released by alkaline hydrolysis of the esters. Unsubstituted, L-alpha amino acids are also more stimulatory than simple peptides. The rule that was established for the taste system of the channel catfish (Caprio, 1978), and for which no exceptions have yet been shown for any species of teleost tested (Kiyohara et al., 1981; Hidaka and Ishida, 1985; Marui and Kiyohara, 1987), is that a peptide is less effective than its most stimulatory amino acid residue; furthermore, the order of the amino acid residues in a simple peptide does not significantly affect its ability to stimulate the taste system (Caprio, 1978; Hidaka and Ishida, 1985; Marui and Kiyohara, 1987).

As indicated by Hara and Zielinski (1989), among the teleosts whose taste responses were examined systematically, two general groups were identified: (1) those whose facial taste systems respond to many different types of amino acids (i.e. 'wide response range' in Table 9.1), with varying stimulatory effectiveness depending upon the particular amino acid tested, and (2) those whose facial taste systems are highly selective and respond to few of the amino acids tested (i.e. 'limited response range'). So far, of 19 species whose facial taste systems have been tested with amino acids, 9 species consisting of both freshwater and marine organisms are indicated to be of the limited response range (Table 9.1). In general, regardless of whether fish had taste systems with wide or limited response ranges, neutral amino acids containing two or fewer carbon atoms having unbranched and uncharged side chains and the imino acid, L-proline, were highly stimulatory. With few exceptions, acidic amino acids were poor gustatory stimuli. Taste responses to the basic amino acids, generally L-arginine, were quite variable depending upon the species. In the channel catfish (Caprio, 1975), the Japanese eel (Yoshii et al., 1979), the red sea bream, and mullet, *Mugil cephalus* (Goh and Tamura, 1980a), the topmouth minnow (Kiyohara et al., 1981), the maaji jack mackerel, *Trachurus japonicus* (Ishida and Hidaka, 1987), and an African cichlid, *Tilapia nilotica* (Marui, unpublished), at least one of the basic amino acids was highly effective. In other species, such as the common carp (Marui et al., 1983a), the rainbow trout (Marui et al., 1983b), the brook charr, *Salvelinus fontinalis* (Hara, unpublished), the aigo rabbitfish, *Siganus fuscescens* (Ishida and Hidaka, 1987) and the isaki grunt, *Parapristipoma trilineatum* (Ishida and Hidaka, 1987), the basic amino acids were non-stimulatory or resulted in minimal taste activity. It was reported (Marui et al., 1983b), however, for the rainbow trout that L-arginine and its derivatives became quite active at basic pH (> 8.5). It is most probable that the differences in amino acid specificity noted above for the different teleosts is related to the particular feeding niche of the respective species.

Aliphatic acids

Aliphatic acids have also been shown to be stimulatory to the taste system of the Japanese eel (Yoshii et al., 1979), the Atlantic salmon (Sutterlin and Sutterlin, 1970) and the common carp (Marui, unpublished; Table 9.2). The magnitude of the integrated taste response of the eel is greater for carboxylic acids than for amino acids at 1 mM. The gustatory threshold for carboxylic acids on the palatine taste buds of the eel ranged between 100 nM and 0.1 mM, and the D–R curves for some carboxylic acids were steeper than for the corresponding amino acids. Treatment of the taste epithelium of the eel with papain (a protease used to inactivate membrane bound taste receptor proteins) caused taste responses to three of four amino acids tested to be eliminated, but this treatment had little effect on taste activity to acetic acid. Also, there was little cross-adaptation between amino acids and carboxylic acids having the same carbon chain length. These last two results suggested that carboxylic acids stimulate receptors independent from those that bind amino acids.

The magnitude of the taste responses of the Atlantic salmon and Japanese eel generally increased with increasing carbon chain length of the aliphatic acid. In the carp, both aliphatic and aromatic carboxylic acids stimulated the facial taste system (Table 9.2). Although compounds consisting of two carbon atoms were more stimulatory than those containing one carbon atom for both groups of acids, the taste responses to the monocarboxylic acids peaked at compounds containing between three and five carbons,

Table 9.2 Relative stimulatory effectiveness of carboxylic acids tested at 1 mM on the taste receptors of carp, *Cyprinus carpio*, as a percentage of the response to the standard, 1 mM L-alanine

Chemicals*	Structural formula	Stimulatory effectiveness (%) (mean ± SD)	Number of tests
Aliphatic carboxylic acid			
Monocarboxylic acid			
Formic acid	HCOOH	22.5 ± 12.2	5
Acetic acid	CH_3COOH	97.4 ± 24.4	7
Propionic acid	CH_3CH_2COOH	128.0 ± 18.7	7
n-Butyric acid	$CH_3(CH_2)_2COOH$	113.8 ± 18.9	4
n-Valeric acid	$CH_3(CH_2)_3COOH$	135.6 ± 9.8	5
Caproic acid	$CH_3(CH_2)_4COOH$	99.8 ± 41.7	4
Dicarboxylic acid			
Oxalic acid	HOOCCOOH	0	3
Malonic acid	$HOOCCH_2COOH$	93.2 ± 46.4	2
Succinic acid	$HOOC(CH_2)_2COOH$	259.2 ± 40.8	2
Glutaric acid	$HOOC(CH_2)_3COOH$	350.6 ± 86.9	2
Adipic acid	$HOOC(CH_2)_4COOH$	308.5 ± 54.0	2
Pimelic acid	$HOOC(CH_2)_5COOH$	128.4 ± 46.0	2
Tartaric acid	HOOCCHOHCHOHCOOH	59.0 ± 28.2	6
Tricarboxylic acid			
Citric acid	$HOOCCH_2C(OH)(COOH)CH_2COOH$	245.1 ± 72.6	4
Aromatic carboxylic acid			
Benzoic acid	C_6H_5COOH	32.5 ± 7.8	2
Salicylic acid	HOC_6H_4COOH	5.0 ± 7.0	2
Phthalic acid	$C_6H_4(COOH)_2$	25.0 ± 2.8	2

*All chemicals were adjusted in neutral range (pH 6.8–7.1); see text for explanation.

whereas for dicarboxylic acids, taste responses were maximal for compounds containing five to six carbon atoms. It is important to note that all stimuli in Table 9.2 were pH adjusted with 0.1 N sodium hydroxide and that 1 mM sodium chloride was not an effective taste stimulus in the carp (Marui et al., 1983a); therefore, the taste activity recorded in response to the compounds listed was due primarily to the carboxylic anions rather than to the sodium ion.

Nucleotides

Kiyohara et al. (1975) first indicated that nucleotides might be important gustatory stimuli to teleosts. Nucleotides, such as adenosine-5'-monophosphate (AMP), inosine-5'-monophosphate (IMP), uridine-5'-monophosphate (UMP), and adenosine-5'-diphosphate (ADP), were stimulatory to the lip taste buds of the puffer, *Fugu pardalis*. Single fibre analyses revealed that facial taste fibres in the puffer were divided into three groups: (1) those responding well to hydrochloric acid, (2) those responding well to nucleotides, and (3) those responding well to amino acids. Also, a preliminary study in the channel catfish (Michel et al., 1987) implied that nucleotide taste information is transmitted to the central nervous system by fibres that are not especially sensitive to amino acids. Since the work on the puffer, nucleotides have been found to be excitatory gustatory stimulants for the following species of teleosts: oriental weatherfish, *Misgurnus anguillicaudatus* (Harada, 1986); yellowtail, *Seriola quinqueradiata* (Harada, 1986; Hidaka et al., 1985); amberjack, *Seriola dumerili* (Ishida and Hidaka, 1987); jack mackerel, *Trachurus japonicus* and club mackerel, *Scomber japonicus* (Ishida and Hidaka, 1987); rabbit fish, *Siganus fuscescens* (Ishida and Hidaka, 1987); tiger fish, *Therapon oxyrhynchus* (Hidaka and Ishida, 1985); grunt, *Parapristipoma trilineatum* (Ishida and Hidaka, 1987); carp (Marui, unpublished), and channel catfish (Littleton et al., 1989). The ability to detect nucleotides is not limited to the taste system of fishes, but has also been found for the olfactory system of the channel catfish (Michel et al., 1987) and for gustatory and olfactory systems of crustaceans, such as shrimp, *Palaemonetes pugio* (Carr and Thompson, 1983; Carr and Derby, 1986a,b), lobster, *Homarus americanus* (Derby and Atema, 1982a,b), crayfish, *Austropotamobius torrentium* (Hatt, 1984) and spiny lobster, *Panulirus argus* (Derby et al., 1984; Carr et al., 1986). Interestingly, the threshold range for nucleotides in the taste system of teleosts (10–100 μM) is similar to that reported for the rat (Yoshii et al., 1986), and is significantly higher than that determined for amino acids in the facial taste systems of fishes.

Bile salts

The detection of bile salts, which are potent gustatory and olfactory stimuli to salmonids, is proposed to be important in homing migration (Döving et al.,

1980; Selset and Döving, 1980). These compounds, however, are quite different from other gustatory stimuli, such as amino acids, aliphatic acids and nucleotides, in that bile salts have been shown to stimulate the facial taste system of only salmonids (Hara et al., 1984). Unpublished data (Marui) indicate that the taste systems neither of the common carp, nor of the sea catfish, *Plotosus lineatus*, nor of an African cichlid, *Tilapia nilotica*, are stimulated by up to 0.1 mM bile salts. Integrated taste responses recorded from the palatine nerve in the rainbow trout to taurolithocholate, the most effective bile salt tested, increased linearly with logarithmic increase in concentration (Hara et al., 1984). The threshold ranged between 1 and 10 pM, which is approximately four log units lower than that for L-proline, the most potent amino acid taste stimulant for that species (Marui et al., 1983b). Taurodeoxycholic and cholic acids, less effective stimuli than taurolithocholic acid, were nevertheless far more potent taste stimuli than L-proline. The functional significance for taste systems of salmonids to be sensitive to bile salts is unknown; possibly, both olfaction and taste are involved in the detection of key stimulus compounds that are involved in homing behaviour.

9.4 RECEPTOR SITE TYPES FOR AMINO ACIDS

Although studies published since 1975 established conclusively that the taste system of teleosts detects amino acids with high sensitivity, little was known concerning the types (defined by their specificities) of amino acid taste receptor sites that exist on the taste cells. Information on receptor site types for amino acids in teleosts has primarily been obtained from electrophysiological cross-adaptation and biochemical competitive binding experiments performed primarily in the channel catfish. Both the electrophysiological and biochemical experiments indicated the existence of a multiplicity of relatively independent amino acid receptor site types; further, the electrophysiological studies indicated that the types and specificities of these receptor sites varied across species (Table 9.3; Brand and Bruch, Chapter 7 herein).

Two different electrophysiological cross-adaptation paradigms were used for determining the types of amino acid receptors that exist for the species listed in Table 9.3. In the first method, the 'test' stimulus is presented just after the response to a single presentation of the 'adapting stimulus' has returned to baseline (for the eel and puffer); in the second scheme, the 'test' stimulus is injected into an 'adapting stimulus' that continuously bathes the sensory epithelium (for the carp, channel catfish and rainbow trout). Test amino acids whose responses were suppressed completely (to control levels) by the adapting stimulus are presumed to share a common receptive pathway and possibly the same receptor site. Those test amino acids that remain stimulatory are considered to have at least some portion of their

Table 9.3 Classification of receptor types by cross-adaptation experiments

Species*	Receptor types†	
Carp[1]	(1)	L-Ala (L-Pro, L-Ser, Gly, L-α-Abu, β-Ala)
	(2)	L-CysH
	(3)	L-His
	(4)	L-Glu
	(5)	**L-Asp**
	(6)	**Bet**
Japanese eel[2]	(1)	L-Ala (L-Lys, L-Arg, L-Ser, Gly)
	(2)	L-Pro
	(3)	L-His
	(4)	**Bet**
Channel catfish[3]	(1)	L-Ala (Gly, L-Ser, L-Met)
	(2)	L-Arg
	(3)	L-Pro
	(4)	L-Lys
	(5)	L-His
	(6)	D-Ala[3,4]
	(7)	D-Arg
	(8)	D-Pro
Rainbow trout[5]	(1)	L-Pro
	(2)	L-Leu
	(3)	L-AGPA (Bet)
Puffer[6]	(1)	L-Ala (Gly, L-Ser, Sarcosine)
	(2)	L-Pro (Trimethylglycine)
	(3)	**Bet** (Dimethylglycine)

*Sources: 1, Marui *et al.* (1987); 2, Yoshii *et al.* (1979); 3, Caprio, (1982), Wegert and Caprio (1991); 4, Brand *et al.* (1987); 5, Hara and Marui (1982); 6, Kiyohara *et al.* (1991).

†Chemicals set in **bold** enhance the responses to amino acids. Abbreviations: Abu, alpha-aminobutyric acid; AGPA, alpha-amino-beta-guanidino-propionic acid; Ala, alanine; Arg, arginine; Asp, aspartic acid; Bet, betaine (N-trimethyl-glycine); CysH, cysteine; Glu, glutamic acid; Gly, glycine; His, histidine; Hpr, hydroxyproline; Hyp, hydroxyproline; Leu, leucine; Lys, lysine; Met, methionine; Phe, phenylalanine; Pro, proline; Sarcosine, N-methyl glycine; Ser, serine; Try, tryptophan; Val, valine.

Note: the compounds included with a parenthesis are thought to bind to the same receptor sites as the initially listed amino acid in each row.

receptive pathways (and therefore receptor sites) independent from that of the adapting stimulus.

It is important to note, however, that in addition to differences in how the adapting stimulus was presented as noted above, an additional difference in the testing paradigm still remains and may be a factor in the results obtained. The concentrations of the adapting and test stimuli vary across the species tested. For the majority of the experiments, the stimuli were tested at a single concentration, usually 10^{-4} M or 10^{-3} M. For the channel catfish (Wegert and Caprio, 1991), however, all stimuli for each fish tested were adjusted in concentration to be equipotent stimuli. This concentration adjustment allows each stimulus, independent of its relative stimulatory effictiveness, to have equal effect in eliciting a similar magnitude of action potential activity along the gustatory neurones. For relatively independent receptor sites, this also allows a poor agonist, because of its higher concentration, to have a chance of competing for the receptor site approximately equal to that of a more stimulatory agonist at lower concentration. Using this testing paradigm, eight different receptor site types for amino acids were indicated in the facial taste system of the channel catfish (Wegert and Caprio, 1991) (Table 9.3). Based on the magnitude of the taste responses to the eight different amino acid stimuli, the binding sites for L-alanine, L-arginine and L-proline appear to be in relatively large numbers. Recently, experiments utilizing membrane vesicles isolated from the cutaneous taste epithelium of the channel catfish that were incorporated into phospholipid bilayers on the tips of patch pipettes (see Chapter 7), have confirmed the existence of both the arginine (Teeter *et al.*, 1990) and proline (Kumazawa *et al.*, 1990) receptor sites. These and other recent advances concerning receptor site identification and the biochemical processes of transduction in the taste system of the channel catfish are summarized by Brand and Bruch (Chapter 7 herein).

There are few generalities concerning receptor site types for amino acids in teleosts that can be summarized from the limited number of species tested electrophysiologically (Table 9.3). Other than the indication of multiple receptor site types for amino acids in the same species, there is an indication in the carp (Marui *et al.*, 1987), Japanese eel (Yoshii *et al.*, 1979), channel catfish (Wegert and Caprio, 1991) and puffer (Yamashita *et al.*, 1987) that some of the neutral amino acids may share the same or highly cross-reactive receptive pathways. In contrast, the receptor sites for the basic amino acids may be different from that (or those) indicated for the neutral amino acids, in that the L-isomers of arginine, lysine and histidine in the channel catfish were all indicated to have relatively independent receptor sites (Wegert and Caprio, 1991). A similar finding occurred in the carp for the acidic amino acids, where the facial taste receptor sites for the L-isomers of aspartic and glutamic acids were indicated to be relatively independent (Marui *et al.*, 1987). Also, from data obtained from the channel catfish, it is evident that receptor sites for the stereoisomers of particular amino acids may be different (Brand *et al.*, 1987; Wegert and Caprio, 1991). Evidence from four of the five species listed in Table 9.3 also indicated that betaine binds to receptor sites independent from those to other amino acids. However, from unpublished experiments in the sea catfish, *Plotosus lineatus* (Caprio, Marui and Kiyohara), L-proline and betaine bind to the same receptor site with similar affinities.

9.5 TASTE RESPONSES OF SINGLE FACIAL TASTE FIBRES

Although much is known concerning multiunit taste responses of certain species of teleosts to amino acids, quantitative studies of the responses of single facial taste fibres to these stimuli are rare. A major question related to quality coding of amino acid taste information is how that information is represented in the output of single peripheral taste neurones. A knowledge of the 'tuning' (i.e. response specificity) of individual neurones to amino acid stimuli is critical to a better understanding of how the organism may identify the stimulus.

The first quantitative study of single fibre taste activity in a teleost to amino acid stimuli was performed in the Japanese puffer, *Fugu pardalis* (Kiyohara et al., 1975). That study indicated that amino acids, nucleotides and hydrochloric acid activated different facial taste fibres, although betaine stimulated both the amino acid and hydrochloric acid fibres. Another study of the responses of individual facial taste fibres in the puffer reported that only 16 of the 106 fibres isolated responded to L-proline, the only amino acid tested. The remaining fibres responded to nucleotides, hydrochloric acid or mechanical stimulation (Kiyohara et al., 1985). Two more recent reports on the puffer indicated that although L-alanine and L-proline activate different receptor sites (based on electrophysiological, cross-adaptation data) (Kiyohara et al., 1989), these stimuli excite the same taste fibres (Marui and Kiyohara, 1987).

The responses of individual facial taste fibres to amino acids were also reported in the North American, freshwater channel catfish, *Ictalurus punctatus* (Davenport and Caprio, 1982). Since the facial taste system of the channel catfish responded to all the common amino acids (Caprio, 1975, 1978, 1982), it was of interest to determine whether the amino acid receptor sites and taste cells were organized in such a manner to allow for the existence of different fibre types. Nine of 15 single recurrent facial taste fibres that innervated taste cells on the flank of the animal were stimulated best by L-alanine, whereas the remaining six fibres were stimulated best by L-arginine. A more recent quantitative study indicated that of the 94 facial taste fibres analysed that innervated maxillary barbel taste buds in *I. punctatus* 49 were L-alanine-best and 42 were L-arginine-best taste fibres (Kohbara and Caprio, unpublished). The responses from each of the two fibre groups clearly showed that the L-arginine fibre type was narrowly tuned; one portion of the arginine-best fibres responded with high frequency to only L-arginine, whereas the remaining portion responded well both to L-arginine and to L-proline. The L-alanine fibre type was responsive primarily to a number of neutral amino acids. It is rather interesting that transduction of L-alanine taste information, transmitted primarily by the L-alanine neural channel (i.e. fibre type) to the facial lobe of the medulla (the primary gustatory centre), probably involves G-proteins and second-messenger systems (Bruch and Kalinoski, 1987, 1988), whereas the transduction of arginine (Teeter et al., 1990) and proline (Kumazawa et al., 1990) information, transmitted by the L-arginine neural channel, is thought to occur through direct ligand-operated ion channels (see also Brand and Bruch, Chapter 7 herein). Thus, not only were the taste receptor sites for the L-isomers of alanine, arginine and proline relatively independent from one another in the taste system of *Ictalurus* (Section 9.4), but there was a differential innervation of these taste cells by the two major types of facial taste fibres. In addition, a small number of D-alanine-best fibres was indicated from recordings from small twigs of the facial nerve in the channel catfish (Wegert and Caprio, 1991). Also, as found for the puffer, nucleotide-sensitive fibres are different from the major amino-acid-sensitive, facial taste fibres in *Ictalurus* (Kohbara, Michel and Caprio, unpublished).

An analysis of responses from single facial taste fibres in the North American sea catfish, *Arius felis*, also indicated the presence of two major types of amino-acid-sensitive fibres (Michel and Caprio, 1991). Similar to that found in *Ictalurus*, the major fibre type present in *Arius* (22 of 42 fibres analysed) was highly sensitive to L-alanine. However, glycine, which in *Arius* was approximately equally stimulatory to L-alanine, was much less stimulatory to the taste system of *Ictalurus*. In addition, the sensitivity of the *Arius* facial taste system to L-arginine was much lower than that indicated for *Ictalurus*, and the second most common fibre type in *Arius* was most responsive to D-alanine. For these D-alanine units, L-alanine was the second most effective amino acid, whereas in general the L-alanine/glycine units responded poorly to glycine. Thus, it appears that the detection of D-alanine and possibly other D-isomers might be important to the feeding behaviour of *Arius*. These findings are reasonable in that D-alanine was recently shown to be in measurable levels in 18 of 43 invertebrate species tested, and that D-amino acids were more abundant than their L-isomers in six species of bivalve molluscs (Preston, 1987; Felbeck and Wiley, 1987).

The major amino-acid-sensitive, facial taste fibres of the Indo–Pacific sea catfish, *Plotosus lineatus*, also consisted of two types of fibres, those that were most stimulated by L-proline and betaine, and those that were most stimulated by L-alanine and glycine (Caprio, Marui and Kiyohara, unpublished). Thus, the input of amino acid taste information to the central nervous system (i.e. the facial lobe of the medulla) in *I. punctatus*, *A. felis* and *P. lineatus*, each species being a member of a different family of catfishes, occurs primarily through only two major types of neural channel (i.e. fibre types). Whether this will be similar for other species of catfish and non-siluriform teleostean species, and how this organization is used by the organism to determine stimulus quality, are unknown. Additional studies of the response specificity of individual taste fibres (facial, glossopharyngeal and vagal) in other teleostean species are necessary for a better understanding of the organization of amino acid neural channels in teleosts.

9.6 ENHANCED TASTE ACTIVITY

Chemoreceptors in nature rarely, if ever, are presented with single pure stimuli. As a rule, chemical stimuli in nature are rather complex mixtures, and the peripheral taste and olfactory systems must be organized to detect and process this information centrally. As such, animal extracts and certain mixtures of amino acids were shown to be potent stimuli to fishes (Konosu et al., 1968; Funakoshi et al., 1981; Carr, 1976; Carr and Chaney, 1976; Hidaka et al., 1976, 1978; Adron and Mackie, 1978; Yoshii et al., 1979; Goh and Tamura, 1980a,b; Mackie et al., 1980; Ellingsen and Döving, 1986; Marui and Kiyohara, 1987; Marui et al., 1987; Kiyohara et al., 1989). However, although particular animal extracts or mixtures of amino acids were shown to be stimulatory, individual components of some of these mixtures were shown to be inactive or weak stimulants. So, it is suggested that the stimulatory ability of certain mixtures is due to potentiation or synergistic interactions among the components.

In electrophysiological experiments, synergistic effects were observed qualitatively for mixtures of betaine and amino acids in the taste systems of several species of fish (Hidaka et al., 1976; Yoshii et al., 1979). A betaine and L-alanine mixture resulted in enhancement of taste activity in the puffer (Hidaka et al., 1976). Also in the same species, single unit responses to alanine and glycine were potentiated by the presence of betaine in cross-adaptation experiments (Marui and Kiyohara, 1987; Kiyohara et al., 1989). Mixtures of particular amino acids and betaine also enhanced feeding behaviour in some fishes (Hidaka et al., 1978; Goh and Tamura, 1980b; Mackie et al., 1980). In a quantitative electrophysiological study of the effects of amino acid cross-adaptation in the carp, an enhancement of gustatory responses to amino acids with betaine and with L-aspartate-Na was clearly shown (Marui et al., 1987); however, the enhancement observed for some pairs of stimuli might be fibre specific because the variability in the multiunit taste responses recorded was considerable. In the facial taste system of the carp, cross-adaptation with 1 mM betaine resulted in a shift in the D–R function for L-alanine to a lower concentration range (about 4 log units), whereas betaine alone at stimulus concentrations between 10^{-8} and 10^{-13} M was ineffective (Marui et al., 1987). Conversely, the D–R function for betaine during adaptation with 1 mM L-alanine was also shifted to a lower concentration range. Although synergism was observed in the taste system of the Japanese eel in response to binary mixtures of betaine with an amino acid, this same effect was also observed with other binary amino acid mixtures not containing betaine (Yoshii et al., 1979).

The mechanisms proposed for the taste enhancement that occurs with certain mixtures are varied. It was suggested for the Japanese eel that synergism might be caused by the weakening of the negative cooperativity between specific amino acid receptor sites (Yoshii et al., 1979). Because of a linear relationship between the integrated taste response and the log stimulus dose over a wide concentration range, it was postulated that this occurred as a result of negative cooperativity between amino acid receptor sites (i.e. it required increasingly large elevations in stimulus concentration to provide the same relative increase in the height of the integrated taste response). Thus, lessening of the negative cooperativity between sites would result in greater taste activity. Similarly, positive cooperativity between certain sites would also result in enhanced neural activity. A mechanism offered for the synergism observed electrophysiologically between 5'-nucleotides and L-amino acids in taste responses in the rat was an increase in affinity of the stimulus for the receptor site (Yoshii et al., 1986); however, biochemical evidence suggested that an increased number of available binding sites, and not an increased affinity for the ligands, was responsible for the enhanced taste activity (Torii and Cagan, 1980; Cagan, 1987).

Enhancement of olfactory (Caprio et al., 1989; Kang and Caprio, 1991) and gustatory (Kohbara and Caprio, unpublished) neural activity in catfish to particular mixtures of amino acids, presented simultaneously rather than sequentially, as in the previously described cross-adaptation experiments, has also been documented. Here, the mechanism suggested for the resulting enhanced neural activity is the simultaneous activation of multiple receptor site types by the components in the mixtures.

9.7 BEHAVIOUR TO CHEMICAL STIMULI

A number of studies have shown conclusively that amino acids acting singly and in combination stimulate feeding behaviour in fishes (Hashimoto et al., 1968; Carr, 1982; Hidaka, 1982; Mackie, 1982; Sutterlin et al., 1982; Little, 1983; Takeda et al., 1984; Harada, 1985; Mearns, 1985, 1986; Olsén et al., 1986; Jones, 1989 and Chapter 14 herein). Although given the extreme paucity of chemosensory information concerning all but a few species of teleosts, a correlation between the potent amino acids determined electrophysiologically and those determined behaviourally is indicated for the channel catfish, Japanese eel, red sea bream and puffer (Table 9.4). In the red sea bream, a correlation of 0.78 was found between the amino acids that were effective in initiating feeding behaviour and those that stimulated gustatory neural activity, whereas a correlation of only 0.01 occurred between the behaviourally effective amino acids and those that were potent olfactory stimuli determined electrophysiologically (Goh and Tamura, 1980a,b). In the puffer, the amino acids that stimulated lip taste buds electrophysiologically were also effective in stimulating feeding behaviour, but the concentration of the amino acids necessary to release the behaviour was greater by 2–3 orders of magnitude than that required to elicit gustatory neural activity (Hidaka et al., 1978). In the channel catfish, taste thresholds for the more stimulatory amino acids tested electrophysiologically

Table 9.4 Potent amino acids determined by electrophysiology and behaviour

Species	Electrophysiology*†	Feeding behaviour*†
Channel catfish, *Ictalurus punctatus*	L-Ala, L-Arg, L-Pro[1,2]	L-CysH[3], L-Arg, L-Met[4], L-Ala, L-Arg, L-Pro[5]
Japanese eel, *Anguilla japonica*	L-Arg, Gly, L-Ala, L-Pro[6]	L-Arg, L-Ala, Gly[7]
Puffer, *Fugu pardalis*	Gly, L-Pro, Bet, L-Ala[8]	L-Pro, Bet, L-Ala, Gly[9]
Rainbow trout, *Oncorhynchus mykiss*	L-Pro, L-Hyp, L-Leu, L-Ala[10]	(L-Tyr, L-Phe and L-His) or (L-Tyr, L-Phe and L-Lys)[11]
Red sea bream, *Chrysophrys major*	L-Ala, Gly, L-Arg, L-Ser[12]	L-Ala, Gly, L-Ser[13]

*Sources: 1, Caprio (1975); 2, Kanwal and Caprio (1983); 3, Little (1977); 4, Holland and Teeter, (1981); 5, Valentincic and Caprio (unpublished); 6, Yoshii *et al.* (1979); 7, Hashimoto *et al.* (1968); 8, Hidaka *et al.* (1976); 9, Hidaka *et al.* (1978); 10, Marui *et al.* (1983b); 11, Adron and Mackie (1978); 12, Goh and Tamura (1980a); 13, Goh and Tamura (1980b).
†Abbreviations as in Table 9.3.

(Caprio, 1978; Davenport and Caprio, 1982) and behaviourally (Holland and Teeter, 1981; Valentincic and Caprio, unpublished) were reported to be in fair agreement.

A comparison of the results obtained electrophysiologically (Marui et al., 1983b) and behaviourally (Adron and Mackie, 1978) in the rainbow trout, however, is confusing. The L-isomers of tyrosine, phenylalanine and either lysine or histidine were the constituents of the simplest mixture tested that stimulated feeding activity. When this mixture was subdivided into two fractions, neither fraction was active. When these same compounds were omitted from the effective mixture (i.e. the mixture which applied to casein food pellets caused a significantly greater number of daily actuations of the food dispenser trigger than did the unflavoured casein pellets) of L-amino acids constituting an artificial squid extract, the resulting mixture was not reduced in its stimulatory ability. Amazingly, lysine, tyrosine and histidine were inactive taste stimuli determined electrophysiologically (Marui et al., 1983b). Of the four amino acids identified above, electrophysiological studies indicated histidine and phenylalanine to be an effective olfactory (Hara, 1973) and gustatory (Marui et al., 1983b) stimulus, respectively. In addition, proline, which was in highest concentration in the synthetic squid extract, was inactive as a feeding stimulus (Adron and Mackie, 1978), but was a potent taste stimulus determined electrophysiologically (Marui et al., 1983b).

Nucleotides and nucleosides are also recognized as effective chemicals in the feeding behaviour of teleosts. In the turbot, *Scophthalmus maximus*, inosine and a few closely related purine nucleotides and nucleosides at low concentrations were specific feeding stimulants (Mackie and Adron, 1978). A weak activity of 5'-IMP and 5'-UMP, and a rather high activity of 5'-GMP were observed in the Japanese eel, although a repellent activity of 5'-AMP was recognized (Takeda et al., 1984). Adult oriental weatherfish and juvenile yellowtail were also stimulated behaviourally by nucleotides (Harada, 1986). In contrast, although nucleotides were shown to be effective physiological taste stimuli in the puffer (Hidaka et al., 1977), these compounds did not evoke feeding behaviour in that species (Hidaka, 1982).

Additional species of teleosts need to be tested both physiologically and behaviourally before a definitive statement can be made concerning the correlation of electrophysiological and behavioural studies of chemoreception. It does appear, however, that in some species there is a stronger relation between the sense of taste and feeding behaviour than that found for olfaction (Goh and Tamura, 1980b). This is consistent with behavioural evidence obtained from bullhead catfish, *Ictalurus nebulosus*, in which the gustatory sense was involved primarily in stimulus localization and swallowing or rejection during feeding (Atema, 1971), whereas the olfactory system served as a general alerting sense involved in social interactions (Todd et al., 1967). Further experiments are required to determine whether this functional distinction between olfaction and taste indicated for catfish is common among other teleostean species.

9.8 TACTILE SENSITIVITY OF PERIPHERAL NEURONES

Mechanosensation and gustation are thought to be intricately involved in food ingestion in fish (Herrick, 1904, 1906; Ariens-Kappers et al., 1960; Bardach et al., 1967; Bartheld and Meyer, 1985). Similarly, in mammals, mechanosensory (trigeminal) information contributes to both the sensorimotor and the motivational control of ingestive behaviour (Zeigler et al., 1984), and the interaction between somatosensation and gustation is involved in palatability (Berridge and Fentress, 1985). The high percentage of bimodal (taste/touch) neurones within the primary gustatory nucleus of the medulla in carp (Marui, 1977) and catfish (Marui and Caprio, 1982; Marui et al., 1988; Hayama and Caprio, 1989) is indicative that somatosensation is at least as important in the feeding processes of fishes as it appears to be for mammals (Kanwal and Finger, Chapter 5 herein).

The possible origins of bimodal (taste and tactile) units in the primary gustatory nucleus of teleosts are numerous. (1) The finding of trigeminal input to the primary gustatory nucleus of the medulla in the puffer (Kiyohara et al., 1985) and the sea catfish (Kiyohara et al., 1986), suggests the possibility of the convergence of mechanically sensitive trigeminal and taste-sensitive facial nerve fibres in the facial lobe (FL). Electrophysiological experiments clearly showed that sectioning the trigeminal and facial nerves, respectively, during microelectrode recordings from bimodal (taste and tactile) FL neurones in the carp resulted in the loss of tactile and taste responses, respectively (Marui and Funakoshi, 1979). (2) The occurrence in the puffer (Kiyohara et al., 1985) and channel catfish (Davenport and Caprio, 1982) of mechanically sensitive facial nerve fibres suggests the possibility of the convergence of mechanically sensitive and taste-sensitive facial fibres, respectively, in the FL. In addition, glossopharyngeal and vagal nerves innervating the oropharyngeal cavity in the channel catfish also contain a large population of mechanosensory fibres (Kanwal and Caprio, 1983). (3) Peripheral nerve fibres responsive to both mechanical and chemical stimulation were also indicated in the channel catfish (Davenport and Caprio, 1982).

9.9 SUMMARY

The gustatory sense in the majority of teleosts studied rivals the olfactory sense in terms of sensitivity for particular chemical stimuli. Although a number of different classes of stimuli are known to excite the gustatory receptors of teleosts (e.g. aliphatic acids, bile salts, and nucleotides), research concerning the taste of amino acids has dominated the field since 1975. For the facial taste system, electrophysiologically determined gustatory thresholds for amino acids are generally reported between micromolar and nanomolar concentrations. Taste buds innervated by glossopharyngeal and vagal nerves appear to have amino acid specificities similar to those of taste buds innervated by facial nerves, but are somewhat less sensitive. Two general types of patterns of taste responses to amino acids occur in teleosts. Some taste systems have a 'wide response range' and detect many of the amino acids, while other species have a 'limited response range' and respond well to only few of the amino acids. Competition experiments indicate that multiple types of amino acid receptors exist within taste receptor cell membranes, and that the organization of these sites is not random, as particular types of taste fibres within the same species have been identified. Generally, L-amino acids are more stimulatory than their stereoisomers, but recent electrophysiological data indicate the existence of D-amino acid taste receptor sites and D-alanine-best facial taste fibres. Certain mixtures of amino acids may result in enhanced (synergistic) taste activity. One mechanism for this enhancement is the simultaneous activation of different receptor site types by the components of the mixture.

ACKNOWLEDGEMENTS

This work is supported by NSF grant BNS8819772 and NIH grant NS14819 to J.C.

REFERENCES

Adron, J.W. and Mackie, A.M. (1978) Studies on the chemical nature of feeding stimulants for rainbow trout, *Salmo gairdneri*. *J. Fish Biol.*, **12**, 303–10.

Ariens-Kappers, C.U., Huber, G.C. and Crosby, E.C. (1960) *The Comparative Anatomy of the Central Nervous System of Vertebrates, Including Man*, Hanfer, New York, 1845 pp.

Atema, J. (1971) Structures and functions of the sense of taste in the catfish (*Ictalurus natalis*). *Brain Behav. Evol.*, **4**, 273–94.

Baatrup, E. (1985) Physiological studies on the pharyngeal terminal buds in the larval brook lamprey, *Lampetra planeri* (Bloch). *Chem. Senses*, **10**, 549–58.

Bardach, J.E. and Atema, J. (1971) The sense of taste in fishes, in *Handbook of Sensory Physiology*. Vol. IV (ed. L.M. Beidler), Springer-Verlag, Berlin, pp. 293–336.

Bardach, J.E., Fujiya, M. and Holl, A. (1967) Investigations of external chemoreceptors of fishes, in *Olfaction and Taste II* (ed. T. Hayashi), Pergamon Press, Oxford, pp. 647–65.

Bartheld, C.S. von and Meyer, D.L. (1985) Trigeminal and facial innervation of cirri in three teleost species. *Cell Tissue Res.*, **241**, 615–22.

Berridge, K.C. and Fentress, J.C. (1985). Trigeminal–taste interaction in palatability processing. *Science, Wash. D.C.*, **228**, 747–50.

Brand, J.G., Bryant, B.P., Cagan, R.H. and Kalinoski, D.L. (1987) Biochemical studies of taste sensation. XIII. Enantiomeric specificity of alanine taste receptor sites in catfish, *Ictalurus punctatus*. *Brain Res. Amst.*, **416**, 119–28.

References

Bruch, R.C. and Kalinoski, D.L. (1987) Interaction of GTP-binding regulatory proteins with chemosensory receptors. *J. biol. Chem.*, **262**, 2401–4.

Bruch, R.C. and Kalinoski, D.L. (1988) Biochemistry of vertebrate olfaction and taste. *Ann. Rev. Nutr.*, **8**, 21–42.

Bryant, B.P. and Leftheris, K. (1991) Structure/activity relationships in the L-alanine taste receptor system of the channel catfish, *Ictalurus punctatus*. *Physiol. Behav.*, **49**, 891–8.

Bryant, B.P., Harpaz, S. and Brand, J.G. (1989) Structure/activity relationships in the arginine taste pathway of the channel catfish. *Chem. Senses*, **14**, 805–15.

Cagan, R.H. (1987) Allosteric regulation of glutamate taste receptor function in *Umami: A Basic Taste* (eds Y. Kawamura and M.R. Kare), Marcel Dekker, New York, pp. 155–72.

Caprio, J. (1975) High sensitivity of catfish taste receptors to amino acids. *Comp. Biochem. Physiol.*, **52A**, 247–51.

Caprio, J. (1978) Olfaction and taste in the channel catfish: an electrophysiological study of the responses to amino acids and derivatives. *J. comp. Physiol.*, **123A**, 357–71.

Caprio, J. (1982) High sensitivity and specificity of olfactory and gustatory receptors of catfish to amino acids, in *Chemoreception in Fishes* (ed. T.J. Hara), Elsevier, Amsterdam. pp. 109–34.

Caprio, J. (1984) Olfaction and taste in fish, in *Comparative Physiology of Sensory Systems* (eds L. Bolis, R.D. Keynes and S.H.P. Madrell), Cambridge University Press, Cambridge, pp. 257–83.

Caprio, J. (1988) Peripheral filters and chemoreceptor cells in fishes, in *Sensory Biology of Aquatic Animals* (eds J. Atema, R.R. Fay, A.N. Popper and W.N. Tavolga), Springer-Verlag, Berlin, pp. 313–38.

Caprio, J., Dudek, J. and Robinson, J.J., II (1989) Electro-olfactogram and multiunit olfactory receptor responses to binary and trinary mixtures of amino acids in the channel catfish, *Ictalurus punctatus*. *J. gen. Physiol.*, **93**, 245–62.

Carr, W.E.S. (1976) Chemoreception and feeding behavior in the pigfish, *Orthopristis chrysopterus*: characterization and identification of stimulatory substances in a shrimp extract. *Comp. Biochem. Physiol.*, **55A**, 153–7.

Carr, W.E.S. (1982) Chemical stimulation of feeding behavior, in *Chemoreception in Fishes* (ed. T.J. Hara), Elsevier, Amsterdam, pp. 259–74.

Carr, W.E.S. and Chaney, T.B. (1976) Chemical stimulation of feeding behavior in the pinfish, *Lagodon rhomboides*: characterization and identification of stimulatory substances extracted from shrimp. *Comp. Biochem. Physiol.*, **54A**, 437–41.

Carr, W.E.S. and Derby, C.D. (1986a) Behavioral chemoattractants for the shrimp, *Palaemonetes pugio*: identification of active components in food extracts and evidence of synergistic interactions. *Chem. Senses*, **11**, 49–64.

Carr, W.E.S. and Derby, C.D. (1986b) Chemically stimulated feeding behavior in marine animals. Importance of chemical mixtures and involvement of mixture interactions. *J. Chem. Ecol.*, **12**, 989–1011.

Carr, W.E.S. and Thompson, H.W. (1983) Adenosine 5'-monophosphate, an internal regulatory agent, is a potent chemoattractant for a marine shrimp. *J. comp. Physiol.*, **153**, 47–53.

Carr, W.E.S., Gleeson, R.A., Ache, B.W. and Milstead, M.L. (1986) Olfactory receptors of the spiny lobster: ATP-sensitive cells with similarities to P2-type purinoceptors of vertebrates. *J. comp. Physiol.*, **158A**, 331–8.

Davenport, C.J. and Caprio, J. (1982) Taste and tactile recordings from the ramus recurrens facialis innervating flank taste buds in the catfish. *J. comp. Physiol.*, **147**, 217–29.

Derby, C.D. and Atema, J. (1982a) Chemosensitivity of walking legs of the lobster *Homarus americanus*: neurophysiological response spectrum and thresholds. *J. exp. Biol.*, **98**, 303–16.

Derby, C.D. and Atema, J. (1982b) Narrow-spectum chemoreceptor cells in the walking legs of the lobster *Homarus americanus*: taste specialists. *J. comp. Physiol.*, **146**, 181–9.

Derby, C.D., Carr, W.E.S. and Ache, B.W. (1984) Purinergic olfactory cells of crustaceans: response characteristics and similarities to internal purinergic cells of vertebrates. *J. comp. Physiol.*, **155A**, 341–9.

Döving, K.B., Selset, R. and Thommesen, G. (1980) Olfactory sensitivity to bile acids in salmonid fishes. *Acta physiol. scand.*, **108**, 123–31.

Ellingsen, O.F. and Döving, K.B. (1986) Chemical fractionation of shrimp extracts inducing bottom food search behavior in cod (*Gadus morhua* L.). *J. Chem. Ecol.*, **12**, 155–68.

Felbeck, H. and Wiley, S. (1987) Free D-amino acids in the tissues of marine bivalves. *Biol. Bull. mar. biol. Lab., Woods Hole*, **173**, 252–9.

Finger, T.E. and Morita, Y. (1985) Two gustatory systems: facial and vagal gustatory nuclei have different brainstem connections. *Science, Wash. D.C.*, **227**, 776–8.

Funakoshi, M., Kawakita, K. and Marui, T. (1981) Taste responses in the facial nerve of the carp, *Cyprinus carpio* L. *Jap. J. Physiol.*, **31**, 381–90.

Goh, Y. and Tamura, T. (1980a) Olfactory and gustatory responses to amino acids in two marine teleosts—red sea bream and mullet. *Comp. Biochem. Physiol.*, **66C**, 217–24.

Goh, Y. and Tamura, T. (1980b) Effect of amino acids on the feeding behaviour in red sea bream. *Comp. Biochem. Physiol.*, **66C**, 225–9.

Goh, Y., Tamura, T. and Kobayashi, H. (1979) Olfactory responses to amino acids in marine teleosts. *Comp. Biochem. Physiol.*, **62A**, 863–8.

Gordon, K.D. and Caprio, J. (1985) Taste responses to amino acids in the southern leopard frog, *Rana sphenocephalus*. *Comp. Biochem. Physiol.*, **81A**, 525–30.

Hara, T.J. (1973) Olfactory responses to amino acids in rainbow trout, *Salmo gairdneri*. *Comp. Biochem. Physiol.*, **44A**, 407–16.

Hara, T.J. (1975) Olfaction in fish, in *Progress in Neurobiology*. Vol. 5 (eds G.A. Kerkut and J.W. Phillis), Pergamon Press, Oxford, pp. 271–335.

Hara, T.J. and Mauri, T. (1982) Multiplicity of taste receptors for amino acids in rainbow trout: evidence from cross-adaptation experiments and kinetic analysis. *Chem. Senses*, **8**, 250. abs.

Hara, T.J. and Zielinski, B. (1989) Structural and functional development of the olfactory organ in teleosts. *Trans. Am. Fish. Soc.*, **118**, 183–94.

Hara, T.J., Macdonald, S., Evans, R.E., Marui, T. and Arai, S. (1984) Morpholine, bile acids and skin mucus as possible chemical cues in salmonid homing: electrophysiological re-evaluation, in *Mechanisms of Migration in Fishes* (eds J.D. McCleave, G.P. Arnold, J.J. Dodson and W.H. Neill), Plenum, New York, pp. 363–78.

Harada, K. (1985) Feeding attraction activities for amino acids and nitrogenous bases for oriental weatherfish. *Bull. Jap. Soc. scient. Fish.*, **51**, 461–6.

Harada, K. (1986) Feeding attraction activities of nucleic acids-related compounds for abalone, oriental weatherfish and yellowtail. *Bull. Jap. Soc. scient. Fish.*, **52**, 1961–8.

Harada, S., Marui, T. and Kashara, Y. (1982) Amino acids as taste stimuli in the mouse. *Taste and Smell*, **16**, 123–6.

Harada, S., Marui, T. and Kasahara, Y. (1983) Gustatory stimulatory effectiveness of basic amino acids in rat and mouse. *Taste and Smell*, **17**, 61–4.

Hashimoto, Y., Konosu, S., Fusetani, N. and Nose, T. (1968) Attractants for eels in the extracts of short-necked clam–I. Survey of constituents eliciting feeding behavior by the omission test. *Bull. Jap. Soc. scient. Fish.*, **34**, 78–83.

Hatt, H. (1984) Structural requirements of amino acids and related compounds for stimulation of receptors in crayfish walking leg. *J. comp. Physiol.* **155A**, 219–32.

Hayama, T. and Caprio, J. (1989) Lobule structure and somatotopic organization of the medullary facial lobe in the channel catfish *Ictalurus punctatus*. *J. comp. Neurol.*, **285**, 9–17.

Herrick, C.J. (1904) The organ and sense of taste in fishes. *Bull. U.S. Fish. Comm.*, **22**, 237–72.

Herrick, C.J. (1906) On the centers for taste and touch in the medulla oblongata of fishes. *J. comp. Neurol. Psychol.*, **16**, 403–21.

Hidaka, I. (1982) Taste receptor stimulation and feeding behavior in the puffer, in *Chemoreception in Fishes* (ed. T.J. Hara), Elsevier, Amsterdam, pp. 243–58.

Hidaka, I. and Ishida, Y. (1985) Gustatory response in the shimaisaki (tigerfish) *Therapon oxyrhynchus*. *Bull. Jap. Soc. scient. Fish.*, **51**, 387–91.

Hidaka, I., Nyu, N. and Kiyohara, S. (1976) Gustatory response in the puffer–IV. Effects of mixtures of amino acids and betaine. *Bull. Fac. Fish.*, *Mie Univ.*, **3**, 17–28.

Hidaka, I., Kiyohara, S. and Oda, S. (1977) Gustatory response in the puffer–III. Stimulatory effectiveness of nucleotides and their derivatives. *Bull. Jap. Soc. scient. Fish.*, **43**, 423–8.

Hidaka, I., Ohsugi, T. and Kubomatsu, T. (1978) Taste receptor stimulation and feeding behaviour in the puffer, *Fugu pardalis*. I. Effect of single chemicals. *Chem. Senses Flavor*, **3**, 341–54.

Hidaka, I., Ohsugi, T. and Yamamoto, Y. (1985) Gustatory response in the young yellowtail *Seriola quinqueradiata*. *Bull. Jap. Soc. scient. Fish.*, **51**, 21–4.

Hoagland, H. (1933) Specific nerve impulses from gustatory and tactile receptors in catfish. *J. gen. Physiol.*, **16**, 685–93.

Holland, K.N. and Teeter, J.H. (1981) Behavioral and cardiac reflex assays of the chemosensory acuity of channel catfish to amino acids. *Physiol. Behav.*, **27**, 699–707.

Ishida, Y. and Hidaka, I. (1987) Gustatory response profiles for amino acids, glycinebetaine, and nucleotides in several marine teleosts. *Nippon Suisan Gakkaishi*, **53**, 1391–8.

Jones, K.A. (1989) The palatability of amino acids and related compounds to rainbow trout, *Salmo gairdneri* Richardson. *J. Fish Biol.*, **34**, 149–60.

Kang, J. and Caprio, J. (1991) Electro-olfactogram and multiunit olfactory recpetor responses to complex mixtures of amino acids in the channel catfish, *Ictalurus punctatus*. *J. gen. Physiol.*, **98**, 699–721.

Kanwal, J.S. and Caprio, J. (1983) An electrophysiological investigation of the oro-pharyngeal (IX–X) taste system in the channel catfish, *Ictalurus punctatus*. *J. comp. Physiol.*, **150A**, 345–57.

Kanwal, J.S., Hidaka, I. and Caprio, J. (1987) Taste responses to amino acids from facial nerve branches innervating oral and extra-oral taste buds in the channel catfish, *Ictalurus punctatus*. *Brain Res. Amst.*, **406**, 105–12.

Kiyohara, S. and Hidaka, I. (1991) Receptor sites for alanine, proline and betaine in the palatal taste system of the puffer, *Fugu pardalis*. *J. comp. Physiol.*, **169A**, 523–30.

Kiyohara, S., Hidaka, I. and Tamura, T. (1975) Gustatory response in the puffer–II. Single fiber analyses. *Bull. Jap. Soc. scient. Fish.*, **41**, 383–91.

Kiyohara, S., Yamashita, S. and Harada, S. (1981) High sensitivity of minnow gustatory receptors to amino acids. *Physiol. Behav.*, **26**, 1103–8.

Kiyohara, S., Hidaka, I., Kitoh, J. and Yamashita, S. (1985) Mechanical sensitivity of the facial nerve fibers innervating the anterior palate of the puffer, *Fugu pardalis*, and their central projection to the primary taste center. *J. comp. Physiol.*, **157A**, 705–16.

Kiyohara, S., Houman, H. and Yamashita, S., Caprio, J. and Marui, T. (1986) Morphological evidence for a direct projection of trigeminal nerve fibers to the primary gustatory center in the sea catfish *Plotosus anguillaris*. *Brain Res. Amst.*, **379**, 353–7.

Konishi, J., Uchida, M. and Mori, Y. (1966) Gustatory fibers in the sea catfish. *Jpn. J. Physiol.*, **16**, 194–204.

Konosu, S., Fusetani, N., Nose, T. and Hashimoto, Y. (1968) Attractants for eels in the extracts of short-necked clam–II. Survey of constituents eliciting feeding behavior by fractionation of the extracts. *Bull. Jap. Soc. scient. Fish.*, **34**, 84–7.

Kumazawa, T., Teeter, J.H. and Brand, J.G. (1990) L-Proline-activated cation channels in isolated catfish taste epithelial membranes. *Chem. Senses*, **15**, 603–4.

Little, E.E. (1977) Conditioned aversion to amino acid flavors in the catfish, *Ictalurus punctatus*. *Physiol. Behav.*, **19**, 743–7.

Little, E.E. (1983) Behavioral function of olfaction and taste in fish, in *Fish Neurobiology*. Vol. 1 (eds R.C. Northcutt and R.E. Davis), University of Michigan Press, Ann Arbor, pp. 351–76.

Littleton, J.T., Kohbara, J., Michel, W. and Caprio, J. (1989) Gustatory responses of the channel catfish, *Ictalurus punctatus*, to nucleotides and related substances. *Chem. Senses*, **14**, 732.

Mackie, A.M. (1982) Identification of the gustatory feeding stimulants, in *Chemoreception in Fishes* (ed. T.J. Hara), Elsevier, Amsterdam, pp. 275–92.

Mackie, A.M. and Adron, J.W. (1978) Identification of inosine and inosine 5′-monophosphate as the gustatory feeding stimulants for the turbot, *Scophthalmus maximus*. *Comp. Biochem. Physiol.*, **60A**, 79–83.

Mackie, A.M., Adron, J.W. and Grant, P.T. (1980) Chemical nature of feeding stimulants for the juvenile Dover sole, *Solea solea* (L.). *J. Fish Biol.*, **16**, 701–8.

Marui, T. (1977) Taste responses in the facial lobe of the carp, *Cyprinus carpio* L. *Brain Res. Amst*, **130**, 287–98.

Marui, T. and Caprio, J. (1982) Electrophysiological evidence for the topographical arrangement of taste and tactile neurons in the facial lobe of the channel catfish. *Brain Res. Amst.*, **231**, 185–90.

Marui, T. and Funakoshi, M. (1979) Tactile input to the facial lobe of the carp, *Cyprinus carpio* L. *Brain Res. Amst.*, **177**, 479–88.

Marui, T. and Kiyohara, S. (1987) Structure–activity relationships and response features for amino acids in fish taste. *Chem. Senses.*, **12**, 265–75.

Marui, T., Harada, S. and Kasahara, Y. (1983a) Gustatory specificity for amino acids in the facial taste system of the carp, *Cyprinus carpio* L. *J. comp. Physiol.*, **153A**, 299–308.

Marui, T., Harada, S. and Kasahara, Y. (1987) Multiplicity of taste receptor mechanisms for amino acids in the carp, *Cyprinus carpio*, in *Umami: A. Basic Taste* (eds Y. Kawamura and M.R. Kare), Marcel Dekker, New York, pp. 185–99.

Marui, T., Evans, R.E., Zielinski, B. and Hara, T.J. (1983b) Gustatory responses of the rainbow trout (*Salmo gairdneri*) palate to amino acids and derivatives. *J. comp. Physiol.* **153A**, 423–33.

Marui, T., Kiyohara, J., Caprio, S. and Kasahara, Y. (1988) Topographical organization of taste and tactile neurons in the facial lobe of the sea catfish, *Plotosus lineatus*. *Brain Res. Amst.*, **446**, 178–82.

Mearns, K.J. (1985) Response of Atlantic salmon (*Salmo salar* L.) yearlings to individual L-amino acids. *Aquaculture*, **48**, 253–9.

Mearns, K.J. (1986) Sensitivity of brown trout (*Salmo trutta* L.) and Atlantic salmon (*Salmo salar* L.) fry to amino acids at the start of exogenous feeding. *Aquaculture*, **55**, 191–200.

Michel, W. and Caprio, J. (1991) Responses of single facial taste fibers in the sea catfish, *Arius felis*, to amino acids. *J. Neurophysiol.*, **66**, 247–60.

Moore, A. and Cobb, J.L. (1985) Neurophysiological studies on the detection of amino acids by *Ophiura ophiura*. *Comp. Biochem. Physiol.*, **82A**, 395–9.

Morita, Y. and Finger, T.E. (1985) Reflex connections of the facial and vagal gustatory systems in the brainstem of the bullhead catfish, *Ictalurus nebulosus*. *J. comp. Neurol.*, **231**, 547–58.

Olsén, K.H., Karlsson, L. and Helander, A. (1986) Food search behavior in Arctic char, *Salvelinus alpinus* (L.), induced by food extracts and amino acids. *J. Chem. Ecol.*, **12**, 1987–98.

Preston, R.L. (1987) Occurrence of D-amino acids in higher organisms: a survey of the distribution of D-amino acids in marine invertebrates. *Comp. Biochem. Physiol.*, **87D**, 55–62.

Schiffman, S.S., Sennewald, K. and Gagnon, J. (1981) Comparison of taste qualities and thresholds of D- and L-amino acids. *Physiol. Behav.*, **27**, 51–9.

Selset, R. and Döving, K.B. (1980) Behaviour of mature anadromous char (*Salmo alpinus* L.) towards odorants by smolts of their own population. *Acta physiol. scand.*, **108**, 113–22.

Sutterlin, A.M. and Sutterlin, N. (1970) Taste responses in Atlantic salmon (*Salmo salar*). *J. Fish. Res. Bd Can.*, **27**, 1927–42.

Sutterlin, A.M. and Sutterlin, N. (1971) Electrical responses of the olfactory epithelium of Atlantic salmon (*Salmo salar*). *J. Fish. Res. Bd Can.*, **28**, 565–72.

Sutterlin, A.M., Solemdal, P. and Tilseth, S. (1982) Baits in fisheries with emphasis on the North Atlantic cod fishing industry, in *Chemoreception in Fishes* (ed. T.J. Hara), Elsevier, Amsterdam, pp. 293–305.

Suzuki, N. and Tucker, D. (1971) Amino acids as olfactory stimuli in freshwater catfish, *Ictalurus catus* (Linn.). *Comp. Biochem. Physiol.*, **40A**, 399–404.

Takeda, M., Takii, K. and Matsui, K. (1984) Identification of feeding stimulants for juvenile eel. *Bull. Jap. Soc. scient. Fish.*, **50**, 645–51.

Teeter, J.H., Brand, J.G. and Kumazawa, T. (1990) A stimulus-activated conductance in isolated taste epithelial membranes. *Biophys. J.*, **58**, 253–9.

Todd, J.H., Atema, J. and Bardach, J.E. (1967) Chemical communication in social behavior of a fish, the yellow bullhead (*Ictalurus natalis*). *Science, Wash. D.C.*, **158**, 672–3.

Torii, K. and Cagan, R.H. (1980) Biochemical studies of taste sensation, XI. Enhancement of L-[^{3}H]glutamate binding to bovine taste papillae by 5'-ribonucleotides. *Biochim. biophys. Acta*, **627**, 313–23.

Tucker, D. (1983) Fish chemoreception: peripheral anatomy and physiology, in *Fish Neurobiology*. Vol. 1 (eds R.G. Northcutt and R.E. Davis), University of Michigan Press, Ann Arbor, pp. 311–49.

Wegert, S. and Caprio, J. (1991) Structure/activity relations of the L-Proline taste receptor site in the channel catfish. *Chem. Senses*, **16**, 597.

Wegert, S. and Caprio, J. (1991) Receptor sites for amino acids in the facial taste system of the channel catfish. *J. comp. Physiol.*, **168A**, 201–11.

Yamashita, S., Kiyohara, S. and Hidaka, I. (1987) The role of amino and carboxyl groups for the amino acid receptors in the puffer, *Fugu pardalis*. *Taste and Smell*, **21**, 141–4.

Yoshii, K., Kamo, N., Kurihara, K. and Kobatake, Y. (1979). Gustatory responses of eel palatine receptors to amino acids and carboxylic acids. *J. gen. Physiol.*, **74**, 301–17.

Yoshii, K., Yokouchi, C. and Kurihara, K. (1986) Synergistic effects of 5′-nucleotides on rat taste responses to various amino acids. *Brain Res. Amst.*, **367**, 45–51.

Yoshii, K., Yoshii, C., Kobatake, Y. and Kurihara, K. (1982) High sensitivity of *Xenopus* gustatory receptors to amino acids and bitter substances. *Am. J. Physiol.*, **243**, R42–R48.

Zeigler, H.P., Jacquin, M.F. and Miller, M.G. (1984) Trigeminal sensorimotor mechanisms and ingestive behaviour. *Neurosci. biobehav. Rev.*, **8**, 415–24.

Chapter ten

Hormones, pheromones and chemoreception

Peter W. Sorensen

10.1 INTRODUCTION

Living in an aquatic environment often devoid of light but rich in dissolved compounds, teleost fish have evolved highly developed chemosensory and chemical signalling systems. These systems have a variety of functions including species and kin recognition, recognition of predatory risk (fright reaction), orientation, and the promotion of reproductive synchrony (reviews: Bardach and Todd, 1970; Solomon, 1977; Colombo *et al.*, 1982; Liley, 1982; Liley and Stacey, 1983; Stacey *et al.*, 1986, 1987). Although our understanding of these systems is generally poor, dramatic advances in our understanding of sex pheromones have been made over the past 10 years. These strongly suggest that sex hormones and their metabolites function as sex pheromones with distinct, fundamental roles controlling the reproductive physiology and behaviour of fish. This chapter reviews these advances from an historical perspective, paying particular attention to the endocrinological and neural basis of pheromone function. Emphasis is on the goldfish, *Carassius auratus*, because it represents the best-understood model of sex pheromone function.

10.2 HISTORY OF THE DEFINITION OF PHEROMONE

Although the term 'pheromone' is commonly used, its meaning is controversial. It has long been recognized that many animals release specialized chemicals to the environment, where these have specific behavioural and/or physiological effects on conspecifics (Darwin, 1887). Bethe (1932) proposed that these chemicals be called 'ectohormones' and contrasted their effects with those of 'endohormones', a term he suggested to replace 'hormones', coined by Starling (1905) to designate 'chemical messages secreted by one organ in the body and acting on another within the body of the producing organism'. Bethe's terminology failed to gain acceptance, and in 1959 Karlson and Luscher coined the term 'pheromone' (from the Greek *pherin* to transfer; *hormon* to excite) to replace ectohormone. Pheromones were defined as 'substances that are excreted to the outside by an individual and received by a second individual of the same species in which they release a specific reaction, for example a definite behaviour or developmental process'. Ironically, in light of what we now know about goldfish, Karlson and Lüscher (1959) state that 'unlike hormones, the substance is not secreted into the blood but outside the body; it does not serve humoral correlation within the organism but communication between individuals'. They also suggest that 'the principle of minute amounts being effective holds' and that 'species-specificity is not required'.

Karlson and Lüscher's definition has been vigorously criticized by mammalian biologists, who have felt that it is too vague (Drickamer, 1989), and fish biologists, who have questioned whether chemical signals should be considered pheromones if they prove to be metabolic by-products and not specialized communicative signals (Liley, 1982). Although numerous attempts have been made to modify the definition or coin new terminology (Kirschenblatt, 1962; Beauchamp et al., 1976; Brown, 1979), none have gained popular acceptance, and pheromone remains the term of choice. Unfortunately, however, its meaning has been weakened and many scientists now use it to describe all chemical signals of social significance (Drickamer, 1989). Our findings in the goldfish offer compelling reasons for a return to the original definition for fish. Goldfish pheromones have distinct patterns of release, are innately recognized by specific neural systems, and elicit immediate, distinct responses of adaptive significance. Although it is unclear whether the hormonal metabolites used as pheromones by goldfish have a communicative role in the strict sense of the word (Liley, 1982), they do appear to represent a necessary step in the evolution of the kind of specialized signals which characterize 'true' communication (Sorensen and Stacey, 1990; Stacey and Sorensen, 1991). In this chapter, 'pheromone' is used in compliance with its original definition.

10.3 ORIGINS OF THE HYPOTHESIS THAT FISH USE HORMONAL SEX PHEROMONES

Correlations between endocrine state and pheromone release

Although reports that fish commonly employ sex pheromones first appeared in the 1930s (Bardach and Todd, 1970), it was Tavolga (1955, 1956) who conducted the first experiments to test the relationship between reproductive condition and pheromone release. Studying the estuarine gobiid, *Bathygobius soporator*, he observed that mature males establish territories which they vigorously defend against all conspecifics except for mature females, which they court. Wondering how males recognize females Tavolga (1956) found that small quantities of the water in which gravid females had been kept would rapidly elicit male courtship behaviour. Among various bodily fluids, mucous rinses, and gonadal extracts tested, Tavolga found only ovarian fluid from gravid (presumably ovulated) females and egg washings to be active. In the first experiment ever to test which sensory system mediates responses to pheromones, Tavolga (1956) used heat cautery to destroy the male's olfactory sense and established that an intact sense of smell is required. Tavolga (1955) also found that castrated animals will court males and females of all maturities, and speculated the olfactory dysfunction might be responsible.

Studies on the ovoviviparous (live-bearing) guppy, *Poecilia reticulata*, which have ironically proven unrepeatable, provided the next boost for the hypothesis that endocrinological and pheromonal function are closely related in fish. Amouriq (1964, 1965) reported that male guppies become active when exposed to female water and that ovarian extracts elicit a similar response. Speculating that an ovarian hormone might be responsible, she tested hexestrol dipropionate, a synthetic analogue of oestradiol, and found it evoked hyperactivity in males after several hours of exposure (Amouriq 1967a, b). This finding was enthusiastically received and several authors confirmed the pheromonal activity of female water (Gandolfi, 1969; Crow and Liley, 1978). However, when Meyer and Liley (1982) re-tested ovarian fluid and oestradiol for pheromonal activity they were unable to confirm Amouriq's results. Studies by Liley and his colleagues did, however, offer two pieces of evidence supporting Amouriq's contention that the guppy pheromone is synthesized in the ovary under endocrine control. First, female guppies are most attractive during the first few days of their parturition cycle and lose attractiveness when hypophysectomized (Meyer and Liley, 1982) or gonadectomized (Crow and Liley, 1978). Second, hypophysectomized females become attractive when treated with either gonadotrophin or oestradiol (Meyer and Liley, 1982). Recently, the possibility that this pheromone might be oestrogen-like received support from the finding that male preference behaviour changes when they are exposed to a saturated solution of oestradiol and that these changes do not occur if the olfactory sense is blocked (Johansen, 1984).

Studies by Liley and his colleagues on the oviparous (externally-fertilizing) goldfish, provided new clues as to how pheromones function in fish. Noting that males chase recently ovulated (postovulatory) females, which they appear to identify in the absense of visual cues, Partridge *et al.* (1976) hypothesized and tested the possibility that ovulated female goldfish release a pheromone which males recognize by smell. Several experiments lent convincing support to this hypothesis. First, sham-operated males interact much more intensively with ovulated females than do anosmic males (i.e. those lacking a functional sense of smell). Second, when males were tested in a Y-maze they preferred water containing ovulated females or ovarian fluid, but not water from non-ovulated females. Third, spermiated males were much more responsive than non-spermiated individuals to sexual and food odours. Partridge et al. (1976) speculated that 'If the ovary is the primary source of sexual pheromones in fish, presumably such pheromones arose originally in evolution as by-products of endocrine and other changes associated with the maturation and ovulation of eggs....'

In another interesting study, which must be cautiously interpreted because of small sample size ($n=3$), Yamazaki and Watanabe (1979) directly tested the role of sex hormones in goldfish sex behaviour and olfaction. They reported that sexual activity and olfactory epithelial thickness decreased in hypophysectomized goldfish but not in methyl testosterone-treated hypophysectomized fish. Oestradiol-treated hypophysectomized fish were attractive to methyltestosterone-treated fish with an intact olfactory sense, and oestradiol treatment induced changes in kidney structure which Yamazaki and Watanabe (1979) speculated were related to pheromone release.

By the early 1980s, behavioural studies confirmed that ovulation and pheromone release are closely associated in at least 10 species of oviparous fish (Emanuel and Dodson, 1979; Lee and Ingersoll, 1979; Honda, 1979, 1980a,b, 1982a,b; Sorensen and Winn, 1984). Furthermore, when examined (Partridge et al., 1976; Honda, 1980a,b, 1982a,b), an intact olfactory system has been found necessary for pheromonal responsiveness. Although these studies firmly established a link between ovulation and pheromone production, they did not suggest an endocrinological basis for it.

Evolutionary perspectives

Kittredge et al. (1971) were the first to propose that aquatic organisms might commonly use hormones as pheromones. Although Kittredge drew his experimental evidence from studies on crabs, which have since proven controversial (see below), his ideas about the evolution of aquatic pheromones strongly influenced studies of fish pheromones (Doving, 1976; Colombo et al., 1982; Sorensen et al., 1988). Because the criticism levelled at Kittredge presumably reflects the criteria by which claims that fish use hormones as pheromones will be judged, these studies will be reviewed.

Noting that female crustaceans mate soon after moulting and that males exhibit reproductive behaviours when exposed to moulted female water, Kittredge et al. (1971) speculated that crabs release the steroidal moulting hormone crustecdysone to function as a pheromone. They exposed three species of crab to crustecdysone at concentrations as low as 10^{-13} M and found that males exhibited 'typical pre-copulatory behaviour' within 20 min (or longer) of exposure. Kittredge and Takahashi (1972) then proposed that the evolution of pheromonal communication requires the simultaneous appearance of the ability: (1) to produce and release the pheromone, and (2) to detect it via an external receptor mechanism. Arguing that it is unlikely for these events to occur simultaneously *de novo*, they hypothesized that aquatic animals probably evolved to use naturally released hormones as pheromones because this would require only one step, the chance expression of hormonal receptors on chemosensory receptor cells.

Kittredge *et al.*'s (1971) studies were criticized because the behavioural changes elicited by crustecdysone were not clearly related to sexual function and they had a long latency period (Dunham, 1978). Additionally, Kittredge *et al.* (1974) predicted that crustecdysone should have similar pheromonal effects on other crustaceans, but when tested this could not be verified (Atema and Gagosian, 1973; Gagosian and Atema, 1973; Seifert, 1982; Gleeson *et al.*, 1984). Expectations that crustecdysone should be released at specific times in the moult cycle also proved unfounded; one study could not measure it at all (Gleeson *et al.*, 1984), and another documented continuous release (Seifert, 1982). Most recently, however, supporting evidence has emerged from an electrophysiological study which found that the lateral antennule of spiny lobsters responds rapidly and specifically to water-borne crustecdysone (Spencer and Case, 1984), and it certainly bears pointing out that no-one has yet attempted to duplicate Kittredge *et al.*'s findings using the same species they used. In any case, the confusion surrounding Kittredge's studies has made most researchers extremely sceptical about crustecdysone's role as a 'hormonal' pheromone.

Links between hormone metabolism and pheromone release

Although Doving (1976) was the first to suggest that Kittredge's hypothesis might apply to fish, Colombo *et al.* (1979) were the first to test it. Male black gobies, *Gobius jozo*, defend a spawning site which ovulated females approach for courtship and oviposition. Colombo *et al.* (1979) found that ovulated females are attracted to male odour. Speculating that hormonal metabolites might be the attractant, they examined steroid metabolism *in vitro* in the Leydig-cell-rich mesorchial gland associated with the testes. Etiocholanolone glucuronide, a reduced androgenic steroid conjugated with glucuronic acid (a modification which makes it highly soluble), was identified, and when tested, it was found to attract ovulated females and induce oviposition (Colombo *et al.*, 1980). In the only study of how fish might release pheromones, Colombo *et al.* (1982) catheterized the urogenital papillae of females and found urine to be more attractive to males than ovarian fluid. They suggested that urine is more likely to contain small water-soluble hormonal metabolites than ovarian fluid, and noted 'important similarities' between pig (Melrose et al., 1971) and goby pheromones: both are androgen metabolites with no evident hormonal function and high solubility, both are produced by specialized glands, and only ovulated females respond to them.

Dutch endocrinologists have also investigated the possibility that the zebrafish, *Brachydanio rerio*, and African catfish, *Clarias gariepinus*, use specialized steroidal metabolites as pheromones. Because female zebrafish fail to ovulate when isolated from males but recommence when exposed to male water, and because this response is blocked by olfactory ablation, male pheromones are thought to initiate spawning activity (Chen and Martinich, 1975). Van den Hurk et al. (1987) discovered that testicular extracts are as effective as male water at triggering ovulation and that the activity of these extracts is associated with a fraction containing glucuronated steroids. Although seven steroid glucuronides were identified in testicular incubates, only two were found in male water. On ovulation, female zebrafish appear to release another pheromone, which stimulates male reproductive behaviour. Van den Hurk and Lambert (1983) and Lambert et al. (1986) suggest that this postovulatory pheromone is a mixture of oestradiol glucuronide and testosterone glucuronide produced by the ovaries, on the basis of several findings: (1) ovarian extracts containing steroid glucuronides attract males, (2) ovaries incubated with radiolabelled steroidal precursors produce these glucuronides, and (3) a mixture of these compounds (but not the individual compounds) attracts males if their olfactory sense is intact.

The African catfish spawns in the summer in response to flooding. Although evidence suggests that both males and females release pheromones, research has focused on the males, which have large seminal vesicles associated with their testes. Resink et al. (1987) discovered that ovulated *C. gariepinus* prefer male water to female water, and that non-ovulated and olfactory-ablated ovulated fish do not show a preference. Investigating a role for the seminal vesicles, they established that the odour of males lacking seminal vesicles is unattractive and that the odour of castrated males (which develop hypertrophied seminal vesicles) is highly attractive. Separating seminal vesicle fluid into fractions containing free and conjugated steroids, Resink et al. (1989b) found that only the conjugated fraction was active and that it lost its attractiveness when treated with β-glucuronidase, an enzyme which cleaves glucuronic acid. Subsequent *in vitro* incubations established that the seminal vesicles produce at least 17 free and eight glucuronated steroids (Lambert et al., 1986; Schoonen and Lambert, 1986; Schoonen et al., 1988), and Resink et al. (1989b) demonstrated that a mixture of seven of these glucuronides is attractive. Tests of the olfactory potency of these compounds using electrical recording from the olfactory epithelium (electroolfactogram; EOG) (see Hara, Chapter 8 herein, for a description of this technique) confirmed that 5β-pregnan-$3\alpha,17\alpha$-diol-20-one-3α-glucuronide and 5β-androstane-$3\alpha,11\beta$-diol-17-one-3α-glucuronide are strong olfactory stimulants, with detection thresholds of 10^{-11} M and 10^{-9} M respectively (Resink et al., 1989a). Although these studies convincingly demonstrate that the seminal vesicles of male African catfish synthesize steroidal metabolites, and that these compounds are olfactory stimulants, the identities of the chemicals actually released by the animal, and their biological function(s), have yet to be determined.

The goldfish: a neuroendocrine model of hormonal sex pheromone function

Female goldfish ovulate in the spring, in response to a preovulatory surge in gonadotrophin (GtH) which is triggered by rising water temperatures and emerging aquatic vegetation (Stacey, 1987). Females ovulate in the early morning when light levels are low, and as in other externally fertilizing teleosts, sexual receptivity coincides with ovulation (Stacey, 1987). Because females spawn (release) their eggs within a few hours, male–female reproductive physiology and behaviour must be tightly synchronized to ensure reproductive success. The importance of this synchrony is no doubt magnified by the intense male–male competition that characterizes the group-spawning behaviour of goldfish. Collaborative research between Norman Stacey, Toshiaki Hara, Joseph Dulka, Glen Van Der Kraak, Rick Goetz, and myself suggests that this synchrony is mediated by at least two hormone-derived pheromones released in a sequential manner by ovulatory females (Fig. 10.1).

Many fish, including goldfish, synthesize $17\alpha,20\beta$-dihydroxy-4-pregnen-3-one ($17\alpha,20\beta P$) or a closely related progestational steroid hormone (derived from progesterone and having 21 carbons – Fig. 8.2) in response to the ovulatory surge in GtH, to induce final oocyte maturation (germinal vesicle migration and breakdown: Goetz, 1983; Scott and Canario, 1987). Studying whether this steroid also functions as a male hormone stimulating production of milt (sperm and seminal fluid), we accidentally discovered that male goldfish held in the same water as $17\alpha,20\beta P$-injected fish experience large overnight increases in milt volume (Stacey and Sorensen, unpublished). Recalling that male goldfish held in the same water as ovulatory females have a GtH surge (Kobayashi et al., 1986), and that GtH stimulates milt production, we hypothesized that both ovulatory females (whose $17\alpha,20\beta P$ levels are naturally elevated) and $17\alpha,20\beta P$-injected males release unmetabolized $17\alpha,20\beta P$ directly to the water where it functions as a pheromone stimulating GtH release and milt production. To test this we added $17\alpha,20\beta P$ to tanks of males and measured their milt production 8 h later. Concentrations of $17\alpha,20\beta P$ as low as 10^{-10} M stimulated milt increases, and when we tested related progestational steroids we encountered evidence of high specificity; only 17α-hydroxy-4-pregnen-3-one, the precursor of $17\alpha,20\beta P$, elicited notable responses. Furthermore, severing the medial olfactory tracts eliminated responsiveness (Stacey and

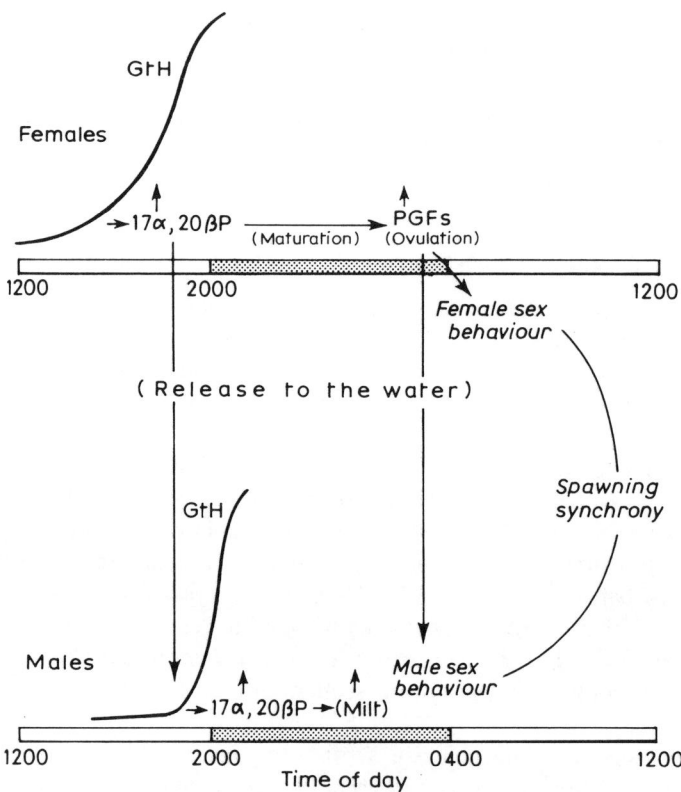

Fig. 10.1 A model of the goldfish sex pheromone system. Environmental cues trigger a surge in gonadotrophin (GtH) in mature females. This hormone in turn stimulates synthesis of the steroidal maturation hormone $17\alpha,20\beta$-dihydroxy-4-pregnen-3-one ($17\alpha,20\beta$P) by the ovary. Hormonal $17\alpha,20\beta$P induces final oocyte maturation and as it increases in the blood (short upward arrows), it is increasingly cleared to the water, where it then functions as a preovulatory 'primer' pheromone. The goldfish olfactory system is acutely sensitive to waterborne $17\alpha,20\beta$P and males exposed to it experience a surge in GtH which stimulates testicular synthesis of $17\alpha,20\beta$P (short upward arrow) which, in turn, stimulates milt (sperm and seminal fluid) production by the time spawning commences. Later, at the time of ovulation, females start producing F prostaglandin (PGF) and $17\alpha,20\beta$P levels fall. PGF plays a role mediating follicular rupture and accumulates in the blood where it travels to the brain to stimulate female sexual receptivity. Ovulated females metabolize PGF and release it to the water where it then functions as a postovulatory 'releaser' pheromone stimulating male sexual behaviour. The female hormone cycle (and hence pheromone release) is regulated by a circadian rhythm so that the GtH surge occurs in the late afternoon and ovulation (spawning) at daybreak. In this figure scotophase is represented by darkened bars on the time axes. (After Sorensen et al., 1988.)

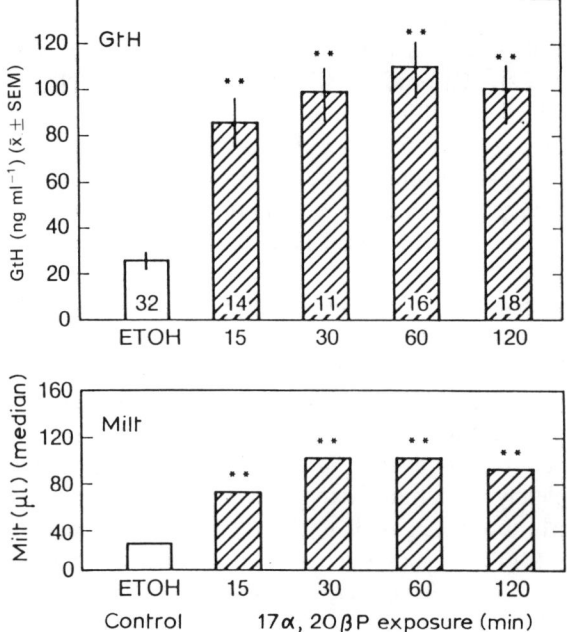

Fig. 10.2 Effects of exposing mature male goldfish to 5×10^{-10} Molar (M) 17α,20βP. Fish were exposed to 17α,20βP for 5 to 120 minutes, their blood was sampled for GtH determination, and then moved to aquaria with fresh clean water where they were held for 8 h before their milt volumes were measured. Control fish were exposed to ethanol carrier (ETOH) for 60 min. Mean values are given for GtH and the numbers within the bars signify sample size for both GtH and milt determinations. Significance of the GtH responses was evaluated using analysis of variance with Newman–Keuls follow-up tests. The milt data was analysed using Kruskal–Wallis. **Significantly different from control ($P<0.01$). (After Dulka et al., 1987.)

Sorensen, 1986). Testing the neuro-endocrine basis of the phenomenon, Dulka et al. (1987) determined that male goldfish exposed to 17α,20βP for as little as 15 min experienced an approximate tripling in GtH, a doubling in circulating 17α,20βP, and, within 6 h, a doubling in milt volume (Fig. 10.2; Dulka et al., 1987). Although the tiny amounts of 17α,20βP tested (tank concentrations were 10% that of male blood) made it seem unlikely that these effects were caused by the absorption of 17α,20βP from the water, we had yet to test directly whether the olfactory system mediates pheromonal responsiveness.

To test whether the goldfish olfactory system detects 17α,20βP we used EOG recording. As measured by electrophysiological means, 17α,20βP is the most stimulatory odorant identified in a fish (Sorensen et al., 1987a). It has a detection threshold of 10^{-12} M to 10^{-13} M, and at a concentration of 10^{-8} M elicits a response three times that of 10^{-5} M L-serine, an amino acid previously recognized as a feeding attractant and potent olfactory stimulant for fish (Fig. 10.3; Hara, 1986). Because many other fish also appear to use $17\alpha,20\beta P$-like hormones (Scott and Canario, 1987) and, like goldfish, probably release them (Van Der Kraak et al., 1989), we reasoned that for $17\alpha,20\beta P$ to function as a useful signal, the goldfish olfactory system must detect it in a highly specific manner. Using EOG recording and assays of endocrine responsiveness to confirm bioactivity, we tested the pheromonal actions of several dozen steroids and established that goldfish detect $17\alpha,20\beta P$ in the most specific manner yet described in a fish (Sorensen et al., 1990). Of 22 progestational steroids tested at a concentration of 10^{-8} M, the concentration at which $17\alpha,20\beta P$ elicits near-maximal responses, only $17\alpha,20\beta,21$-trihydroxy-4-pregnen-3-one, a probable metabolite of $17\alpha,20\beta P$ which has an additional hydroxyl group, evokes a significant response. Seemingly minor variations in molecular structure (e.g. reorientating the position of a hydroxyl group ($17\alpha,20\alpha P$), saturating a double bond ($3\alpha,17\alpha,20\beta P$), or removing a hydroxyl ($17\alpha P$) almost completely eliminate activity. When the olfactory epithelium is adapted (continuously exposed) to $17\alpha,20\beta P$ (theoretically filling receptor sites for $17\alpha,20\beta P$), it does not respond to other progestational steroids, suggesting that a single specific transduction process is responsible for responses to these compounds. Furthermore, because olfactory potency is correlated with similarity to $17\alpha,20\beta P$, specificity is probably attributable to a single class of olfactory receptors. Neither $17\alpha,20\beta P$ glucuronide nor any of the conjugated steroids suggested to function as pheromones in the black goby, zebrafish and African catfish stimulate the goldfish olfactory system (unpublished results). Endocrine responsiveness to homogeneous steroid solutions has confirmed the EOG results (Sorensen et al., 1990). However, because the male endocrine system responds very quickly to pheromonal stimulation (Sorensen et al., 1990), whether the specificity of the goldfish olfactory system to $17\alpha,20\beta P$ allows this compound actually to function as a species-specific cue is probably strongly determined by the manner in which it is released and encountered (Sorensen et al., 1990).

Although we had determined that males detect and respond to $17\alpha,20\beta P$, we still did not know whether females release this steroid in adequate quantities and at appropriate times for it to function as a pheromone. To address this question, ovulatory females were held in the same tanks as males, and the entire group was removed at regular intervals to determine blood GtH and $17\alpha,20\beta P$, female $17\alpha,20\beta P$ release, and milt production (Stacey et al., 1989). Males were maintained either in contact with females or isolated from them by a permeable barrier so that the influence of pheromones could be distinguished from that of behavioural interaction. Blood $17\alpha,20\beta P$ in ovulatory females paralleled their GtH surge, rising rapidly 8 h prior to ovulation, plateauing 5 h before, and falling to pre-surge levels at the time of ovulation and spawning (Fig. 10.4). Release of $17\alpha,20\beta P$ from ovulatory females paralleled blood levels, indicating rapid

Origins of the hypothesis that fish use hormonal sex pheromones   209

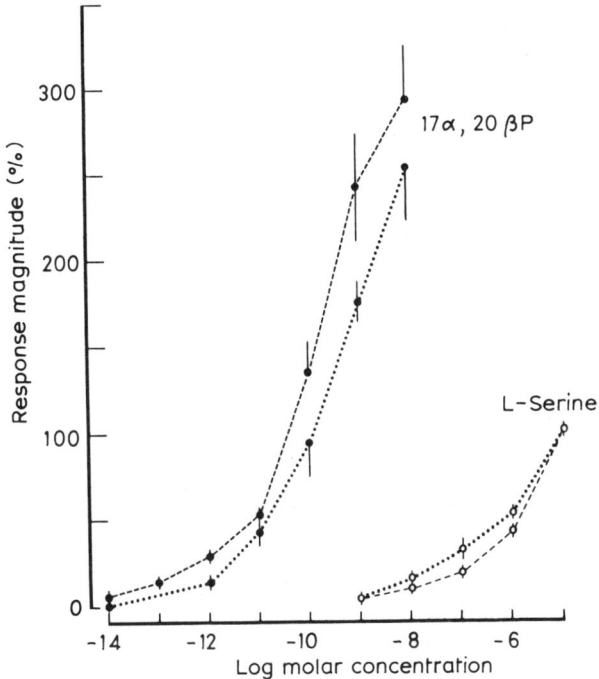

Fig 10.3 Electroolfactogram (EOG) responses of mature and immature goldfish to log molar concentrations of $17\alpha,20\beta$-dihydroxy-4-pregnen-3-one ($17\alpha,20\beta P$) and L-serine. The broken line represents mature males (n = 5) and the dotted line represents immature fish (3 gonadally-regressed males and 3 gonadally-regressed females). Responses are relative to 10^{-5} M L-serine standard and vertical bars represent standard error. (After Sorensen et al., 1987b.)

clearance. The GtH levels of 'isolated' and 'contact' males were similar to each other and closely followed the pattern of female $17\alpha,20\beta P$ release (although perhaps slightly preceding it) until the time of ovulation and spawning, when the levels of 'isolated' fish fell and those of 'contact' fish increased. Milt levels paralleled GtH increases. Although these results established that $17\alpha,20\beta P$ functions as a potent preovulatory pheromone which stimulates the male's endocrine system prior to spawning, they also suggested that male GtH release associated with spawning cannot be attributed to $17\alpha,20\beta P$ because very little is released at this time. In addition, this finding, and the results of pilot studies suggesting that $17\alpha,20\beta P$ has little immediate effect on male behaviour, made it appear extremely unlikely that $17\alpha,20\beta P$ was the postovulatory, behaviourally active pheromone described earlier by Partridge et al. (1976) (see p. 202).

Instead we wondered whether the goldfish postovulatory pheromone might be another hormone more closely associated with ovulation and

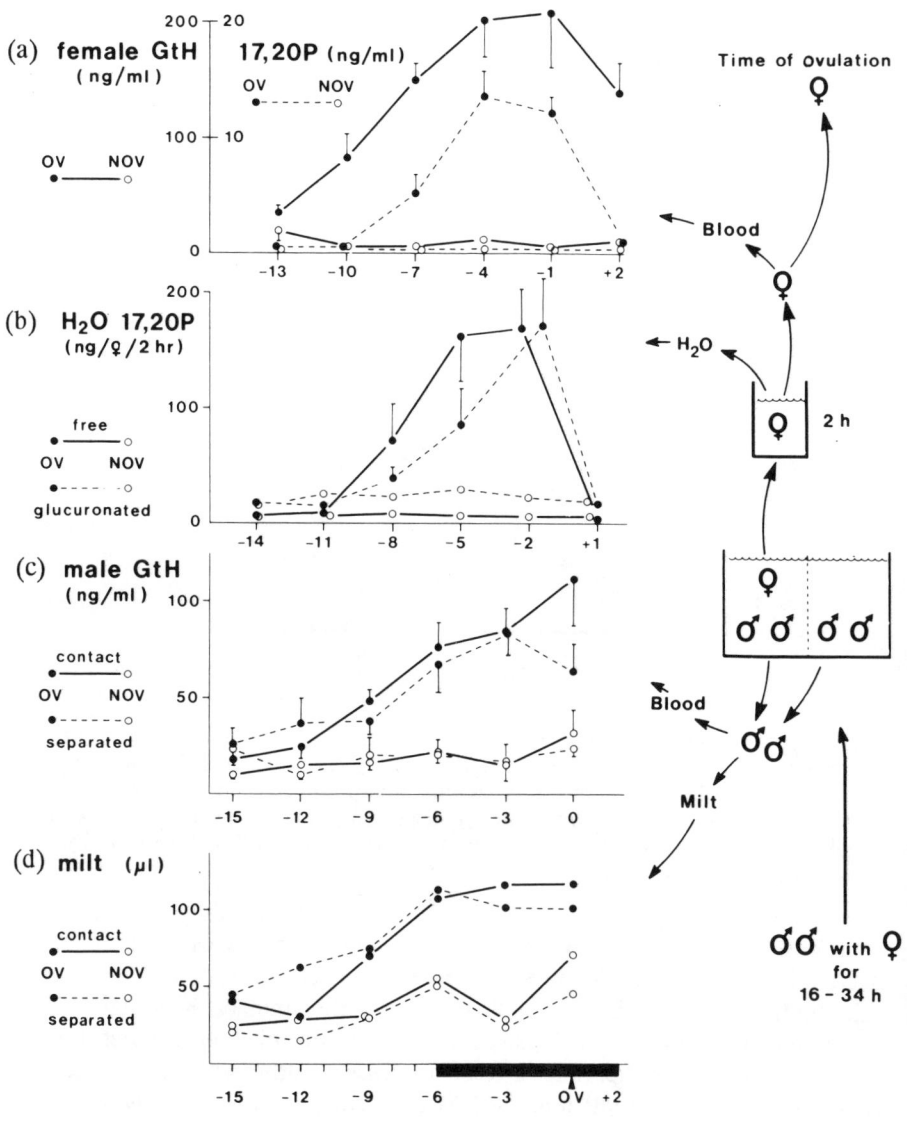

Fig. 10.4 Results of an experiment performed by Stacey *et al.* (1989) which proves that 17α,20βP functions as a preovulatory pheromone in goldfish; i.e. that ovulatory goldfish release 17α,20βP and that males exposed to water containing naturally-released 17α,20βP experience GtH and milt increases. A schematic representation of the experimental protocol which is also described in the text is shown on the right-hand side. Male and female goldfish were held together in aquaria divided by a perforated partition. Each aquaria contained one vitellogenic female and five males; two of these males were with the female ('contact'; solid line) and two were on the other side of the barrier so that they were in contact with the female's water only

spawning. We suspected prostaglandins (PGs) for several reasons. First, levels of circulating F prostaglandins (PGF) increase in goldfish and at least several other fish at the time of ovulation (Bouffard, 1979; Cetta and Goetz, 1982), presumably reflecting a role modulating follicular rupture (Goetz, 1983). Second, circulating PGF appears to function as a hormonal signal triggering female spawning behaviour through direct actions on the brain of goldfish (Stacey and Peter, 1979; Stacey and Goetz, 1982; Stacey, 1987). Third, when non-ovulated goldfish are injected with prostaglandin $F_{2\alpha}$ ($PGF_{2\alpha}$), they become sexually receptive and males spawn normally with them, suggesting that $PGF_{2\alpha}$ injection evokes pheromone release (Stacey, 1981). Furthermore, destroying the olfactory system of male goldfish makes them unresponsive to $PGF_{2\alpha}$-injected fish (Stacey and Kyle, 1983).

To test whether $PGF_{2\alpha}$ injection elicits pheromone release, we exposed male goldfish to the odour of naturally-ovulated and $PGF_{2\alpha}$-injected female goldfish and observed their behaviour. Both odours elicited similar reproductive behaviours (Sorensen et al., 1986). To verify this possibility, we used EOG recording to establish that the odour of $PGF_{2\alpha}$-injected goldfish is much more stimulatory than the odour of saline-injected controls (Sorensen et al., 1988). Wondering whether $PGF_{2\alpha}$ might be directly released like $17\alpha,20\beta P$ to function as the stimulatory odour of $PGF_{2\alpha}$-injected fish, we tested olfactory potency of $PGF_{2\alpha}$ using EOG recording. This experiment established prostaglandins as a new class of olfactory stimulants in fish; of eight PGs tested, $PGF_{2\alpha}$ was the most potent, with a threshold of 10^{-10} M (Sorensen et al., 1988; Hara, Chapter 8 herein). However, as potent as $PGF_{2\alpha}$ was, it was not adequate to explain the odour of $PGF_{2\alpha}$-injected fish. Knowing that mammals metabolize PGs very rapidly and release the metabolites in their urine (Granstrom and Kindahl, 1982), we hypothesized that the goldfish postovulatory pheromone might be a $PGF_{2\alpha}$ metabolite.

('separated', dashed line). Aquatic vegetation was then added to the aquaria to stimulate an ovulatory GtH surge. At 3 h intervals during this surge fish were removed from selected aquaria for sampling. Males were bled for GtH determination and their milt stripped and measured. Females were first placed into containers from which water samples were collected to measure free and glucuronated $17\alpha,20\beta P$ (17,20P in the figure), and then removed for blood sampling 2 h later. Approximately half the 78 females ovulated (OV–open circle) and half did not (NOV–closed circle). 312 males were used in this experiment. Fish were sampled once only with approximately equal numbers being represented at each sample time (i.e. approximately 6 OV and 6 NOV females at each time and 26 contact males and 26 separated males). (a) Mean levels of circulating GtH and $17\alpha,20\beta P$ ($\pm$SEM) in ovulatory(OV) and nonovulatory (NOV) female goldfish. (b) Mean ($\pm$SEM) quantities of free and glucuronated $17\alpha,20\beta P$ released to the water by OV and NOV females. (c) Mean GtH ($\pm$SEM) levels of males in contact with females (Contact) or their water only (Separated). (d) Median volume of milt stripped from males. Darkened region on time axis is scotophase. (Reproduced from Stacey et al. (1989) with the permission of N.E. Stacey and Academic Press, Inc.)

Testing the olfactory potency of mammalian PGF metabolites (PGF metabolism has not been studied in fish), we found 15-keto-prostaglandin $F_{2\alpha}$ (15K-$PGF_{2\alpha}$) to have a threshold of 10^{-12} M. Calculations suggested that if $PGF_{2\alpha}$-injected goldfish excrete half of the injected $PGF_{2\alpha}$ as 15K-$PGF_{2\alpha}$, their potency would be explained. A pheromonal role for a 15K-$PGF_{2\alpha}$-like compound received additional support from cross-adaptation studies, which found that adaptation to 15K-$PGF_{2\alpha}$ greatly reduces responsiveness to $PGF_{2\alpha}$-injected fish water. Interestingly, these experiments also suggest that responses to 15K-$PGF_{2\alpha}$ and $PGF_{2\alpha}$ are transduced by different mechanisms, suggesting in turn that the natural pheromone may have several components. Recent experiments using radiolabelled $PGF_{2\alpha}$ suggest that goldfish metabolize $PGF_{2\alpha}$ to several unknown components resembling, but not identical to, $PGF_{2\alpha}$ and 15K-$PGF_{2\alpha}$ (Sorensen, unpublished); the precise identity of this pheromone is unknown.

To test whether ovulated goldfish release PGF-like compounds to the water, water samples from tanks containing ovulated fish were measured for PGFs using radioimmunoassay. Although the antisera used does not detect $PGF_{2\alpha}$ metabolites, and probably grossly under-estimated immunoreactive PGF release, it clearly established that female goldfish release rapidly-increasing quantities of PGF-like compounds immediately after ovulation (Fig. 10.5; Sorensen *et al.*, 1988), a pattern also described for blood-borne PGFs (Bouffard, 1979). To test whether water-borne PGFs trigger male reproductive behaviour, we exposed groups of male goldfish to water-borne PGFs and observed their behaviour. PGs were pumped into the tanks at a low flow rate to mimic the transient pheromonal wisps which fish probably encounter naturally. Water-borne $PGF_{2\alpha}$ and/or 15K-$PGF_{2\alpha}$ stimulated immediate increases in reproductive behaviour similar to responses elicited by the odour of ovulated fish (Sorensen *et al.*, 1988). These responses only lasted 10–15 min, suggesting the importance of other cues to normal spawning behaviour. EOG and behavioural detection thresholds to PGFs were similar.

Although it was clear that the preovulatory 17α,20βP and postovulatory PGF pheromones have different patterns of release, it was unclear how their actions might complement each other. In particular we wondered whether these pheromones could be categorized according to Wilson and Bossert's (1963) definitions as a 'primer' (pheromones with primarily physiological functions) and 'releaser' (pheromones with immediate behavioural functions). Grouped (socially active) and isolated (without social stimulation) males were exposed to 17α,20βP, or to PGFs, or to various mixtures of these, or to food, or to spawning $PGF_{2\alpha}$-injected females, and their behavioural and endocrinological responses were monitored (Sorensen *et al.*, 1989b). Exposure to 17α,20βP stimulated GtH increases in both groups but had only small effects on behaviour; we consider it a primer (Fig. 10.6). In contrast, grouped males exposed to PGFs exhibited large, immediate increases in

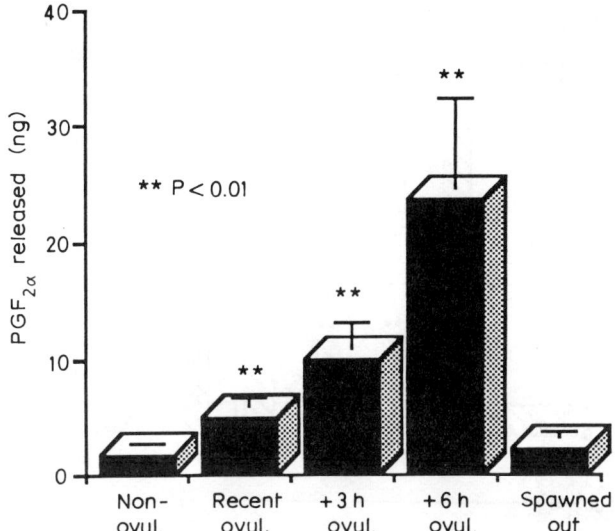

Fig. 10.5 Mean ($\pm$SEM) quantities of immunoreactive F prostaglandin (ng) released by female goldfish during two h periods at different points in their ovulatory cycle: before ovulation ('Non-ovul'), within an hour after ovulating ('Recent ovul'), 3 and 6 h after ovulation, and when 'spawned out' (all eggs released). The antisera used may have underestimated the quantity of PGF metabolites released. A portion of these data was published by Sorensen et al. (1988).

reproductive behaviour and small GtH increases, and the GtH of isolated males was unchanged. We consider the postovulatory PGF pheromone a releaser. Although the PGF pheromone clearly plays an essential role stimulating male sexual behaviour, nevertheless the full expression of male spawning behaviour, and the GtH and milt increases which accompany it (Kyle et al., 1985), appear to require behavioural interaction and other sensory cues associated with it.

We are only beginning to understand pheromonal function in goldfish. For example, females also detect $17\alpha,20\beta P$ (Sorensen et al., 1987a,b), and when exposed to it, often ovulate (Sorensen and Stacey, 1987). $17\alpha,20\beta P$ could be a 'bisexual' pheromone synchronizing reproductive activity in both sexes. Recent studies (unpublished) indicate that $17\alpha,20\beta P$ may also function as a behavioural primer, so that in addition to having more milt, males are more behaviourally competitive by the time of spawning. Whether males release enough $17\alpha,20\beta P$ for it to function as a pheromone has yet to be determined, and the physiological and behavioural significance of increased steroidogenesis and milt production have not been studied. The precise chemical identity of the postovulatory PGF pheromone has yet to be established, and its true biological (behavioural) significance and mechanism of synthesis (metabolism) and release are unknown. The multi-component,

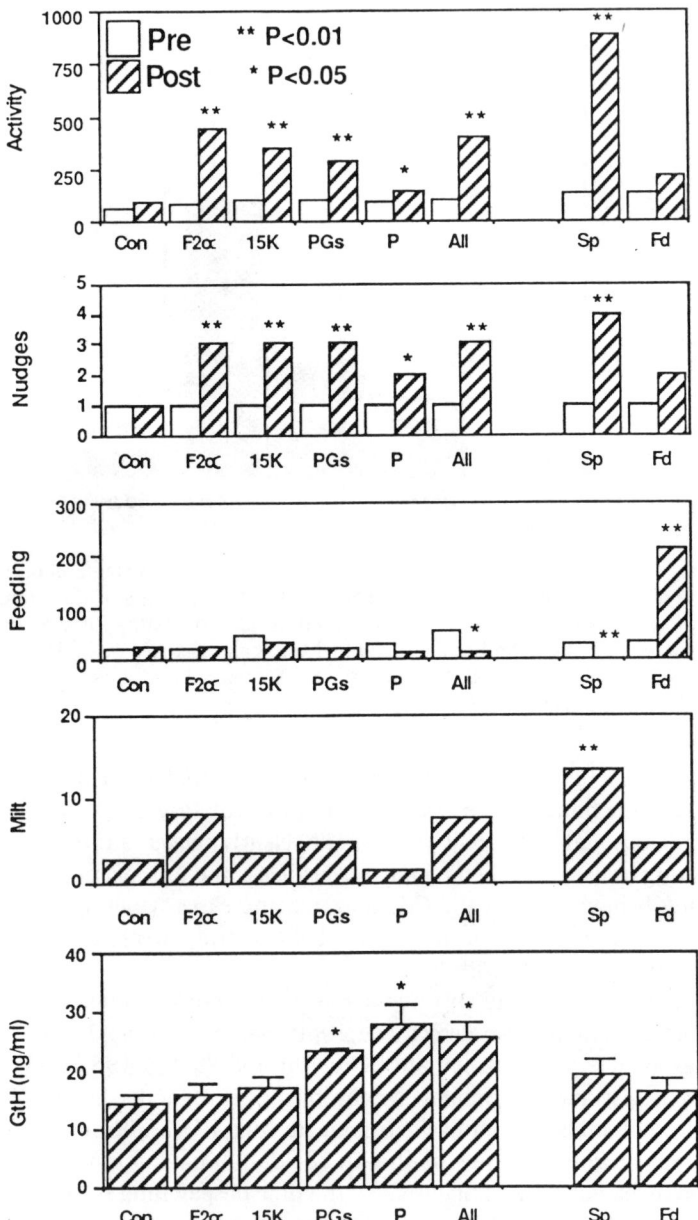

Fig. 10.6 Behavioural and endocrinological responses of groups of 5 male goldfish exposed to various pheromones at a concentration of 10^{-8} M (n = 11 groups of 5 fish tested for each treatment; N = 440). Observation periods prior to odorant addition ('Pre') and during pheromone exposure ('Post') were 15 min long. 'Activity' was a measure of swimming activity. 'Nudges' (physical contact between fish which characterizes sexual arousal) was measured on an ordinal scale. Milt and blood for

metabolite-based PGF pheromone may have more sophisticated functions than $17\alpha,20\beta P$. Evidence is also emerging of other hormonal pheromones in goldfish. In addition to the gender-specific pheromone described by Yamazaki and Watanabe (1979), our recent studies suggest that goldfish release a reproductive inhibitor composed of androgens (unpublished).

10.4 CURRENT STATUS OF THE HYPOTHESIS AND FUTURE CONSIDERATIONS

No teleost sex pheromone system is well understood, but there is convincing evidence that males and females from a number of species use hormones and/or their metabolites as sex pheromones. Because hormone production is synchronized with discrete reproductive events, and the evolution of chemosensory detection mechanisms may be facilitated by the prior existence of hormone receptors, this may be an extremely common phenomenon. Certainly, the finding that the goldfish, a species whose reproductive biology typifies that of many fish, uses common hormones as pheromones strengthens this hypothesis (Stacey and Sorensen, 1991). Still it is essential that we test a variety of species before drawing conclusions. It is important to keep in mind the remarkable behavioural and physiological diversity of fish and the fact that their pheromone systems are likely to be equally diverse. For example, male primers such as $17\alpha,20\beta P$ are not likely to be found in species in which spawning is protracted and/or females form long-term relationships with particular males of their choosing. On the other hand, species found in 'featureless' environments (e.g. the deep sea, turbid tropical rivers, under ice) and/or whose spawning activity is limited to short periods because of unpredictable social or environmental factors, may commonly use bisexual primers and releasers. It seems likely that most externally fertilizing fish use PG postovulatory pheromones, because males that are able to recognize reproductively active females will almost always have a competitive advantage over those without this ability. $PGF_{2\alpha}$ injection has already been shown to elicit seemingly normal spawning activity in several species of fish (Stacey, 1987), and in the fathead minnow, *Pimephales promelas*, it has specifically been shown to evoke pheromone release (Cole and Smith, 1987).

GtH were collected 30 minutes after exposure. Median values are given for all measures except for GtH which are means ($\pm$SEM). Differences represent comparisons between Pre and Post for each treatment and were calculated using Mann-Whitney-U ($P<0.05$) comparisons corrected for multiple comparisons. Abbreviations: Con: ethanol carrier; F2α:PGF2α; 15K:15K-PGF$_{2\alpha}$; PGs:mixture of PGF$_{2\alpha}$ and 15K-PGF$_{2\alpha}$; P:17α,20βP; All:A mixture of 17α,20βP, PGF$_{2\alpha}$, and 15K-PGF$_{2\alpha}$; Sp:Spawning with PGF$_{2\alpha}$-injected female; Fd:food odour. (Reproduced from Sorensen *et al.* (1989) with the permission of Academic Press, Inc.)

Because the diversity of fish exceeds that of their endocrine systems, it is reasonable to ask whether hormonal pheromones should always be expected to be species specific. I believe that this expectation is unwarranted, and evidence for both species-specific and non-species-specific pheromones is well documented (Hunter and Hasler, 1965; Rossi, 1969; Chien, 1973; Chen and Martinich, 1975; Honda, 1982a,b; Liley, 1982; McKinnon and Liley, 1987). As suggested and demonstrated by McKinnon and Liley (1987), closely related species may not come to evolve species-specific pheromonal systems unless there is evolutionary pressure for them to do so, i.e. they are sympatric during their spawning season. Even then, sympatric species may 'share' pheromones if they can use other sensory information to identify the pheromone donor. An example of this may be the anabantid fish *Colisa* spp.; when tested in the absence of other cues, the odour of female *C. lalia* stimulates nest building in male *C. lalia* and *C. labiosa*, but when these species are housed together, only male *C. lalia* build nests (Rossi, 1969). The role of other sensory cues may be particularly relevant when pheromones are used for short-range (within meters) signalling. Although pheromones are generally thought to function over great distances, our goldfish data indicate that this need not always be true. For instance, if one hypothesizes that ovulatory goldfish continuously release $17\alpha,20\beta P$, then the size of the space 'activated' by pheromones is so small (about one litre) and short-lived (lasting only a few seconds) that only proximate conspecifics are likely to encounter it (Sorensen and Stacey, 1990).

Species that use pheromones for long-distance signalling or freely associate with closely related species probably require species-specific pheromone systems. Two mechanisms could lend hormonal pheromone systems this characteristic. First, minor changes in hormone metabolism combined with high olfactory specificity could impart species specificity. It is notable that studies of the olfactory sensitivities of African catfish, goldfish and sculpins to steroidal pheromones constitute the first known demonstration of species specificity in olfactory function in fish, and preliminary studies of cyprinids hint that these differences may be common (Stacey and Sorensen, 1991; unpublished). Second, species-specific information may be conveyed through species-specific odour mixtures. These mixtures could be exclusively hormonal, as suggested for the zebrafish and goldfish (van den Hurk and Lambert, 1983; Sorensen *et al.*, 1988), but they may also be supplemented with mixtures of amino acids and bile acids (Doving *et al.*, 1980; Saglio and Fauconneau, 1985; Bryant and Atema, 1987) which appear to act as social cues in several species. It is interesting to speculate that the unusual fin glands described in the glandulocaudine fish (Nelson, 1964; Weitzman and Fink, 1985) and Mediterranean blenniform fish (Laumen *et al.*, 1974) may produce chemical cues to supplement hormonal metabolites.

Many other fundamental questions about sex pheromone function in fish remain to be addressed. For example, the mechanism(s) by which pheromones are synthesized (metabolized), their relationship to hormones, their complexity, and the means by which they are released are unknown. The behavioural and ecological roles of pheromones and their impact on reproductive success are also not understood in any species. Finally, the basis of neurological responsiveness is poorly understood.

10.5 NEURAL RESPONSIVENESS TO SEX PHEROMONES

Chemosensory systems mediating pheromonal responsiveness

Although olfactory ablation studies indicate that the olfactory system (cranial nerve I) mediates responses to sex pheromones (Stacey and Sorensen, 1986; Stacey et al., 1987), and EOG recording from African catfish (Resink et al., 1989a) and goldfish (Sorensen et al., 1987a, 1990) has verified that the olfactory epithelium is acutely sensitive to pheromones, it has yet to be determined whether other chemosensory systems also detect pheromones. This is an important question because the neurobiology and central connections of the gustatory system (cranial nerves VII, IX, X) and the common chemical sense (principally the trigeminal nerve, cranial nerve V) differ markedly from that of olfaction. They have different receptor cells, their responses are conveyed by different cranial nerves with different terminal fields (Satou, Chapter 3, and Kanwall and Finger, Chapter 5 herein), and they appear to mediate 'simpler' behaviours than olfaction (Atema, 1980). Studies of the threespine stickleback, *Gasterosteus aculeatus* (Segaar et al., 1983), *Mollienesia sphenops* (Zeiske, 1968), and *Trichogaster trichopterus* (Pollack et al., 1978), suggest that gustation plays a limited role in the reproductive behaviour of these species. Intriguingly, studies of olfactory and gustatory sensitivity to amino acids and bile salts (Caprio, 1976, 1982; Hara et al., 1984) have yet to demonstrate a clear difference between these systems.

Although it is clear that the olfactory epithelium is sensitive to pheromones, it cannot be assumed that the olfactory system necessarily mediates these responses. The reason for this is that although most nerve fibres associated with the olfactory epithelium are olfactory, a small number from the trigeminal (cranial nerve V) and terminal nerve (cranial nerve 0) also enter this tissue (Satou, Chapter 3, and Hara, Chapter 8, herein). The trigeminal system is believed to detect noxious stimuli (Silver, 1987), but the function of the terminal nerve (TN) is unknown, and Demski and Northcutt (1983) suggest that it, and not the olfactory system, mediates responses to sex pheromones. They base this hypothesis on the fact that the TN is immunoreactive for gonadotrophin hormone releasing hormone (GnRH) and has terminal fields in brain regions known to mediate reproductive behaviour, and electrical stimulation of the goldfish visual system (which receives TN input) evokes sperm release. Although this hypothesis has generated considerable controversy, it remains unresolved because the TN is intimately associated with the olfactory system and difficult to isolate. Three findings from the goldfish suggest that the terminal nerve is not chemoreceptive. First, male goldfish lacking a TN continue to respond to pheromonal 17α,20βP (Kyle, 1987). Second, although we have been unable to record any change in the firing rate of TN cell bodies of goldfish exposed to pheromones, we have recorded changes in the firing rate of the mitral and granule cells of these fish (Fujita et al., 1991). Third, no-one has demonstrated that the TN reaches the surface of the olfactory epithelium (Kyle, unpublished). In conclusion, although prevailing evidence indicates that the olfactory system exclusively mediates responses to sex pheromones in fish, conclusive evidence is lacking.

Pheromone receptor mechanisms and signal transduction

Next to nothing is known about how the olfactory epithelium of fish detects and transmits pheromonal information. Although the fundamental similarity between EOG responses to pheromones and amino acids (Sorensen et al., 1987a, 1988; Resink et al., 1989a) suggests that they use similar biochemical mechanisms, our poor understanding of the EOG and second messenger function (Lancet, 1988; Breer et al., 1990; Brand and Bruch, Chapter 7 herein) make this a speculative suggestion. More substantive evidence comes from recent ligand binding studies indicating that 17α,20βP, like other olfactory stimulants (Rhein and Cagan, 1983), binds to the membrane fraction of goldfish olfactory epithelial preparations (Rosenblum et al., 1991). Fish olfactory epithelia have at least two types of presumptive sensory cells, and it will be interesting to test whether, as proposed by Thommesen (1982), they are differentially sensitive to pheromones and amino acids (see also Hara, Chapter 8 herein). Pheromone receptors, second messenger systems, deactivation (metabolism) processes, signal transduction systems and the location of these systems all await investigation.

Could pheromonal receptor systems be derived from hormonal systems? The answer is unclear, but it is striking that membrane-bound hormonal receptors for both 17α,20βP and prostaglandins have been described. Although steroid hormone receptors are traditionally thought to be located in the cytoplasm or cell nucleus and evoke slow changes by altering gene expression (Jensen et al., 1982), clear evidence that certain neural tissue has membrane-bound steroid hormone receptors is emerging (Haukkamaa, 1987). Furthermore, these systems have rapid and specific electrical actions: oestradiol rapidly alters neural firing in the rat brain (Kelley et al., 1977; Nabekura et al., 1986), and several steroids bind specifically *in vitro* to synaptic membranes from the rat brain (Towle and Sze, 1983). Especially intriguing is the characterization of 17α,20βP receptors on the surface of goldfish and rainbow trout, *Oncorhynchus mykiss*, oocytes (Nagahama, 1987; Maneckjee *et al.*, 1989) and progestational receptors on *Xenopus* oocytes (Maller, 1985). Membrane-bound $PGF_{2\alpha}$ receptors have also been isolated in sheep corpora lutea (Rao, 1975) and are presumably present in the goldfish brain (Stacey and Peter, 1979). It seems reasonable to speculate that, as first suggested by Kittredge *et al.* (1971), pheromone olfactory receptors have evolutionary links to membrane-bound hormonal receptors. Whether membrane-bound pheromone and hormone receptors need to (or in fact do) share similar transduction mechanisms is pure conjecture, but similar kinds of secondary messenger systems have been described in both systems (Lancet, 1988).

Central mechanisms

Although very little is known about brain mechanisms responsible for pheromonal responsiveness, we do have a cursory understanding of neural processing in the olfactory bulbs and tracts. Fish olfactory bulbs are organized into relatively independent lateral and medial components, which send secondary olfactory afferents to the brain in the lateral and medial olfactory tracts (the LOT and MOT) (Satou, Chapter 3 herein). The MOT also includes efferent fibres from the brain and the TN (Kyle *et al.*, 1987). Several studies indicate that the MOT carries pheromonal information. First, fibres from the MOT project to the ventral telencephalon and preoptic areas–regions of the brain which lesioning and electrical stimulation have shown to control reproductive processes in goldfish (Kyle *et al.*, 1987). Second, electrical stimulation of the MOT stimulates reproductive behaviour in cod, *Gadus morhua* (Doving and Selset, 1980) and sperm release in the goldfish (Demski and Dulka, 1984). Third, lesioning the MOT reduces greatly the courtship behaviour of male goldfish placed with $PGF_{2\alpha}$-injected females (Stacey and Kyle, 1983) and eliminates $17\alpha,20\beta P$-stimulated increases in milt and GtH (Stacey and Sorensen, 1986; Dulka and Stacey, 1991). Fourth, electrical responses to $17\alpha,20\beta P$ and PGFs are recorded from the MOT and not from the LOT (Sorensen *et al.*, 1991). How the olfactory bulbs encode pheromonal information, which component of the MOT carries pheromonal information, what the role of the efferent olfactory tract fibres might be, and how the brain processes this information have yet to be studied.

Hormonal influences on pheromonal responsiveness

Peripheral responsiveness to hormonal sex pheromones appears to be influenced by endocrine state. Testosterone treatment increases the thickness of the olfactory epithelium of hypophysectomized goldfish (Yamazaki and Watanabe, 1979), and EOG recording suggests that hypophysectomy reduces general olfactory sensitivity (Sorensen et al., 1987b). Only mature goldfish respond to $PGF_{2\alpha}$-injected female goldfish, and their olfactory sense is slightly more sensitive to PGs than that of females (Sorensen, unpublished). Recently, Moore and Scott (1991) report that the olfactory system of precocious male Atlantic salmon (*Salmo salar* L.) is highly sensitive to testosterone for a limited period of the year. This dramatic seasonality in olfactory function contrasts with the goldfish which shows only subtle changes. Histological studies are less conclusive; both positive (Schreibman et al., 1986), and negative (Pankhurst and Lythgoe, 1983; Sorensen and Pankhurst, 1988) correlations between maturity and development of the olfactory epithelium have been described in different species.

Steroidal influences on central processes appear complex. Female responsiveness to pheromones appears to be triggered by ovulation in several species (Colombo et al., 1982; Resink et al., 1987), and although immature male and female goldfish detect $17\alpha,20\beta P$, they do not respond to it (unpublished results). Electroencephalogram (EEG) recording from the olfactory bulbs of goldfish indicates that prolonged treatment with either progesterone or oestradiol can potentiate or reduce olfactory responsiveness to NaCl (Oshima and Gorbman, 1968). Hara (1967) reports that oestradiol alters evoked neural activity in the olfactory bulbs in a sex-dependent manner and alters the magnitude of the positive after-potentials. A relationship between thyroid activity and olfactory-mediated learning (imprinting) has been suggested in salmonids, based on correlations between heart rate conditioning experiments and thyrotrophic hormone administration (Hasler and Scholz, 1983) and thyroid cell height (Morin et al., 1989). Oshima and Gorbman (1966) reported that thyroxine potentiates responsiveness of the goldfish olfactory bulb and inhibits centrifugal impulses from the brain. All of these experiments were conducted when fish hormone and pheromone systems were poorly understood, and it is difficult to draw clear conclusions from them. Hopefully more conclusive experiments will be conducted on this fascinating topic in the near future.

10.6 GENERAL CONCLUSIONS

Evidence is rapidly accumulating that pheromonal and endocrinological function are intertwined in teleost fish. Hormones and their metabolites have been found to function as pheromones in several species; this relationship appears likely to be common, although that has yet to be tested. Many basic questions about pheromone identity and synthesis, its relationship to endocrine function, species specificity, neural and endocrine mechanisms regulating responsiveness, and the significance of pheromones to reproductive processes, need to be addressed. There is little doubt that many exciting discoveries await in this rapidly growing field.

ACKNOWLEDGEMENT

Contribution No. 17,853 from the Minnesota Agricultural Experiment Station. I thank T.J. Hara and C. Hollingworth for their careful reviews of this chapter and N.E. Stacey for his enthusiastic collaboration and support.

REFERENCES

Amouriq, L. (1964) L'activité et le phénomène social chez *Lebistes reticulatus* (Poeciliidae; Cyprinidontiformes). *C.r. hebd. Séanc. Acad. Sc. Paris*, **259**, 2701–2.

Amouriq, L. (1965) Origine de la substance dynamogène émise par *Lebistes reticulatus* (Poissonfemelle Poeciliidae; Cyprinidontiforme). *C.r. hebd. Séanc. Acad. Sc. Paris*, **260**, 2334–5.

Amouriq, L. (1967a) L'optimum de sensibilité de *Lebistes reticulatus* (Poisson, Poeciliidae; Cyprinidontiformes) à l'hormone synthétique femelle. *Revue. Comp. Animal*, **3**, 57–60.

Amouriq, L. (1967b) Sensibilité des *Lebistes reticulatus* mâle à la substance dynamogène émise par les femelles de Poeciliidae et de Gasterosteidae. *Revue. Comp. Animal*, **4**, 83–6.

Atema, J. (1980) Smelling and tasting underwater. *Oceanus*, **23**, 4–18.

Atema, J. and Gagosian, R.B. (1973) Behavioral responses of male lobsters to ecdysones. *Mar. Behav. Physiol.*, **2**, 15–20.

Bardach, J.E. and Todd, J.H. (1970) Chemical communication in fish, in *Communication by Chemical Signals* (eds J.W. Johnston, jun., D.G. Moulton and A. Turk), Appelton-Century-Crofts, New York, pp. 205–40.

Beauchamp, G.K., Doty, R.L., Moulton, D.G. and Mugford, R.A. (1976) The pheromone concept in mammalian chemical communication: a critique, in *Mammalian Olfaction, Reproductive Processes, and Behavior* (ed. R. Doty), Academic Press, New York, pp. 143–60.

Bethe, A. (1932) Vernachlassigte hormone. *Naturwissenschaften*, **11**, 177–81.

Bouffard, R.E. (1979) The role of prostaglandins during sexual maturation, ovulation, and spermiation in the goldfish, *Carassius auratus*. MSc thesis, University of British Columbia, Vancouver, 155 pp.

Breer, H. Boekhoff, I. and Tareilus, E. (1990) Rapid kinetics of second messenger formation in olfactory transduction. *Nature, Lond.* **343**, 65–8.

Brown, R.E. (1979) Mammalian social odors: a critical review, in *Advances in the Study of Behavior*. Vol. 10 (eds J.S. Rosenblatt, R.A. Hinde, C. Beer and M.-C. Busnel), Academic Press, New York, pp. 103–62.

Bryant, B.P. and Atema, J. (1987) Diet manipulation affects social behavior of catfish: the importance of food odor. *J. Chem. Ecol.*, **13**, 1645–62.

Caprio, J. (1976) Electrophysiological distinctions between taste and smell of amino acids in catfish. *Nature, Lond.*, **266**, 850–1.

Caprio, J. (1982) High sensitivity and specificity of olfactory and gustatory receptors of catfish to amino acids, in *Chemoreception in Fishes* (ed. T.J. Hara), Elsevier, Amsterdam, pp. 109–34.

Cetta, F. and Goetz, F.Wm. (1982) Ovarian and plasma prostaglandin E and F levels in brook trout (*Salvelinus fontinalis*) during pituitary induced ovulation. *Biol. Reprod.*, **27**, 1216–21.

Chen, L.C. and Martinich, R.L. (1975) Pheromonal stimulation and metabolite inhibition of ovulation in the zebrafish, *Brachydanio rerio*. *Fish. Bull.*, (Natl. Mar. Fish. Serv. U.S.), **73**, 889–94.

Chien, A.K. (1973) Reproductive behavior of the angelfish *Pterophyllum scalare* (Pisces: Cichlidae). II. Influence of male stimuli upon the spawning rate of females. *Anim. Behav.*, **21**, 457–63.

Cole, K.S. and Smith, R.J.F. (1987) Release of chemicals by prostaglandin treated female fathead minnows, *Pimephales promelas*, that stimulates male courtship. *Horm. Behav.*, **21**, 440–56.

Colombo, L., Belvedere, P.C. and Marconato, A. (1979) Biochemical and functional aspects of gonadal biosynthesis of steroid hormones in teleost fishes. *Proc. natn. Acad. Sci. India*, **B45**, 443–51.

Colombo, L., Marconato, A., Belvedere, P.C. and Frisco, C. (1980) Endocrinology of teleost reproducton: a testicular steroid pheromone in the black goby, *Gobius jozo* L. *Boll. Zool.*, **47**, 355–64.

Colombo, L., Belvedere, P.C., Marconato, A. and Bentivegna, F. (1982) Pheromones in teleost fish, in *Proc. Second Int. Symp. Reprod. Physiol. Fish* (eds C.J.J. Richter and H.J.Th. Goos), Pudoc, Wageningen, The Netherlands, pp. 84–94.

Crow, R.T. and Liley, N.R. (1978) A sexual pheromone in the guppy, *Poecilia reticulata* (Peters). *Can. J. Zool.*, **57**, 184–8.

Darwin, C. (1887) *The Descent of Man and Selection in Relation to Sex* (2nd edn, revised). John Murray, London, 688 pp.

Demski, L.S. and Dulka, J.G. (1984) Functional-anatomical studies on sperm release evoked by electrical stimulation of the olfactory tract in goldfish. *Brain Res., Amst.*, **291**, 241–7.

Demski, L.S. and Northcutt, R.G. (1983) The terminal nerve: a new chemosensory system in vertebrates? *Science, Wash., D.C.*, **202**, 435–7.

Døving, K.B. (1976) Evolutionary trends in olfaction, in *The Structure–Activity Relationships in Chemoreception* (ed. G. Benz), IRL Press, London, pp. 149–59.

Døving, K.B. and Selset, R. (1980) Behavioral patterns in cod released by electrical stimulation of olfactory tract bundles. *Science, Wash., D.C.*, **207**, 559–60.

Døving, K.B., Selset, R. and Thommesen, G. (1980) Olfactory sensitivity to bile acids in salmonid fishes. *Acta. physiol. scand.*, **108**, 123–31.

Drickamer, L.C. (1989) Pheromones: behavioral and biochemical aspects, in *Advances in Comparative and Environmental Physiology 3*, (ed. J. Balthazart), Springer-Verlag, New York, pp. 270–356.

Dulka, J.G. and Stacey, N.E. (1991) Effects of olfactory tract lesions on gonadotropin and milt responses to the female pheromone, $17\alpha,20\beta$-dihydroxy-4-pregnen-3-one, in male goldfish. *J. exp. Zool.*, **253**, 223–9.

Dulka, J.G., Stacey, N.E., Sorensen, P.W. and Van Der Kraak, G.J. (1987) Sex steroid pheromone synchronizes male–female spawning readiness in the goldfish. *Nature, Lond.*, **325**, 251–3.

Dunham, P.J. (1978) Sex pheromones in crustacea. *Biol. Rev.*, **53**, 555–83.

Emanuel, M.E. and Dodson, J.J. (1979) Modification of the rheotropic behavior of male rainbow trout (*Salmo gairdneri*) by ovarian fluid. *J. Fish. Res. Bd Can.*, **36**, 63–8.

Fujita, I., Sorensen, P.W., Stacey, N.E. and Hara, T.J. (1991) The olfactory system, not the terminal nerve, functions as the primary chemosensory pathway mediating responses to sex pheromones in goldfish. *Brain Behav. Evol.*, **38**, 313–21.

Gagosian, R.B. and Atema, J. (1973) Behavioral responses of male lobsters to ecdysone metabolites. *Mar. Behav. Physiol.*, **2**, 115–20.

Gandolfi, G. (1969) A chemical sex attractant in the guppy *Poecilia reticulata* (Peters: Poeciliidae). *Monitore zool. ital. (N.S.).*, **3**, 89–98.

Gleeson, R.A., Adams, M.A. and Smith, A.B. (1984) Characterization of a sex pheromone in the blue crab, *Callinectes sapidus*: crustecdysone studies. *J. Chem. Ecol.*, **10**, 913–21.

Goetz, F.Wm. (1983) Hormonal control of oocyte final maturation and ovulation in fishes, in *Fish Physiology* Vol. IXB (eds W.S. Hoar, D.J. Randall and E.M. Donaldson), Academic Press, New York, pp. 117–70.

Granstrom, E. and Kindahl, H. (1982) Species differences in circulating prostaglandin metabolites: relevance for the assay of prostaglandin release. *Biochim. biophys. Acta*, **713**, 555–69.

Hara, T.J. (1967) Electrophysiological studies of the olfactory system of the goldfish, *Carassius auratus* L.—III. Effects of sex hormones on olfactory activity. *Comp. Biochem. Physiol.*, **22**, 209–25.

Hara, T.J. (1986) Role of olfaction in fish behaviour, in *Behaviour of Teleost Fishes* (ed. T.J. Pitcher), Croom Helm, London, pp. 152–76.

Hara, T.J., Macdonald, S., Evans, R.E., Marui, T. and Arai, S. (1984) Morpholine, bile acids, and skin mucus as possible chemical cues in salmonid homing: electrophysiological re-evaluation, in *Mechanisms of Migration in Fishes* (eds J.D. McCleave, G.P. Arnold, J.J. Dodson and W.H. Neill), Plenum Press, New York, pp. 363–78.

Hasler, A.D. and Scholz, A.T. (1983) *Olfactory Imprinting and Homing in Salmon* (Zoophysiology, Vol. 14), Springer-Verlag, Berlin, 134 pp.

Haukkamaa, M. (1987) Membrane-associated steroid hormone receptors, in *Steroid Hormone Receptors* (ed. C.R. Clark), Ellis Horwood, Chichester, England, pp. 155–69.

Honda, H. (1979) Female sex pheromone of the ayu, *Plecoglossus altivelis* involved in courtship behaviour. *Bull. Jap. Soc. scient. Fish.*, **45**, 1375–80.

Honda, H. (1980a) Female sex pheromone of rainbow trout, *Salmo gairdneri*, involved in courtship behaviour. *Bull. Jap. Soc. scient. Fish.*, **46**, 1109–12.

Honda, H. (1980b) Female sex pheromone of the loach, *Misgurnus anguillicaudatus*, involved in courtship behaviour. *Bull. Jap. Soc. scient. Fish.*, **46**, 1223–5.

Honda, H. (1982a) On the female sex pheromones and courtship behaviour in the bitterlings, *Rhodeus ocellatus ocellatus* and *Acheilognathus lanceolatus*. *Bull. Jap. Soc. scient. Fish.*, **48**, 43–6.

Honda, H. (1982b) On the female sex pheromones and courtship behavior in the salmonids, *Oncorhynchus masou* and *O. rhodorus*. *Bull. Jap. Soc. scient. Fish.*, **48**, 47–55.

Hunter, J.R. and Hasler, A.D. (1965) Spawning association of the redfin shiner, *Notropis umbratilis*, and the green sunfish, *Lepomis cyanellus*. *Copeia*, **1965**, 265–81.

Jensen, E.V., Greene, G.L., Closs, L.E., deSombre, E.R. and Nadji, M. (1982) Receptors reconsidered: a 20 year perspective. *Recent Prog. Horm. Res.*, **38**, 1–40.

Johansen, P.H. (1984) Female pheromone and the behaviour of male guppies (*Poecilia reticulata*) in a temperature gradient. *Can. J. Zool.*, **63**, 1211–13.

Karlson, P. and Lüscher, M. (1959) 'Pheromones': a new term for a class of biologically active substances. *Nature, Lond.*, **183**, 55–6.

Kelley, M.J., Moss, R.L. and Dudley, C.A. (1977) Effects of microelectrophoretically applied estrogen, cortisol, and acetylcholine on medial preoptic–septal unit activity throughout the estrus cycle of the female rat. *Exp. Brain Res.*, **30**, 53–64.

Kirschenblatt, J. (1962) Terminology of some biologically active substances and validity of the term 'pheromones'. *Nature, Lond.*, **195**, 916–17.

Kittredge, J.S. and Takahashi, F.T. (1972) The evolution of sex pheromone communication in Arthropoda. *J. theor. Biol.*, **35**, 467–71.

Kittredge, J.S., Terry, M. and Takahashi, F.T. (1971) Sex pheromone activity of the moulting hormone, crustecdysone, on male crabs (*Pachygrapsus crassipes*, *Cancer antennarius*, and *C. anthonyi*). *Fish. Bull.*, (Natl. Mar. Fish Serv. U.S.), **69**, 337–43.

Kittredge, J.S., Takahashi, F.J., Lindsey, J. and Lasker, R. (1974) Chemical signals in the sea: marine allelochemics and evolution. *Fish. Bull.*, (Natl. Mar. Fish Serv. U.S.), **72**, 1–11.

Kobayashi, M., Aida, K. and Hanyu, I. (1986) Pheromone from ovulatory female goldfish induces gonadotropin surge in males. *Gen. comp. Endocrinol.*, **62**, 70–9.

Kyle, A.L. (1987) Effects of the nervus terminalis ablation on gonad weight, response to a sex pheromone, and courtship behaviour in the male goldfish (*Carassius auratus*), in *Proc. Third Int. Symp. Reprod. Physiol. Fish* (eds D.R. Idler, L.W. Crim and J.M. Walsh), Memorial University Press, St. John's, Newfoundland, Canada, p. 161.

Kyle, A.L., Stacey, N.E., Peter, R.E. and Billard, R.E. (1985) Elevations in gonadotropin concentrations and milt volumes as a result of spawning in the goldfish. *Gen. comp. Endocrinol.*, **57**, 10–22.

Kyle, A.L., Sorensen, P.W., Stacey, N.E. and Dulka, J.G. (1987) Medial olfactory tract pathways controlling sexual reflexes and behavior in teleosts, in *The Terminal Nerve* (*nervus terminalis*) *Structure, Function and Evolution* (eds L.S. Demski and M. Schwanzel-Fukuda), *Ann. N.Y. Acad. Sci.*, **519**, 97–107.

Lambert, J.G.D., van den, R. Hurk, Schoonen, W.G.E.J., Resink, J.W. and van Oordt, P.G.W.J. (1986) Gonadal steroidogenesis and the possible role of steroid glucuronides as sex pheromones in two species of teleosts. *Fish Physiol. Biochem.*, **2**, 101–7.

Lancet, D. (1988) Molecular components of olfactory reception and transduction, in *Molecular Neurobiology of the Olfactory System* (eds F.L. Margolis and T.V. Getchell), Plenum Press, New York, pp. 25–50.

Laumen, J., Pern, U. and Blum, V. (1974) Investigations on the function and hormonal regulation of the anal appendices in *Blennius pavo* (Risso). *J. exp. Zool.*, **190**, 47–56.

Lee, C.-T. and Ingersoll, D.W. (1979) Social chemosignals in five Belontiidae (Pisces) species. *J. comp. Physiol. Psychol.*, **93**, 117–18.

Liley, N.R. (1982) Chemical communication in fish. *Can. J. Fish. Aquat. Sci.*, **39**, 22–35.

Liley, N.R. and Stacey, N.E. (1983) Hormones, pheromones and reproductive behavior, in *Fish Physiology*. Vol. IX *Reproduction* Part B *Fertility Control* (eds W.S. Hoar, D.G. Randall and E.M. Donaldson), Academic Press, New York, pp. 1–49.

McKinnon, J.S. and Liley, N.R. (1987) Asymmetric species specificity in responses to female sexual pheromone by males of two species of *Trichogaster* (Pisces: Belontiidae). *Can. J. Zool.*, **65**, 1129–34.

Maller, J.L. (1985) Regulation of amphibian oocyte maturation. *Cell. Differ.*, **16**, 211–21.

Maneckjee, A., Weisbart, M. and Idler, D.R. (1989) The presence of 17α,20β-dihydroxy-4-pregnen-3-one receptor activity in the ovary of the brook trout, *Salvelinus fontinalis*, during terminal stages of oocyte maturation. *Fish Physiol. Biochem.*, **6**, 19–38.

Melrose, D.R., Reed, H.C.B. and Patterson, R.L.S. (1971) Androgen steroids associated with boar odour as an aid to detection of oestrus in pig artificial insemination. *Br. vet. J.*, **127**, 497–502.

Meyer, J.H. and Liley, N.R. (1982) The control of production of a sexual pheromone in the female guppy, *Poecilia reticulata*. *Can. J. Zool.*, **60**, 1505–10.

Moore, A. and Scott, A.P. (1991) Testosterone is a potent odorant in precocious male Atlantic salmon (*Salmo salar* L.) parr. *Phil. Trans. R. Soc. Lond.* B., **332**, 241–4.

Morin, P.-P., Dodson, J.J. and Doret, F.Y. (1989) Thyroid activity concomitant with olfactory learning and heart rate changes in Atlantic salmon, *Salmo salar*, during smoltification. *Can. J. Fish. Aquat. Sci.*, **46**, 131–6.

Nabekura, J., Oomura, Y., Minami, T., Mizuno, Y. and Fukuda, A. (1986) Mechanism of the rapid effect of 17β-estradiol on medial amygdala neurons. *Science, Wash., D.C.*, **233**, 226–8.

Nagahama, Y. (1987) $17\alpha,20\beta$-dihydroxy-4-pregnen-3-one: a teleost maturation-inducing hormone. *Dev. Growth and Differ.*, **29**, 1–12.

Nelson, K. (1964) Behavior and morphology in the glandulocaudine fishes (Ostariophysi, Characidae). *Univ. Calif., Berkeley, Publs Zool.*, **75**, 55–152.

Oshima, K. and Gorbman, A. (1966) Influence of thyroxine and steroid hormones on spontaneous and evoked unitary activity in the olfactory bulb of goldfish. *Gen. comp. Endocrinol.*, **7**, 482–91.

Oshima, K. and Gorbman, A. (1968) Modification by sex hormones of the spontaneous and evoked bulbar activity in goldfish. *J. Endocrinol.*, **40**, 409–20.

Pankhurst, N.E. and Lythgoe, J.N. (1983) Changes in vision and olfaction during sexual maturation in the European eel *Anguilla anguilla* (L.). *J. Fish Biol.*, **23**, 229–40.

Partridge, B.L., Liley, N.R. and Stacey, N.E. (1976) The role of pheromones in the sexual behaviour of the goldfish. *Anim. Behav.*, **24**, 291–9.

Pollack, E.I., Becker, L.R. and Haynes, K. (1978) Sensory control of mating in blue gourami, *Trichogaster trichopterus* (Pisces, Belontiidae). *Behav. Biol.*, **22**, 92–103.

Rao, Ch.V. (1975) The presence of discrete receptors for prostaglandin $F_{2\alpha}$ in the cell membranes of bovine corpora lutea. *Biochem. biophys. Res. Commun.*, **64**, 416–24.

Resink, J.W., van den Hurk, R., Groenix Van Zoelen, R.F.O. and Huisman, E.A. (1987) The seminal vesicle as source of sex attracting substances in the African catfish, *Clarias gariepinus*. *Aquaculture*, **63**, 115–27.

Resink, J.W., Voorthius, P.K., van den Hurk, R., Peters, R.C. and Van Oordt, P.G.W.J. (1989a) Steroid glucuronide of the seminal vesicle as olfactory stimuli in African catfish, *Clarias gariepinus*. *Aquaculture*, **83**, 153–66.

Resink, J.W., Schoonen, W.G.E.J., Albers, P.C.H., File, D.M., Notenboom, C.D., van den Hurk, and van Oordt, P.G.W.J. (1989b) The chemical nature of sex attracting pheromones from the seminal vesicle of the African catfish, *Clarias gariepinus*. *Aquaculture*, **83**, 137–51.

Rhein, L.D. and Cagan, R.H. (1983) Biochemical studies of olfaction: binding specificity of odorants to a cilia preparation from rainbow trout olfactory rosettes. *J. Neurochem.*, **41**, 569–77.

Rosenblum, P.M., Sorensen, P.W., Stacey, N.E. and Peter, R.E. (1991) Binding of the steroidal pheromone $17\alpha,20\beta$-dihydroxy-4-pregnen-3-one to goldfish, *Carassius auratus*, olfactory epithelium preparations. *Chem. Senses*, **16**, 143–54.

Rossi, A.C. (1969) Chemical signals and nest building in two species of *Colisa* (Pisces, Anabantidae). *Monitore zool. ital. (N.S.)*, **3**, 225–37.

Saglio, P. and Fauconneau, B. (1985) The amino acid content in the skin mucus of goldfish, *Carassius auratus*: influences of feeding. *Comp. Biochem. Physiol.*, **82A**, 67–70.

Schoonen, W.G.E.J. and Lambert, J.G.D. (1986) Steroid metabolism in the seminal vesicles of the African catfish *Clarias gariepinus* (Burchell), during the spawning season, under natural conditions, and kept in ponds. *Gen. comp. Endocrinol.*, **61**, 355–67.

Schoonen, W.G.E.J., Lambert, J.G.D. and van Oordt, P.G.W.J. (1988) Quantitative analysis of steroids and steroid glucuronides in the seminal vesicle fluid of feral spawning and feral and cultivated nonspawning African catfish, *Clarias gariepinus* (Burchell). *Gen. comp. Endocrinol.*, **70**, 91–100.

Schreibman, M.P., Margolis-Nunno, H. and Halpern-Sebold, L. (1986) The structural and functional relationships between olfactory and reproductive systems from birth to old age in fish, in *Chemical Signals in Vertebrates. 4. Ecology, Evolution, and Comparative Biology* (eds D. Duvall, D. Muller-Schwarze and R.D. Silverstein), Plenum Press, New York, pp. 155–72.

Scott, A.P. and Canario, A.V.M. (1987) Status of oocyte maturation-inducing steroids in teleosts, in *Proc. Third Int. Symp. Reprod. Physiol. Fishes* (eds D.R. Idler, L.W. Crim and J.W. Walsh), Memorial University Press, St. John's, Newfoundland, Canada, pp. 224–34.

Segaar, J., deBruin, J.P.C., and van der Meche-Jacobi, M.E. (1983) Influence of chemical receptivity on reproductive behavior of the male three spined stickleback (*Gasterosteus aculeatus* L.). *Behaviour*, **86**, 100–66.

Seifert, P. (1982) Studies on the sex pheromone of the shore crab, *Carcinus maenas*, with special regard to ecdysone excretion. *Ophelia*, **21**, 147–58.

Silver, W.L. (1987) The common chemical sense, in *Neurobiology of Taste and Smell* (eds T.E. Finger and W.L. Silver), Wiley, New York, pp. 65–88.

Solomon, D.J. (1977) A review of chemical communication in freshwater fish. *J. Fish Biol.*, **11**, 363–76.

Sorensen, P.W. and Pankhurst, N.E. (1988) Histological changes in the gonad, skin, intestine and olfactory epithelium of artificially-matured male American eels, *Anguilla rostrata* (LeSueur). *J. Fish Biol.*, **32**, 297–307.

Sorensen, P.W. and Stacey, N.E. (1987) $17\alpha,20\beta$-dihydroxy-4-pregnen-3-one functions as a bisexual priming pheromone in goldfish. *Am. Zool.*, **27**, 412.

Sorensen, P.W. and Stacey, N.E. (1990) Identified hormonal pheromones in the goldfish: a model for sex pheromone function in teleost fish, in *Chemical Signals in Vertebrates V* (eds D.W. MacDonald, D. Muller-Schwarze and S.E. Natynczuk), Oxford University Press, Oxford, pp. 302–14.

Sorensen, P.W. and Winn, H.E (1984) The induction of maturation and ovulation in American eels, *Anguilla rostrata* (LeSueur), and the relevance of chemical and visual cues to male spawning behaviour. *J. Fish Biol.*, **25**, 261–8.

Sorensen, P.W., Hara, T.J. and Stacey, N.E. (1987a) Extreme olfactory sensitivity of mature and gonadally-regressed goldfish to a potent steroidal pheromone, $17\alpha,20\beta$-dihydroxy-4-pregnen-3-one. *J. comp. Physiol.*, **160A**, 305–13.

Sorensen, P.W., Hara, T.J. and Stacey, N.E. (1987b) Peripheral olfactory responses of mature, regressed, and hypophysectomized goldfish to a steroidal sex pheromone, L-serine, and taurocholic acid, in *Olfaction and Taste IX* (eds S.D. Roper and J. Atema), *Ann. N.Y. Acad. Sci.*, **510**, 635–7.

Sorensen, P.W., Hara, T.J. and Stacey, N.E. (1991) Sex pheromones selectively stimulate the medial olfactory tracts of male goldfish. *Brain Res. Amst.*, **558**, 343–7.

Sorensen, P.W., Hara, T.J., Stacey, N.E. and Goetz, F.Wm. (1988) F prostaglandins function as potent olfactory stimulants that comprise the postovulatory female sex pheromone in goldfish. *Biol. Reprod.*, **39**, 1039–50.

Sorensen, P.W., Stacey, N.E. and Chamberlain, K.J. (1989b) Differing behavioral and endocrinological effects of two female sex pheromones on male goldfish. *Horm. Behav.*, **23**, 317–32.

Sorensen, P.W., Stacey, N.E. and Naidu, P. (1986) Release of spawning pheromone(s) by naturally ovulated and prostaglandin injected, nonovulated female goldfish, in *Chemical Signals in Vertebrates. 4. Ecology, Evolution and Comparative Biology* (eds D. Duvall, D. Muller-Schwarze and R.D. Silverstein), Plenum Press, New York, pp. 149–54.

Sorensen, P.W., Hara, T.J., Stacey, N.E. and Dulka, J.G. (1990) Extreme olfactory specificity of male goldfish to the preovulatory steroidal pheromone $17\alpha,20\beta$-dihydroxy-4-pregnen-3-one. *J. comp. Physiol.*, **166A**, 373–83.

Spencer, M. and Case, J.F. (1984) Exogenous ecdysteroids elicit low-threshold sensory responses in spiny lobsters. *J. exp. Zool.*, **229**, 163–6.

Stacey, N.E. (1981) Hormonal regulation of female reproductive behavior in fish. *Am. Zool.*, **21**, 305–16.

Stacey, N.E. (1987) Roles of hormones and pheromones in fish reproductive behavior, in *Psychobiology of Reproductive Behavior* (ed. D. Crews), Prentice-Hall, Englewood Cliffs, N.Y. pp. 28–69.

Stacey, N.E. and Goetz, F.Wm. (1982) Role of prostaglandins in fish reproduction. *Can. J. Fish. Aquat. Sci.*, **39**, 92–8.

Stacey, N.E. and Kyle, A.L. (1983) Effects of olfactory tract lesions on sexual and feeding behavior in the goldfish. *Physiol. Behav.*, **30**, 621–8.

Stacey, N.E. and Peter, R.E. (1979) Central action of prostaglandins in spawning behavior of female goldfish. *Physiol. Behav.*, **22**, 1191–6.

Stacey, N.E. and Sorensen, P.W. (1986) $17\alpha,20\beta$-dihydroxy-4-pregnen-3-one: a steroidal primer pheromone which increases milt volume in the goldfish, *Carassius auratus*. *Can. J. Zool.*, **64**, 2412–17.

Stacey, N.E. and Sorensen, P.W. (1991) Function and evolution of fish hormonal pheromones, in *Biochemistry and Molecular Biology of Fishes*. Vol. 1 (eds P.W. Hochachka and T.P. Mommsen), Elsevier Press, New York, pp. 109–35.

Stacey, N.E., Kyle, A.L. and Liley, N.R. (1986) Fish reproductive pheromones, in *Chemical Signals in Vertebrates 4 Ecology, Evolution, and Comparative Biology* (eds D. Duvall, D. Muller-Schwarze and R.D. Silverstein), Plenum Press, New York, pp. 117–33.

Stacey, N.E., Sorensen, P.W., Dulka, J.G., Van Der Kraak, G.J. and Hara, T.J. (1987) Teleosts sex pheromones: recent studies on identity and function, in *Proc. Third Int. Symp. Reprod. Physiol. Fish* (eds D.R. Idler, L.W. Crim and J.M. Walsh), Memorial University Press, St. John's, Newfoundland, Canada, pp. 150–4.

Stacey, N.E., Sorensen, P.W., Dulka, J.G. and Van Der Kraak, G.J. (1989) Direct evidence that $17\alpha,20\beta$-dihydroxy-4-pregnen-3-one functions as the preovulatory pheromone in goldfish. *Gen. comp. Endocrinol*, **75**, 62–70.

Starling, E.H. (1905) The chemical correlations of the functions of the body. *Lancet*, **i**, 340–1.

Tavolga, W.N. (1955) Effects of gonadectomy and hypophysectomy on pre-spawning behavior in males of the gobiid fish, *Bathygobius soporator*. *Physiol. Zool.*, **28**, 218–33.

Tavolga, W.N. (1956) Visual, chemical, and sound stimuli as cues in the sex discriminatory behavior of the gobiid fish, *Bathygobius soporator*. *Zoologica, N.Y.* **41**, 49–65.

Thommesen, G. (1982) Specificity and distribution of receptor cells in the olfactory mucosa of char (*Salmo alpinus* L.). *Acta. physiol. scand.*, **115**, 47–56.

Towle, A.C. and Sze, P.Y. (1983) Steroid binding to synaptic plasma membrane: differential binding of glucocorticoids and gonadal steroids. *J. Steroid Biochem.*, **18**, 135–43.

van den Hurk, R. and Lambert, J.G.D. (1983) Ovarian steroid glucuronides function as sex pheromones for male zebrafish, *Brachydanio rerio*. *Can. J. Zool.*, **61**, 2381–7.

van den Hurk, R., Schoonen, W.G.E.J., Zoelen, G.A. van and Lambert, J.G.D. (1987) The biosynthesis of steroid glucuronides in the testis of the zebrafish *Brachydanio rerio*, and their pheromonal function as ovulation inducers. *Gen. comp. Endocrinol.*, **68**, 179–88.

Van Der Kraak, G.J., Sorensen, P.W., Stacey, N.E. and Dulka, J.G. (1989) Periovulatory female goldfish release three potential pheromones: 17α,20β-dihydroxy-progesterone, 17α,20β-dihydroxy-progesterone glucuronide and 17α-hydroxy-progesterone. *Gen. comp. Endocrinol.*, **73**, 452–7.

Yamazaki, F. and Watanabe, K. (1979) The role of steroid hormones in sex recognition during spawning behaviour of the goldfish, *Carassius auratus* L. *Proc. natn. Acad. Sci. India,*, **45B**, 505–11.

Weitzman, S.H. and Fink, S.V. (1985) Xenurobryconin phylogeny and putative pheromone pumps in glandulocaudine fishes (Teleostei: Characidae). *Smithsonian Contrib. Zool.*, **421**, 1–121.

Wilson, E.O. and Bossert, W.H. (1963) Chemical communication among animals, in *Recent Progress in Hormone Research*. Vol. 19 (ed. G. Pincus), Academic Press, New York, pp. 673–716.

Zeiske, E. (1968) Pradispositionen bei *Mollienesia sphenops* (Pisces, Poeciliidae) für einen Ubergang zum Lebeb in subterranen Gewassern. *Z. vergl. Physiol.*, **58**, 190–222.

Chapter eleven

Kin recognition in fish mediated by chemical cues

K. Håkan Olsén

11.1 INTRODUCTION

There is an increasing number of studies which show that social organisms frequently have the ability to distinguish relatives from non-relatives, and determine their degree of relatedness (reviews: Holmes and Sherman, 1983; Blaustein and O'Hara, 1986; Fletcher and Michener, 1987; Waldman, 1988). Kin recognition is shown by differential behaviour to traits 'labels', often based on odours) of conspecifics according to their genetic relatedness. The ability to recognize kin may decrease the risk of close inbreeding. It may also, by way of altruistic behaviour directed towards genetic relatives, increase an individual's inclusive fitness (Hamilton, 1964). An individual's 'inclusive fitness' is not only determined by its own reproductive success, but also by the survival and reproductive success of closely related individuals, which increase the probability that genes identical to its own pass on to the next generation (Hamilton, 1964). It has been proposed that there are four possible mechanisms for kin recognition (Blaustein, 1983; Holmes and Sherman, 1983; Waldman, 1988).

1. Spatial distribution: Relatives are distributed predictably in space, and altruistic behaviours are directed to individuals in a particular place, e.g. home site, territory (indirect recognition).
2. Familiarity and previous association: Individuals of the same litter or clutch may become familiar with features of the other members.
3. Phenotypic matching: An individual learns the phenotypes of relatives or of itself (self-matching). The individual later recalls the phenotype and is able to determine its relationship with other individuals by comparing the learned phenotype with that of the unfamiliar conspecific.

4. Recognition alleles: No learning is involved. Both the phenotype and its recognition have genetic bases.

Odours may be important in recognizing relatives in all four cases, but in phenotypic matching and recognition alleles, in contrast to spatial distribution and familiarity, the chemical cues acting as phenotypic markers should have inherited components.

Very little is known about the ability of fish to distinguish between individuals or groups of different degree of relationship. Most studies have focused on the ability of salmonids to recognize kin. These studies have been initiated to test the possible existence of intraspecific odours of importance for homing to the natal river during the spawning migration, as suggested by Nordeng (1971, 1977). In this chapter, the evidence for kin recognition in salmonids and its possible function will be discussed, along with some data obtained in other taxonomic groups.

11.2 SALMONIDS

Local populations or stocks are important ecological units in salmonids because mature individuals return with high precision to the natal river and spawning ground of the stock (Ricker, 1972; Bams, 1976; Stabell, 1984; Stabell, Chapter 12 herein). This concept is valid both in populations with mature fish returning from the sea (anadromous populations) and in land-locked populations which live all their life in fresh water (potamodromous populations). Stock-specific chemical cues may evolve in consequence of salmonids' precise homing to their natal river and spawning ground. Such homing behaviour appears in the process of segregation of spawning groups, and subsequently yields genetically different stocks or local populations adapted to their particular environment (Ryman et al., 1979; Child, 1984).

The young salmonid spends the embryonic stage within the egg and, after hatching, as an alevin buried in the gravel of a stream or a lake. Development in the gravel nest, which for the most part contains siblings and half siblings, takes several months, depending upon water temperature. After emerging from the gravel nests the fry remain in fresh water, or in the case of pink salmon, *Oncorhynchus gorbuscha*, and chum salmon, *O. keta*, the fry immediately move downstream into the estuary (Hoar, 1976). After 2–3 years in fresh water the young fish undergoes major changes in physiology, appearance and behaviour. It turns from a highly territorial fish with parr marks into a silvery schooling smolt prepared for a life in the marine environment (or in a lake for land-locked populations). Most body growth takes place in the sea in anadromous populations or in lakes in land-locked populations. Some species undergo long migrations in the open sea, but others remain in coastal waters during the summer before returning to fresh water for wintering, or in the case of mature fish both for spawning and wintering. After at least a year in the sea or in the lake, the mature fish returns to its natal river to spawn (Chapter 12).

Recent studies have shown that juvenile salmonids are able to discriminate between siblings and non-siblings from the same population (Quinn and Busack, 1985; Olsén, 1989). These results raise the question of whether both population- and sibling-specific chemical cues exist in salmonids. In the following sections, the possible existence of population-specific chemical cues will be discussed, based on experiments with sibling groups.

Sibling recognition

Quinn and Busack (1985) demonstrated that juvenile coho salmon, *Oncorhynchus kisutch*, distinguish water conditioned by either familiar or unfamiliar siblings from water conditioned by non-siblings of the same population, but if the two sibling groups were reared together, the test fishes showed preference for neither (Quinn and Hara, 1986). The results do not, however, exclude the possibility that the test fishes are able to distinguish familiar siblings from familiar non-siblings. The short time periods for adaptation and tests, i.e. 3 and 10 min respectively, might, however, have affected the ability of the test fish to make a correct choice. The test fish were familiar with both chemical cues, and the differences may have had no substantial importance during the instantaneous behavioural response studied. Quinn and Busack (1985) suggested that sibling recognition might be important during shoaling, and that information about kinship can later reduce the risk of inbreeding (Hamilton, 1964; Blaustein and O'Hara, 1982).

Behavioural experiments have recently shown that also juvenile Arctic charr, *Salvelinus alpinus*, are able to discriminate between siblings and non-siblings from the same population (Olsén, 1989). The experiments were conducted using a Y-maze (Olsén, 1985b) and a modified fluviarium (Fig. 11.1). The test fishes preferred water conditioned by siblings to that by non-siblings, even though the test fish and the siblings conditioning the water were taken from separate aquaria. No preference was observed when charr were given a choice between water scented by siblings from the same and a separate aquarium (Figs 11.2 and 11.3). The ability or motivation to choose their own siblings was more obvious in the large standard fluviarium, 3–6 months after the tests in the Y-maze. This may be due to an ontogenetic process or to learning. An ontogenetic change in ability to discriminate or in motivation may also be interpreted on the basis of results obtained with juvenile coho salmon by Quinn and Busack (1985). During the initial part of the experiment, three test families out of 10 preferred their own family scent to that of a strange family, but during the final part of the experiment all test families preferred their own scent.

Recently performed experiments in the standard fluviarium with juvenile Arctic charr indicate that learning is a part of the sibling-recognition

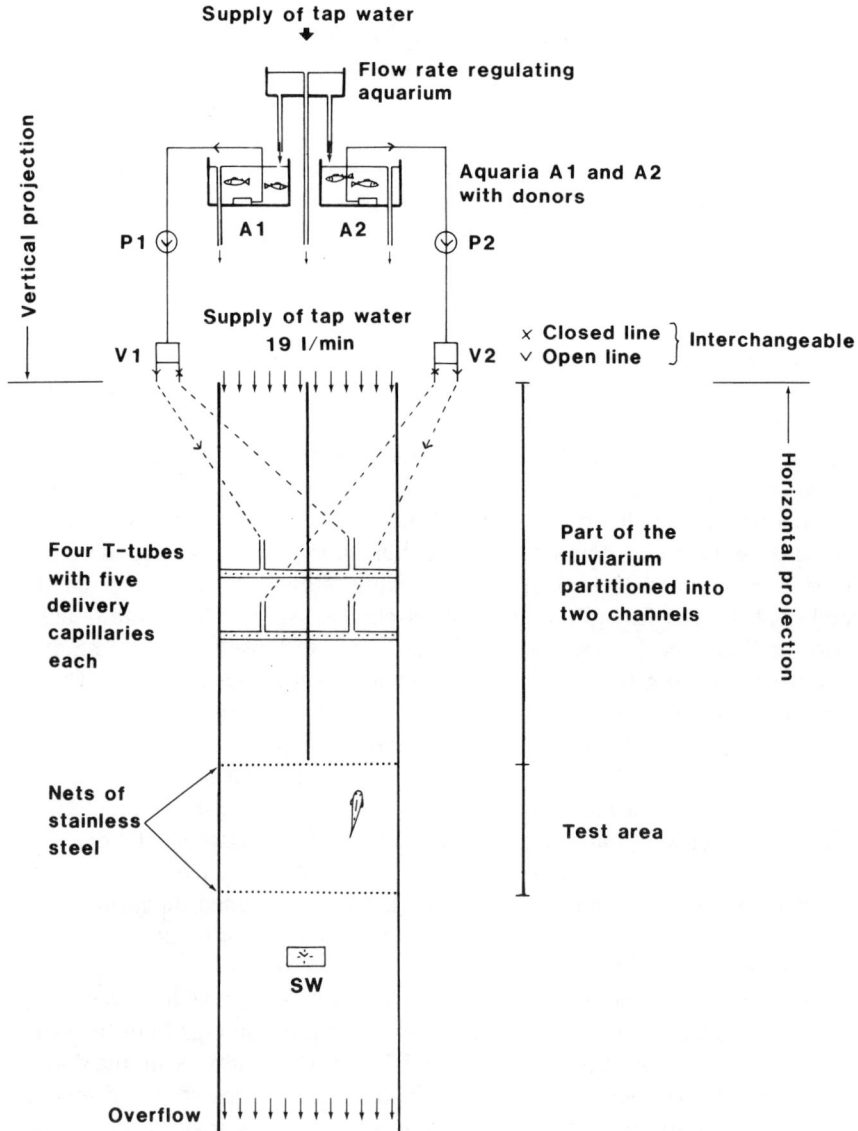

Fig. 11.1 The modified standard fluviarium. The test area of the fluviarium measured 23.5 cm long or 46 cm (depending on the size of the fish) × 33 cm wide × 11.5 cm deep. A1 and A2 are two identical 12-1 donor aquaria supplied with equal amounts of tap water (400–460 ml/min); P1 and P2, peristaltic pumps supplying equal amounts of water from the two donor aquaria (300–360 ml/min) to opposite halves of the fluviarium; V1 and V2, electromagnetic valves switching the supply of the two water qualities from A1 and A2; SW, switch and time recorder connected to V1 and V2. For further details of the fluviarium, see Höglund (1961). Reproduced from Olsén (1989) with permission from Academic Press.

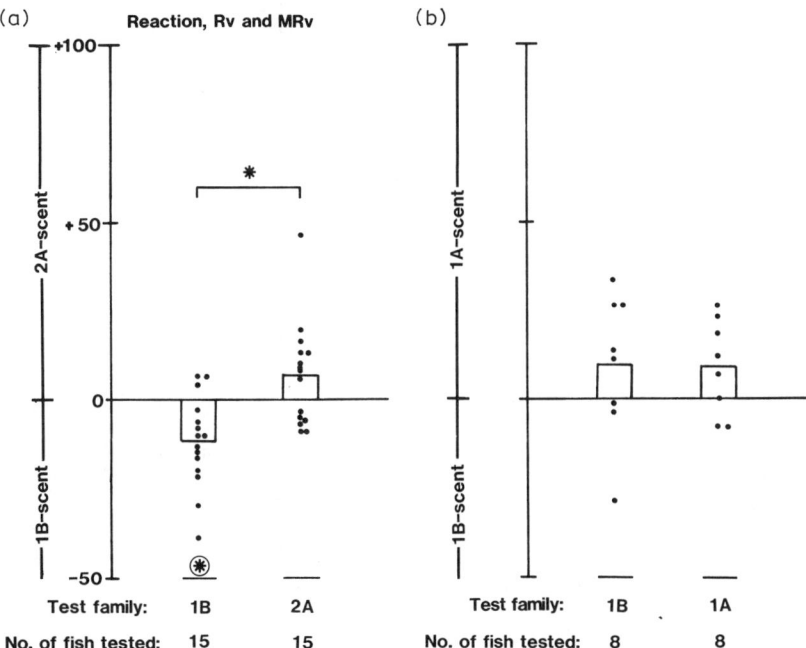

Fig. 11.2 Choice of single specimens of Arctic charr between water conditioned by (a) siblings and unfamiliar non-siblings, and (b) siblings from the same aquarium and from a separate aquarium, in a small Y-maze fluviarium. Artificial fertilization of eggs from female 1 with sperms from male 1 gave rise to sibling group 1 which was divided into two groups, 1A and 1B. Sibling group 2 from male 2 and female 2 was divided into 2A and 2B. The positions taken in darkness by a test fish in the test area (18 cm long × 7 cm wide × 3 cm deep) were recorded every 30 s by time-lapse photography with a 16 mm film camera and a flashlight filtered through a Kodak Wratten filter 89B (no transm. < 670 nm). Each test was divided into two test periods, 35 and 45 min, respectively. After the first test period the two odour qualities tested were switched from one side of the Y-maze to the other. A fish's preference for either odour quality was determined on the basis of the number of observations (positions of the snout) in each.

A reaction value (Rv) for a test was calculated from the following equation: $Rv = ((N1 - N2)/(N1 + N2)) \times 100$; where N1 and N2 represent the number of observations in each half of the test area supplied with water scented by different groups of fish. A dot denotes a reaction value, Rv, for a fish, and a column represents the mean reaction value, MRv., which is the arithmetic mean of Rvs from identical tests. Each Rv was based on 120 observations. A significant preference (t-test, $P < 0.05$, two-tailed), i.e. MRv deviating from zero, is shown by an encircled asterisk. A significant difference in MRv of sibling groups is denoted by bracket and asterisk (Wilcoxon's two sample test, $P < 0.05$, two-tailed). Reproduced from Olsén (1989) with permission from Academic Press.

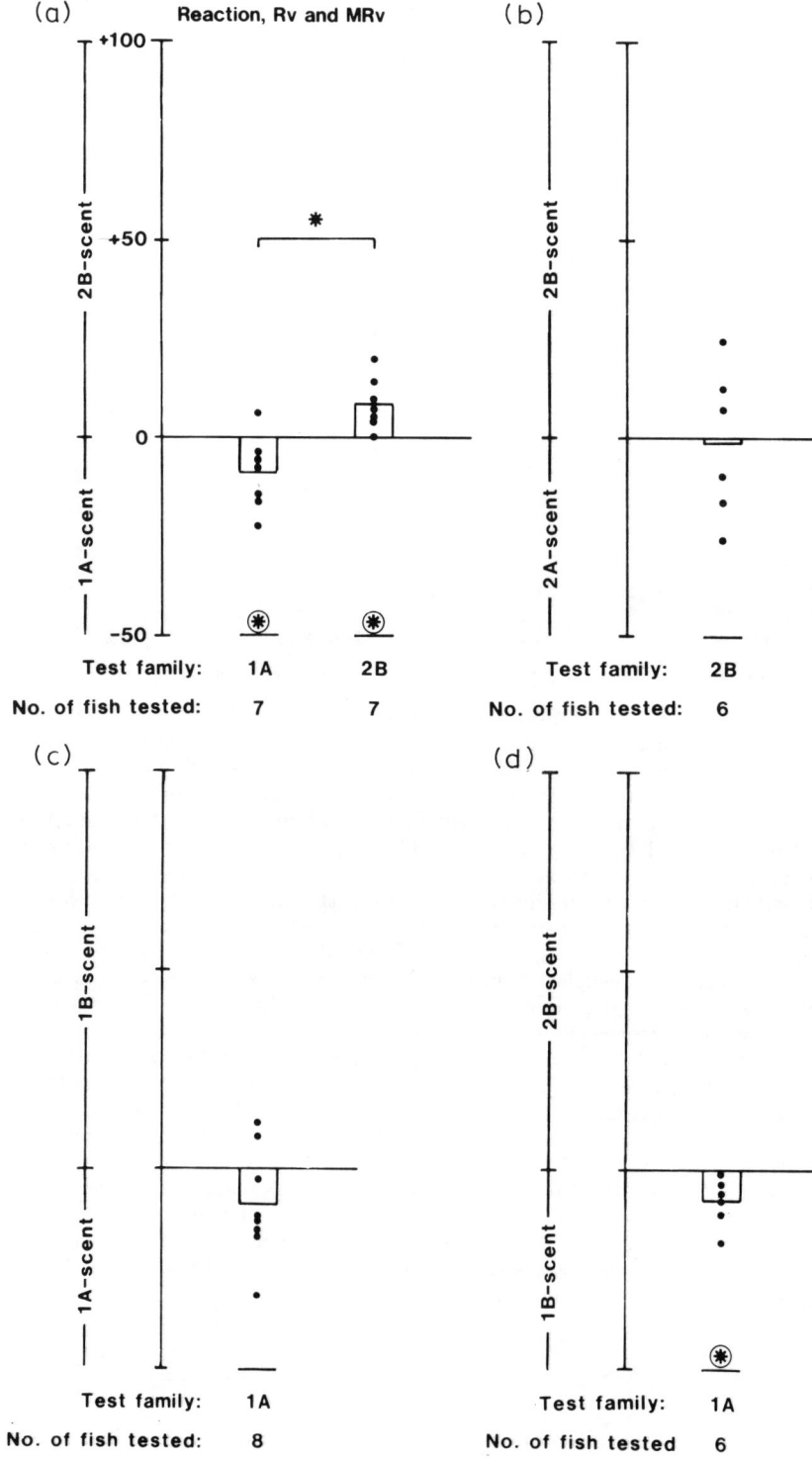

Fig. 11.3

mechanism in this species (Winberg and Olsén, unpublished). Individuals reared in isolation from the day eggs were fertilized did not discriminate between siblings and non-siblings from the same population, though they preferred water scented by siblings to pure water. But fish reared with siblings discriminated between unknown siblings and unknown non-siblings. They preferred water conditioned by siblings. Fish reared together with siblings and non-siblings varied in their ability to discriminate. Individuals from one of the two different sibling groups were still attracted to water from siblings, but the other group which originated from another pair of parents did not discriminate. The results indicate that juvenile Arctic charr learn the odour quality of siblings and also the odour from other sibling groups, but they do not learn their own smell or they do not use the information, i.e. there is no self-matching.

Population-specific chemical cues

Döving et al. (1974) studied the electrical response of single olfactory bulbar neurones of Arctic charr to stimulation with odours from specimens of different populations. The excitation and inhibition pattern of the activity of the olfactory neurones differed between populations. However, whether the fish showed behavioural discrimination between the odours was not examined.

Selset and Döving (1980) showed in behavioural experiments that hatchery-raised mature charr of an anadromous population originating from Lake Storvatnet, Hammerfest (Norway) were attracted to water scented by intestinal contents and bile from conspecific smolt. Only scent from smolt of the same population as the test fishes was used in these tests. When given the choice between intestinal contents of fish of the same and of a strange population, the water conditioned by the former was significantly preferred. These results must be interpreted cautiously, however, because the intestinal contents from the other population were not as fresh as those from their own population. Adults from only one of the two populations were used as

Fig. 11.3 Choice of single specimens of Arctic charr between water conditioned by (a) siblings and unfamiliar non-siblings; (b) siblings from the same aquarium and from a separate aquarium; (c) siblings from the same aquarium and from a separate aquarium; (d) siblings from a separate aquarium and unfamiliar non-siblings. The experiments were performed in the standard fluviarium (Fig. 11.1). One charr was tested at a time in darkness from the evening to the next morning and its positions were recorded every 150 s by time-lapse photography and filtered flash light (no transmittance < 670 nm). Each test was divided into eight 90-min test periods and after each period the supply of charr scented water was switched automatically from one side of the fluviarium to the other. A dot represents a reaction value (Rv) for a fish, and each is based on 192 observations. A column represents the mean reaction value, MRv. See Fig. 11.2 for further explanation of statistical treatment. Reproduced from Olsén (1989) with permission from Academic Press.

test fish, so it cannot be excluded that the preference was due to attractive substances not specific to the population.

Stabell (1982) studied preference behaviour combined with rheotactic responses of Atlantic salmon, *Salmo salar*, parr from the River Imsa, Norway, when given the choice between chemical cues emitted by their own population and that from a different river. Siblings of the Imsa population discriminated between water scented by their own and another sibling group originating from the River Opo, Norway. The parr preferred the water current scented by their siblings. No reciprocal test with the other sibling group as test fish was done. The results do not indicate population differences in odour quality, as both the test fish and the odour donors of the River Imsa population were taken from the same sibling group. The test fishes presumably recognized chemical cues from their own siblings.

More recently, Stabell (1987) studied Atlantic salmon parr from several spawners. The test fishes, which originated from the River Sandvikselv, were able to discriminate between chemical cues emitted by their own and by a different population. No reciprocal tests were, however, done.

Behavioural experiments were performed with two sympatric populations of Arctic charr from Lake Torrön, Sweden (Olsén, 1986a). The charr spawning in the Holderströmmen inlet can be distinguished from others spawning on reefs and shores in the lake by differences in spawning time, spawning age, growth and gene frequency for serum esterases (Hammar, 1980; Fürst et al., 1981; Hammar, 1984). The Holderströmmem charr (HO) and the charr from the village of Åbränna (AB) should accordingly represent two discrete sympatric populations or stocks. Previous studies have shown that juvenile Arctic charr preferred to stay in a water current scented by conspecifics (but not in water scented by other salmonid species) compared with pure water, and that an intact olfactory sense was essential for the discrimination (Höglund and Åstrand, 1973; Höglund et al., 1975). This has been confirmed by recent experiments (Olsén, 1986c). Attraction to conspecific chemical cues is already present in newly free-swimming fry of Arctic charr (Olsén, 1985a,b, 1990). Offspring of specimens caught at the two spawning locations in Lake Torrön (Holderströmmen and Åbränna) were studied in the standard fluviarium (Fig. 11.1). Both populations were attracted to water scented by either type of charr. However, the AB individuals were attracted significantly more strongly to their own scent. No such difference was observed using HO as test fish (Fig. 11.4). The same results were obtained when the test fish were given a choice between water scented by the two populations. AB specimens were, in contrast to HO specimens, significantly attracted to their own scent (Fig. 11.5). A gene frequency analysis of serum esterase indicated that the HO sample, in contrast to AB, was heterogeneous, i.e. they were not offspring of a discrete population. The lack of discriminating ability in the HO specimens, therefore,

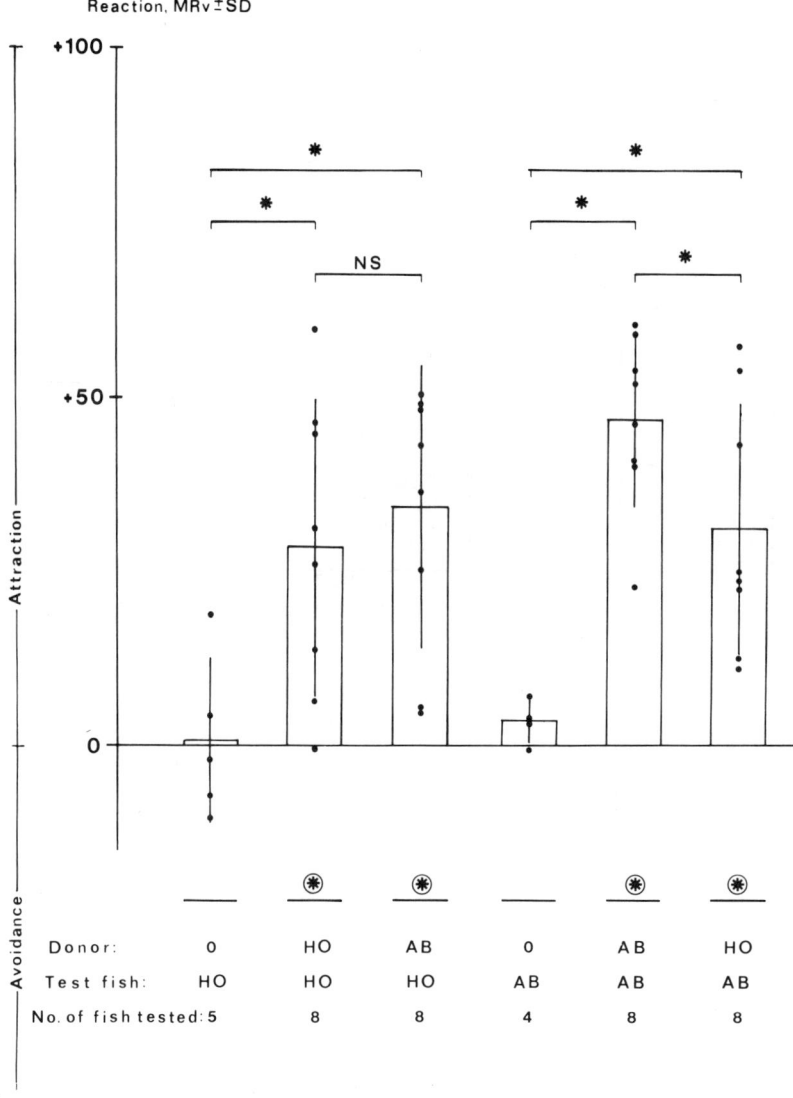

Fig. 11.4 Reaction to conspecific odour by the offspring of two sympatric populations of Arctic charr spawning in Lake Torrön, Sweden, at Åbränna (AB) and at Holderströmmen (HO). The test fish, which was always a single specimen, chose between pure water (no fish in one of the two donor aquaria) and charr scented water in the standard fluviarium. In control tests only pure tap water was supplied from donor aquaria, A1 and A2, at the same rate as with donors present. See Figs 11.1 and 11.3 for further details of the experimental design and Fig. 11.2 for information of statistical treatment. Reproduced from Olsén (1986a) with permission from Academic Press.

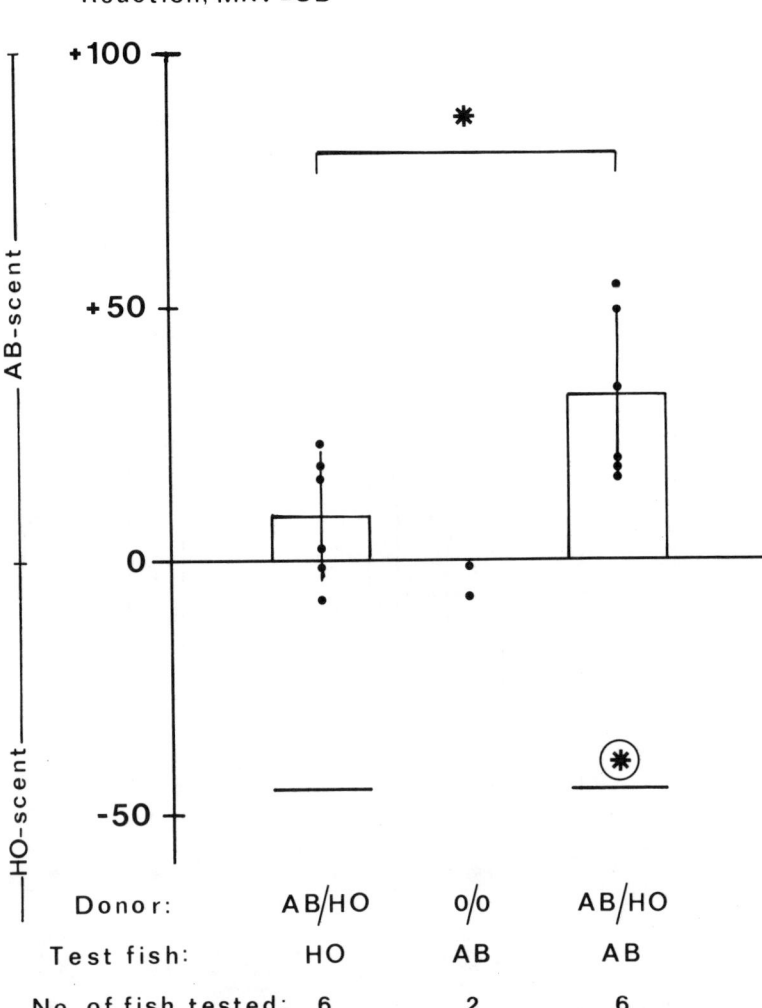

Fig. 11.5 Choice of single Åbränna (AB) or Holderströmmen (HO) specimen between water scented by equal number of charr of each population (AB and HO) supplied to opposite halves of the test area. See Figs 11.1 and 11.3 for information of experimental design and Fig. 11.2 for further information of statistical treatment. Reproduced from Olsén (1986a) with permission from Academic Press.

may have arisen because they were genetically less homogeneous. The results should not be due to sibling recognition, as several individuals from each stock were used as parents (about 100 of each sex, 2 females × 2 males fertilization) and all fertilized eggs within the same stock were mixed in the same container.

In experiments performed with hatchery-raised juvenile and wild adult coho salmon, *Oncorhynchus kisutch* (Quinn and Tolson, 1986) and with adult wild sockeye salmon, *O. nerka* (Groot *et al.*, 1986), no fish in the donor tank belonged to the same sibling group as the test fish. Both species were attracted to water conditioned by conspecifics, but only one of the two populations tested showed significant attraction to water scented by their own population. Individuals of the other population were indifferent to the two waters in experiments with sockeye, in which adult fish were used both as test fish and as donors. The experiment with adult coho salmon males as test fish and juveniles as donors indicated that water from one of the two populations (Quinsam River, British Columbia) was in general more attractive.

A recent study with young coho salmon gives strong support for the contention that population-specific chemical cues exist. Fish from the populations of Quinsam River (30 females, 1 female × 1 male fertilization), Big Qualicum River (British Columbia) (110 females, 1 female × 3 males) and Puntledge River (British Columbia) (40 females, 4 females × 4 males) preferred water scented by their own population when given a choice between two populations (Courtenay, 1989). It is, however, uncertain whether the 2000 eyed coho eggs obtained from the hatcheries were evenly sampled among the parental groups stated, or if some family groups were in a majority. Furthermore, recognition by familiarity cannot be completely ruled out as the test fish and donors of the same population were tankmates.

Summary and recommendations

The results of the experiments discussed indicate that chemical cues emitted by salmonids vary at the population level. But final proof of the existence of population- or stock-specific odours has yet to be presented. Previous studies have used only fish from one of the populations as test fish, and in the cases with reciprocal crosses of test fish, only one of the two populations has shown the ability to discriminate. The only exception is the study performed by Courtenay (1989), but sibling recognition and preference for tankmates cannot be completely ruled out.

The genetical background of fish that are designated as coming from the same population is in most cases not known. Do parental fish that are designated as individuals from the same population originate from the same spawning ground, or do they belong to more than one reproductive unit? The lack of the ability to discriminate may be due to mixture of gametes from individuals belonging to different, genetically discrete, stocks or local populations (Stabell, 1984).

The experiments with siblings emphasize that in designing future experiments, it is important to keep in mind that sibling recognition exists and that learning is one part of the mechanisms involved. It may be practicable to use sibling groups from different stocks as working units to answer the question about the existence of population-specific chemical cues. Several sibling groups within each stock should be mixed to ensure that if individuals from other stocks (strayers) were used as parents by mistake, their contribution to the blend would be small. If population- or stock-specific chemical cues exist, then the test fish should be able to discriminate between (a) an unfamiliar blend from non-siblings of the same local population, and (b) a blend from unfamiliar conspecifics which belong to another local population. The risk that the observed discrimination ability might be due to familiarity with the scent of tankmates would also be eliminated by this procedure.

Initial attempts to identify intraspecific chemical cues in salmonids

There are several possible sources for the emission of the substances. Fish emit amino acids, ammonia, steroids and several other substances with urine, faeces, mucus and through the gills. To examine the significance of free amino acids and ammonia for the attraction to conspecific odour in juvenile Arctic charr, fluviarium tests were performed.

Amino acids were examined for four reasons.

1. The olfactory sense in many species of fish, including Arctic charr and other salmonids, is sensitive to amino acids (e.g. Sutterlin and Sutterlin, 1971; Belghaug and Döving, 1977; Hara, 1982 and Chapter 8 herein).
2. Free amino acids are present in urine and skin mucus (e.g. Uskova *et al.*, 1971; Stabell and Selset, 1980; Ogata *et al.*, 1983; Hara *et al*, 1984), which are potent olfactory stimuli in salmonids (Döving *et al.*, 1974; Hara and Macdonald, 1976; Fisknes and Döving, 1982).
3. Urine and skin mucus of some teleosts contain unidentified substances of importance for chemical communication (Todd *et al.*, 1967; Richards, 1974; Barnett, 1981). Hara and collaborators (1984) demonstrated that free amino acids are mainly responsible for the electrophysiological response of rainbow trout, *Oncorhynchus mykiss*, to its own skin mucus.
4. Bryant and Atema (1987) showed that diet manipulation may change the body odour of bullheads, *Ictalurus nebulosus*. They suggested that amino acids and other nonspecific metabolites are important parts of the body odour which carry information to other members of the social group.

Ammonia was tested because it is the main nitrogen excretion product in fish (Brett and Zala, 1975) and excess ammonia may affect attraction of fish to conspecific odour.

Juvenile Arctic charr were shown to emit at least 17 of the amino acids commonly found in proteins (Olsén, 1986b). The charr avoided a mixture of synthetic amino acids in the same rank order and relative concentrations as was found in water conditioned by conspecifics. The response was dependent upon an intact olfactory sense. The charr also avoided L-alanine at the same concentration as the total concentration of the mixture. The charr also avoided ammonium chloride. Both L-alanine and ammonium chloride were shown to affect the attractiveness to water scented by conspecifics. The attraction of charr to scented water became repulsed or indifferent during addition of L-alanine or ammonium chloride respectively (Olsén, 1986c). These observations support the view that amino acids are not responsible for the attractive effect of water conditioned by intact conspecifics or by urine and intestinal content, respectively (Olsén, 1987). Amino acids may instead induce food search (Olsén et al., 1986) in the same way as shown in other species of fish (Carr and Chaney, 1976; Pawson, 1977).

Selset and Døving (1980) observed that adult anadromous Arctic charr, when offered a choice between pure and conditioned water, were attracted to water scented by intestinal contents or bile from conspecifics. No significant preference was obtained with skin mucus, which had been suggested as a important source for population-specific odours in salmonids (Døving et al., 1973, 1974). Selset and Døving (1980) also tested fractions of intestinal contents against pure water. The charr were attracted to the fraction containing polar lipids, which, according to Selset (1980), are likely to be polar steroid molecules. His assumption was supported by an electrophysiological study which showed that bile acids are potent olfactory stimuli in salmonid fishes, more potent than any amino acid previously tested (Døving et al., 1980). Selset and Døving (1980) suggested that the great variability in structure of bile acids and of their derivatives might be valuable sources of information for fishes, e.g. steroids excreted by salmon parr might guide the mature fish to their home river. Selset (1980) discussed the possibility that these substances are adsorbed to organic matter and released slowly to the water.

Groot et al. (1986) also showed in cross-adaptation experiments using the electroolfactogram (EOG) that the major stimulatory effect of water scented by sockeye salmon could be bile acids or bile-acid-like substances, and not amino acids.

Conclusion and recommendations

Intraspecific variation in chemical cues exists in salmonids. Recent experiments have shown that juvenile coho salmon and juvenile Arctic charr can discriminate between water scented by siblings and by non-siblings from the same population, even if the donors are unknown. The results of other studies also support the idea of variation in chemical cues at the population level. But the final proof of their existence has yet to be presented.

The mechanism of sibling recognition must be examined. Is information about kinship of significance during social interactions? Quinn and Busack (1985) suggested that sibling recognition might be important during shoaling and that information about kinship can later reduce the risk of inbreeding (Hamilton, 1964; Blaustein and O'Hara, 1982). This may also be the case among individuals of Arctic charr, as they are associated in shoals. Very young fish may learn the odours of related fish while still buried in the gravel, as olfactory neurones are already differentiated before hatching (Brannon, 1972; Jahn, 1972; Zielinski and Hara, 1988). Learning of close relatives' traits should be facilitated by proximity with siblings and half siblings in the same nest during the embryonic and alevin stages. Recent results with coho salmon indicate that chemical cues from conspecifics are learned at least as early as during the alevin stage (Courtnay, 1989). Discrimination between unfamiliar siblings and unfamiliar non-siblings in Arctic charr indicates that kin recognition may be close to phenotypic matching, suggesting that the chemical cues are inherited.

There are interesting studies with tadpoles which may reflect the same mechanisms as in the Arctic charr and other shoaling salmonids. In comprehensive studies by Blaustein and O'Hara (1986) and Waldman (1986), chemical stimuli associated with kinship information have been demonstrated in some anuran species. In general, these authors proposed that chemical cues bearing information about kinship facilitate shoaling with related individuals. Tadpoles of the woodfrog, *Rana sylvatica*, which are able to discriminate between familiar siblings and familiar non-siblings, associate preferentially with familiar siblings, both in a laboratory test pool and in natural ponds (Waldman, 1984).

In future research it will be important to make efforts to isolate intraspecific chemical cues with information about kinship, and to clarify their chemical identity. Knowledge of their identity should make it possible to determine their implication in social behaviour and mate choice as well as in homing behaviour (Stabell, Chapter 12 herein).

Chemical cues acting in sibling recognition may at least to some degree contain specific proteinaceous materials, though Hara (1977) has shown that simple peptides have low stimulatory effects on the olfactory sense of rainbow trout. Proteins, peptides and other water-soluble macromolecules do have important functions in the chemical ecology of aquatic invertebrates (Rittschof *et al.*, 1984; Rittschof and Bonaventura, 1986; Forward *et al.*, 1987).

The significance of the high olfactory sensitivity to bile acids in salmonids (Döving *et al.*, 1980) remains to be discovered. It may be that conjugated bile acids and other steroids carry enough information for discrimination at the stock and the species levels.

Nothing is known about the identity and origin of the substances which give a river its characteristic odour. Is the whole mixture of odours from conspecifics an integral part of the home river blend? These conspecific odours, not only from close relatives, may be learned just before leaving the spawning ground and natal river, and this memory may later be retrieved during homing.

11.3 STUDIES WITH NONSALMONID FISHES

There are few studies on nonsalmonid species that present evidence of kin recognition. Several authors have studied the complex and diversified parental care behaviour of cichlid fishes regarding learning of, and 'imprinting' to, features of the offspring or the parents (review: Keenleyside, 1979). It would be advantageous for cichlid parents to distinguish between their own offspring and other conspecific fry, so that they might focus their efforts to increase the chance of survival of their own offspring. Some authors have shown that cichlid parents of substratum-spawning species are able to discriminate between conspecific and heterospecific fry by visual and/or chemical cues (Kühme, 1963, 1964; Myrberg, 1966, 1975; McKaye and Barlow, 1976). Chemical cues were sufficient to keep maternal individuals behaving parentally. Kühme (1963, 1964) also demonstrated that jewel fish, *Hemichromis bimaculatus*, are capable of discriminating between their own brood and others of unknown conspecifics. This capability is the result of a learning process. 'Foster brooding' due to brood adoption or 'cuckoo' behaviour of both conspecific and heterospecific fry in mouthbrooding cichlids indicate that neither visual nor chemical recognition of kin are functioning or of importance in some species (Ribbink, 1977; Mrowka, 1987a,b).

Fry of the Midas cichlid, *Cichlasoma citrinellum*, preferred water scented by their mother to plain water, but they could not distinguish between water scented by their own mother and water scented by another mother (Barnett, 1977). Water scented by the father could also be attractive, depending upon the rearing conditions (with mother or father or with both) (Barnett, 1986).

Cooperative joint care of young has been observed in a cichlid fish, *Neolamprologus brichardi* (formerly *Lamprologus brichardi*), from Lake Tanganyika, both in captivity and in the lake. Individuals from earlier spawns are accepted by the parents as helpers (Taborsky, 1984), and the fish in the family group can recognize each other individually (Hert, 1985). But because parents may accept strange young as helpers, and because young stay and continue as helpers after replacement of both breeders (Taborsky and Limberger, 1981), it is doubtful whether they can recognize kin directly.

Experiments performed by Van Havre and FitzGerald (1988) showed that adult threespine sticklebacks, *Gasterosteus aculeatus*, spent more time in water conditioned by kin, i.e. water conditioned by specimens from the same pond, than in water conditioned by specimens from another pond. These authors also showed that stickleback fry raised in captivity were able to discriminate between siblings and unfamiliar or familiar non-siblings, which were offspring of a pair from another pond. Also, specimens raised alone from the egg stage and isolated from odours of other individuals preferred chemical cues from siblings. Van Havre and FitzGerald concluded that sticklebacks prefer to shoal with kin rather than with non-kin, and that the kin recognition has a genetic basis. Female threespine sticklebacks are also able to distinguish between chemical cues from their own eggs and those from others, and so avoid eating their own offspring (FitzGerald and Van Havre, 1987; Smith and Whoriskey, 1988). Females often destroy nests and eat eggs (Whoriskey and FitzGerald, 1985).

Kin recognition has also been suggested to occur in poeciliids. Loekle *et al.* (1982) observed that females preferentially cannibalized conspecific fry of other females in preference to their own. Whether the female used chemical or visual cues was not investigated.

Future research should focus on investigation into how frequently kin recognition occurs among fishes, and the significance of such a capability in fish behaviour. The mechanisms behind kin recognition should reflect the species' life history pattern and the habitat occupied (Blaustein and O'Hara, 1986; Waldman, 1988).

ACKNOWLEDGEMENT

This work was supported by The Bank of Sweden Tercentenary Foundation.

REFERENCES

Bams, R.A. (1976) Survival and propensity for homing as affected by presence or absence of locally adapted paternal genes in two transplanted populations of pink salmon (*Oncorhynchus gorbuscha*). *J. Fish. Res. Bd Can.*, **33**, 2716–25.

Barnett, C. (1977) Chemical recognition of the mother by the young of the cichlid fish, *Cichlasoma citrinellum*. *J. Chem. Ecol.*, **3**, 461–6.

Barnett, C. (1981) The role of urine in parent–offspring communication in a cichlid fish. *Z. Tierpsychol.*, **55**, 173–82.

Barnett, C. (1986) Rearing conditions affect chemosensory preferences in young cichlid fish. *Ethology*, **72**, 227–35.

Blaustein, A.R. (1983) Kin recognition mechanisms: phenotypic matching or recognition alleles? *Am. Nat.*, **121**, 749–54.

Blaustein, A.R. and O'Hara, R.K. (1982) Kin recognition cues in *Rana cascadae* tadpoles. *Behav. neural Biol.*, **36**, 77–87.

Blaustein, A.R. and O'Hara, R.K. (1986) Kin recognition in tadpoles. *Scient. Am.*, **254**, 90–6.

Belghaug, T. and Döving, K.B. (1977) Odour threshold determined by studies of the induced waves in the olfactory bulb of the char (*Salmo alpinus* L.). *Comp. Biochem. Physiol.*, **54A**, 327–30.

References

Brannon, E.L. (1972) Mechanisms controlling migration of sockeye salmon fry. *Int. Pac. Salmon Fish. Comm. Bull.*, **21**, 1–86.

Brett, J.R. and Zala, C.A. (1975) Daily pattern of nitrogen excretion and oxygen consumption of sockeye salmon (*Oncorhynchus nerka*) under control conditions. *J. Fish. Res. Bd Can.*, **32**, 2479–86.

Bryant, B.P. and Atema, J. (1987) Diet manipulation affects social behavior of catfish: importance of body odor. *J. Chem. Ecol.*, **13**, 1645–61.

Carr, W.E.S. and Chaney, T.B. (1976) Chemical stimulation of feeding behavior in the pinfish, *Lagodon romboides*: characterization and identification of stimulatory substances extracted from shrimp. *Comp. Biochem. Physiol.*, **54A**, 437–41.

Child, A.R. (1984) Bichemical polymorphism in charr (*Salvelinus alpinus* L.) from the Cumbrian lakes. *Heredity*, **53**, 249–57.

Courtenay, S.C. (1989) Learning and memory of chemosensory stimuli by underyearling coho salmon *Oncorhynchus kisutch* (Walbaum), PhD thesis, Department of Zoology, University of British Columbia, Vancouver, Canada, 231 pp.

Døving, K.B., Enger, P.S. and Nordeng, H. (1973) Electrophysiological studies on the olfactory sense in char (*Salmo alpinus* L.). *Comp. Biochem. Physiol.*, **45A**, 21–4.

Døving, K.B., Nordeng, H. and Oakley, B. (1974) Single unit discrimination of fish odours released by char (*Salmo alpinus* L.) populations. *Comp. Biochem. Physiol.*, **47A**, 1051–63.

Døving, K.B., Selset, R. and Thommesen, G. (1980) Olfactory sensitivity to bile acids in salmonid fishes. *Acta physiol. scand.*, **108**, 123–31.

Fisknes, B. and Döving, K.B. (1982) The olfactory sensitivity to group specific substances in Atlantic salmon (*Salmo salar* L.). *J. Chem. Ecol.*, **8**, 1083–91.

FitzGerald, G.J. and Van Havre, N. (1987) The adaptive significance of cannibalism in sticklebacks (Gasterosteidae: Pisces), *Behav. Ecol. Sociobiol.*, **20**, 125–8.

Fletcher, D.J. and Michener, C.D. (eds) (1987) *Kin Recognition in Animals*, Wiley, Chichester, 453 pp.

Forward, R.B., jun., Rittschof, D and De Vries, M.C. (1987) Peptide pheromones synchronize crustacean egg hatching and larval release. *Chem. Senses*, **12**, 491–8.

Fürst, M., Boström, U. and Hammar, J. (1981) Effects of introduced *Mysis relicta* on fish in Lake Torrön. *Inform. Inst. Freshw. Res. Drottningholm*, (1), 1–48. (In Swedish with English summary.)

Groot, C., Quinn, T.P. and Hara, T.J. (1986) Responses of migrating adult sockeye salmon (*Oncorhynchus nerka*) to population specific odours. *Can. J. Zool.*, **64**, 926–32.

Hamilton, W.D. (1964) The genetic evolution of social behaviour, I and II. *J. Theor. Biol.*, **7**, 1–52.

Hammar, J. (1980) The ecology and taxonomy of Arctic char in lake reservoirs in Sweden. *Internat. Soc. Arctic Char Fanatics, ISACF, Inform. Ser.* (1) 18–28.

Hammar, J. (1984) Ecological characters of different combinations of sympatric populations of Arctic charr in Sweden, in *Biology of the Arctic Charr* (eds L. Johnson and B.L. Burns), Univ. Manitoba Press, Winnipeg, pp. 35–63.

Hara, T.J. (1977) Further studies on the structure–activity relationships of amino acids in fish olfaction. *Comp. Biochem. Physiol.*, **56A**, 559–65.

Hara, T.J. (1982) Structure–activity relationships of amino acids as olfactory stimuli, in *Chemoreception in Fishes* (ed. T.J. Hara), Elsevier, Amsterdam, pp. 135–57.

Hara, T.J. and MacDonald, S. (1976) Olfactory responses to skin mucus substances in rainbow trout *Salmo gairdneri*. *Comp. Biochem. Physiol.*, **54A**, 41–4.

Hara, T.J., Macdonald, S., Evans, R.E., Marui, T. and Arai, S. (1984) Morpholin, bile acids and skin mucus as possible chemical cues in salmonid homing: electrophysiological re-evaluation, in *Mechanisms of Migration in Fishes* (eds J.D. McCleave, G.P. Arnold, J.D. Dodson, and W.H. Neill), Plenum, New York, pp. 363–77.

Hert, E. (1985) Individual recognition of helpers by breeders in *Lamprologus brichardi* (Cichlidae). *Z. Tierpsychol.*, **68**, 313–25.

Hoar, W.S. (1976) Smolt transformation: evolution, behavior, and physiology. *J. Fish. Res. Bd Can.*, **33**, 1324–52.

Höglund, L.B. (1961) The reaction of fish in concentration gradients. *Rep. Inst. Freshwat. Res. Drottningholm*, **43**, 1–147.

Höglund, L.B. and Åstrand, M. (1973) Preferences among juvenile char (*Salvelinus alpinus* L.) to intraspecific odours and water currents studied with the fluviarium technique. *Rep. Inst. Freshwat. Res. Drottningholm*, **53**, 21–30.

Höglund, L.B., Bohman, A. and Nilsson, N.-A. (1975) Possible odor responses of juvenile Arctic char (*Salvelinus alpinus* (L.)) to three other species of subarctic fish. *Rep. Inst. Freshwat. Res. Drottningholm*, **54**, 21–35.

Holmes, W.G. and Sherman, P.W. (1983) Kin recognition in animals. *Am. Scient.*, **71**, 46–55.

Jahn, L.A. (1972) Development of the olfactory apparatus of the cutthroat trout. *Trans. Am. Fish. Soc.*, **101**, 284–9.

Keenleyside, M.H.A. (1979) *Diversity and Adaptation in Fish Behaviour*, Springer-Verlag, Berlin, 208 pp.

Kühme, W. (1963) Chemisch ausgelöste Brutflege und Schwarmreaktionen bei *Hemichromis bimaculatus* (Pisces). *Z. Tierpsychol.*, **20**, 688–704.

Kühme, W. (1964) Ein chemisch ausgelöste Brutpflegeraktion bei Cichliden (Pisces). *Naturwissenschaften*, **51**, 20–1.

Loekle, D.M., Madison, D.M. and Christian, J.J. (1982) Time dependency and kin recognition of cannibalistic behavior among Poeciliid fishes. *Behav. neural Biol.*, **35**, 315–18.

McKaye, K.R. and Barlow, G.W. (1976) Chemical recognition of young by the Midas cichlid, *Cichlasoma citrinellum*. *Copeia*, **1976**, 276–82.

Mrowka, W. (1987a) Brood adoption in a mouthbrooding cichlid fish: experiments and a hypothesis. *Anim. Behav.*, **35**, 922–3.

Mrowka, W. (1987b) Egg stealing in a mouthbrooding cichlid fish. *Anim. Behav.*, **35**, 923–5.

Myrberg, A.A., jun. (1966) Parental recognition of young in cichlid fishes. *Anim. Behav.*, **14**, 565–71.

Myrberg, A.A., jun. (1975) The role of chemical and visual stimuli in the preferential discrimination of young by the cichlid fish *Cichlasoma nigrofasciatum* (Gunther). *Z. Tierpsychol.*, **37**, 274–97.

Nordeng, H. (1971) Is the local orientation of anadromous fishes determined by pheromones? *Nature, Lond.*, **233**, 411–13.

Nordeng, H. (1977) A pheromone hypothesis for homeward migration in anadromous salmonids. *Oikos*, **28**, 155–9.

Ogata, H., Murai, T. and Nose, T. (1983) Free amino acid composition in urine of carp and channel catfish. *Bull. Jap. Soc. scient. Fish.*, **49**, 1471.

Olsén, K.H. (1985a) Chemoreceptive behaviour in Arctic charr – Response to conspecific scent and nitrogenous metabolites, PhD thesis, Department of Zoophysiology, Uppsala University; *Acta Universitatis Upsaliensis*, **9**, 1–43.

Olsén, K.H. (1985b) Chemoattraction between fry of Arctic charr (*Salvelinus alpinus* (L.)) studied in a Y-maze fluviarium. *J. Chem. Ecol.*, **11**, 1009–17.

References

Olsén, K.H. (1986a) Chemoattraction between juveniles of two sympatric stocks of Arctic charr (*Salvelinus alpinus* (L.)) and their gene frequency of serum esterases. *J. Fish Biol.*, **28**, 221–31.

Olsén, K.H. (1986b) Emission rate of amino acids and ammonia and their role in olfactory preference behaviour of juvenile Arctic charr, *Salvelinus alpinus*. *J. Fish Biol.*, **28**, 255–65.

Olsén, K.H. (1986c) Modification of conspecific chemoattraction in Arctic charr (*Salvelinus alpinus* (L.)) by nitrogenous excretory products. *Comp. Biochem. Physiol.*, **85A**, 77–81.

Olsén, K.H. (1987) Chemoattraction of juvenile Arctic charr, *Salvelinus alpinus* (L.), to water scented by conspecific intestinal content and urine. *Comp. Biochem. Physiol.*, **87A**, 641–3.

Olsén, K.H. (1989) Sibling recognition in juvenile Arctic charr (*Salvelinus alpinus* (L.)). *J. Fish Biol.*, **34**, 571–81.

Olsén, K.H. (1990) Further studies concerning chemoattraction among fry of Arctic charr (*Salvelinus alpinus* (L.)) to water conditioned by conspecifics. *J. Chem. Ecol.*, **16**, 2081–90.

Olsén, K.H., Karlsson, L. and Helander, A. (1986) Food search behaviour in Arctic charr (*Salvelinus alpinus* (L.)) induced by food extract and amino acids. *J. Chem. Ecol.*, **12**, 1997–8.

Pawson, M.G. (1977) Analysis of a natural chemical attractant for whiting *Merlangius merlangius* L. and cod *Gadus morhua* L. using a behavioural bioassay. *Comp. Biochem. Physiol.*, **56A**, 129–35.

Quinn, T.P. and Busack, C.A. (1985) Chemosensory recognition of siblings in juvenile coho salmon (*Oncorhynchus kisutch*). *Anim. Behav.*, **33**, 51–6.

Quinn, T.P. and Hara, T.J. (1986) Sibling recognition and olfactory sensitivity in juvenile coho salmon (*Oncorhynchus kisutch*). *Can. J. Zool.*, **64**, 921–5.

Quinn, T.P. and Tolson, G.M. (1986) Evidence of chemically mediated population recognition in coho salmon (*Oncorhynchus kisutch*). *Can. J. Zool.*, **64**, 84–7.

Ribbink, A.J. (1977) Cuckoo among Lake Malawi cichlid fish. *Nature, Lond.*, **267**, 243–4.

Richards, I.S. (1974) Caudal neurosecretory system: possible role in pheromone production. *J. exp. Zool.*, **187**, 405–8.

Ricker, W.E. (1972) Hereditary and environmental factors affecting certain salmonid populations, in *The Stock Concept in Pacific Salmon, H.R. McMillan Lectures in Fisheries* (eds R.C. Simon and P.A. Larkin), Univ. British Columbia, Vancouver, B.C., pp. 27–160.

Rittschof, D. and Bonaventura, J. (1986) Macromolecules in marine chemical ecology. *J. Chem. Ecol.*, **12**, 1013–23.

Rittschof, D., Shepherd, R. and Williams, L.G. (1984) Concentration and preliminary characterization of chemical attractant of the oyster drill *Urosalpinx cinerea*. *J. Chem. Ecol.*, **10**, 63–79.

Ryman, N., Allendorf, F.W. and Ståhl, G. (1979) Reproduction isolation with little genetic divergence in sympatric populations of brown trout (*Salmo trutta*). *Genetics*, **92**, 247–62.

Selset, R. (1980) Chemical methods for fractionation of odorants produced by char smolts and tentative suggestions for pheromone origins. *Acta physiol. scand.*, **108**, 97–103.

Selset, R. and Döving, K.B. (1980) Behaviour of mature anadromous char (*Salmo alpinus* L.) towards odorants by smolts of their own population. *Acta physiol. scand.*, **108**, 113–22.

Smith, R.S. and Whoriskey, F.G., jun. (1988) Multiple clutches: female threespine sticklebacks lose the ability to recognize their own eggs. *Anim. Behav.*, **36**, 1838–9.

Stabell, O.B. (1982) Detection of natural odorants by Atlantic salmon parr using positive rheotaxis olfactometry, in *Proc. Salmon and Trout Migratory Behavior Symp.*, (eds E.L. Brannon and E.O. Salo), Univ. Washington Press, Seattle, pp. 71–8.

Stabell, O.B. (1984) Homing and olfaction in salmonids: a critical review with special reference to the Atlantic salmon. *Biol. Rev.*, **59**, 333–88.

Stabell, O.B. (1987) Intraspecific pheromone discrimination and substrate marking by Atlantic salmon parr. *J. Chem. Ecol.*, **13**, 1625–43.

Stabell, O.B. and Selset, R. (1980) Comparison of mucus collecting methods in fish olfaction. *Acta physiol. scand.*, **108**, 91–6.

Sutterlin, A.M. and Sutterlin, N. (1971) Electrical responses of the olfactory epithelium of Atlantic salmon (*Salmo salar*). *J. Fish. Res. Bd Can.*, **28**, 565–72.

Taborsky, M. (1984) Broodcare helpers in the cichlid fish *Lamprologus brichardi*: their costs and benefits. *Anim. Behav.*, **32**, 1236–52.

Taborsky, M. and Limberger, D. (1981) Helpers in fish. *Behav. Ecol. Sociobiol.*, **8**, 143–5.

Todd, J.H., Atema, J. and Bardach, J.E. (1967) Chemical communication in social behavior of a fish, the yellow bullhead (*Ictalurus natalis*). *Science, Wash., D.C.*, **158**, 672–3.

Uskova, Y.T., Chaykovskaya, A.V. and Uskov, I.A. (1971) Composition of amino acids in the mucus of fish skin. *Hydrobiol. J.* **7**, 83–5.

Van Havre, N. and FitzGerald, G.J. (1988) Shoaling and kin recognition in the threespine stickleback (*Gasterosteus aculeatus* L.). *Biol. Behav.*, **13**, 190–201.

Waldman, B. (1984) Kin recognition and sibling association among wood frog (*Rana sylvatica*) tadpoles. *Behav. Ecol. Sociobiol.*, **14**, 171–80.

Waldman, B. (1986) Preference for unfamiliar siblings over familiar nonsiblings in American toad (*Bufo americanus*) tadpoles. *Anim. Behav.*, **34**, 48–53.

Waldman, B. (1988) The ecology of kin recognition. *A. Rev. Ecol. Syst.*, **19**, 543–71.

Whoriskey, F.G. and FitzGerald, G.J. (1985) Sex, cannibalism and sticklebacks. *Behav. Ecol. Sociobiol.*, **18**, 15–18.

Zielinski, B. and Hara, T.J. (1988) Morphological and physiological development of olfactory receptor cells in rainbow trout (*Salmo gairdneri*) embryos. *J. comp. Neurol.*, **271**, 300–11.

Chapter twelve

Olfactory control of homing behaviour in salmonids

Ole B. Stabell

12.1 INTRODUCTION

A connection between homing* behaviour and olfaction in fishes is now a well-established fact, but interpretations and concepts remain controversial. While speculations about a possible involvement of the olfactory sense with homing in fishes appeared at least a century ago (Buckland, 1880), direct scientific evidence for such a coupling has been obtained only during the course of the past four decades.

Salmonid fishes are ectothermic animals, which display a high degree of plasticity in functional characteristics and life-history patterns (Nordeng, 1983; Jonsson, 1985). The process of homing is a behavioural feature that serves to link the various life stages and, as such, homing behaviour must be viewed as an integral part of salmonid ecology. If studies of homing behaviour are carried out without this important point in mind, incorrect conclusions may be reached which may have serious implications for fisheries management practices. Thus, in reviewing olfactory control of homing behaviour in salmonid fishes, a brief introduction to their natural ecology and migratory systems is necessary.

---

*The term 'homing' is here used in the general sense of Gerking (1959), to mean 'the return of fish (after migrating, accidental or experimental displacement) to a place formerly occupied instead of going to other equally probable places'.

12.2 ECOLOGICAL FRAMEWORK

Population structures

Several salmonid species of the genera *Salmo*, *Salvelinus* and *Oncorhynchus* have been studied with respect to homing behaviour (Table 12.1). These species are usually termed **anadromous** (Table 12.2; Myers, 1949), meaning that they undertake migrations from fresh water to the sea and return to fresh water for spawning. However, many of these species also have land-locked stocks, representing freshwater forms that never migrate to sea. Freshwater forms may also undertake migrations, using lakes as their 'sea' and performing homing to streams for spawning. Such forms, in which all migration takes place only within fresh water, are denoted **potamodromous**.

The complexity of life-history patterns within the various species of anadromous salmonids has been described in detail by Nordeng (1977, 1983, 1989b) and Jonsson (1985). Within a population of fish, several year classes and life-history stages may be present sympatrically. In Arctic charr, *Salvelinus alpinus*, and brown trout, *Salmo trutta* (system I, Fig. 12.1), for instance, all life stages may be found in the freshwater habitat. Smolts, as well as immature and sexually maturing fish, display an annual anadromous migratory pattern. Strictly speaking, the immature 'anadromous' Arctic charr and brown trout should be termed **amphidromus** (Table 12.2), since their return to freshwater is not for the purpose of spawning but is possibly related to an inability to survive in seawater during winter.

Table 12.1 Species of salmonid fishes in which homing behaviour has been studied

Genus	Species	Popular name
Salmo	*salar*	Atlantic salmon
	trutta	Brown trout
Salvelinus	*alpinus*	Arctic charr
	malma	Dolly Varden charr
	namaycush	Lake 'trout'
	fontinalis	Brook 'trout'
Oncorhynchus	*masou*	Masu salmon
	nerka	Sockeye salmon
	kisutch	Coho salmon
	tshawytscha	Chinook salmon
	keta	Chum salmon
	gorbuscha	Pink salmon
	*mykiss**	Rainbow (steelhead) trout
	clarki†	Cutthroat trout

*Formerly *Salmo gairdneri* (Kendall, 1989).
†Formerly *Salmo clarki* (Kendall, 1989).

Table 12.2 Terms for migratory fishes*

Term	Definition
Diadromous	Migratory fishes that migrate between the sea and fresh water
Anadromous	Diadromous fishes that spend most of their lives in the sea and migrate to fresh water to spawn
Catadromous	Diadromous fishes that spend most of their lives in fresh water and migrate to the sea to spawn
Amphidromous	Diadromous fishes whose migration from fresh water to the sea, or vice versa, is not for the purpose of spawning, but occurs regularly at some other definite stage of the life cycle
Potamodromous	Migratory fishes whose migration occurs wholly within fresh water
Oceanodromous	Migratory fishes that live and migrate wholly in the sea

*After Myers (1949).

To understand fully the complexity of life-history patterns within distinct populations of anadromous salmonids, it is important to realize that progeny of both anadromous and resident parents during the parr stage may develop either into early-maturing resident individuals or into smolts. In addition, some of the resident early-maturing individuals may later transform into smolts and undertake an anadromous life style (Nordeng, 1983; Fig. 12.1).

The population structures of anadromous charrs and trout (system I, Fig. 12.1) comprise a single base in the freshwater habitat consisting of fry, juveniles and early-maturing resident individuals of both sexes. Winter survival of the anadromous individuals of both the immature and mature stages is dependent on fresh water. On the other hand, the population structures of the various salmon species (systems II–V) comprise two bases, one in fresh water and the other in the sea. In addition to fry and parr, the freshwater bases of masu and sockeye salmon, *Oncorhynchus masou* and *O. nerka* (system II) include early-maturing individuals of both sexes. In Atlantic salmon, *Salmo salar*, and in coho and chinook salmon, *O. kisutch* and *O. tskawytscha*, it is generally only the males that mature early and are in the freshwater base (system III), but resident individuals appear to be absent in chum salmon, *O. keta* (system IV) and pink salmon, *O. gorbuscha* (system V). The marine bases of the salmon migratory systems (II–V) consist of immature individuals plus maturing individuals that are ready for homeward migration.

Stock characteristics

Different salmonid species display intraspecific variations in several morphological and functional characters (Scheer, 1939; Ricker, 1972). These

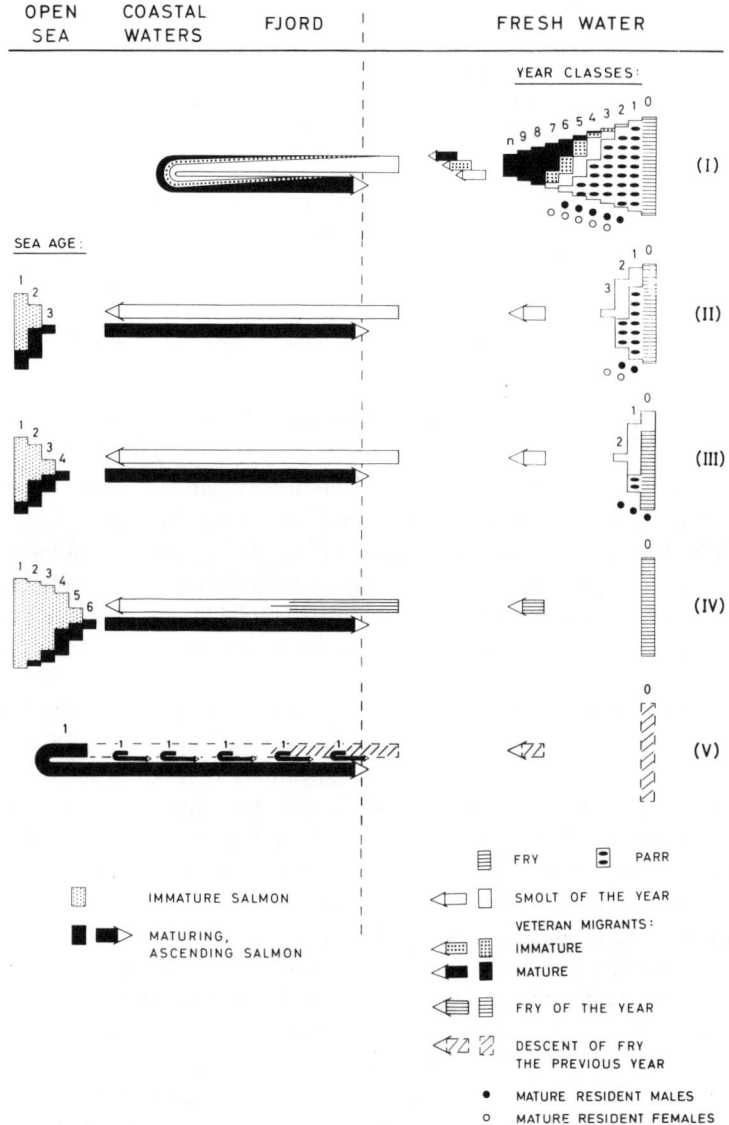

Fig. 12.1 Schematic representation of the migratory systems of anadromous salmonids, including the various life stages that may be present in fresh water and in the sea. The illustration shows the five major systems, representing charr and trout (I), masu and sockeye salmon (II), Atlantic, coho and chinook salmon (III), chum salmon (IV), and pink salmon (V) (species names, Table 12.1). The systems are arranged according to decreasing strength of the freshwater bases. (From Nordeng, 1989b; reproduced by permission of H. Nordeng, and *Physiology and Ecology*.)

variations are usually referred to as stock- (or population-) specific characters, and may reflect an adjustment by a local population to environmental variables, owing to long-term selection. Physical, chemical and biological factors that may have influenced the development of local population characteristics are length of seasons, seasonal temperature regimes, stream size and seasonal changes in water flow, presence of various parasites and predators, as well as food abundance and diversity. Among the morphological and functional characteristics that display population-specific differences in salmonid fishes are fecundity and egg size, oxygen requirement, growth rate, age at smolting, age at sexual maturity including early maturation, time of spawning, and disease resistance. Variation in population-specific characters can be detected by means of biochemical genetics (Allendorf et al., 1976, 1987). It should especially be noted that within a river system, several local populations of the same species may be present sympatrically (Cope, 1957; Saunders and Bailey, 1980; Ryman and Ståhl, 1981). The genetical differences found between sympatric populations may represent a framework for population-specific chemical cues to be used in migration.

Migratory systems

The discrete characteristics of a local salmonid population are maintained through specific homing for reproduction. The process of homing takes place within a neatly tuned migratory system, which can be described in relation to the freshwater base occupied by the resident individuals of the population (Fig. 12.1; Nordeng, 1977, 1989b). For charr and trout (system I), seaward migration starts with the descent of veteran migrants that are the forthcoming spawners, then by immature veteran migrants and smolts of the year. The migratory cycle is completed by the return to fresh water by the descending fish the same year. For all salmon species (systems I–V), the migratory system comprises two bases, one in the freshwater habitat and the other in the sea. The migration season starts with the descent of smolts (masu, sockeye, chinook, and Atlantic salmon) or fry (chum and pink) to the sea, and culminates with the maturing adults ascending into fresh water at a later date. In these species the annual migratory cycle is complete when the descending smolts or fry and ascending mature adults have exchanged habitats.

A characteristic feature of the migratory systems of anadromous salmonids is that smolts descend downstream for a certain period of time before the first ascending fish enters the river. Downstream migration takes place in schools and occurs during spring and summer. In Atlantic salmon from the Salangen River system, North Norway, downstream migration may commence on different dates during May, depending upon the time at which water temperature reaches approximately 6 °C. Despite this variation on a year-to-year basis, the time lag between the initial descent of smolts and the first entry of ascending salmon into fresh water has been found to be the

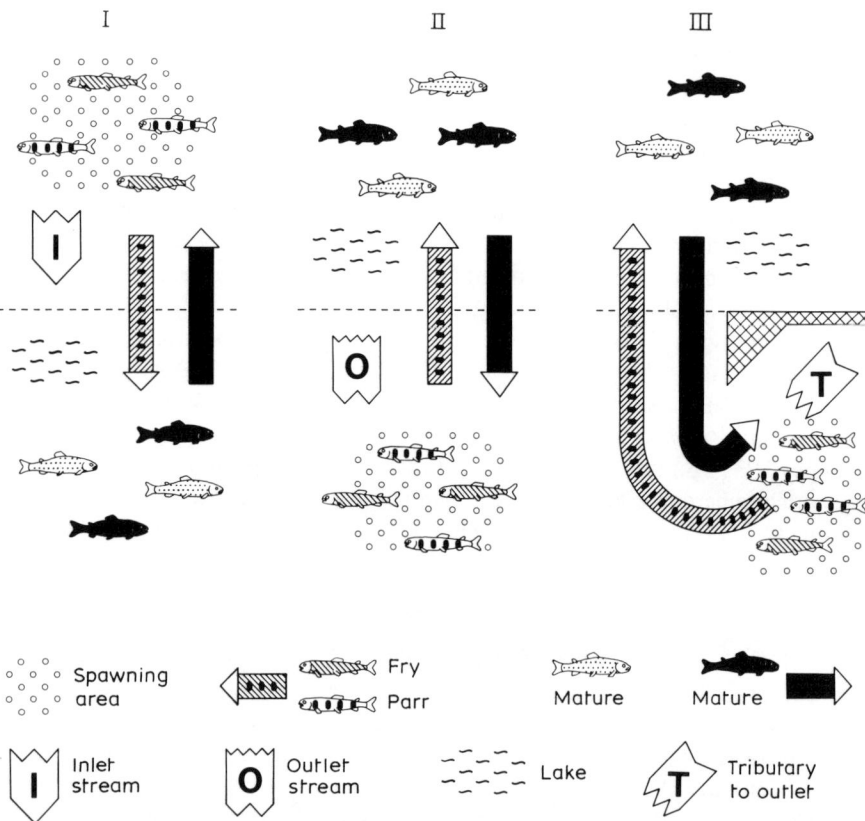

Fig. 12.2 Schematic diagram presenting the migratory systems of potamodromous salmonids, together with the different life stages involved. The direction of migration by juvenile and mature fish denotes the system: I, a 'juvenile downstream–mature upstream' system; II, 'juvenile upstream–mature downstream' system; III, 'combined juvenile downstream/upstream–mature downstream/upstream' system. Migration of juveniles may take place as fry or as fingerling or as parr; the life stage of initial movement within each system is species or stock dependent.

same (around 15 days) each year for several years (Nordeng, 1977; H. Nordeng, 1991; pers. comm.). This indicates that a close connection exists between downstream migration of smolts and return of mature salmon. A similar connection between outward migration of smolts or fry and homeward return of adults also seems to be present in the salmon species of the genus *Oncorhynchus* (Nordeng, 1989b).

Potamodromous populations of salmonids also demonstrate specific features in relation to their freshwater migrations. In addition to freshwater stocks of charr and trout, some of the salmon species also have pure 'land-locked' forms that perform migrations either between rivers and lakes, or within river systems. Examples of such freshwater forms of salmon are yamame (masu), kokanee (sockeye), and 'blege' or 'småblank' (Atlantic salmon), the latter two being local names for land-locked forms in Norway (Dahl, 1928; Berg, 1953; Lorz and Northcote, 1965; Tanaka, 1965; Machidori and Kato, 1984).

Potamodromous salmonids may perform either upstream or downstream spawning migrations from the rearing lake to the area of spawning and progeny incubation in inlet or outlet streams (Stuart, 1957; Lindsey et al., 1959; Tilzey, 1977; Kaasa, 1980). Three types of migratory systems have been described in potamodromous systems (systems I–III, Fig. 12.2):(1) system I, in which the juveniles migrate downstream from the incubation area in an inlet stream to their rearing lake, followed by upstream spawning migration by mature adults (Stuart, 1957; Lindsey et al., 1959); (2) system II, in which juveniles undertake an upstream migration from the incubation area in outlet streams to the rearing lake, followed by a downstream spawning migration by mature adults (Lindsey et al., 1959; Kaasa, 1980); and (3) system III, in which juveniles undertake a combined downstream–upstream migration from the incubation area in a tributary stream to reach the rearing lake, i.e. downstream to the outlet stream and then upstream into the lake. Spawning migration by adults from the lake is the reverse, i.e. downstream in the outlet stream followed by an upstream migration into the tributary to the spawning site (Kaasa, 1980).

The various types of lakeward migration shown in potamodromous systems may also be seen in the freshwater migration by fry, fingerlings or parr before smolt transformation and seaward migration in anadromous systems (Northcote, 1962, 1969; Brannon, 1967, 1972; Raleigh, 1967, 1971; Raleigh and Chapman, 1971; Hutchings, 1986). The specific migrations of juveniles may also be under olfactory influence (Brannon, 1972).

12.3 OLFACTION AND HOMING BEHAVIOUR

Homing for reproduction

As mentioned above, salmonid fishes return to their home stream, and within the stream to a specific area, for spawning. The existence of distinct stocks of salmon has been suspected for at least 400 years (Stabell, 1984; Nordeng, 1989a), but not until the beginning of this century were tagging experiments started in order to verify this (Scheer, 1939; Table 12.3).

Among several proposals for the underlying sensory mechanisms involved in home stream detection, Buckland (1880) and later Chidester (1924) tentatively suggested olfactory participation in the process. Craigie (1926) conducted a preliminary study in order to investigate the role of olfaction in

Table 12.3 Homing experiments with salmonids in natural systems, demonstrating an ability for specific return to natal streams or spawning grounds. In several of the experiments an evaluation of the underlying sensory mechanisms has also been carried out

Behaviour and Species	Life stage	Sensory evaluation*	Source
Anadromous			
Sockeye salmon	Mature	Olfactory	Craigie (1926)
Atlantic salmon	Mature	–	White (1936)
Chinook salmon	Mature	–	Taft and Shapovalov (1938)
Steelhead trout	Mature	–	Taft and Shapovalov (1938)
Sockeye salmon	Mature	Olfactory	Clemens et al. (1939)
Pink salmon	Mature	–	Clemens et al. (1939)
Coho salmon	Mature	Olfactory	Wisby and Hasler (1954)
Chum salmon	Mature	Olfactory/visual	Hiyama et al. (1967)
Chinook salmon	Mature	Olfactory/visual	Groves et al. (1968)
Atlantic salmon	Mature	Olfactory	Bertmar and Toft (1969)
Chinook salmon	Mature	Olfactory	DeLacy et al. (1969)
Arctic charr	Mature	–	Nordeng (1971)
Dolly Varden	Mature/immature	–	Armstrong (1974)
Atlantic salmon	Mature	Olfactory/visual	Toft (1975)
Pink salmon	Mature	–	Bams (1976)
Brown trout	Mature	Olfactory/visual	Bertmar (1979)
Potamodromous			
Brown trout	Mature/immature	Olfactory	Stuart (1957)
Rainbow trout	Mature	–	Lindsey et al. (1959)
Arctic charr	Mature	–	Frost (1963)
Kokanee (sockeye)	Mature	Olfactory	Lorz and Northcote (1965)
Cutthroat trout	Mature	Olfactory/visual	McCleave (1967)
Cutthroat trout	Mature	Olfactory/visual	Jahn (1969)
Cutthroat trout	Mature	Olfactory/visual	LaBar (1971)
Brook trout	Mature	–	O'Connor and Power (1973)
Brown trout	Mature	–	Tilzey (1977)
Stationary			
Cutthroat trout	Immature	Olfactory	Miller (1954)
Atlantic salmon	Parr	–	Saunders and Gee (1964)
Brown trout	Immature	–	Harcup et al. (1984)
Brown trout	Mature/immature	Olfactory	Halvorsen and Stabell (1990)

homing by sockeye salmon from coastal regions to the home river. Although some reduction in homing performance was seemingly revealed by elimination of the olfactory sense, the experiment suffered from inadequate design and uncertain origin of the experimental fish.

There was no further progress before a classical experiment was performed by Wisby and Hasler (1954). They captured silver (coho) salmon that had returned to Issaquah Creek, Washington, and its tributary, East Fork. Following olfactory occlusion by plugging the nostrils with cotton and vaseline, the anosmic fish, together with untreated controls, were displaced one mile (1.6 km) below the junction of the two streams. The control salmon repeated their original choices at the stream junction, whilst the anosmic fish distributed themselves randomly between the two streams. This strongly suggested that olfaction was important for home stream detection.

Several homing experiments have been carried out to evaluate factors affecting stream location in potamodromous salmonids (Table 12.3). Using kokanee (resident sockeye), Lorz and Northcote (1965) suggested that olfaction may play a major part in finding the home stream. McCleave (1967) concluded that blockage of either the olfactory or the visual sense did not affect the percentage of cutthroat trout, *O. clarki*, homing to or straying from Clear Creek, Yellowstone Lake, Wyoming. Access to visual clues, however, increased the speed of homing, since homing times for blinded trout were much longer than for olfactory-occluded fish. It is worth noting that several of the anosmic trout reported as homers in this experiment had actually lost their latex nasal plugs. Contrary to McCleave, Jahn (1969) concluded that fewer anosmic as well as blind anosmic cutthroat trout homed than did controls. Olfactory occlusion was carried out by injection of petroleum jelly into the nares. When using the same olfactory occlusion technique, LaBar (1971) obtained an olfactory-dependent within-stream homing of cutthroat trout. Results from other sensory occlusion experiments also strongly suggest that olfaction is the key sense that directs the return of the fish during their freshwater migration (Groves *et al.*, 1968; DeLacy *et al.*, 1969).

The same trends as found in fresh water are also seen when sensory-deprived anadromous fish are displaced from rivers and back out into the sea (Table 12.3). Blinded fish are able to repeat their original choice by returning to the river, whereas olfactory-occluded fish appear to lose their ability to perform such behaviour, and anosmic fish demonstrate a high rate of straying (Hiyama *et al.*, 1967; Bertmar and Toft, 1969; Toft, 1975; Bertmar, 1979).

In experiments carried out by Bertmar and Toft (1969), Toft (1975) and Bertmar (1979), Atlantic salmon and brown (sea) trout were rendered anosmic either by cutting the olfactory nerves or by removal of the olfactory rosettes by heat cauterization. In both cases, reliable controls were produced by sham operations. Vision was impaired by heat cauterization of the cornea. Blinded Atlantic salmon were able to return from a distance of 100 km from the river mouth, whilst distances of 45–60 km were covered by blinded brown trout. Fish with the olfactory sense destroyed, however, did not return to their native river even when displaced with controls. Thus, olfaction seems to be an important factor for return from release points even far away from the home river.

Non-reproductive homing

Mature salmonids that have homed to a tributary or a spawning site will return there if displaced, irrespective of whether the displacement is in a downstream or an upstream direction (Groves et al., 1968; DeLacy et al., 1969). Similar behaviour has been seen in the immature stages of a number of freshwater fish species (Gunning, 1959; Stott et al., 1963). This homing behaviour appears to be a general feature related to the phenomenon of 'home-area' preference and stationary behaviour in stream-dwelling fishes (Gerking, 1959).

Stream-dwelling salmonids occupy territories, but movements between bordering territories may occur. Thus, an extended area may be covered with time, but the fish will still be restricted to a limited area of the stream. This restricted area is usually termed the 'home range' or 'home area' of the fish, and may be up to 18 m of the stream length for resident brown and cutthroat trout (Miller, 1957; Harcup et al., 1984).

When resident salmonids are displaced from their home area, they follow the general pattern by returning to the home area from both up- and downstream displacements. A slightly higher rate of homing may, however, be seen following downstream displacement (Miller, 1954; Saunders and Gee, 1964; Harcup et al., 1984; Halvorsen and Stabell, 1990). If brown trout are displaced from a tributary into the main stream, and released upstream from the tributary–stream junction, they are still able to home. In homing, such fish must first perform a downstream movement to the tributary–stream junction, followed by an upstream movement in the tributary to the home area (Halvorsen and Stabell, 1990).

The homing of stream-dwelling trout after displacement is dependent on the olfactory sense, since fish rendered anosmic by heat cauterization are unable to home whereas sham-burned controls can do so (Halvorsen and Stabell, 1990). The olfactory control of homing behaviour in resident salmonids agrees with observations made on another species of stream-dwelling fish by Gunning (1959). In his experiments with longear sunfish, *Lepomis megalotis megalotis* Raf., anosmic fish moved at random in the stream after displacement while blinded individuals were found to home as quickly and accurately as controls. The resident immature fishes seem therefore to rely on the same sensory mechanisms for homing behaviour as found for homing mature anadromous and potamodromous salmonids.

12.4 OLFACTORY HYPOTHESES IN SALMONID HOMING

Learning of clues: the imprinting hypothesis

Following conditioning experiments with bluntnose minnows, *Hyborhynchus notatus* Raf., which demonstrated that the fish could discriminate between creek waters by olfaction, Hasler and Wisby (1951) proposed the imprinting hypothesis for homing by anadromous salmonids. This hypothesis postulates that waters from home streams contain certain characteristic odours, to which the young salmon become conditioned (or imprinted) while in the stream, and which they recognize and which they orientate to upon reaching their parent stream as mature migrants. Aquatic plants and soil have been suggested as being the source of these stream odours. The sensitive period, during which the juveniles may become imprinted, has been suggested to be the period of smolt transformation, when the young fish undergo metamorphic processes including morphological, physiological and behavioural changes prior to seaward migration (Hasler *et al.*, 1978; Hasler and Scholz, 1983).

Since odours from a home area will undergo dilution along a large river system, Harden Jones (1968) suggested that the descending smolts could be imprinted to characteristic odorant features along their route during the seaward migration. This idea has been further extended by Hartt and Dell (1986), who suggested that juvenile salmon may experience a sequence of sensory cues to which they are imprinted also whilst proceeding along the coast after entering the sea. Although it is difficult to imagine how fixed reference points could be obtained by olfaction in large bodies of moving water, the 'sequential imprinting' hypothesis provides an integration of the model explaining homing orientation of salmonids by means of acquired clues.

Innate responses: the pheromone hypothesis

Based on an analysis of the population structures and migratory systems displayed by Arctic charr, brown trout and Atlantic salmon in the Salangen River system, North Norway, Nordeng (1977) proposed that the entire homeward orientation (i.e. in open sea, coastal waters, fjord and home river) of adult salmonids may be directed by population-specific pheromones. The pheromones were proposed to be derived from the smolts during their downstream migration in the river and sea during spring and summer. Thus, the pheromone hypothesis for the homeward orientation of anadromous salmonids incorporates the entire homeward migration and is based on the migratory systems of local populations (Nordeng 1977, 1983). In this context, the term 'pheromones' should be understood as strain-specific substances which are detected by the olfactory sense, and that release an attraction behaviour of the recipient. It is suggested that orientation using population-specific pheromones may be a common feature for all species of anadromous fishes (Fig. 12.1; Nordeng, 1971, 1989b).

The downstream migration of smolts from a local population in a river system takes place over a specific period prior to the time at which the first mature fish ascend the river. It is proposed, therefore, that during the migrating season, the downstream migration of smolts and the upstream migration of ascending adults overlap to form a two-way 'traffic' system (Nordeng, 1977). With reference to all species of anadromous salmon, it is suggested that the homeward migration of adults is initiated and directed by innate responses to population-specific pheromone 'trails' secreted by smolts along the route (Fig. 12.1). During the final stage of migration, the adult fish may also detect pheromones secreted by fry, juveniles and early-maturing individuals in their freshwater home localities.

Behavioural evidence for the presence of strain-specific chemical signals that release an attractive behaviour in sexually mature fishes has been demonstrated in several species of salmonids (Selset and Døving, 1980; Groot et al., 1986; Quinn and Tolson, 1986). Pheromones used in migration are believed to be produced in the liver, and released into the environment via the bile and the faeces (Selset, 1980; Stabell, 1982, 1987).

Pro aut contra: acquired or innate responses?

Evidence often presented in support for imprinting as the basic mechanism for homeward orientation is based on observations that when salmonid fishes are transplanted to foreign river systems as parr or smolt, they generally return as mature adults to the river of release and not to the river of origin. Since the river of release, in most cases, has contained native fishes of the same species, representing established natural migratory systems, Nordeng (1977) suggested that return to the river of release seemed to depend on association of the transplanted fish with migratory systems already existing within the river of release.

The percentage return of fish from non-native releases is often much lower than that of native fishes in natural systems, and considerable straying often takes place from non-native releases. The low rates of return from non-native releases appear not to be related to poor survival at sea of transplanted, hatchery-raised fish compared with wild fish, since hatchery-raised fish are captured as adults at sea at the same rate as the wild, native fish (Stabell, 1984). Accordingly, the low rates of return and the high incidence of straying have been taken to indicate a diminished navigational ability in transplanted fishes.

A reduced homing ability to the river of release following transplantation is not consistent with theories based upon learning of, or imprinting to, environmental clues. According to the imprinting hypothesis, fishes should imprint to olfactory clues, and presumably respond to these clues equally well irrespective of the site of release. Results of the transplantations of fishes to non-native rivers, therefore, suggest that orientation based on imprinting may be inadequate to explain homing.

Why, then, do any fish return to the stream of release following transplantation? A possible answer is that they school with conspecifics that are native to the releasing site, and follow the adopted school throughout the journey. Since attractive features of non-native pheromone trails must be expected to be suboptimal, many transplanted fishes may drop off en route. This suggestion may also explain the increased straying of transplanted fish. Further, the reason that transplanted fishes do not return to their native river system may simply be that they do not find any correct odour 'trail'.

Even when artificially reared fish are released within their native river system, a low return and a high rate of straying may occur (Stabell, 1984). The fact that such 'native' hatchery-raised fishes may behave in a manner similar to that of 'non-native' fishes does not necessarily contradict the concept of pheromone-guided navigation. As previously mentioned, a river system may contain several local populations of the same species. If mature adults are taken during stream migration and artificial fertilization is conducted for management purposes, population hybrids may be inadvertently produced. Any population of hybrid fishes should be expected to secrete, and/or be 'hard-wired' to respond to, hybrid pheromones (Stabell, 1984). From the point of view of pheromone navigation, therefore, population hybrids may not be truly 'native' with regard to the pheromone systems present. On the other hand, this may open the possibility of producing new 'homogeneous' hatchery populations in sea-ranching programmes by careful selection of returning migrants for breeding purposes.

The chemicals morpholine and phenethyl alcohol have been used as 'odorants' in order to demonstrate imprinting (Scholz et al., 1976; Hasler et al., 1978). Whilst the results of the behavioural experiments suggested that the fish displayed an impressive accuracy in discrimination of these chemicals, it has been questioned whether the substances can be detected at sufficiently low levels by the olfactory sense of fishes (Hara, 1974; Hara and Macdonald, 1975; Hara and Brown, 1979; Tucker, 1983; Stabell, 1984). It should also be noted that the odorants derived from soil and vegetation, to which the fish are suggested to imprint, have not been classified by chemical means. The questions can be raised as to whether soil and vegetation adjoining small stream branches running through similar terrain (Stuart, 1957), are likely to produce highly distinct cocktails of odorants, and if so, are they likely to be more specific to the fish than any odorants produced by closely related conspecifics?

In most of the recent reviews of homing studies, evidence for imprinting has generally been considered conclusive (Thorpe, 1988; Quinn, 1990; Quinn and Dittman, 1990). In general, the experiments cited are one of two types. In experiments of the first category, adults have been captured in one river, the eggs fertilized artificially, and the progeny raised in a hatchery adjacent to a non-native river, prior to being transported and released at two sites some distance from each other. The releasing sites may or may not have been within the 'native' river, but both have been within the range of detection for the fish upon return from the sea. Upon return from the sea, adults have returned to their site of release, and have not strayed to the other site (Donaldson and Allen, 1957; Johannesson, 1987; Gunnerød et al., 1988). Since eggs are usually fertilized in batches, and fry raised and transported in batches, the fish released at the alternative sites may simply not have been closely enough related to each other. Combined with the possibility that many silvering 'smolts' are actually precociously maturing male parr (Jones and Orton, 1940; Jones, 1959) that remain in the area of release as odour donors, this may explain the separate return of fish.

In the second category of experiments, adult fish have been captured at a specific site in a watercourse, the eggs fertilized artificially and the progeny raised in a hatchery within that watercourse. The progeny have then been released either from the hatchery or from a third locality within the watercourse prior to seaward migration. The adults have returned to the hatchery where they were raised, and not to the releasing site or to the site of capture of their parents (Sutterlin and Gray, 1973; Brannon, 1982; Quinn et al., 1989; Brannon and Quinn, 1990).

Unintended mixing of genetically distinct strains of fish by artificial fertilization might explain results obtained in homing studies. Sibling groups of fishes within a single population can discriminate between each other by olfaction (Olsén, Chapter 11 herein) indicating that local populations may express pheromone profiles where family and possibly individual variations are present. In the above cases, arguments based on either the presence of close relatives left in the hatchery, and/or on a possible scent marking of the substratum by the fish in the hatchery prior to release (Stabell, 1987), may be evoked to explain why there was homing to the hatchery and not to the site of capture of the parents. In other words, hybrid progeny produced by artificial fertilization of eggs may have displayed kin recognition behaviour in accordance with their pheromone profile.

Can the 'imprinting' hypothesis then be used at all as a model to explain homing behaviour? A positive answer to that question seems doubtful, and two series of arguments can be presented in support of that conclusion. Firstly, the imprinting theory is inconsistent in the sense that no specific critical period for imprinting has been demonstrated. Although the period of smolt transformation was initially suggested as the criticial stage, it is indisputable that homing takes place also in potamodromous systems where smolt transformation, in its strict sense, does not occur (Stuart, 1957; Halvorsen and Stabell, 1990). In addition, in sockeye salmon the fry move upstream or downstream from the incubation site and into a rearing lake, where they eventually smoltify for seaward migration. Upon return as adults they aim specifically for the spawning area (i.e. the incubation site of fry), and not the lake where smolt transformation (and presumably imprinting) took place (Quinn, 1985; Quinn and Ditman, 1990). Any suggestion of 'sensitive' periods additional to that of smolt transformation must necessarily be speculative, and are not in accordance with the basic definition for the imprinting theory (Stabell, 1984). This casts doubt on the imprinting hypothesis as a universal theory that fits with natural migratory systems in general, in contrast to the theory based on intraspecific pheromones as a basis for homing behaviour.

12.5 POSSIBLE MECHANISMS OF OLFACTORY CONTROL

Salmonids as well as other fish species display clearly orientated movements towards water currents. The behaviour is commonly known as a rheotactic response (Arnold, 1974). In the basic form of rheotaxis the fish face upstream and maintain station. In the presence of chemical stimuli, both the direction of rheotaxis and the kinetic component may be affected. When attractive odours are present, the fish may display an upstream movement, i.e. positive rheotaxis towards the stimuli. In the absence of attractants, or in the presence of repellent odours, a downstream movement may be released, i.e. a negative rheotaxis in response to the stimuli (De Lacy et al., 1969; Halvorsen and Stabell, 1990).

When the fish perform orientation movements, vision, magnetic sensing and the lateral line organ may all be used in combination with olfaction for directional sensing. Thus, the direction of currents will be registered by senses that detect the physical environment, whilst the olfactory sense will detect the chemical quality of the currents and determine the type of response that will be elicited. Regardless of homing theories, this model can be used to explain how fish in a river system may orientate by a combination of positive and negative rheotaxis in the presence or absence of attractive chemical stimuli.

As regards navigation in the sea, Westerberg (1982) demonstrated by telemetric studies that orientation of Atlantic salmon involved movements in a vertical direction, and that the fish tended to follow fine-structured gradient layers in the quasi-mixed surface layer or in the thermocline. Anosmic salmon were found to swim more haphazardly in depths that were unrelated to these microstructure layers. Accordingly, Westerberg suggested that the vertical dives that the salmon perform in the sea are exploratory searches for the vertical distribution of a home stream odour, and that it uses information found in the vertical to choose horizontal direction. The behaviour of anosmic and sham-operated salmon in relation to the microstructure layers was confirmed by Døving et al. (1985). Water samples taken from different depths that showed layering, and to which migrating salmon showed behavioural preferences, were tested by electrophysiological means. Neurones in the olfactory bulb showed differential responses to the water samples, indicating that odour differences were present between the stratified layers found in the ocean, and that the olfactory system of salmon may be able to discriminate between these layers.

An intriguing speculation is that, if conclusive evidence can be given for detection of current directions within microstructure layers by fishes, the model based on olfactory-controlled rheotactic behaviour may also be extended to apply to open waters of lakes and sea. Thus, the homeward orientation of anadromous salmonids might be integrated into one single type of behaviour that may control the entire homeward migration.

12.6 A MODEL SYSTEM AND ITS ECOLOGICAL IMPLICATIONS

It has been suggested that homing behaviour may in some way be under genetic influence (White and Huntsman, 1938; Bams, 1976; McIsaac and Quinn, 1989), but the basic mechanisms by which a genetic influence on homing behaviour may occur have generally not been addressed.

Adult salmonid fishes undertake specific homing to their native spawning site, seemingly by a process that is in some way under genetic influence. Furthermore, the process of homing relies highly on olfactory mechanisms (Section 12.3). Homing to specific sites would be expected to maintain genetic homogeneity within the native population. Accordingly, a homing theory based on olfaction and related to genetic considerations should be given priority. The pheromone hypothesis proposed by Nordeng (1971, 1977) fulfils the necessary requirements for such a theory by its integration of behavioural, sensory and genetic aspects of fish migration. As such, the pheromone hypothesis represents a coherent model system within the field of chemical ecology.

Extensive releases of Atlantic salmon have been made in order to attempt to compensate for the effects of hydroelectric development in Scandinavia during the last three to four decades. In Norway, releases have mostly been of non-native fish. Despite the extensive stocking programmes, a decline in number of returning spawners has been reported in recent years, both in Norway (Atlantic coast) and in Sweden (Baltic Sea). Mixing of different genetic strains of fish may often give unintended results in the management of salmon stocks (Wilkins, 1986), and the term 'genetic pollution' has been introduced (Mills, 1989). Releases of artificially fertilized fish may result in increased numbers of strays, and this may further increase the chances of producing hybrids. A possible scenario would be the occurrence of large numbers of strays among natural spawners that may interfere with the spawning process, with the consequence of an eventual breakdown of the natural migratory systems via a disruption of a genetically based, pheromonally controlled homing behaviour. Therefore, population-specific pheromones should be considered amongst the spectrum of population-specific characters within salmonid ecology.

12.7 SUMMARY AND CONCLUSIONS

The process of homing in salmonid fishes has generally been associated with a return to spawning sites for reproduction. However, following both natural and artificial displacements, salmonids display an ability for homing also within other life stages. Four different life-history stages have been shown to possess an ability for homing behaviour: the adult anadromous and the adult potamodromous stages, where individuals are homing for reproduction, and the resident immature and mature stages, where individuals display homing in relation to stationarity and home-area preferences. In all cases, olfaction has been shown to be an indispensable sense for homing to occur. The behavioural mechanism underlying homing ability in streams is probably based on a behavioural balance between positive and negative rheotaxis towards hitherto unknown odours from the home area. As regards olfactory-controlled behavioural mechanisms in sea and lake orientation, the field is still open to speculation.

Since homing may take place at different life stages, seemingly by mechanisms of similar types, it should be viewed as an integral part of salmonid ecology. Salmonid ecology includes a segregation of fish into genetically distinct intraspecific populations, often occurring sympatrically within waterchains. The process of homing functions to uphold this genetic homogeneity.

Learning of environmental cues during possible sensitive periods (i.e. imprinting), and innate responses to pheromones from local populations, are opposing theories seeking to explain the homing phenomenon. Fish stocked into foreign environments demonstrate a reduced homing performance, which should not occur if imprinting during seaward migration were the ruling mechanism for home stream detection. While the existence of strain-specific pheromones has been demonstrated in behavioural experiments, confirmation of any specific odorants from soil and vegetation necessary to support an imprinting theory is still lacking. On the basis of this circumstantial evidence, it is concluded that the pheromone theory should be given priority.

The possibility of genetic contamination of pheromones in natural systems, resulting from hatchery escapees or poorly designed stocking programmes, may seriously interfere with homing performance and population structures. Accordingly, the importance of including the concept of chemical ecology into management practices is stressed.

REFERENCES

Allendorf, F., Ryman, N., Stennek, A. and Ståhl, G. (1976) Genetic variation in Scandinavian brown trout (*Salmo trutta* L.): evidence of distinct sympatric populations. *Hereditas*, **83**, 73–82.

Allendorf, F.W., Ryman, N. and Utter, F.M. (1987) Genetics and fishery management, in *Population Genetics and Fishery Management* (eds N. Ryman and F.Utter), Univ. Washington Press, Seattle, pp. 1–19.

Armstrong, R.H. (1974) Migration of anadromous Dolly Varden (*Salvelinus malma*) in Southeastern Alaska. *J. Fish. Res. Bd Can.*, **31**, 435–44.

Arnold, G.P. (1974) Rheotropism in fishes. *Biol. Rev.*, **49**, 515–76.

Bams, R.A. (1976) Survival and propensity for homing as affected by presence or absence of locally adapted paternal genes in two transplanted populations of pink salmon (*Oncorhynchus gorbuscha*). *J. Fish. Res. Bd Can.*, **33**, 2716–25.

Berg, M. (1953) A relict salmon, *Salmo salar* L., called 'småblank' from the River Namsen, North-Trøndelag. *Acta borealia, A. Sci.*, **6**, 1–17.

Bertmar, G. (1979) Home range, migrations and orientation mechanisms of the River Indalsälven trout, *Salmo trutta* L. *Fish. Bd Swed. Inst. Freshwat. Res. Drottningholm Rep.*, **58**, 5–26.

Bertmar, G. and Toft, R. (1969) Sensory mechanisms of homing in salmonid fish I. Introductory experiments on the olfactory sense in grilse of Baltic salmon (*Salmo salar*). *Behaviour*, **35**, 235–41.

Brannon, E.L. (1967) Genetic control of migrating behavior of newly emerged sockeye salmon fry. *Int. Pac. Salm. Fish. Comm. Prog. Rep.* **16**, 1–31.

Brannon, E.L. (1972) Mechanisms controlling migration of sockeye salmon fry. *Int. Pac. Salm. Fish. Comm. Bull.*, **21**, 1–86.

Brannon, E.L. (1982) Orientation mechanisms of homing salmonids, in *Proc. Salmon Trout Migr. Behav. Symp.*, (eds E.L. Brannon and E.O. Salo), Univ. Washington Press, Seattle, pp. 219–27.

Brannon, E.L. and Quinn, T.P. (1990) Field test of the pheromone hypothesis for homing by Pacific salmon. *J. Chem. Ecol.*, **16**, 603–9.

Buckland, F. (1880) *Natural History of British Fishes*, Unwin, London, pp. 300–3.

Chidester, F.E. (1924) A critical examination of the evidence for physical and chemical influences on fish migration. *Br. J. exp. Biol.*, **2**, 79–118.

Clemens, W.A., Foerster, R.E. and Pritchard, A.L. (1939) The migration of Pacific salmon in British Columbia waters. *Publs. Am. Ass. Advmt. Sci.*, **8**, 51–9.

Cope, O.B. (1957) The choice of spawning sites by cutthroat trout. *Proc. Utah Acad. Sci. Arts Lett.*, **34**, 73–9.

Craigie, E.H. (1926) A preliminary experiment upon the relation of the olfactory sense to the migration of the sockeye salmon (*Oncorhynchus nerka*, Walbaum). *Trans. R. Soc. Can.*, **5**, 215–24.

Dahl, K. (1928) The dwarf salmon of Lake Byglands fjord. *Salm. Trout Mag.*, **51**, 108–12.

DeLacy, A.C., Donaldson, L.R. and Brannon, E.L. (1969) Homing behavior of chinook salmon. *Univ. Wash. Res. Fish. Seattle, Contrib.*, **300**, 59–60.

Donaldson, L.R. and Allen, G.H. (1957) Return of silver salmon *O. kisutch* to point of release. *Trans. Am. Fish. Soc.*, **87**, 13–22.

Døving, K.B., Westerberg, H. and Johnsen, P.B. (1985) Role of olfaction in the behavioral and neural responses of Atlantic salmon, *Salmo salar*, to hydrographic stratification. *Can. J. Fish. Aquat. Sci.*, **42**, 1658–67.

Frost, W.E. (1963) The homing of charr *Salvelinus willughbii* (Gunther) in Windermere. *Anim. Behav.*, **11**, 74–82.

Gerking, S.D. (1959) The restricted movement of fish populations. *Biol. Rev.*, **34**, 221–42.

Groot, C., Quinn, T.P. and Hara, T.J. (1986) Responses of migrating adult sockeye salmon (*Oncorhynchus nerka*) to population specific odours. *Can. J. Zool.*, **64**, 926–32.

Groves, A.B., Collins, G.B. and Trefethen, P.S. (1968) Roles of olfaction and vision in choice of spawning site of homing adult chinook salmon (*Oncorhynchus tshawytscha*). *J. Fish. Res. Bd Can.*, **25**, 867–76.

Gunnerød, T.B., Hvidsten, N.A. and Heggberget, T. (1988) Open sea releases of Atlantic salmon smolts, *Salmo salar*, in central Norway, 1973–83. *Can. J. Fish. Aquat. Sci.*, **45**, 1340–5.

Gunning, G.E. (1959) The sensory basis for homing in the longear sunfish, *Lepomis megalotis megalotis* (Rafinesque). *Invest. Indiana Lakes Streams*, **5**, 103–30.

Halvorsen, M. and Stabell, O.B. (1990) Homing behaviour of displaced stream-dwelling brown trout. *Anim. Behav.*, **39**, 1089–97.

Hara, T.J. (1974) Is morpholine an effective olfactory stimulant in fish? *J. Fish. Res. Bd Can.*, **31**, 1547–50.

Hara, T.J. and Brown, S.B. (1979) Olfactory bulbar electrical responses of rainbow trout (*Salmo gairdneri*) exposed to morpholine during smoltification. *J. Fish. Res. Bd Can.*, **36**, 1186–90.

Hara, T.J. and Macdonald, S. (1975) Morpholine as olfactory stimulus in fish. *Science, Wash., D.C.*, **187**, 81–2.

Harcup, M.F., Williams, R. and Ellis, D.M, (1984) Movements of brown trout, *Salmo trutta* L., in the River Gwyddon, South Wales. *J. Fish Biol.*, **24**, 415–26.

Harden Jones, F.R. (1968) *Fish Migration*, Edward Arnold, London, 325 pp.

Hartt, A.C. and Dell, M.B. (1986) Early oceanic migrations and growth of juvenile Pacific salmon and steelhead trout. *Int. North Pac. Fish. Comm. Bull.*, **46**, 105 pp.

Hasler, A.D. and Scholz, A.T. (1983) *Olfactory Imprinting and Homing in Salmon. Investigations into the Mechanism of the Imprinting Process*, Springer-Verlag, Berlin, 134 pp.

Hasler, A.D. and Wisby, W.J. (1951) Discrimination of stream odors by fishes and relation to parent stream behavior. *Am. Nat.*, **85**, 223–38.

Hasler, A.D., Scholz, A.T. and Horrall, R.M. (1978) Olfactory imprinting and homing in salmon. *Am. Scient.*, **66**, 347–55.

Hiyama, Y., Taniuchi, T., Suyama, K., Ishioka, K., Sato, R. Kajihara, T. and Maiwa, T. (1967) A preliminary experiment on the return of tagged chum salmon to the Otsuchi River, Japan. *Bull. Jap. Soc. scient. Fish.*, **33**, 18–19.

Hutchings, J.A. (1986) Lakeward migrations by juvenile Atlantic salmon, *Salmo salar. Can. J. Fish. Aquat. Sci.*, **43**, 732–41.

Jahn, L.A. (1969) Movement and homing of Cutthroat trout (*Salmo clarki*) from open-water areas in Yellowstone Lake. *J. Fish. Res. Bd Can.*, **26**, 1243–61.

Johannesson, B. (1987) Observations related to the homing instinct of Atlantic salmon (*Salmo salar* L.). *Aquaculture*, **64**, 339–41.

Jones, J.W. (1959) *The Salmon*. Collins, London. 192 pp.

Jones, J.W. and Orton, H. (1940). The paedogenic cycle in *Salmo salar* L., *Proc. Roy. Soc. Lond.* (B), 128: 485–499.

Jonsson, B. (1985) Life history patterns of freshwater resident and sea-run migrant brown trout in Norway. *Trans. Am. Fish. Soc.*, **114**, 182–94.

Kaasa, H. (1980) Stock characteristics and population-specific migration in resident brown trout, *Salmo trutta* L., in the Eika watershed, Bø, Telemark County. Cand. real. thesis, Inst. Zoology, Univ. Oslo, Norway, 96 pp. (In Norwegian)

Kendall, R.L. (1989) Taxonomic changes in North American trout names. *Trans. Am. Fish. Soc.*, **117**, 321.

LaBar, G.W. (1971) Movement and homing of cutthroat trout (*Salmo clarki*) in Clear and Bridge Creeks, Yellowstone National Park. *Trans. Am. Fish. Soc.*, **100**, 41–9.

Lindsey, C.C., Northcote, T.G. and Hartman, G.F. (1959) Homing of rainbow trout to inlet and outlet spawning streams at Loon Lake, British Columbia. *J. Fish. Res. Bd Can.*, **16**, 695–719.

Lorz, H.W. and Northcote, T.G. (1965) Factors affecting stream location, and timing and intensity of entry by spawning Kokanee (*Oncorhynchus nerka*) into an inlet of Nicola Lake, British Columbia. *J. Fish. Res. Bd Can.*, **23**, 665–87.

McCleave, J.D. (1967) Homing and orientation of cutthroat trout (*Salmo clarki*) in Yellowstone Lake, with special reference to olfaction and vision. *J. Fish. Res. Bd Can.*, **24**, 2011–44.

Machidori, S. and Kato, F. (1984) Spawning populations and marine life of masu salmon (*Oncorhynchus masou*). *Int. North Pac. Fish. Comm. Bull.*, **43**, 1–131.

McIsaac, D.O. and Quinn, T.P. (1989) Evidence for a hereditary component in homing behavior of Chinook salmon (*Oncorhynchus tshawytscha*). *Can. J. Fish. Aquat. Sci.*, **45**, 2201–5.

Miller, R.B. (1954) Movements of cutthroat trout after different periods of retention upstream and downstream from their homes. *J. Fish. Res. Bd Can.*, **11**, 550–8.

Miller, R.B. (1957) Permanence and size of home territory in stream-dwelling cutthroat trout. *J. Fish. Res. Bd Can.*, **14**, 687–91.

Mills, D. (1989) *Ecology and Management of Atlantic Salmon*, Chapman and Hall, London, 351 pp.

Myers, G.S. (1949) Usage of anadromous, catadromous and allied terms for migratory fishes. *Copeia*, **1949**, 89–97.

Nordeng, H. (1971) Is the local orientation of anadromous fishes determined by pheromones? *Nature, Lond.*, **233**, 411–13.

Nordeng, H. (1977) A pheromone hypothesis for homeward migration in anadromous salmonids. *Oikos*, **28**, 155–9.

Nordeng, H. (1983) Solution to the 'Char Problem' based on Arctic char (*Salvelinus alpinus*) in Norway. *Can. J. Fish. Aquat. Sci.*, **40**, 1372–87.

Nordeng, H. (1989a) Salmonid migration: hypotheses and principles, in *Proc. Salm. Migr. Distr. Symp.* (eds E.L. Brannon and B. Jonsson), Univ. Washington Press, Seattle, pp. 1–8.

Nordeng, H. (1989b) Migratory systems in anadromous salmonids. *Physiol. Ecol. Japan.* Spec. Vol. **1**, 167–8.

Northcote, T.G. (1962) Migratory behaviour of juvenile rainbow trout, *Salmo gairdneri*, in outlet and inlet spawning streams of Loon Lake, British Columbia. *J. Fish. Res. Bd Can.*, **18**, 201–70.

Northcote, T.G. (1969) Patterns and mechanisms in the lakeward migratory behaviour of juvenile trout, in *Symposium on Salmon and Trout in Streams* (ed. T.G. Northcote), Univ. British Columbia Press, Vancouver, pp. 183–203.

O'Connor, J.F. and Power, G. (1973) Homing of brook trout (Salvelinus fontinalis) in Motamek Lake, Quebec. *J. Fish. Res. Bd Can.*, **30**, 1012–14.

Quinn, T.P. (1985) Salmon homing: is the puzzle complete? *Env. Biol. Fishes*, **12**, 315–17.

Quinn, T.P. (1990) Current controversies in the study of salmon homing. *Ethol. Ecol. Evol.* **2**, 49–64.

Quinn, T.P. and Dittman, A.H. (1990) Pacific salmon migrations and homing: mechanisms and adaptive significance. *Trends Ecol. Evol.*, **5**, 174–7.
Quinn, T.P. and Tolson, G.M. (1986) Evidence of chemically mediated population recognition in coho salmon (*Oncorhynchus kisutch*). *Can. J. Zool.*, **64**, 84–7.
Quinn, T.P., Brannon, E.L. and Dittman, A.H. (1989) Spatial aspects of imprinting and homing in coho salmon, *Oncorhynchus kisutch. Fishery Bulletin, U.S.* **87**, 769–74.
Raleigh, R.F. (1967) Genetic control in the lakeward migration of sockeye salmon (*Oncorhynchus nerka*) fry. *J. Fish. Res. Bd Can.*, **24**, 2613–22.
Raleigh, R.F. (1971) Innate control of migrations of salmon and trout fry from natal gravels to rearing areas. *Ecology*, **52**, 291–7.
Raleigh, R.F. and Chapman, D.W. (1971) Genetic control in lakeward migrations of cutthroat trout fry. *Trans. Am. Fish. Soc.*, **100**, 33–40.
Ricker, W.E. (1972) Hereditary and environmental factors affecting certain salmonid populations, in *The Stock Concept in Pacific Salmon* (eds R.C. Simon and P.A. Larkin) Univ. British Columbia Press, Vancouver, pp. 19–160.
Ryman, N. and Ståhl, G. (1981) Genetic perspectives of the identification and conservation of Scandinavian stocks of fish. *Can. J. Fish. Aquat. Sci.*, **38**, 1562–75.
Saunders, R.L. and Bailey, J.K. (1980) The role of genetics in Atlantic salmon management, in *Atlantic Salmon: its Future* (ed. A.E.J. Went), Fishing News Books, Farnham, England, pp. 182–200.
Saunders, R.L. and Gee, J.H. (1964) Movements of young Atlantic salmon in a small stream. *J. Fish. Res. Bd Can.*, **21**, 27–36.
Scheer, B.T. (1939) Homing instinct in salmon (*Oncorhynchus*). *Q. Rev. Biol.*, **14**, 408–30.
Scholz, A.T., Horall, R.M., Cooper, J.C. and Hasler, A.D. (1976) Imprinting to chemical cues: the basis for home stream selection in salmon. *Science, Wash., D.C.*, **192**, 1247–9.
Selset, R. (1980) Chemical methods for fractionation of odorants produced by char smolts and tentative suggestions for pheromone origins. *Acta physiol. scand.*, **108**, 97–103.
Selset, R. and Døving, K.B. (1980) Behaviour of mature anadromous char (*Salmo alpinus* L.) towards odorants produced by smolts of their own population. *Acta physiol. scand.*, **108**, 113–22.
Stabell, O.B. (1982) A comparative chemical study on population-specific odorants from Atlantic salmon. *J. Chem. Ecol.*, **8**, 201–17.
Stabell, O.B. (1984) Homing and olfaction in salmonids: a critical review with special reference to the Atlantic salmon. *Biol. Rev.*, **59**, 333–88.
Stabell, O.B. (1987) Intraspecific pheromone discrimination and substrate marking by Atlantic salmon parr. *J. Chem. Ecol.*, **13**, 1625–43.
Stott, B., Elsdon, J.W.V. and Johnston, J.A.A. (1963) Homing behaviour in gudgeon (*Gobio gobio* L.). *Anim. Behav.*, **11**, 93–6.
Stuart, T.A. (1957) The migrations and homing behaviour of brown trout (*Salmo trutta* L.). *Freshwat. Salm. Fish. Res.*, **18**, 27 pp.
Sutterlin, A.M. and Gray, R. (1973) Chemical basis for homing of Atlantic salmon (*Salmo salar*) to a hatchery. *J. Fish. Res. Bd Can.*, **30**, 985–9.
Taft, A.C. and Shapovalov, L. (1938) Homing instinct and straying among steelhead trout (*Salmo gairdneri*) and silver salmon (*Oncorhynchus kisutch*). *Calif. Fish Game*, **24**, 118–25.
Tanaka, S. (1965) Salmon of the North Pacific Ocean–part IX. Coho, chinook and masu salmon in offshore waters. 3. A review of the biological information on

masu salmon (*Oncorhynchus masou*). *Int. North Pac. Fish. Comm. Bull.*, **16**, 75–135.

Thorpe, J.E. (1988) Salmon migration. *Sci. Prog. Oxf.*, **72**, 345–70.

Tilzey, R.D.J. (1977) Repeat homing of brown trout (*Salmo trutta*) in Lake Eucumbene, New South Wales, Australia. *J. Fish. Res. Bd Can.*, **34**, 1085–94.

Toft, R. (1975) The significance of the olfactory and visual senses in the behaviour of spawning migration in Baltic salmon. *Swed. Salm. Res. Inst. Rep.* (LFI Medd, 10/1975), pp. 79. (In Swedish)

Tucker, D. (1983) Fish chemoreception: peripheral anatomy and physiology, in *Fish Neurobiology*. Vol. 1, *Brain Stem and Sense Organs* (eds R.G. Northcutt and R.E. Davis), Univ. Michigan Press, Ann Arbor, pp. 311–49.

Westerberg, H. (1982) Ultrasonic tracking of Atlantic salmon (*Salmo salar* L.)–II. Swimming depth and temperature stratification. *Fish. Bd Swed. Inst. Freshwat. Res. Drottningholm Rep.*, **60**, 102–20.

White, H.C. (1936) The homing of salmon in Apple river, N.S. *J. Biol. Bd Can.*, **2**, 391–400.

White, H.C. and Huntsman, A.G. (1938) Is the local behavior in salmon heritable? *J. Fish. Res. Bd Can.*, **4**, 1–18.

Wilkins, N.P. (1986) *Salmon Stocks: a Genetic Perspective*, Atlantic Salmon Trust, Pitlochry, 30 pp.

Wisby, W.J. and Hasler, A.D. (1954) Effect of olfactory occlusion in migrating silver salmon (*Oncorhynchus kisutch*). *J. Fish. Res. Bd Can.*, **11**, 472–8.

Chapter thirteen

Gustation and nutrition in fishes: application to aquaculture

Masahiko Takeda and Kenji Takii

13.1 INTRODUCTION

Numerous studies on feeding stimulants for aquatic animals have been conducted in the last decade. Although fundamental knowledge has been accumulated (Jones, Chapter 14, this volume), there is little information on the relationship between gustation and nutrition in fishes and possible application of feeding stimulants to aquaculture. This chapter deals with the effects of feeding stimulants on fish nutrition and feeding, with special emphasis on their application to aquaculture. Previous studies on feeding stimulants have been reviewed by Atema (1980), Takeda (1980a, b), Bromley and Sykes (1985), Mackie and Mitchell (1985) and Smith (1989).

13.2 IDENTIFICATION OF FEEDING STIMULANTS

Although a variety of feeding stimulants have been identified from experiments conducted with various fish species, most belong to a small group of chemicals: (1) free amino acids, (2) nucleotides and nucleosides, and (3) quaternary ammonium bases (Takeda, 1980a,b; Mackie and Mitchell, 1985).

Free amino acids

L-Alanine, glycine and L-proline appear to be the major components of feeding stimulants for many fish species. Some variabilities exist in the composition of the active amino acid mixtures for different fish species (Takeda, 1980a,b). L-Tryptophan, L-valine, and a mixture of L-tyrosine, L-phenylalanine, L-lysine and L-histidine have also been identified as feeding stimulants for jack mackerel, *Trachurus japonica* (Ikeda et al., 1988), red sea bream, *Pagrus major* (Ina and Matsui, 1980; Fuke et al., 1981) and rainbow trout, *Oncorhynchus mykiss* (Adron and Mackie, 1978), respectively.

Nucleotides and nucleosides

Inosine-5'-monophosphate (IMP), inosine, adenosine-5'-diphosphate (ADP), guanosine-5'-monophosphate (GMP) and uridine-5'-monophosphate (UMP) have been identified as feeding stimulants for fish. IMP showed marked feeding stimulant activity for yellowtail, *Seriola quinqueradiata* (Takeda, 1980a, b), turbot, *Scophthalmus maximus*, brill, *Scophthalmus rhombus* (Mackie and Mitchell, 1985), jack mackerel (Ikeda et al., 1988) and marbled rockfish, *Sebasticus marmoratus* (Takaoka et al., 1990). In these fish, feeding activity was further enhanced by addition of some amino acids to the diet. Mackie and Adron (1978) indicated that the stimulatory activity of IMP to turbot is related to its chemical structure, e.g. having a purine base and a phosphatidic base at the 5'-position. Japanese eel, *Anguilla japonica*, preferred a diet supplemented with a mixture of L-alanine, glycine, L-proline, L-histidine and UMP, to one supplemented with the amino acids alone. UMP, with a pyrimidine base, showed only minimal feeding stimulant activity to eel (Takeda et al., 1984). Inosine has been found to act as a feeding stimulant for turbot (Mackie and Adron, 1978) and marbled rockfish (Takaoka et al., 1990), with the effect potentiated by amino acids.

Quaternary ammonium bases

Glycine betaine, trimethylglycine, has been reported to act as a feeding stimulant for benthos feeders such as puffer, *Fugu pardalis* (Hidaka, 1982), pinfish, *Lagodon rhomboides* (Carr et al., 1976), pigfish, *Orthopristis chrysopterus* (Carr et al., 1977), red sea bream (Goh and Tamura, 1980) and Dover sole, *Solea solea* (Mackie et al., 1980). The effects are also potentiated by the addition of some amino acids to the diet. The sulphur analogue of glycine betaine, dimethylthetin, has also been reported to be an effective stimulant for Dover sole (Mackie and Mitchell, 1982).

13.3 RELEVANCE OF FEEDING STIMULANTS TO NUTRITION AND FEEDING

Feeding activities of fish change with dietary acceptability and palatability associated with their chemical and physical properties, as well as surrounding environmental conditions (Atema, 1980; Takeda, 1980a; Fletcher, 1984). When food is given to fish, some of the soluble nutrients may leach out into the surrounding water. Fish nutritionists and culturists have consequently developed formula diets that reduce feeding loss. An increase in dietary palatability results in an increase of satiation amounts and a reduction of feeding time (Ishiwata, 1968a,b). The incorporation of feeding stimulants is a useful means to increase the palatability of formula diets for fish, particularly where alternative proteins, such as plant proteins, are substituted for widely used fish meal. The nutritional value of formula diets is normally evaluated by how the composition of essential nutrients meets the requirements of fish. Unpalatable and/or indigestible diets that meet fish nutritional requirements are of little value to fish nutritionists and culturists.

The digestive mechanism in mammals is thought to consist of the cephalic, gastric and intestinal phases. The cephalic phase is concerned with gustation, and the gastric and intestinal phases with nutrition. Olbe (1963) reported that in the alimentary canal of a dog, one-third of the gastric secretion during nutrient digestion results from the cephalic phase. Giduck et al. (1987) indicated that oropharyngeally stimulated responses induced by the smell and taste of palatable foods initiate the cephalic reflex, included in the cephalic phase. This reflex elicits salivation, gastric secretion, and pancreatic exocrine and endocrine secretions stimulated via the vagus nerve. These secretory activities prepare the alimentary canal and other organs for digestion, transport, and utilization of ingested nutrients. Fange and Grove (1979) suggested that the cephalic reflex occurs in fish as well as mammals. Therefore it is conceivable that more palatable diets will enhance fish growth as a result of efficient activation of the cephalic reflex.

We have recently investigated the direct effect of dietary feeding stimulant supplements on food intake, and indirect effects on digestion, absorption and metabolism of ingested nutrients in some fish species. Based primarily on these results, the following sections will demonstrate that the palatability of diets is one of the important factors influencing dietary nutritive value and growth performance, as well as feeding activity.

13.4 DIETARY APPLICATION OF FEEDING STIMULANTS

Feeding stimulants have not frequently been applied to fish formula diets. However, attention has recently been paid to their application in solving problems facing fish culture, such as water pollution due to excessive feeding, and the utilization of unpalatable plant protein as a substitute for fish meal protein (Takeda, 1981).

Starter diets

Effective starter diets have been formulated and widely used for salmonid culture based on knowledge accumulated since the 1940s (National Research Council, 1981). However, the starter formula diets for marine species, both larval and juvenile, are not as convenient as those for freshwater species, and a large amount of live food (mainly rotifers and brine shrimp) is used as starter diets. Therefore it is necessary for each hatchery to maintain large tanks for rearing live food in addition to tanks holding larval and juvenile fish. Practical use of starter formula diets for marine fishes would allow culturists to raise the fish more easily and economically.

Meteiller et al. (1983) showed that larval Dover sole grew faster when reared on a formula diet flavoured with a mixture of glycine betaine, glycine, L-alanine, L-glutamic acid and inosine. They also achieved better survival, specific growth rate and feed conversion rate, compared with fish fed an unflavoured diet.

In Japan, seed stock for yellowtail and eel cultures are mainly obtained by capturing wild juveniles, which are usually reared on minced fish and live small earthworm, *Tubifex* sp., respectively. According to our unpublished data, juvenile yellowtail (mean body weight 1.3 g), fed a fish-meal-based diet supplemented with a mixture of scallop and skipjack extracts for flavour, showed better food intake, weight gain and survival rate than fish raised on an unflavoured diet. However, fish on the flavoured diet gained less weight than fish fed minced sand lance, *Ammodytes personatus*. In contrast, a higher survival rate was found in the fish provided with the flavoured diet (Table 13.1) (Y. Nakao, M. Urayasu, and M. Takeda, Oral presentation in the autumn meeting of Japanese Society of Scientific Fisheries, 4 Oct. 1986). Takii et al. (1984) reported a similar effect of supplemental dietary feeding stimulants on feeding activity of juvenile Japanese eel. The basal diet used for these experiments was a commercial formula diet containing a large amount of fish meal and a small amount of krill meal for flavour. The basal diet was supplemented with (1) a feeding stimulant mixture of L-alanine, glycine, L-proline, L-histidine and UMP, or (2) an extract of *Tubifex* sp. (*Tubifex* is an excellent live starter food traditionally employed in Japanese eel culture), or (3) deionized water as control. Eels weighing an average of 150 mg, a total of 100 g were selectively reared on one of the test diets, fed *ad libitum* for 30 min, three times a day, for two days. Food intake of the fish group given the diet supplemented with the feeding stimulant mixture was 1.78 and 1.34 times as high as values for the control and *Tubifex* extract diets, respectively (Table 13.2).

It is obvious from these results that the identified feeding stimulants are responsible, to some extent, for increasing growth performance as well as for stimulating feeding behaviour of larval and juvenile fish. However, more suitable feeding stimulant mixtures for starter diets must be formulated because it has been reported that the appeal of some amino acids to herring, *Clupea harengus*, vary as they grow from first-feeding larval to juvenile stages (Dempsey, 1978). Therefore, the identification of feeding stimulants for weaning larval and juvenile fish is very important for the development of

Table 13.1 Effect of supplemental dietary extract on survival rate, weight gain and feeding activity of juvenile yellowtail, *Seriola quinqueradiata*

Variable	Flavoured diet*	Unflavoured diet†	Minced sand lance
Number of yellowtail	50	50	50
Survival rate (%)	90	72	78
Mean body weight (g)			
Initial	1.30	1.25	1.35
Final	1.67	1.26	1.91
Weight gain (%)	28.5	0.8	41.5
Food intake over 6 days (g)‡	43.7	15.7	69.0
Relative feeding activity (%)¶	63	23	100

*Commercial skipjack extract (1.34 g) and scallop extract (6.31 g) were added to 100 g of basal diet which was composed of brown fish meal (70), wheat gluten (6), dextrin (5), pollack liver oil (10), vitamin mix (3), mineral mix (2.5), CMC.Na, (2), Ethoxyquin (0.03), cellulose (1.47).
†Deionized water was added to the basal diet.
‡Food intake as g dry weight per 100 g of body weight.
¶The rate of food intake for each dietary supplement, relative to that of minced sand lance.
Source: Nakao *et al.* (unpublished).

Table 13.2 Effect of supplemental dietary feeding stimulants* on feeding activity of juvenile Japanese eel, *Anguilla japonica*

Dietary supplement	Total body weight (g)	Food intake† (g dry weight)	Relative food intake
Deionized water	100	4.54	1
Feeding stimulants	100	8.08	1.78
Extract of *Tubifex* sp.‡	100	6.01	1
Feeding stimulants	100	8.05	1.34

*L-Alanine, 75.7 mg; glycine, 63.8 mg; L-proline, 163 mg; L-histidine, 39.8 mg; UMP·2Na, 626 mg were supplemented to 100 g of commercial diet, contained the following constituents (%): white fish meal, 72.5; α-starch, 5.0; others, 22.5, produced by Nihon Nosan Kogyo Co. Ltd., Yokohama.
†Total amount of food consumed by each group weighing about 100 g, 3 times a day for 2 days.
‡Alcohol extract (Konosu *et al.*, 1965).
Source: Takii *et al.* (1984).

acceptable microparticulate formula diets essential for mass production of fish larvae in aquaculture.

Grower diets

The effects of supplemental dietary feeding stimulants on immature fish will be considered in this section. In a study with Japanese eel, Takii *et al.* (1986a) prepared two fish-meal-based test diets. One diet was supplemented with a feeding stimulant mixture of L-alanine, glycine, L-proline, L-histidine and UMP. The other diet was left intact as control. Two groups of 30 young eel, each having initial mean body weight of 78 g, were fed the test diets, *ad libitum* for 30 min once a day, for 25 days. Enhanced growth performance (percent weight gain), feed efficiency, protein efficiency ratio, and protein and energy retention rates were observed in the fish fed the flavoured diet, compared with those of the control group, in spite of almost the same daily feeding rate between dietary groups (Table 13.3).

Kumai *et al.* (1989) investigated the effects of scallop extract supplemented in fish-meal-casein-based diets on feeding and growth of young tiger puffer, *Fugu rubripes*, which lacks a stomach. Fish reared on diets flavoured with 1% and 2% extract for 3 weeks showed weight gain, feed efficiency and protein efficiency ratios superior to those of fish reared on a diet supplemented with tap water (Table 13.4).

The physiological and nutritional mechanisms underlying the stimulatory effects of dietary feeding stimulants on growth performance of these young fish are not clear. However, these effects may be attributed to enhanced digestion and metabolic processes, as well as to increased food intake. As previously stated, better growth performances were obtained in eel fed on the flavoured diet, despite the fact that the daily feeding rate was almost the

Table 13.3 Effect of supplemental dietary feeding stimulants* on growth performances of Japanese eel reared for 25 days

Variable	Flavoured diet	Unflavoured diet
Number of eels	30	30
Mean body weight (g)		
Initial	77.6	77.7
Final	93.3	80.6
Weight gain (%)	20.2	3.7
Survival rate (%)	100	93.3
Daily feeding rate (%)†	0.70	0.66
Feed efficiency (%)‡	108.2	50.2
Protein efficiency ratio¶	1.96	0.91
Protein retained (%)#	34.4	16.0
Energy retained (%)§	78.6	35.6

*L-Alanine, 285 mg; glycine, 508 mg; L-proline, 217 mg; L-histidine, 39 mg; UMP·2Na, 63 mg were supplemented to 100 g of basal diet.
†The percentage of daily food intake (dry weight) for midterm body weight. Midterm body weight was calculated by the average value of initial and final total body weights.
‡The percentage of weight gain for food intake (dry weight).
¶The ratio of weight gain to protein intake.
#The percentage of body protein gained for protein intake.
§The percentage of body gross energy gained for digestible energy intake.
Source: Takii *et al.* (1986a).

Table 13.4 Effect of supplemental dietary scallop extract* on growth performances of young tiger puffer, *Fugu rubripes*, reared for 21 days

Variable	Flavoured diet		Unflavoured diet
	1% extract	2% extract	(Tap water added)
Number of puffers	50	50	50
Mean body weight (g)			
Initial	127.5	130.0	127.5
Final	178.4	170.0	162.0
Weight gain (%)	39.9	30.8	27.1
Survival rate (%)	100	100	98.0
Daily feeding rate (%)†	1.89	1.63	1.64
Feed efficiency (%)†	84.0	77.9	63.4
Protein efficiency ratio†	1.75	1.59	1.30

*Commercially available (Riken Vitamin Co. Ltd., Tokyo).
†Explained in Table 13.3.
Source: Kumai *et al.* (1989).

same as that for fish fed the unflavoured diet. This result suggests that the chemical cues originating from the dietary feeding stimulants may enhance the cephalic reflex response, resulting in an improved nutritional status of the fish. Alternatively, the nutritional value of formula diets may be improved as a result of feeding stimulant supplementation. However, this is unlikely because nucleotides and betaine, despite their feeding stimulant activity for many fish species, are non-essential nutrients, and amino acids at low supplementation levels may not influence the nutritional balance of formula diets. Thus, the former hypothesis is most likely, i.e. the desirable chemical cues may stimulate the cephalic reflex responses and promote the digestive and metabolic functions for ingesting nutrients, as in mammals (Giduck *et al.*, 1987).

13.5 FEEDING STIMULANT EFFECTS ON NUTRITION

Digestion, absorption and growth

The hypothesis proposed above was tested in the following investigations. Two groups of 30 eel, with initial mean body weight of 78 g, were fed twice daily for 21 days on a diet with or without a feeding stimulant mixture. The feeding stimulant mixture consisted of L-alanine, glycine, L-proline, L-histidine and UMP. The basal diet was composed of white fish meal, pollack liver oil, α-starch, and vitamin and mineral mixtures. Growth performance, feed efficiency, protein efficiency ratio, and protein and energy retention rates were superior in the fish reared on the flavoured diet. In order to define possible causes of the higher performance obtained, postprandial changes in digesta weight, digestive enzyme activities, and the efficiency of dietary protein and carbohydrate digestion were measured on the final day of the feeding trial (Takii et al., 1986a). The gastric digesta weight decreased linearly with time in both dietary groups, whereas the intestinal digesta weight was lower in the group fed the flavoured diet. The total gastrointestinal digesta weight in the flavoured diet group decreased immediately after feeding, but decreased only 3 h after feeding in the unflavoured diet group (Fig. 13.1). Pepsin-like enzyme activity in the gastric digesta was twice as high in the flavoured diet group as in the unflavoured diet group 3 h after feeding, and was equivalent 6 h after feeding. Trypsin-like enzyme activities in the intestinal digesta 3 h and 6 h after feeding were higher in the unflavoured diet group (Table 13.5). Digestibility coefficients of protein and carbohydrate, assayed by the indirect method (Furukawa and Tsukahara, 1966) using total intestinal digesta 5 h after feeding, were higher by 3% and 10% respectively, in the flavoured diet group (Table 13.6).

It is clear from these results that the digestive activity soon after eating in eels fed the flavoured diet was higher than in the unflavoured diet group. The enhanced secretion of gastric juices in eels that were provided the flavoured diet may be the result of the cephalic reflex response to chemical stimulation by L-alanine, glycine, L-proline, L-histidine and UMP, as in dogs (Olbe, 1963). The higher trypsin-like enzyme activities observed in intestinal digesta of eels fed the unflavoured diet (Table 13.5) suggest that pancreatic secretion of the enzyme might increase in compensation for the lower pepsin-like enzyme activity in gastric digesta. It is conceivable that enhanced growth in eels reared on the flavoured diet is the result of increased function

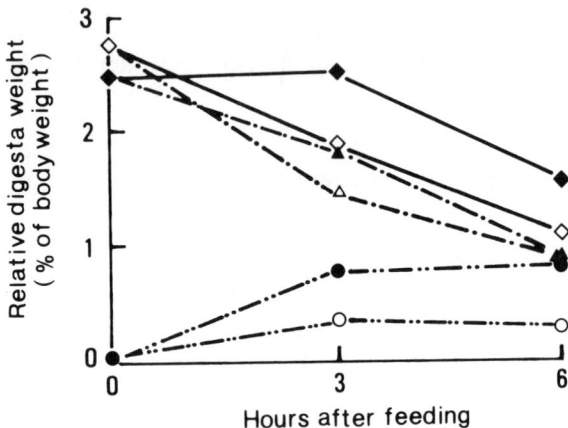

Fig. 13.1 Changes in gastrointestinal digesta weight in Japanese eel, *Anguilla japonica*, after consuming flavoured and unflavoured diets. Flavoured diet: ◇, total gastrointestinal digesta; △, gastric digesta; ○, intestinal digesta. Unflavoured diet: ◆, total gastrointestinal digesta; ▲, gastric digesta; ●, intestinal digesta. (After Takii et al. 1986a.)

Table 13.5 Activities of pepsin-like and trypsin-like enzymes in gastrointestinal digesta of Japanese eel after consuming flavoured and unflavoured diets

Interval (h)	Diet type	Pepsin-like enzyme activity*		Trypsin-like enzyme activity*	
		per g gastric digesta	per 100 g body weight	per g intestinal digesta	per 100 g body weight
3	Flavoured	0.20±0.09†	0.30±0.20	49.5±14.7	17.5 ± 8.33
	Unflavoured	0.10±0.07	0.19±0.17	64.6±14.0	44.6 ±12.7
6	Flavoured	0.24±0.12	0.17±0.18	36.7±17.8	7.51± 3.53
	Unflavoured	0.26±0.06	0.17±0.08	40.5± 2.76	32.9 ±10.7

*Units expressed as μmol of tyrosine-liberated min^{-1} under assay conditions.
†All entries are mean ± standard deviation (n=5).
Source: Takii et al. (1986a).

Table 13.6 Apparent digestibilities (%) of protein and carbohydrate in Japanese eel 5 h after consuming flavoured and unflavoured diets

Diet	Protein	Carbohydrate
Flavoured	88.4	87.8
Unflavoured	85.3	77.1

Source: Takii et al. (1986a).

of digestion and absorption through the induction of the cephalic reflex by chemosensory stimulation.

Hepatic enzyme activity

The enhanced growth of eels reared on the flavoured diet is thought to be indirectly attributable to more efficient nutrient metabolism as well as to increased digestive and absorptive activities soon after eating. Therefore, the effects of dietary feeding stimulants on the metabolic enzyme activities relating to amino acids and carbohydrates were examined (Takii et al., 1986b). The activity levels of six hepatic enzymes, alanine aminotransferase (GPT), aspartate aminotransferase (GOT), glucose-6-phosphatase (G6Pase), phosphoglucose isomerase (PGI), glucose-6-phosphate dehydrogenase (G6PDH) and phosphogluconate dehydrogenase (PGDH), along with the levels of blood glucose and plasma free amino acid nitrogen (PAA), are summarized in Tables 13.7 and 13.8.

Generally, PAA levels were lower, and both GOT and GPT activities were higher, in the flavoured diet group than in the unflavoured diet group (Table 13.7). The blood glucose level 3 h after feeding was higher in the flavoured diet group, but it was significantly higher, at $P<0.05$ on the Student's t-test (Mizushima, 1952), in the unflavoured group 6 h after feeding. There was no significant difference ($P>0.05$) in the activity of G6Pase associated with gluconeogenesis between groups. However, the activities of PGI, relating to glycolysis, and G6PDH and PGDH relating to the hexose monophosphate shunt were maintained significantly higher ($P<0.05$) in eels fed the flavoured diet (Table 13.8). These data indicate that metabolism of protein and carbohydrate is enhanced at a relatively early stage of feeding in eels reared on the flavoured diet as compared with those in the control group. In particular, high activities of hepatic enzymes relating to glycolysis and the hexose monophosphate shunt are observed in fish provided with the flavoured diet. Therefore, dietary feeding stimulants seem to have enhancing effects on the utilization of dietary carbohydrate as an energy source through glycolysis, and on the synthesis of body protein and fat by the supply of nucleic acid and nicotinamide adenine dinucleotide phosphate

Table 13.7 Plasma amino acid nitrogen level (PAA) and hepatic transaminase activities of Japanese eel after consuming flavoured and unflavoured diets

Variable	After 3 h		After 6 h	
	Flavoured diet	Unflavoured diet	Flavoured diet	Unflavoured diet
PAA (mg per 100 ml)	5.40 ± 0.48*	6.36 ± 2.26	4.70 ± 0.88	5.32 ± 1.05
Alanine aminotransferase†				
per g tissue	13.3 ± 4.90	11.3 ± 2.13	18.2 ± 2.96	16.4 ± 5.03
per 100 g body weight	20.8 ± 7.70	17.6 ± 5.45	25.4 ± 3.01	24.7 ± 7.85
Aspartate aminotransferase†				
per g tissue	27.3 ± 3.35	26.4 ± 3.43	30.5 ± 2.04	25.7 ± 2.60
per 100 g body weight	44.6 ± 10.9	40.5 ± 9.07	43.0 ± 6.17	39.0 ± 6.47

*All entries are mean ± standard deviation (n = 5).
†Units expressed as μmol of pyruvate liberated min^{-1} under assay conditions.
Source: Takii et al. (1986b).

Table 13.8 Blood glucose levels and activities of hepatic enzymes relating to carbohydrate metabolism in eel after consuming flavoured and unflavoured diets

Variable	After 3 h		After 6 h	
	Flavoured diet	Unflavoured diet	Flavoured diet	Unflavoured diet
Blood glucose (mg per 100 ml)	202 ± 25.9*	175 ± 113	174 ± 48.1	283 ± 46.7
Glucose-6-phosphatase†				
per g tissue	6.26 ± 1.12	6.15 ± 0.70	6.36 ± 1.03	6.45 ± 1.54
per 100 g body weight	10.0 ± 1.39	9.39 ± 1.55	8.79 ± 1.95	9.55 ± 0.56
Phosphoglucose isomerase†				
per g tissue	124 ± 20	127 ± 15	139 ± 11	116 ± 7
per 100 g body weight	204 ± 60	199 ± 57	197 ± 33	178 ± 41
Glucose-6-phosphate dehydrogenase†				
per g tissue	56.2 ± 4.3	42.8 ± 12.3	60.2 ± 8.3	45.3 ± 6.5
per 100 g body weight	91.1 ± 15.1	68.8 ± 29.1	85.7 ± 18.7	70.7 ± 23.3
Phosphogluconate dehydrogenase†				
per g tissue	22.2 ± 4.7	20.8 ± 3.8	23.1 ± 2.4	17.0 ± 4.4
per 100 g body weight	35.8 ± 8.8	33.4 ± 12.5	32.4 ± 3.2	26.6 ± 11.4

*All entries are mean ± standard deviation (n = 5).
†Units expressed as μmol of substrate converted min^{-1} under assay conditions.
Source: Takii et al. (1986b).

(NADPH) through the hexose monophosphate shunt, resulting in better growth performance. Since L-alanine, L-proline and glycine are typical glucogenic amino acids, their feeding stimulant properties might be related to the preferential utilization of these amino acids for energy production in carnivorous fishes, as suggested by Torii et al. (1987).

In mammals, Giduck et al. (1987) indicated that the cephalic reflex initiated by chemosensory stimulation has an important role in digestion, absorption and metabolism of ingested nutrients. Also, sweet taste stimulants such as sucrose and glucose activate the endogenous hormone secretions from the pancreas and increase cardiac pulse, resulting in increased blood flow. Gastrointestinal hormones secreted during the cephalic reflex are also involved in regulating the digestive process. Cholecystokinin (CCK), a gastrointestinal hormone, stimulates endogenous insulin secretion from the pancreas (Itoh, 1980). CCK or CCK-like substances have already been identified from the intestinal tissue of eels and other fish species (Nilsson, 1970, 1973; Barrington, 1972; Barrington and Dockray, 1972). Therefore, the increased activities of hepatic enzymes involving carbohydrate and amino acid metabolism in the eels receiving a feeding stimulant might be related to the increased secretion of insulin during the early postprandial period. This view is supported by the fact that PAA and blood glucose levels 6 h after feeding remain lower in the flavoured diet group (Tables 13.7 and 13.8). This may be the result of insulin-activated cellular absorption of these free nutrients. Much research has been conducted on fish metabolism relating to insulin, and some authors have concluded that fish are like diabetic mammals (Hepher, 1988). Furuichi (1983) found that plasma insulin levels in carp, *Cyprinus carpio*, red sea bream and yellowtail do not increase until 3 h after feeding and are lower than the level in mammals. This indicates that nutrients absorbed soon after feeding may be excreted and not sufficiently utilized until about 3 h after feeding, when serum insulin levels increase. The enhanced growth performances detected in eels reared on the flavoured diet might be partly attributed to increased secretion of insulin early in the postprandial period. Further research should be conducted to clarify the function of feeding stimulants in evoking insulin secretion in fish.

13.6 CONCLUSIONS AND PROSPECTS

Information on chemoreception and feeding stimulants in fish advances our understanding of chemical and physiological processes applicable to fisheries and aquaculture. The application of feeding stimulants to fish formula diets and their effects on fish nutrition, particularly digestion and metabolism, may be summarized as follows.

1. Supplemental dietary feeding stimulants increase the feeding activity, survival rate and feed efficiency of larval and juvenile fish.
2. Supplemental dietary feeding stimulants increase the growth performance of young fish in terms of feed efficiency, protein efficiency ratio and retention of nutrients.
3. A diet flavoured with feeding stimulants enhances secretion of gastric juices early in the postprandial period and maintains a greater potential for protein and carbohydrate digestion in Japanese eels.
4. Activities of hepatic enzymes relating to glycolysis and the hexose monophosphate shunt are higher in Japanese eels fed a flavoured diet.

It is therefore clear that supplemental dietary feeding stimulants are effective in enhancing not only the feeding activity but also the growth performance of fish. The effect of feeding stimulants on growth performance may be attributed primarily to improved food intake, digestion and absorption, and secondarily to increased activity of hepatic enzymes relating to protein and carbohydrate metabolism, resulting in effective utilization of ingested nutrients. It is likely that chemosensory stimulation by dietary feeding stimulants induces the cephalic reflex, which serves to increase the activities of digestion, absorption and metabolism soon after eating. It is well documented that gustatory functions to initiate these processes following food intake in humans (Yumikari and Torii, 1981).

In view of the significance of optimal diet development in aquaculture, more research is necessary on dietary supplementation with feeding stimulants to improve the palatability and nutritional value of diets. The use of alternative plant protein sources in grower formula diets may require dietary flavours and specific dietary ingredients, rich in compound that stimulate feeding, to acclimatize fish to less palatable protein sources. Although the use of flavours in domestic animal feeds has increased since 1980 along with the number of flavour products and sweeteners (Bradley, 1983a,b), those in practical fish feeds are scarce. Therefore, further investigations are required for the development of potential flavour products before they can be commercially available for fish feeds. Potentially valuable flavour products include extracts from shrimp, clam or fish generally discarded during processing (Mayers, 1986).

ACKNOWLEDGEMENTS

The authors express their sincere thanks to Dr T.J. Hara, Department of Fisheries and Oceans, 501 University Crescent, Winnipeg, Canada, and Dr C. Hollingworth, School of Biological Sciences, University College of North Wales, Bangor, United Kingdom, who kindly read the manuscript and gave valuable suggestions and careful revision.

REFERENCES

Adron, J.W. and Mackie, A.M. (1978) Studies on the chemical nature of feeding stimulants for rainbow trout, *Salmo gairdneri*. *J. Fish Biol.*, **12**, 303–10.
Atema, J. (1980) Chemical senses, chemical signals and feeding behavior in fishes, in *Fish Behavior and its Use in the Capture and Culture of Fishes* (eds J.E. Bardach, J.J. Magnuson, R.C. May and J.M. Reinhart), ICLARM, Manila, pp. 57–101.
Barrington, E.J.W. (1972) The pancreas and intestine, in *The Biology of Lampreys* (eds M.W. Hardisty and J. Potter), Academic Press, New York, pp. 135–69.
Barrington, E.J.W. and Dockray, G.J. (1972) Cholecystokinin-pancreozymin like activity in the eel (*Anguilla anguilla* L.). *Gen. comp. Endocrinol.*, **19**, 80–7.
Bradley, B. (1983a) Flavor utilization in feed. Part I. The nature and purpose of feed flavors. *Feed Management*, **34**(2), 52–5.
Bradley, B. (1983b) Flavor utilization in feed. Part II. Flavor selection, handling and admixing. *Feed Management*, **34**(3), 12–22.
Bromley, P.G. and Sykes, P.A. (1985) Weaning diets for turbot (*Scophthalmus maximus* L.), sole (*Solea solea* L.) and cod (*Gadus morhua* L.), in *Nutrition and Feeding in Fish* (eds C.B. Cowey, A.M. Mackie and J.G. Bell), Academic Press, London, pp. 191–212.
Carr, W.E.S. and Chaney, T.B. (1976) Chemical stimulation of feeding behavior in the pinfish, *Lagodon rhomboides*: characterization and identification of stimulatory substances extracted from shrimp. *Comp. Biochem. Physiol.*, **54A**, 437–41.
Carr, W.E.S. Blumenthal, K.M. and Netherton, J.C. (1977) Chemoreception in the pigfish, *Orthopristis chrysopterus*: the contribution of amino acids and betaine to stimulation of feeding behavior by various extracts. *Comp. Biochem. Physiol.*, **58A**, 69–73.
Dempsey, C.H. (1978) Chemical stimuli as a factor in feeding and intraspecific behaviour of herring larvae. *J. mar. biol. Ass. U.K.*, **58**, 739–47.
Fange, R. and Grove, D. (1979) Digestion, in *Fish Physiology*. Vol. VIII (eds W.S. Hoar, D.J. Randall and J.R. Brett), Academic Press, New York, pp. 162–260.
Fletcher, D.J. (1984) The physiological control of appetite in fish. *Comp. Biochem. Physiol.*, **78A**, 617–28.
Fuke, S., Konosu, S. and Ina, K. (1981) Identification of feeding stimulants for red sea bream in the extract of marine worm, *Perinereis brevicirrus*. *Bull. Jap. Soc. scient. Fish.*, **41**, 1631–5.
Furuichi, M. (1983) Studies on the utilization of carbohydrate by fishes. *Rep. Fish. Res. Lab., Kyushu Univ.*, **6**, 1–59.
Furukawa, A. and Tsukahara, H. (1966) On the acid digestion method for the determination of chromic oxide as an index substance in the study of digestibility of fish feeds. *Bull. Jap. Soc. scient. Fish.*, **32**, 502–6. (In Japanese)
Giduck, S.A., Threatte, R.M. and Kare, M.R. (1987) Cephalic reflexes: their role in digestion and possible roles in absorption and metabolism. *J. Nutr.*, **117**, 1191–6.
Goh, Y. and Tamura, T. (1980) Olfactory and gustatory responses to amino acids in two marine teleosts – red sea bream and mullet. *Comp. Biochem. Physiol.*, **66C**, 217–24.
Hepher, B. (1988) *Nutrition of Pond Fishes*, Cambridge University Press, Cambridge, pp. 64–88.
Hidaka, I. (1982) Taste receptor stimulation and feeding behavior in the puffer, in *Chemoreception in Fishes* (ed. T.J. Hara), Elsevier, Amsterdam, pp. 243–58.
Ikeda, I., Hosokawa, H., Shimeno, S. and Takeda, M. (1988) Identification of feeding stimulant for jack mackerel in its muscle extract. *Nippon Suisan Gakkaishi*, **54**, 229–33. (In Japanese)

Ina, K. and Matsui, H. (1980) Survey of feeding stimulants for red sea bream (*Chrysophrys major*) in the marine worm (*Perinereis vancaurica tetradentata*). *Nippon Nogeikagaku Kaishi*, **54**, 7–12. (In Japanese)

Ishiwata, N. (1968a) Ecological studies on the feeding of fishes – V. Size of fish and satiation amount. *Bull. Jap. Soc. scient. Fish.*, **34**, 781–4. (In Japanese)

Ishiwata, N. (1968b) Ecological studies on the feeding of fishes – VI. Factors affecting satiation amount (1). *Bull. Jap. Soc. scient. Fish.*, **34**, 785–91. (In Japanese)

Itoh, Z. (1980) Gastrin and its group, in *Gastrointestinal Hormones* (ed. Kodansha Scientific), Kodansha, Tokyo, pp. 211–20. (In Japanese)

Konosu, S., Fujimoto, K., Takashima, R., Matsushita, T. and Hashimoto, Y. (1965) Constituents of the extract and amino acid composition of the protein of short-necked clam. *Bull. Jap. Soc. scient. Fish.*, **31**, 680–6. (In Japanese)

Kumai, H., Kimura, I., Nakamura, M., Takii, K. and Ishida, H. (1989) Studies on digestive system and assimilation of a flavored diet in ocellate puffer. *Nippon Suisan Gakkaishi*, **55**, 1035–43.

Mackie, A.M. and Adron, J.W. (1978) Identification of inosine and inosine 5′-monophosphate as the gustatory feeding stimulants for turbot, *Scophthalmus maximus*. *Comp. Biochem. Physiol.*, **60A**, 79–83.

Mackie, A.M. and Mitchell, A.I. (1982) Further studies on the chemical control of feeding behaviour in the Dover sole, *Solea solea*. *Comp. Biochem. Physiol.*, **73A**, 89–93.

Mackie, A.M. and Mitchell, A.I. (1985) Identification of gustatory feeding stimulants for fish – application in aquaculture, in *Nutrition and Feeding in Fish* (eds C.B. Cowey, A.M. Mackie and J.G. Bell), Academic Press, London, pp. 177–90.

Mackie, A.M., Adron J.W. and Grant, P.T. (1980) Chemical nature of feeding stimulants for the juvenile Dover sole, *Solea solea* (L.). *J. Fish Biol.*, **16**, 701–8.

Mayers, S.P. (1986) Attractants, aquatic diet development examined. *Feedstuffs* (29 December), 11–12.

Meteiller, R., Cadena-Roa, M. and Ruyet, J.P. (1983) Attractive chemical substances for the weaning of Dover sole (*Solea vulgaris*). Qualitative and quantitative approach. *J. World Maricul. Soc.*, **14**, 679–84.

Mizushima, U. (1952) *Introduction to Statistical Analysis in Agricultural Experiments*, Yokendo, Tokyo, pp. 73–85. (In Japanese)

National Research Council (1981) *Nutrient Requirements of Coldwater Fishes*. National Academy Press, Washington, D.C., p. 63.

Nilsson, A. (1970) Gastrointestinal hormones in the holocephalian fish *Chimaera monstrosa*. *Comp. Biochem. Physiol.*, **32**, 387–90.

Nilsson, A. (1973) Secretin-like and cholecystokinin-like activity in *Myxine glutinosa* L. *Acta Regiae Soc. sci. litt. Gothob. Zool.*, **8**, 30–2.

Olbe, L. (1963) Significance of vagal release of gastrin during the nervous phase. *Gastroenterology*, **44**, 463–8.

Smith, L.S. (1989) Digestive function in teleost fishes, in *Fish Nutrition*, 2nd edn (ed. J.E. Halver), Academic Press, New York, pp. 322–423.

Takaoka, O., Takii, K., Nakamura, M., Kumai, H. and Takeda, M. (1990) Identification of feeding stimulants for marbled rockfish. *Nippon Suisan Gakkaishi*, **56**, 345–51. (In Japanese)

Takeda, M. (1980a) Feeding in fish, in *Nutrition and Diets of Fishes* (ed. C. Ogino), Koseisha-Koseikaku, Tokyo, pp. 12–26. (In Japanese)

Takeda, M. (1980b) Feeding stimulants for fish. *Iden (Heredity)*, **34**, 45–52. (In Japanese)

Takeda, M. (1981) Application of feeding stimulants for capture and culture of fish, in *Chemical Sense of Fish and Feeding Stimulants* (ed. Japanese Society of Scientific Fisheries), Koseisha-Koseikaku, Tokyo, pp. 109–18. (In Japanese)

Takeda, M., Takii, K. and Matsui, K. (1984) Identification of feeding stimulants for juvenile eel. *Bull. Jap. Soc. scient. Fish.*, **50**, 645–51.

Takii, K., Shimeno, S., Takeda, M. and Kamekawa, S. (1986a) The effect of feeding stimulants in diet on digestive enzyme activities of eel. *Bull. Jap. Soc. scient. Fish.*, **52**, 1449–54.

Takii, K. Shimeno, S. and Takeda, M. (1986b) The effect of feeding stimulants in diet on some hepatic enzyme activities of eel. *Bull. Jap. Soc. scient. Fish.*, **52**, 2131–4.

Takii, K., Takeda, M. and Nakao, Y. (1984) Effects of supplement of feeding stimulants to formulated feeds on feeding activity and growth of juvenile eel. *Bull. Jap. Soc. scient. Fish.*, **50**, 1039–43. (In Japanese)

Torii, K., Mimura, T. and Yugari, Y. (1987) Biochemical mechanism of umami taste perception and effect of dietary protein on taste preference for amino acids and sodium chloride in rats, in *Umami: A Basic Taste* (eds Y. Kawamura and M.R. Kare), Marcel Dekker, New York, pp. 513–64.

Yumikari, K. and Torii, K. (1981) Gustation and nutrition, in *Science of Gustation* (ed. M. Sato), Asakura Shoten, Tokyo, pp. 244–66. (In Japanese)

Chapter fourteen

Food search behaviour in fish and the use of chemical lures in commercial and sports fishing

Keith A. Jones

14.1 INTRODUCTION

When exposed to chemical stimuli associated with food (e.g. extracts or washings of prey organisms), fish initiate behavioural patterns which enhance their chances of locating and consuming the food source (Bateson, 1890). The contents of these food search patterns vary from species to species, but in many cases they are easily discerned, even by the untrained observer. Indeed, in some instances, as during feeding frenzies, food search behaviour can be quite spectacular (Greene, 1925; Tester, 1963b). By no surprise, the overt response of fish to food odours and flavours has not escaped Man's commercial eye, and for centuries commercial and sports fishermen alike have employed 'chemical lures' in many diverse ways to augment their catch. The historical persistence of these strategies attests to their effectiveness.

The intent of this chapter is to describe various aspects of food search behaviour in fish, its external and internal modifiers, the chemicals that evoke it, and the strategies used by the commercial and sport fisheries in employing chemical lures.

14.2 FOOD SEARCH BEHAVIOUR

Wunder (1927) differentiated three phases of food search: (1) an initial period of arousal or excitement, wherein the fish is alerted to the presence of the stimulus; (2) a subsequent search or exploratory phase to locate the source; and (3) a consummation phase, in which the fish seeks to ingest the potential food. Atema (1971) further separates the final phase into food uptake and food ingestion. It should be emphasized that while it is functionally possible, on the basis of their behavioural elements and neural controls, to separate the three phases, they are in reality a continuum without necessarily distinct transitions.

Arousal phase

The release of food substances into the water stimulates food search behaviour only after the concentration exceeds the animal's threshold for perception. The first signs of detection are often quite subtle, but with continued low-level stimulation, more overt behavioural changes eventually surface. These are best observed in sedentary species because of their quiet life styles. Ictalurid catfish, for example, normally rest on the bottom during the daytime, in shelter if available. Upon stimulation with a food extract, they quickly increase their magnitude and rate of opercular pumping, twitch their maxillary barbels, sway their heads to-and-fro in a pronounced fashion, take one or more large gulps, and finally lunge forward to initiate their search (Fig. 14.1; Olmsted, 1918; Bardach *et al.*, 1967; Atema, 1971; Holland and Teeter, 1981). Other species show additional traits, including extensions and flickings of the fins (especially those which serve as specialized chemoreceptors), muscular jerks, twitches and body quivers, rapid shifts of the eyes, yawning or coughing, changes in body posture, and exaggerated lateral movements of the head and tail (Hoese and Hoese, 1967; Marusov, 1975; Pawson, 1977; Hodgson and Mathewson, 1978; Holland, 1978; Mearns, 1985, 1986). These signs of arousal show considerable plasticity. A single individual may display several simultaneously or in rapid succession, the entire display lasting from a few seconds to several minutes. If the level of stimulation rises rapidly, or if an individual has repeatedly experienced a given stimulus, then some or all of these behaviours can be abbreviated or bypassed altogether. Black bullheads, *Ictalurus melas*, that have learned to feed in a particular corner of their aquarium may respond to a food odour by moving directly to that corner without any preceding sign of awareness (pers. obs.).

Continued stimulation almost always leads to a change in locomotor activity, most often seen as a change in the rate of movement. These initial movements are appetitive in nature and are not necessarily directed towards the source. Thus, they are distinguishable from the more directed locomotor

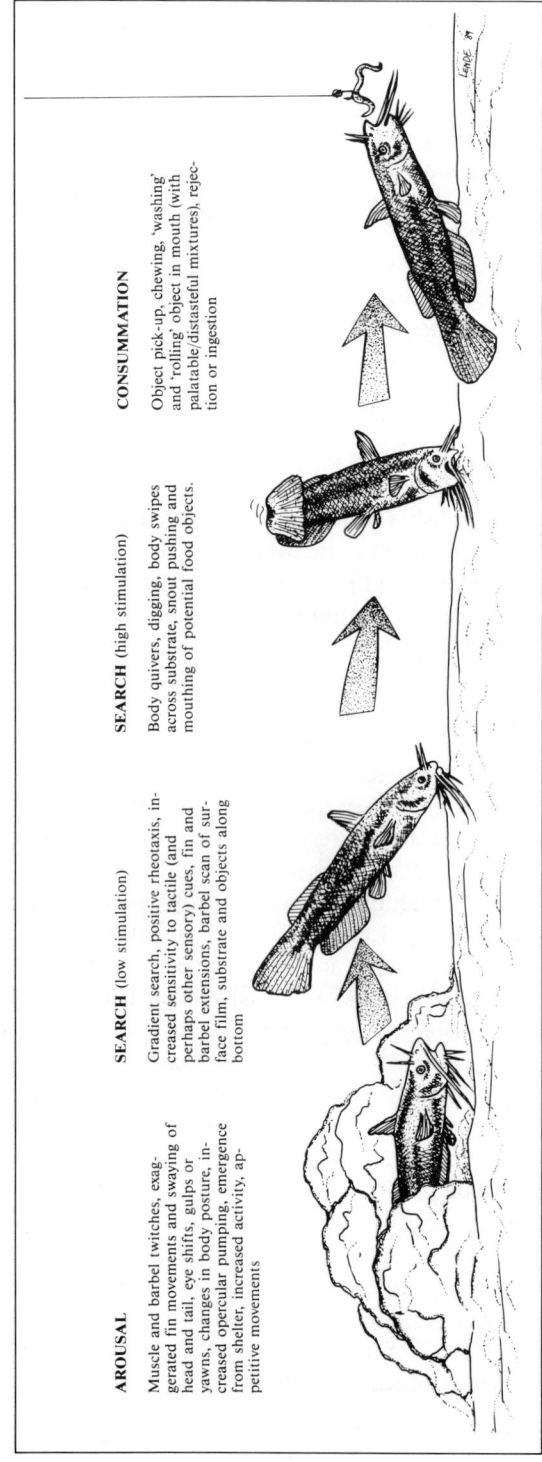

Fig. 14.1 Elements of food search behaviour in the black bullhead, *Ictalurus melas*.

patterns typical of the search phase. Some sedentary species display particularly dramatic outbursts of activity. The sole, *Solea*, for example, erupts from hiding in the sand into a series of violent jumps across the substratum before finally settling down to search for food (Bateson, 1890). In social species, the presence of conspecifics can intensify the locomotor response all the more. Left undisturbed, a group of hagfish, *Polistotrema stouti*, will lie on the bottom, each fish coiled up like a wound spring. Introducing a slurry of fish into their environment causes the group to explode into action, the contagious excitement of those hagfish nearest the source being rapidly transmitted to those further away (Greene, 1925). Such outbreaks of activity are not limited to those times in which fish are normally spontaneously active. Lampreys, *Petromyzon marinus*, for instance, maintain a normal diurnal periodicity in activity (Kleerekoper *et al.*, 1961). Exposure to 'trout water' disrupts the rhythm and causes the lampreys to begin immediate active swimming. Species that are already active (e.g. the fast-swimming oceanic scombrids) respond to food odours by increasing or decreasing their swimming speeds (Steven, 1959).

Changes in locomotion are not limited to activity. Chemical stimulation appears to override those neural mechanisms controlling all aspects of locomotory behaviour (Kleerekoper, 1967). Some complex behaviours may be overridden as well. Stimulation by food extracts causes photonegative species to enter the light (McBride *et al.*, 1962) and disrupts schooling behaviour (Tester *et al.*, 1955; Steven, 1959; McBride *et al.*, 1962; McMahon and Tash, 1970; Atema *et al.*, 1980; Scarfe *et al.*, 1985).

Search phase

Once aroused, the fish attempts to locate the food source. This is by no means a certainty, and it is here that the greatest variability in ways and means of locating food is seen among different species.

There are three common types of chemical distributions: (1) chemicals released as a cloud into relatively calm water, (2) chemicals released as a plume into a current, and (3) chemicals laid along a linear path, or trail. In none of these cases is the chemical gradient likely to be smooth. Currents and turbulence, present in even the calmest water, generate chemical patches varying in concentration often unrelated to the distance from the source. Consequently, a fish swimming within a chemical field and having to rely upon the gradient alone to locate the source must extract directional information from a complex and unstable chemical picture with vague borders. Fish moving through a series of chemical patches must somehow decipher the fluctuations to determine gradient direction; those swimming in a plume can likely detect a sharp boundary between the plume and the surrounding untainted water, but within the plume the gradient may be indiscernably shallow; and those encountering a trail must determine the trail's polarity and overcome occasional breaks if they are to track down the source.

Despite these complexities, some species appear capable of true gradient search. Bardach et al. (1967) found that the yellow bullhead, *Ictalurus natalis*, could readily locate infusions of liver juice or 0.01 M cysteine hydrochloride from a distance of at least 25 body lengths. The bullheads accomplished this feat not only in flowing water (which adds a directional vector to the stimulus) but were even more adept at localizing the chemical source in stagnant water. Nurse sharks, *Ginglymostoma cirratum* (Hodgson and Mathewson, 1978) and sea robins, *Prionotus* sp. (Bardach and Case, 1965) move up odour gradients by zigzag or S-shaped search patterns, suggestive of a constant monitoring of chemical concentration. The former species is also able grossly to locate a point source infusion of shrimp extract in stagnant water (Kleerekoper et al., 1975). Descriptions of the food search patterns in tomtates, *Bathystoma rimator* (Steven, 1959), whitings, *Merlangius merlangus*, and cod, *Gadus morhua* (Pawson, 1977), suggest that these species may be capable of gradient search as well. Some examples of odour trail following have also been reported (Teichmann, 1959; Hobson, 1963).

There has been extensive debate on the mechanisms that might enable fish and other organisms to orientate within chemical gradients. Both kineses and taxes have been proposed (Bell and Tobin, 1982). Although kineses are used by single-cell organisms and some invertebrates (Bell and Tobin, 1982), they are considered inefficient, and their validity in fish has been questioned (Neill, 1979). Taxes (directive orientation mechanisms) seem more plausible. Two types have received the majority of attention: successive temporal comparisons of stimulus intensity at different points within the gradient (klinotaxis), and instantaneous comparisons of stimulus intensity between two or more distinct receptor organs, e.g. bilateral chemosensory organs (tropotaxis). In klinotaxis a fish seeking food would continue its present course as long as the chemical concentration increased between successive samplings, e.g. between olfactory sniffs. A concentration decrease would cause the fish to turn to one side or the other. In tropotaxis the fish would seek to balance stimulus intensity between its bilateral sensory organs. An imbalance would cause the fish to turn towards the side of greater stimulation.

There is experimental support for the tropotactic mechanism. Parker (1912) observed that normal dogfish, *Mustelus canis*, searched for hidden pieces of food by swimming in a figure-of-eight pattern, making an equal number of right and left turns. When he plugged one nostril, the dogfish adopted a more circular path during which the animal turned predominantly towards the side still receiving stimulation. The gradient search of bullheads also suggests the use of tropotaxis. In this case, however, the response is mediated by the bullhead's extraordinary sense of taste. As with the other ictalurids, almost the entire body surface of a bullhead is densely covered with taste buds (Atema, 1971). Yellow bullheads deprived of their sense of smell can still locate distant chemical sources, seemingly orientating by taste alone (Bardach et al., 1967). When, however, the taste buds on the barbels and along the flank of one side are made inoperable, the fish turns more towards the intact side. This suggests that they maintain their headings within chemical gradients by balancing taste bud stimulation between their flanks. The demonstration by Johnsen and Teeter (1980) that bullheads can indeed discern concentration differentials of 10% between the two maxillary barbels supports this hypothesis. Some of the morphologically exaggerated chemosensory structures (e.g. catfish barbels, the long chemoreceptive pectoral fins of the hake, *Merluccius* spp., and the wide heads of hammerhead sharks, *Sphyrna* sp.), as well as the exaggerated swaying movements of fish swimming along a gradient, may be chemoorientation adaptations for emphasizing concentration differences over space and time (Bardach, 1967).

Fish can also draw upon directional information provided by other environmental cues. In this strategy the chemical stimulus acts to release appropriate behavioural responses to these secondary cues (Kleerekoper, 1972). For species that inhabit flowing water, the cue is most often the current; the chemical signals cause the fish to become positively rheotactic. Pelagic sharks have been observed to approach a fish struggling at the end of a line almost exclusively from the downstream direction (Hobson, 1963). Similarly, flounders, *Pseudopleuronectes americanus*, approached crushed clams placed in a tidal brook primarily from downstream, regardless of the direction of flow (Sutterlin, 1975). Kleerekoper (1967) reported that several species of fish tested in a compartmentalized laboratory tank were unable to locate infusions of food extract unless they were associated with a positive flow differential.

Mathewson and Hodgson (1972) studied odour-induced rheotropism in nurse sharks, and lemon sharks, *Negaprion brevirostris*, held captive in a pen containing currents of various strengths. When lemon sharks were stimulated with a mixture of glutamic acid and trimethylamine released into an area of relative calm, they responded by moving into the strongest current present and swimming upstream—even when this manoeuvre carried them away from the source. Thus, in lemon sharks, chemical stimulation seems merely to trigger rheotaxis, which under natural conditions would normally have led them directly to the food source. In contrast, the nurse sharks located the delivery point by swimming in and out of the chemical field in a manner suggestive of klinotaxis. However, when presented with food extracts in a hydrodynamic tunnel, the nurse sharks invariably responded by swimming upstream. Thus, nurse sharks are capable of gradient search but utilize the rheotactic strategy whenever possible. Under controlled laboratory conditions, nurse sharks could indeed localize, to a coarse degree, a point diffusion of food extract in stagnant water, but localization was more precise in the presence of a current (Kleerekoper et al., 1975; Fig. 14.2).

Ginglymostoma cirratum response to shrimp extract in stagnant water

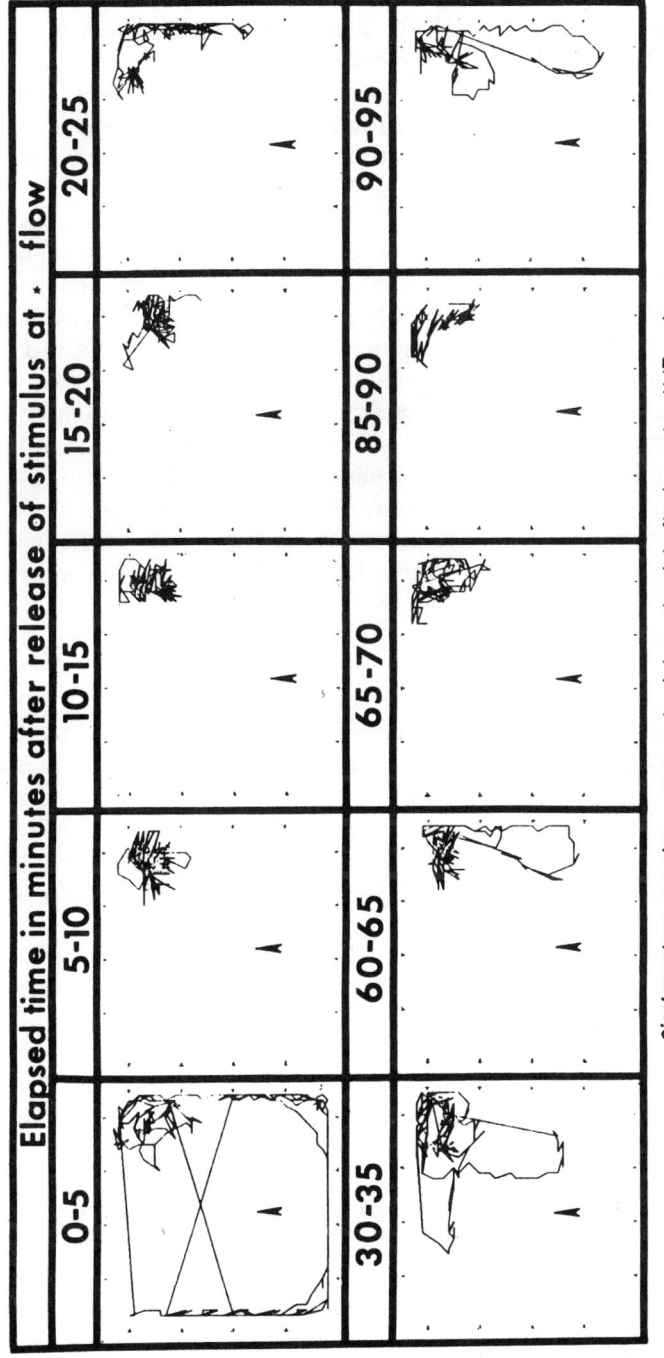

Fig. 14.2 Localization of a point source infusion of shrimp extract (denoted by a star at upper right of each cell) by a nurse shark, *Ginglymostoma cirratum*, in (A) stagnant water and (B) a 1.17 cm s^{-1} current. Note the improved localization in flowing water, the shark often hovering directly above the odour source. (From Kleerekoper et al., 1975; reproduced with permission from Springer-Verlag.)

Chemically induced rheotaxis is sufficient to bring fish within close quarters of an odour source, but the final stages of orientation are often accomplished by other means, principally vision. Nevertheless, even in visual hunting, chemoreception plays a role by serving as a sensory 'primer' to increase the attentiveness to visual cues associated with feeding. Once aroused, sight feeders will repeatedly strike at small objects on the bottom, in mid-water, and on the surface (Steven, 1959; McBride et al., 1962; Carr et al., 1976; McMahon and Tash, 1979). Chemically aroused tuna, *Euthynnus affinis*, will bite at bubbles, small leaves, and virtually any other small object within a food extract field (Tester et al., 1955). Tomtates are motivated to renew their attacks on zooplankton behind a glass barrier when odours from the zooplankton are introduced on the side of the fish. Similarly, visual attacks by darters, *Etheostoma nigrum*, on tubifex worms within glass vials could be slightly enhanced if holes were made in the vial caps (Roberts and Winn, 1962). Chemically aroused riverine salmonids strike at drifting objects which they would ordinarily ignore (Marusov, 1975). When blinded, visual feeders rely strongly on chemical cues for food localization, clearly demonstrating that smell and taste have not lost their functional role in food search within these fish (Steven, 1959; Hoese and Hoese, 1967; Gruber and Cahn, 1975).

Food odours also increase attentiveness to tactile cues, especially in benthic feeders. Once aroused, dogfish, eels, (*Anguilla*), and congers, (*Conger*), feel about for food with their snouts; soles and flounders search for food by creeping along the bottom, patting the substratum, digging, and investigating surface objects (Bateson, 1890; Sutterlin, 1975). Species possessing specialized external taste appendages (e.g. cod, hake, sturgeons, *Acipenser* sp., and catfish) all typically search for food by dragging their chemoreceptive appendages along the bottom. When these structures touch an object, the animal either stops and moves backward to investigate or snaps at the object in a pick-up reflex (Atema, 1971; Pearson et al., 1980).

For some species, chemical stimulation and benthic food search go hand in glove. The red hake, *Urophycis chuss*, will visually attack food objects in mid-water but, when chemically stimulated, drops immediately to the bottom to conduct its search there (Bardach and Case, 1965; Pearson et al., 1980). Sole steadfastly continue their benthic search even when food is suspended only a short distance above their heads (Bateson, 1890). Blinded visual feeders, such as tomtates (Steven, 1959), resort to bottom searches when chemically aroused.

Finally, chemical stimulation releases specific behavioural patterns integral to food search. Chemically aroused cod move to the bottom and perform a stereotypic behaviour termed the 'benthic search behaviour', wherein the head is held down while the fish swims backward, trailing its chemically sensitive chin barbel and fin rays along the bottom (Ellingsen and Doving, 1986). Perfusing the chemoreceptive fin rays of the sea robin with a food extract releases digging behaviour (Bardach and Case, 1965). Chemically aroused lungfish, *Protopterus annectens*, violently whip the substratum with their filamentous pectoral and pelvic fins until they strike food (Bateson, 1890). Bonnethead sharks, *Sphyrna tiburo*, perform a 'looping' behaviour after a whiff of crab extract; the distinctive locomotor pattern persists long after the chemical stimulation subsides (Johnsen and Teeter, 1985).

Consummation phase

Upon locating the chemical source, a fish begins the final phase of food search–taking the object into the mouth and assessing its acceptability. During the search phase a fish can have little or no indication of how acceptable a potential food object may be. Sudden decisive intakes, though efficient food capture mechanisms, can place the fish in jeopardy. Inedible and even harmful objects can be taken into the mouth along with food, particularly by benthic feeders rooting in the substratum. Moreover, a few prey organisms contain distasteful, nauseating, or toxic substances as feeding deterrents (Carr, 1988). Thus, there is a definite need for discrimination before the final stages of swallowing if a fish is to maximize its feeding efforts while protecting itself. That discriminatory role is deferred to the chemical and tactile receptors within the oral cavity. Objects that are flavourless (Olmsted, 1918; Tester *et al.*, 1955; Hobson, 1963), unpalatable (Bateson, 1890), or have a flavour associated with negative experiences (Little, 1977) are quickly spat out, as are those which have the 'wrong' texture. Those passing either or both tests–chemosensory and tactile–may be swallowed, the final 'decision' hinging on the balance between the two and the motivational status of the individual.

14.3 MODIFIERS OF FOOD SEARCH

Exogenous factors

Numerous environmental factors affect a fish's readiness to initiate food search as well as the intensity and content of that behaviour. Water quality parameters such as temperature, pH, and salinity have a broad impact on physiological state and hence the motivation to feed. Within physiologically comfortable limits, food search may be relatively independent of a given parameter, providing that the fish has sufficient time to acclimate following change. However, when a parameter approaches or exceeds stressful levels, feeding behaviour, as well as the proper functioning of the chemoreceptive organs, may be disturbed. For example, goldfish *Carassius auratus*, acclimated to their final thermal preferendum (28 °C) are highly attracted to an infusion of food extract; they are increasingly less responsive to the same extract presented at higher temperatures (Jones, 1984). High acidity can also disrupt food search. The fathead minnow, *Pimephales promelas*, responds strongly to a liquid tissue culture medium at pH 6.5 and higher, but not at pH 6.0 and lower (Lemly and Smith, 1987). Arctic charr, *Salvelinus alpinus* maintain their normal food search behaviour down to pH 5.5. However, food search is moderately inhibited at pH 5.0, and totally eliminated in most individuals at pH 4.75 and lower (Fig. 14.3; Jones *et al.*, 1985, 1987). The effects of other types of pollution on chemically mediated behaviours has been reviewed by Hara *et al.* (1983) and Klaprat *et al.* (Chapter 15, this volume). In many cases the disruption of food search is one of the most sensitive behavioural indicators of physiological stress (Jones *et al.*, 1987).

Changes in external factors rapid enough to stimulate exteroceptors (sensory organs which respond to external stimuli), can themselves act as sensory stimuli, and in that capacity have the potential of interacting with chemoreceptive information at the peripheral or central level. When free to choose, goldfish are more attracted to a food extract presented alone than when combined with a water flow (Rand *et al.*, 1975). However, goldfish prefer a combination of food extract and warm water to either of these stimuli alone; the relative attractiveness of the extract/warm water combination increases with the magnitude of the thermal input (Jones, 1984). The visual presence of conspecifics has a calming effect on some social species and thereby enhances food search (Parker, 1912; Olmsted, 1918; Greene, 1925; Haynes *et al.*, 1967; Hoese and Hoese, 1967). In contrast, certain unique combinations of chemical, visual, mechanical, and tactile cues drive some species into feeding frenzies. The sensory basis for these interactions is not well understood, but the olfactory and gustatory systems seem particularly susceptible to multimodal interactions because of their multiple sensitivities to chemical, mechanical, and thermal stimuli.

Endogenous factors

Food search is also subject to the influence of endogenous factors, although little is known about the physiological mechanisms involved. Hunger has long been known to play a key role in the receptiveness of fish to food (Kleerekoper, 1969). Starved fish have lower behavioural thresholds to chemical feeding cues, are less discriminatory in food selection, and feed more intensively than satiated fish (Tester, 1963a; Atema, 1980). That both of these extremes are graded in degree and change with time suggests an internal control mechanism which monitors one or more blood factors to determine the appropriate level of response. Supporting this concept is the fact that fluctuations in some plasma metabolites have been correlated with changes in food search intensity. During acid stress, Arctic charr display a nonspecific physiological stress syndrome, as well as a sharply reduced attraction to food extracts (Jones *et al.*, 1987). There is a negative correlation between an individual charr's food search intensity and its plasma levels of

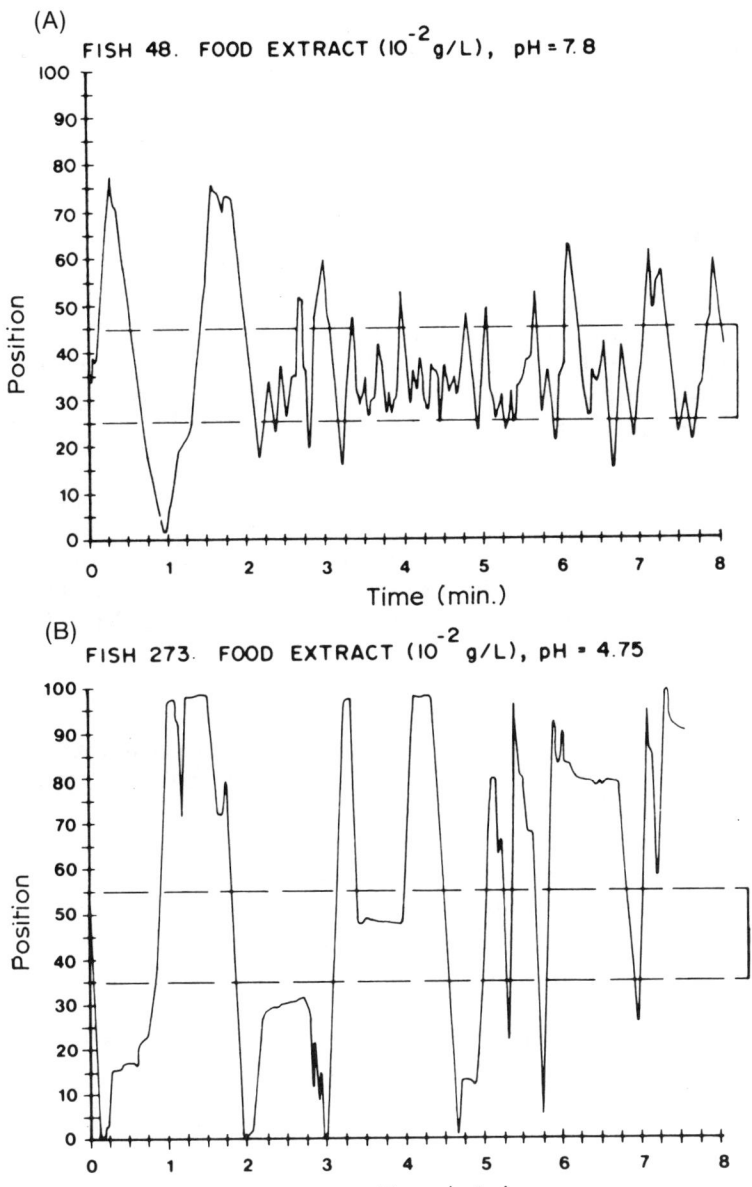

Fig. 14.3 Stress-induced loss of attraction to a food extract by Arctic charr, *Salvelinus alpinus*, tested within a 100-unit horizontal trough. (A) An unstressed charr acclimated to pH 7.8 is strongly attracted to the extract, once detected, restricting its movements to the immediate area of the chemical plume (broken rectangle). (B) In contrast, a charr acclimated to pH 4.7 – a physiologically tolerable but stressful level of acid – fails to orientate positively to the stimulus and even avoids it somewhat during the latter part of the trial. (Reproduced with permission from Jones *et al.*, 1985.)

glucose, cortisol, and protein. These correlations provide indirect evidence that hunger in fish – and hence food search intensity – may be controlled by the circulating levels of one or more of these particular blood factors. Fletcher (1984) lists other possibilities while offering the sombre reminder that work on appetite control in fish lags well behind that in the higher vertebrates.

As with temperature, hormones can have potentially broad effects on chemically mediated behaviours by modulating various aspects of olfaction and taste. In goldfish, sex hormones and thyroxine alter the olfactory response to infusions of NaCl, probably via centrifugal control (Oshima and Gorbman, 1966a,b; Hara, 1967). Godet and Dupé (1965) suggest that olfactory-induced arousal in the telencephalon of the lungfish depends upon the animal's thyroidal state.

Experience

Chemically induced food search, at least in its rudimentary form, is an instinctive behaviour requiring no experience for its release. This is best seen in the feeding responses of immature forms, perhaps most elegantly in the lamprey. The olfactory organ of this animal elaborates developmentally only after the nonparasitic sedentary larvae undergo metamorphosis to the adult parasitic stage (Kleerekoper and Mogensen, 1963). When exposed to 'trout water' for the first time, young naïve transformers immediately release their suction hold on vertical structures and begin searching the environment for potential prey.

As with food search behaviour itself, the preference for specific chemical feeding stimuli can also be innate. Naïve herring larvae, *Clupea harengus*, for example, react positively to extracts of *Balanus* but not to *Artemia*, though both are crustaceans (Dempsey, 1978). Yet the preferences for particular foods, and their odours and flavours, are not permanently fixed. Instead, experience can enhance or detract from the response to specific chemical cues. Odours and tastes associated with positive food rewards tend to receive increasingly greater attention. Juvenile sockeye salmon, *Oncorhynchus nerka*, responded only to extracts of those foods that constituted their diet in captivity (McBride *et al.*, 1962). After gradually being weaned over to a new diet, the salmon made a concomitant change in their extract preferences, now responding to extracts from the new food while ignoring those of the old. Very similar results have been obtained with herring (Dempsey, 1978) and yellowfin tuna, *Thunnus albacares* (Atema *et al.*, 1980). Presumably, such conditioning enables the animal to maximize its feeding efforts in environments where food sources change with season or location.

At its most sophisticated level, positive experience may well lead to the formation of a 'chemical search image' wherein an odour or flavour is associated with a mental impression of a specific prey or food source, similar to pattern recognition known to exist in vision (Atema, 1977). In this way a fish could associate specific chemical images with past experiences. Atema *et al.* (1969) attribute this chemical imagery to the olfactory system because of its plasticity. However, the gustatory system can also retain positive associations for future use in food search selection (Holland and Teeter, 1981).

Negative experiences are retained as well. A single ill experience with a specific food can cause a fish to become 'bait shy', whereby it avoids all further contact with that food and others perceived to be like it (Little, 1977). Some prey organisms have taken advantage of bait shyness by employing chemical feeding deterrents (often located in the epidermis) against their would-be predators (Carr, 1988). Largemouth bass, *Micropterus salmoides*, quickly learn to avoid contact with the noxious larvae of the bullfrog, *Rana catesbeiana*, and those of the toad genus *Bufo* (Kruse and Francis, 1977; Kruse and Stone, 1984). When starved, bass will feed on these larvae only at a maintenance level. In contrast, bullheads choose starvation rather than feed on bullfrog larvae.

14.4 EFFECTIVENESS OF NATURAL MATERIALS AS FOOD SEARCH INDUCERS

Although fish, especially those in captivity, can be entrained to a narrow diet (and its odour and flavour), generalist feeders typically respond to a wide range of chemical feeding stimuli. Indeed, compounds inducing food search are widely distributed throughout nature (Tester, 1963a; Haynes *et al.*, 1967; Carr *et al.*, 1976; Harada, 1982). However, fish do not respond to all food odours and flavours equally: each species has its own preference order. For example, pinfish, *Lagodon rhomboides*, show the greatest response to extracts of shrimp and lesser responses to crab, mullet, oyster, whelk, clam, and sea urchin, whereas the pigfish, *Orthopristis chrysopterus*, is most responsive to crab (Carr, 1982). Generalizations about the efficacy of various natural materials are, therefore, difficult to make. Nevertheless, a few appear to hold true for all but specialist feeders.

1. Some animal phyla are more chemically attractive than others. In comparative behavioural tests, extracts derived from arthropods and annelids frequently score high, whereas those from echinoderms and mammals, including, with some exceptions (Tester, 1963a), various materials from humans (blood, serum, urine, and sweat), tend to score low, and some instances are repulsive (Olmsted, 1918; Brett and MacKinnon, 1954; Tester, 1963a; Hodgson and Mathewson, 1978; Carr, 1982; Harada, 1982, 1983). Fish and mollusc extracts vary from very high to moderately low.
2. Carnivores favour animal extracts over plant extracts, whereas the opposite is true of planktivores (Harada, 1982, 1983; Harada *et al.*, 1982; Adams and Johnsen, 1986b).

3. Within attractive animal phyla there can be substantial variability in the effectiveness of different species (Tester, 1963a; Harada, 1982, 1983), and within the same animal there can be considerable variability in the effectiveness of different tissues (Tester et al., 1955; Haynes et al., 1967; Harada, 1982).
4. Attractiveness generally decreases with the age of the food material, unless preserved. Extracts of decayed meat or extracts that have been allowed to spoil are usually repulsive, the degree of repellence increasing with the extent of fouling (Olmsted, 1981; Tester, 1963a; Harada et al., 1986).
5. There is evidence that the odour of a prey becomes markedly more attractive when the prey is agitated but otherwise unharmed (Olmsted, 1918; Tester, 1963a).

14.5 CHEMICAL NATURE OF FOOD SEARCH INDUCERS

As aquatic organisms, fish require water-soluble olfactory and gustatory stimuli. The aquatic medium serves as a filter of sorts, limiting those compounds which could potentially serve as chemical releasers of food search. However, a wide variety of compounds, from small molecules to large polymers, are water soluble to some degree. More restrictive are the limits imposed by the fish itself: (1) at the peripheral level, where the molecule must fit one or more stereo-specific receptor sites to generate a polarity change within the receptor cell; and (2) at more central levels, where the stimulus perception must be associated, instinctively or experientially, with feeding. Consequently, when confronted with complex chemical mixtures, such as those derived from prey extracts, a fish actually responds to only a narrow portion of the chemical spectrum available. The active portion of the spectrum is not the same for all species.

It seems that the first investigators successfully to identify individual compounds which evoke food search were Kleerekoper and Mogensen (1963). By analysing large volumes of trout water, they showed that lamprey respond to a single amine, originally code-named amine F but later identified as isoleucine methyl ester. Since then several investigators have identified, in whole or in part, the active compounds within various living tissues. In almost all cases, amino acids have been found in the active fractions (Table 14.1). Tests employing authentic (commercial, off-the-shelf chemicals) amino acids have confirmed that many species rely upon these compounds in one or more aspects of food search. However, the particular amino acids involved are not the same in all cases; different species utilize different amino acids, and those with similar amino acid spectra display different preference orders. Killifish, *Fundulus heteroclitus*, and rainbow trout, *Oncorhynchus mykiss*, for example, show a broad attraction to the amino acids (Sutterlin, 1975; Jones, 1989). Silversides, *Menidia menidia*, on the other hand, are more narrowly 'tuned', being attracted only to beta-alanine, threonine, arginine, and methionine

Table 14.1 Chemical stimulants for evoking food search behaviour in different species of fish, listed in taxonomic order

Order/Species*	Source	Observed behaviour	Authentic or isolated chemical stimulant†‡
Petromyzontiformes			
Petromyzon marinus (lamprey)	Kleerekoper and Mogensen (1963)	Increased activity, attraction	Isoleucine methyl ester
Lamniformes			
Ginglymostoma cirratum (nurse shark)	Hodgson and Mathewson (1978)	Attraction, feeding movements	Gly, Glu, Bet, Cys, Ser, Met, TMA, TMAO
Negaprion brevirostris (lemon shark)			
Anguilliformes			
Anguilla japonica (Japanese eel)	Hashimoto et al. (1968)	Attraction	Arg, Ala, Gly
Clupeiformes			
Clupea harengus (herring)	Dempsey (1978)	Increased activity	Glu, Asp, Gly, Met, Ala, Pro
Cypriniformes			
Misgurnus anguillicaudatus (oriental weatherfish)	Harada (1985b, 1986), Harada et al. (1987)	Attraction	His, Arg, Gly, Lys, Ala, Phe, Tyr, Ile, Thr, Cys, Asn, Val, β-Ala, Hcy, Hpr, Orn, GABA, TMA, TMAO, ammonia, pyrrolidine, choline, histamine, monomethylamine, dimethylamine, monoethanolamine, adenine, guanosine, GMP, guanidinoacetic acid
Brachydanio rerio (zebrafish)	Scarfe et al. (1985)	Attraction	Ala

Table 14.1 —Contd.

Order/Species*	Observed behaviour	Source	Authentic or isolated chemical stimulant†‡
Siluriformes			
Ictalurus nebulosus and *I. natalis* (brown and yellow bullheads)	Attraction	Bardach et al. (1967)	Cys-HCl
Salmoniformes			
Oncorhynchus mykiss (rainbow trout)	Prolonged retention of neutral carrier	Jones (1989, 1990)	Leu, Ile, Pro, Met, Glu, Try, Arg, Phe, Tau, norvaline, norleucine, ε-aminocaproic acid, urea, acetamide, valeramide, caproic amide, hexanol, octanol, D-ribose, D-glucose, D-fructose, sucrose
Salmo salar (Atlantic salmon)	Increased activity, snapping, darting, yawning	Mearns (1985, 1986, 1989)	Ala, Arg, Glu, His, Leu, Ile, Lys, Met, Pro, Gly, Val
Salmo trutta (brown trout)	Increased activity, snapping, darting, yawning	Mearns (1986, 1989)	Ala, Arg, Pro, Gly, His
Salvelinus alpinus (Arctic charr)	Attraction, biting	Jones and Hara (1985)	Ala, Ser
Gadiformes			
Gadus morhua (cod)	Attraction / Bottom food search	Pawson (1977) / Ellingsen and Døving (1986)	Gly, Ala, Ser, Glu, Thr, Val / Gly, Ala, Arg, Pro, dimethylthetin
Cyprinodontiformes			
Fundulus heteroclitus (killifish)	Attraction	Gruber and Cahn (1975), Sutterlin (1975)	Cys, Ser, Ala, His, Gly, β-Ala, Thr, Leu, Ile, Val, Glu, Tau, Arg, Met, Asn, Lys, Trp, Asp, GABA, Hpr

Species	Reference	Response	Inducers
Atheriniformes *Menidia menidia* (silverside)	Sutterlin (1975)	Attraction	Ala, Thr, Met, β-Ala
Scorpaeniformes *Prionotus carolinus* and *P. evolans* (sea robins)	Bardach and Case (1965)	Feeding movements	Phe, Pro, Asp, Try, Glu, Gly, TMA, skatol, indole, histamine, sucrose, acetic acid
Sebasticus marmoratus (marbled rockfish)	Takaoka et al. (1990)	Enhanced consumption of bland diet	Ala, Met, Ser, Pro, inosine
Perciformes *Gobiosoma bosci* (goby)	Hoese and Hoese (1967)	Attraction and biting	Gly, Ala, Glu, β-Ala, Asp, Cys, Met, TMA, 2-aminobutyric acid, 4-aminobutyric acid, putrescine, 5-aminovaleric acid
Seriola quinqueradiata (yellowtail)	Harada (1985a, 1986), Harada et al. (1987)	Attraction	His, Arg, Lys, Gly, Val, Thr, Met, Gln, Asn, Orn, cytosine, dGMP, UMP, AMP, dAMP, stearic acid, palmitic acid, monostearin, phosphatidyl choline, ceramide, cardiolipin, sphingomyeline, phosphatidyl ethanolamine, phosphatidic acid
Tilapia zillii (tilapia)	Adams and Johnsen (1986a), Johnsen and Adams (1986), Adams et al. (1988)	Enhanced consumption of bland diet	Glu, Asp, Lys, Ala, Ser, citric acid, acetic acid
Stizostedion vitreum (walleye)	Rottiers and Lemm (1985)	Attraction	NaCl, Arg, Bet, sucrose, glutathione, vitamin B_{12}

Table 14.1—Contd.

Order/Species*	Source	Observed behaviour	Authentic or isolated chemical stimulant†‡
Chrysophrys major (red sea bream)	Goh and Tamura (1980)	Enhanced consumption of bland diet	Ala, Val, Gly, Ser, Pro, Arg, Gln
	Ina and Matsui (1980), Murofushi and Ina (1981), Murofushi et al. (1982)	Increased biting	Ala, Val, Gly, Glu, norvaline, L-α-aminobutyric acid
Orthopristis chrysopterus (pigfish)	Carr (1976), Carr et al. (1977)	Increased striking	Bet
Pleuronectiformes			
Scophthalmus maximus (turbot)	Mackie and Adron (1978)	Enhanced consumption of bland diet	Inosine, 1-methyl inosine, 1-benzylinosine, IMP, IDP, guanosine, 1-methyl guanosine, GMP, ATP, XMP
Solea solea (sole)	Mackie and Mitchell (1982)	Enhanced consumption of bland diet	Bet, dimethylthetin
Pseudopleuronectes americanus (flounder)	Sutterlin (1975)	Attraction, biting, digging	Gly, Ala, Met, Asn, Cys, Glu, Leu, Asp, Ser, Bet, GABA
Tetraodontiformes			
Fugu pardalis (puffer)	Hidaka et al. (1978), Ohsugi et al. (1978)	Enhanced consumption of bland diet	Ala, Gly, Pro, Ser, Bet

*From Nelson (1984).
†Abbreviations: Ala, L-alanine; Arg, L-arginine; Asn, L-asparagine; Asp, L-aspartic acid; βAla, beta alanine; Bet, betaine; Cys, L-cysteine; Cys-HCl, cysteine hydrochloride; GABA, gamma amino butyric acid; Gln, L-glutamine; Glu, L-glutamic acid; Gly, glycine; Hcy, homocysteine; His, L-histidine; Hpr, hydroxyproline; Ile, L-isoleucine; Leu, L-leucine; Lys, L-lysine; Met, L-methionine; Orn, L-ornithine; Phe, L-phenylalanine; Pro, L-proline; Ser, L-serine; Tau, taurine; Thr, L-threonine; TMA, trimethylamine; TMAO, trimethylamine oxide; Trp, L-tryptophan; Try, L-tyrosine; Val, L-valine; AMP, adenosine 5'phosphate; ATP, adenosine 5'triphosphate; GMP, guanosine 5'phosphate; IDP, inosine 5'diphosphate; IMP, inosine 5'phosphate; UMP, uridine 5'phosphate; XMP, xanthosine 5'phosphate.

(Sutterlin, 1975). Ina and Matsui (1980) showed that the sea bream, *Chrysophrys major*, responds only to three amino acids-valine, alanine, and glycine. Small departures from the structures of any of these three compounds are sufficient to negate their effectiveness (Murofushi et al., 1982). At the extreme of this trend, neither tomtates (Haynes et al., 1967), nor lampreys (Kleerekoper and Morgensen, 1963), nor hagfish, *Myxine glutinosa* (Sutterlin, 1975) are activated by any single amino acid.

The list of active compounds is by no means limited to amino acids (Table 14.1). Arousal, attraction, and feeding have been elicited by such diverse substances as amines, amides, alcohols, nucleotides, quaternary ammonium compounds, carbohydrates, ammonia, long chain fatty acids, neutral lipids, phospholipids, glycolipids, vitamin B_{12}, and such xenobiotic compounds as the dipeptide arcamine (hypotauryl-2-carboxyglycine) and strombine (C-methyl-imino diacetic acid). None of these compounds seems to have the universal activity of the amino acids. Compounds that have failed to induce food search include short aliphatic organic acids (Steven, 1959; Bardach and Case, 1965; Sutterlin, 1975; Harada, 1985a; Rottiers and Lemm, 1985), aminophosphonic acids (Haynes et al., 1967), sugar alcohols (Hodgson and Mathewson, 1978), the bitter substance quinine (Bardach and Case, 1965; Jones and Hara, 1985), anethol (the active ingredient of oil and anise), and furfural mercaptan (Tester et al., 1955).

Attempts to isolate the active components of food extracts have often been frustrated by a loss of potency as the active fraction was increasingly simplified, especially during the later stages of fractionation (Tester et al., 1955; Steven, 1959; Hashimoto et al., 1968; Konosu et al., 1968; Pawson, 1977; Ina and Matsui, 1980; Murofushi and Ina, 1981; Olsen et al., 1986). This, together with the fact that mixtures are almost always more potent than any individual component, indicates that the active compounds somehow interact additively or synergistically, whether at the stimulus receptor sites or more centrally. In a clear example, Carr and Chaney (1976) demonstrated that pinfish respond maximally to a synthetic mixture of 20 amino acids plus betaine. However, when either the betaine or the amino acids were absent from the mixture, the effectiveness of the stimulant decreased dramatically. Interactions of this type can occur between compounds of the same class (Harada, 1986; Harada et al., 1987), but the strongest interactions appear to result from the combination of different classes of compounds (Harada, 1985a; Harada et al., 1987).

14.6 CHEMICAL LURES

The use of chemical feeding lures, based almost exclusively on natural materials, has enjoyed a long history in both the commercial and sports fisheries. Although some applications are more common to one fishery than the other, in general the two industries have differed more in their magnitudes of application than in technique.

The requirements of a successful chemical lure in fishing are largely dependent on the application. Some place only minimal demands on the fish; others require the fish to perform very specific tasks. Regardless, chemical lures must be properly applied in a manner that agrees with the target species' own peculiar style of food search behaviour. Chemically stimulating obligate bottom feeders, for example, while angling in mid-water would obviously yield unsatisfactory results. Likewise, expecting a species incapable of following odour gradients to locate an odorous bait, or attempting chemically to arouse a sight-feeder to attack an artificial lure that failed to meet that species' visual criteria for feeding, would also prove fruitless.

In principle, chemical feeding lures for fish carry the same advantages inherent to chemical signals in general. These are the ability to treat a substantial volume of the medium with a minimum investment of energy, a long-lasting stimulus that persists after the source has ceased to generate the signal, an effective signal-to-noise ratio, and the potential for species specificity. The primary disadvantage lies in the inefficiency of the chemical lure in conveying direction. Hence, the greatest potential for error lies in the active search phase.

Three types of chemical lures are recognized: (1) chumming (and the related practice of seeding), (2) odorous baits, and (3) dip baits, or attractants. In practice, the distinction between these three is not always clear.

Chumming and seeding

Chumming – the release of chemical attractants into the water to draw fish to a given locale or to arouse those present to a more aggressive feeding state – was mentioned in scientific literature as early as the turn of the century (Parker, 1912). It is known to be an effective strategy for a number of species and is frequently used on marine forms. Species with a well-developed sense of smell, such as sharks, can be drawn to the chemical source from as far away as several hundred metres.

The quantity of chum used is largely dependent on the application and on the style of food search behaviour exhibited by the target species. Simply raking over barnacle beds is sufficient briefly to attract numerous species of marine fish (McNally, 1986). Similarly, freshwater anglers have been known to chum streams by dropping in a few perforated cans of dog food. In its more classic form, marine charterboat fishermen attract large numbers of fish by means of chemical trails, or 'chum lines', which may extend considerable distances behind the vessel. The trails are generated either by ladling out substantial volumes of fish blood and entrails into a current, or by suspending blocks of frozen fish offal over the stern while slowly trolling.

Since the intent of chumming is merely to cause fish to aggregate within a general area or to evoke their food search, the behavioural demands placed upon them are not particularly stringent. Provided that the target species is capable of detecting the signal and, if necessary, of rudimentary localization, the probability of success is high. Moreover, it can be improved further by incorporating visual stimuli, e.g. small bits of flesh, into the sensory picture. Charterboat sports fishermen along the Florida coast employ the strategy of intermittently releasing clouds of boiled pasta within the chum line and then fishing artificial lures within the clouds (Sosin, 1987). The abundance of small bits of food reinforces the presentation by providing visual targets, which enhance the acceptability of the artificial lures by positive association.

Akin to chumming is seeding, the practice of dispersing food within a confined area to draw in those species which locate their food by chemoreception. Catfishermen, for instance, increase the concentration of catfish within an area of interest by seeding corn along the bottom (Tinsley, 1986). After the catfish have had sufficient time to congregate, the anglers fish within the seeded area, using a more preferred food material as bait.

Odorous baits

As probably the most common chemical lure, odorous baits are used commercially on long- and trot-lines, as well as in fish traps. They are also popular among sports fishermen. Natural prey organisms, either cut or whole, are typically the bait of choice. Sports fishermen in North America alone are estimated to purchase over 800 million dollars worth of live and cut bait annually (U.S. Fish and Wildlife Service, 1988). Commercial consumptions are likely to be much higher.

There are exceptions to the use of live or cut bait, however. In one novel approach, commercial fishermen along the Mississippi River 'bait' their traps with live, sexually ripe adult channel catfish, *Ictalurus punctatus*. The ripe catfish release potent, species-specific sex pheromones which consequently draw in other adults (Timms and Kleerekoper, 1972). Evidently the chemical lures are sex specific: female catfish draw in predominantly males, whereas males draw in largely females (Bailey and Harrison, 1945; Appelget and Smith, 1950). Still stranger baits exist. In certain parts of Louisiana, catfishermen bait their trot-lines with bars of soap wrapped in nylon hosiery for structural support.

Success with natural baits depends partly on the choice of bait. Chemical lures derived from natural prey organisms tend to have broad appeal among different species of predatory fish. Thus, using a natural bait, it is difficult to target a single species to the exclusion of all nuisance species which compete for the offering. Nevertheless, because each species exhibits its own preference order to different natural chemical lures, it is possible to improve a target species' catch rate by offering its most preferred foods as bait.

Anecdotal stories of bait specificity abound in both the commercial and sports fisheries. As described earlier, bait specificity may be due as much to conditioning from positive feeding experiences as to innate chemical preferences. The suggestion by Atema (1977) that fish form chemical search images associated with specific prey could explain why at only certain times of the year given species can be readily taken on one bait but not on another. However, not all instances of bait specificity can be explained in this manner. Fish traps baited with horseshoe crabs (*Limulus*) regularly catch several times as many eels, as do more conventional baits, despite the fact that horseshoe crab is not known to be a natural food source for eels (P. Sorensen, pers. comm., 1988).

The effectiveness of odourous baits as chemical lures is intrinsically linked to the physical properties of the foundational structure from which the stimulants emanate. In many cases the tissue of a prey organism lacks the structural integrity needed for adequate hook retention, controlled release of the stimulant, or some other physical property of a good bait (Sutterlin *et al.*, 1982). Some efforts have gone into producing artificial baits which combine natural foodstuffs as the chemical lure with an artificial material as the carrier. The earliest was that of Bateson (1890), who tried combining extracts of pilchard and squid with gelatine. More recent efforts have focused on artificial carriers made from polyvinyl alcohol (Koyama *et al.*, 1971) and water-insoluble but leachable matrices (Carr, 1984). Pre-formed visual lures of polyvinyl chloride, 'scented' with various substances to augment attraction and bait retention, are becoming increasingly numerous in sports fishing (Fig. 14.4(A)).

Attractants

The practice of chemically enhancing baits and lures by applying attractive substances was encouraged as early as 1653 by the noted naturalist, Izaak Walton, in his book *The Compleat Angler* (cited by Johnson, 1984). Today, from the common practice of spitting on a bait to pre-soaking artificial lures in liquid 'dip baits', chemical enhancement is practised world-wide in many different cultures.

The variety of attractants, particularly homemade brews, is seemingly endless, although the active ingredients are almost always derived entirely from natural sources. North American anglers apply concoctions which range from thick molasses-like mixtures of brown sugar and salt even to some based upon human excretion (Johnson, 1984). In Genio Scott's 1869 book, *Fishing in American Waters* (also cited by Johnson, 1984), reference is made to an apothecary to King Louis XIV who formulated fish attractants composed of cat fat, heron grease, aniseed, camphor, turpentine, civet, and other ingredients.

Regional variations in formulations are common, and some attractants are unique to specific cultures. In the Mayan Indian tribes of Central America, the local witchdoctor, or shaman, is expected to provide fish when meat is scarce. Before fishing, the shaman applies a traditional mixture of local roots and herbs to whatever bait is available, or sometimes simply to strips of cloth (Gibbs, 1984).

Commercial attractants have appeared sporadically, especially in the last two decades with the increasing recognition by anglers of the chemoreceptive powers of fish. Typically consisting of the oily fraction of abundant natural products (e.g. menhaden, *Brevoortia* sp., herring, shad, *Alosa* sp., or shrimp), at times reinforced with the oil of anise or fruity flavourings, the commercial versions have historically performed poorly in laboratory behavioural tests, as compared with fresh extracts (Tester *et al.*, 1955; McBride *et al.*, 1962). More recent versions have generally fared no better (Jones, unpublished).

The most common use of attractants by anglers today is as an external treatment for artificial lures which are repeatedly cast and retrieved. Three reasons are commonly cited for these treatments: (1) to chemically arouse and attract nearby fish, (2) to increase lure retention time once taken into the mouth, and (3) to mask potential feeding deterrents present on the lure. Ideally, the attractant is intended to bleed off the lure steadily during the retrieval, to form a continuous chemical trail without exhausting the limited supply of active ingredients midway. The difficulty of accomplishing this feat under the almost infinite variety of fishing conditions and styles has led to various strategies for improving chemical delivery, such as lures with internal attractant reservoirs and the practice of attaching various absorbent materials (e.g. sponge, yarn, or chamois) to a lure's exterior (Fig. 14.4 (B) and (C)). Prochnow (1989) developed a novel method for simulating the natural release of chemical stimulants by a swimming baitfish by mixing a dry attractant with a polymer which converts to slime upon contact with water. The attractant bleeds off the lure at a rate which approximates the dissolution of mucus from a fish as it swims (Fig. 14.4(D)).

Future developments

In view of the acute sensitivity of many species of fish to water-borne chemicals, the potential for the further development of chemical feeding lures appears considerable. Much of this potential lies in the direction of wholly synthetic lures. Synthetic chemical lures could be more easily preserved, could be more predictably mass produced, would have a more uniform appearance, and would likely be more compatible with a wider variety of synthetic vehicles than would natural chemical lures. One of their greatest potentials, though, lies in the area of specificity. Once the fundamental smell and taste requirements for evoking food search in a particular species are identified, chemical lures could be designed better to fit the preferences of that species. Except in the case of two or more co-occurring species with broadly overlapping chemical spectra, using more species-specific lures would

Fig. 14.4 Vehicles for delivering chemical lures in sports fishing. (a) Scented plastic lures in the shapes of common food items. The 'scent' could include oil of anise, blood chloride, salt, animal protein, the essence of prey organisms or their amino acid equivalents. (b) Representative of lures with internal cavities, tube jigs can be stuffed with solid or gel attractants. The stuffing milks out of the rear of the tube as it is

Chemical lures 313

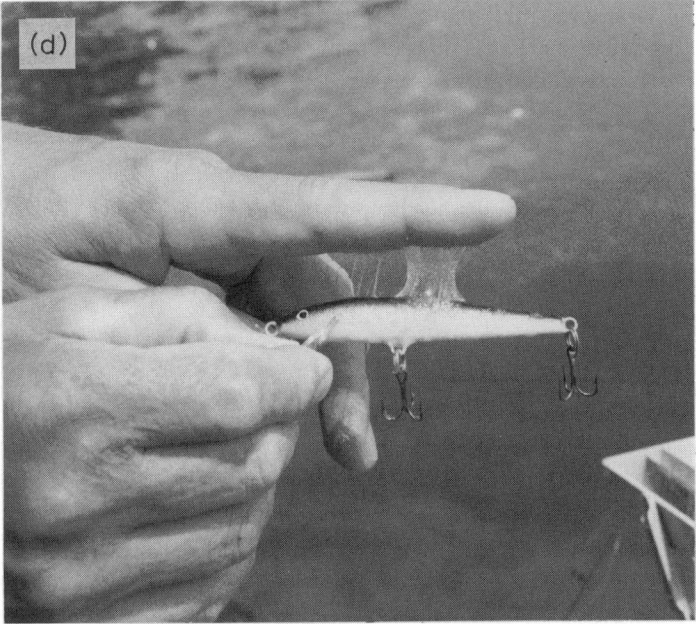

retrieved through the water. (c) A large jig fitted with a strip of attractant-soaked chamois. (d) A hard-bodied lure treated with a water-soluble, mucous-like polymer impregnated with a dry attractant. As the lure is cast and retrieved the polymer, and with it the attractant, disperses at a rate intended to simulate the release of odours by a swimming baitfish.

minimize competition between target and nuisance species, thereby increasing the catch per effort.

A second area of promise for wholly synthetic chemical lures lies in their potential potency. By characterizing the unique chemical requirements of a species, the potency of a suitable chemical lure could be increased in three ways. First, some substances in natural materials act as feeding deterrents (Ellingsen and Doving, 1986; Harada and Fukuda, 1987; Harada et al., 1988). The diluting or negative effects of these substances could be avoided by including in the chemical lure only those elements that have positive effects on feeding. Second, these elements could be incorporated at concentrations higher than their natural levels. Amino acids and nucleotides, for example, are present in living tissue at limited concentrations (Carr, 1988). It follows that synthetic baits containing the appropriate amino acids or nucleotides at levels above these natural ranges would have a stronger appeal than lures made wholly from natural products. They would constitute, in essence, supranatural lures. Third, if the structure–activity relationships of the active compounds were sufficiently defined, it would be theoretically possible to design chemical stimulants that would accentuate those molecular features necessary for activity. Again, the end result could be the development of chemical lures having potencies beyond those attainable using natural materials. Increases in potency by several fold would not be unrealistic. Similar advances have already been achieved in the pharmaceutical industry in the synthesis of 'designer drugs' (Baum, 1985).

ACKNOWLEDGEMENTS

I thank Peter Johnsen, Craig Steele, and the copy editor, Chuck Hollingworth, for their thorough and helpful reviews of the manuscript.

REFERENCES

Adams, M.A. and Johnsen, P.B. (1986a) A solid matrix for determining chemical feeding stimulants. *Progve Fish Cult.*, **48**, 147–9.

Adams, M.A. and Johnsen, P.B. (1986b) Chemical control of feeding in herbivorous and carnivorous fish, in *Chemical Signals in Vertebrates 4.* (eds D. Duvall, D. Muller-Schwarze and R.M. Silverstein), Plenum, New York, pp. 45–61.

Adams, M.A., Johnsen, P.B. and Hong-qi, Z. (1988) Chemical enhancement of feeding for the herbivorous fish *Tilapia zillii. Aquaculture*, **72**, 95–107.

Appelget, J. and Smith, L.L., jun. (1950) The determination of age and rate of growth from vertebrae of the channel catfish, *Ictalurus lacustris punctatus. Trans. Am. Fish. Soc.*, **80**, 119–39.

Atema, J. (1971) Structures and functions of the sense of taste in catfish (*Ictalurus natalis*). *Brain Behav. Evol.*, **4**, 273–94.

Atema, J. (1977) Functional separation of smell and taste in fish and crustacea, in *Olfaction and Taste VI* (eds J. LeMagnen and P. MacLeod), Information Retrieval, London, pp. 165–74.

Atema, J. (1980) Chemical senses, chemical signals, and feeding behaviour in fishes, in *Fish Behavior and Its use in the Capture and Culture of Fishes (ICLARM Conference Proceedings*, Vol. 5) (eds J.E. Bardach, J.J. Magnuson, R.C. May and J.M. Reinhart), International Center for Living Aquatic Resources Management, Manila, Philippines, pp. 57–94.
Atema, J., Todd, J.H. and Bardach, J.E. (1969) Olfaction and behavioral sophistication in fish, in *Olfaction and Taste III* (ed. C. Pfaffman), Rockefeller Univ. Press, N.Y., pp. 241–51.
Atema, J., Holland, K. and Ikehara, W. (1980) Olfactory responses to yellowfin tuna (*Thunnus albacares*) to prey odors: chemical search image. *J. Chem. Ecol.*, 6, 457–65.
Bailey, R.M. and Harrison, H.M., jun. (1945) Food habits of the southern channel catfish (*Ictalurus lacustris punctatus*) in the Des Moines River, Iowa. *Trans. Am. Fish. Soc.*, 75, 110–38.
Bardach, J.E. (1967) The chemical senses and food intake in the lower vertebrates, in *The Chemical Senses and Nutrition* (eds M.R. Kare and O. Maller), The Johns Hopkins Press, Baltimore, MD, pp. 19–43.
Bardach, J.E. and Case, J. (1965) Sensory capabilities of the modified fins of squirrel hake (*Urophycis chuss*) and searobins (*Prionotus carolinus* and *P. evolans*). *Copeia*, 2, 194–206.
Bardach, J.E., Todd, J.H. and Crickmer, R. (1967) Orientation by taste in fish of the genus *Ictalurus*. *Science*, Wash., D.C., 155, 1276–8.
Bateson, W. (1890) The sense-organs and perceptions of fishes; with remarks on the supply of bait. *J. mar. biol. Ass. U.K.*, 1, 225–56.
Baum, R.M. (1985) New variety of street drugs poses growing problem. *Chem. Eng. News*, 63, 7–16.
Bell, W.J. and Tobin, T.R. (1982) Chemo-orientation. *Biol. Rev.*, 57, 219–60.
Brett, J.R. and MacKinnon, D. (1954) Some aspects of olfactory perception in migrating adult coho and spring salmon. *J. Fish. Res. Bd Can.*, 11, 310–18.
Carr, W.E.S. (1976) Chemoreception and feeding behavior in the pigfish, *Orthopristis chrysopterus*: characterization and identification of stimulatory substances in a shrimp extract. *Comp. Biochem. Physiol.*, 55A, 153–7.
Carr, W.E.S. (1982) Chemical stimulation of feeding behavior, in *Chemoreception in Fishes* (ed. T.J. Hara), Elsevier, Amsterdam, pp. 259–73.
Carr, W.E.S. (1984) Artificial bait for aquatic species. U.S. Patent no. 4,463,018.
Carr, W.E.S. (1988) The molecular nature of chemical stimuli in the aquatic environment, in *Sensory Biology of Aquatic Animals* (eds J. Atema, R.R. Fay, A.N. Popper and W.N. Tavolga), Springer-Verlag, N.Y., pp. 3–27.
Carr, W.E.S. and Chaney, T.B. (1976) Chemical stimulation of feeding behavior in the pinfish, *Lagodon rhomboides*: characterization and identification of stimulatory substances extracted from shrimp. *Comp. Biochem. Physiol.*, 54A, 437–41.
Carr, W.E.S., Gondeck, A.R. and Delanoy, R. (1976) Chemical stimulation of feeding behavior in the pinfish, *Lagodon rhomboides*: a new approach to an old problem. *Comp. Biochem. Physiol.*, 54A, 161–6.
Carr, W.E.S., Blumenthal, K.M. and Netherton, J.C., III (1977) Chemoreception in the pigfish, *Orthopristis chrysopterus*: the contribution of amino acids and betaine to stimulation of feeding behavior by various extracts. *Comp. Biochem. Physiol.*, 58A, 69–73.
Dempsey, C.H. (1978) Chemical stimuli as a factor in feeding and intraspecific behaviour of herring larvae. *J. mar. biol. Ass. U.K.*, 58, 739–47.
Ellingsen, O.F. and Doving, K.B. (1986) Chemical fractionation of shrimp extracts inducing bottom food search behavior in cod (*Gadus morhua* L.). *J. Chem. Ecol.*, 12, 155–68.

↯ Fletcher, D.J. (1984) The physiological control of appetite in fish. *Comp. Biochem. Physiol.*, **78A**, 617–28.
Gibbs, J. (1984) Dr. Juice and his amazing elixir. *Outdoor Life*, **173**, 53–5.
Godet, R. and Dupé, M. (1965) Quelques aspects des relations neuro-endocriniennes chez *Protopterus annectens* (Poisson Dipneuste). *Arch. Anat. microsc. Morph. exp.*, **54**, 319–30.
Goh, Y. and Tamura, T. (1980) Effect of amino acids on the feeding behaviour in red sea bream. *Comp. Biochem. Physiol.*, **66C**, 225–9.
Greene, C.W. (1925) Notes on the olfactory and other physiological reactions of the California hagfish. *Science, Wash., D.C.*, **61**, 68–70.
Gruber, D. and Cahn, P.H. (1975) Localization of chemical sources by blinded mummichogs, *Fundulus heteroclitus*. *Am. Midl. Nat.*, **94**, 500–3.
Hara, T.J. (1967) Electrophysiological studies of the olfactory system of the goldfish, *Carassius auratus* L.–III. Effects of sex hormones on olfactory activity. *Comp. Biochem. Physiol.*, **22**, 209–25.
Hara, T.J., Brown, S.B. and Evans, R.E. (1983) Pollutants and chemoreception in aquatic organisms, in *Aquatic Toxicology* (ed. J.O. Nriagu), Wiley, N.Y., pp. 247–306.
Harada, K. (1982) The attractive effect of food based on the behavioral responses of juvenile yellowtail *Seriola quinqueradiata*. *Bull. Jap. Soc. scient. Fish.*, **48**, 1047–54.
Harada, K. (1983) Statistical approach to finding probable feeding attractants for oriental weatherfish. *Bull. Jap. Soc. scient. Fish.*, **49**, 521–6.
Harada, K. (1985a) Feeding attraction activities of amino acids and lipids for juvenile yellowtail. *Bull. Jap. Soc. scient. Fish.*, **51**, 453–9.
Harada, K. (1985b) Feeding attraction activities of amino acids and nitrogenous bases for oriental weatherfish. *Bull. Jap. Soc. scient. Fish.*, **51**, 461–6.
Harada, K. (1986) Feeding attraction activities of nucleic acid-related compounds for abalone, oriental weatherfish and yellowtail. *Bull. Jap. Soc. scient., Fish.*, **52**, 1961–8.
Harada, K. and Fukuda, K. (1987) Chemorepellents derived from food for oriental weatherfish. *Nippon Suisan Gakkaishi*, **53**, 169.
Harada, K., Maeda, H. and Kobayashi, K. (1982) Application of logistic curve as an index of the attractive effects of food for oriental weatherfish. *Bull. Jap. Soc. scient. Fish.*, **48**, 21–9. (In Japanese)
Harada, K., Fukuda, K. and Sugihara, M. (1986) Statistical evaluation of chemotaxes, with special reference to repellence activity of food for oriental weatherfish. *Bull. Jap. Soc. scient. Fish.*, **52**, 1215–24.
Harada, K., Eguchi, A. and Kurosaki, Y. (1987) Feeding attraction activities in the combinations of amino acids and other compounds for abalone, oriental weatherfish and yellowtail. *Nippon Suisan Gakkaishi*, **53**, 1483–9.
Harada, K., Abe, Y. and Sugiyama, T. (1988) Activities of carboxylic acids and acidic amino acids as repellents in the oriental weatherfish *Misgurnus anguillicaudatus*. *Nippon Suisan Gakkaishi*, **54**, 2135–8.
Hashimoto, Y., Konosu, S., Fusetani, N. and Nose, T. (1968) Attractants for eels in the extracts of short-necked clam–I. Survey of constituents eliciting feeding behavior by the omission test. *Bull. Jap. Soc. scient. Fish.*, **34**, 78–83. (In Japanese)
↯ Haynes, L.J., Sangster, A.W., Steven, D.M. and Thomas, S. (1967) Chemical factors inducing exploratory feeding behaviour in fish–E.F.B.-inducing properties of marine invertebrates. *Comp. Biochem. Physiol.*, **20**, 755–65.
Hidaka, I., Ohsugi, T. and Kubomatsu, T. (1978) Taste receptor stimulation and feeding behaviour in the puffer, *Fugu pardalis* I. Effect of single chemicals. *Chem. Senses Flavour*, **3**, 341–54.

Hobson, E.S. (1963) Feeding behavior in three species of sharks. *Pacif. Sci.*, **17**, 145-70.
Hodgson, E.S. and Mathewson, R.F. (1978) Electrophysiological studies of chemoreception in elasmobranchs, in *Sensory Biology of Sharks, Skates, and Rays* (eds E.S. Hodgson and R.F. Mathewson), Off. Naval. Res., Arlington, VA, pp. 227-67.
Hoese, H.D. and Hoese, D. (1967) Studies on the biology of the feeding reaction in *Gobiosoma bosci. Tulane Stud. Zool.*, **14**, 55-62.
Holland, K. (1978) Chemosensory orientation to food by a Hawaiian goatfish (*Parupeneus porphyreus*, Mullidae). *J. Chem. Ecol.*, **4**, 173-86.
Holland, K.N. and Teeter, J.H. (1981) Behavioral and cardiac reflex assays of the chemosensory acuity of channel catfish to amino acids. *Physiol. Behav.*, **27**, 699-707.
Ina, K. and Matsui, H. (1980) Survey of feeding stimulants for the sea bream (*Chrysophrys major*) in the marine worm (*Perinereis vancaurica tetradentata*). *Nippon Nogeikagaku Kaishi*, **54**, 7-12. (In Japanese)
Johnsen, P.B. and Adams, M.A. (1986) Chemical feeding stimulants for the herbivorous fish, *Tilapia zillii. Comp. Biochem. Physiol.*, **83A**, 109-12.
Johnsen, P.B. and Teeter, J.H. (1980) Spatial gradient detection of chemical cues by catfish. *J. comp. Physiol.*, **140**, 95-9.
Johnsen, P.B. and Teeter, J.H. (1985) Behavioral responses of bonnethead sharks (*Sphyrna tiburo*) to controlled olfactory stimulation. *Mar. Behav. Physiol.*, **11**, 283-91.
Johnson, P.C. (1984) *The Scientific Angler*. Scribner's, N.Y., 289 pp.
Jones, K.A. (1984) Temperature dependent attraction by goldfish to a chemical feeding cue presented alone and in combination with heated water. *Physiol. Behav.*, **33**, 509-15.
Jones, K.A. (1989) The palatability of amino acids and related compounds to rainbow trout, *Salmo gairdneri* Richardson. *J. Fish Biol.*, **34**, 149-60.
Jones, K.A. (1990) Chemical requirements of feeding in rainbow trout, *Oncorhynchus mykiss* (Walbum); palatability studies on amino acids, amides, amines, alcohols, aldehydes, saccharides, and other compounds. *J. Fish Biol.*, **37**, 413-23.
Jones, K.A. and Hara, T.J. (1985) Behavioural responses of fish to chemical cues: results from a new bioassay. *J. Fish Biol.*, **27**, 495-504.
Jones, K.A., Hara, T.J. and Scherer, E. (1985) Behavioral modifications in Arctic char (*Salvelinus alpinus*) chronically exposed to sublethal pH. *Physiol. Zool.*, **58**, 400-412.
Jones, K.A., Brown, S.B. and Hara, T.J. (1987) Behavioral and biochemical studies of onset and recovery from acid stress in Arctic char (*Salvelinus alpinus*). *Can. J. Fish. Aquat. Sci.*, **44**, 373-81.
Kleerekoper, H. (1967) Some aspects of olfaction in fishes, with special reference to orientation. *Am. Zool.*, **7**, 385-95.
Kleerekoper, H. (1969) *Olfaction in Fishes*, Indiana Univ. Press, Bloomington, 222 pp.
Kleerekoper, H. (1972) Orientation through chemo-reception in fishes. *NASA Spec. Publ.*, **262**, 459-68.
Kleerekoper, H. and Mogensen, J. (1963) Role of olfaction in the orientation of *Petromyzon marinus*. I. Response to a single amine in prey's body odor. *Physiol. Zool.*, **36**, 347-60.
Kleerekoper, H., Taylor, G. and Wilton, R. (1961) Diurnal periodicity in the activity of *Petromyzon marinus* and the effects of chemical stimulation. *Trans. Am. Fish. Soc.*, **90**, 73-8.

Kleerekoper, H., Gruber, D. and Matis, J. (1975) Accuracy of localization of a chemical stimulus in flowing and stagnant water by the nurse shark, *Ginglymostoma cirratum*. *J. comp. Physiol.*, **98**, 257–75.

Konosu, S., Fusetani, N., Nose, T. and Hashimoto, Y. (1968) Attractants for eels in the extracts of short-necked clam – II. Survey of constituents eliciting feeding behavior by fractionation of the extracts. *Bull. Jap. Soc. scient. Fish.*, **34**, 84–7. (In Japanese)

Koyama, T., Saruya, R., Misono, M., Inone, T. and Shibata, T. (1971) Studies on artificial baits for fishing. I. Artificial bait for tuna longlining. *Bull. Tokai Reg. Fish. Res. Lab.*, **69**, 89–97.

Kruse, K.C. and Francis, M.G. (1977) A predation deterrent in larvae of the bullfrog, *Rana catesbeiana*. *Trans. Am. Fish. Soc.*, **106**, 248–52.

Kruse, K.C. and Stone, B.M. (1984) Largemouth bass (*Micropterus salmoides*) learn to avoid feeding on toad (*Bufo*) tadpoles. *Anim. Behav.*, **32**, 1035–9.

Lemly, A.D. and Smith, R.J.F. (1987) Effects of chronic exposure to acidified water on chemoreception of feeding stimuli in fathead minnows (*Pimephales promelas*): mechanisms and ecological implications. *Environ. Toxicol. Chem.*, **6**, 225–38.

Little, E.E. (1977) Conditioned aversion to amino acid flavors in the catfish, *Ictalurus punctatus*. *Physiol. Behav.*, **19**, 743–7.

McBride, J.R., Idler, D.R., Jonas, R.E.E. and Tomlinson, N. (1962) Olfactory perception in juvenile salmon. I. Observations on response of juvenile sockeye to extracts of food. *J. Fish. Res. Bd Can.*, **19**, 327–34.

Mackie, A.M. and Adron, J.W. (1978) Identification of inosine and inosine 5'-monophosphate as the gustatory feeding stimulants for the turbot, *Scophthalmus maximus*. *Comp. Biochem. Physiol.*, **60A**, 79–83.

Mackie, A.M. and Mitchell, A.I. (1982) Further studies on the chemical control of feeding behaviour in the Dover sole, *Solea solea*. *Comp. Biochem. Physiol.*, **73A**, 89–93.

McMahon, T.E. and Tash, J.C. (1979) The use of chemosenses by threadfin shad, *Dorosoma petenense*, to detect conspecifics, predators and food. *J. Fish Biol.*, **14**, 289–96.

McNally, B. (1986) The art of chum fishing. *Southern Outdoors*, **34**, 38–9.

Marusov, Y.A. (1975) Reactions of young salmonids to certain natural chemical stimuli. *J. Ichthyol.*, **15**, 341–3.

Mathewson, R.F. and Hodgson, E.S. (1972) Klinotaxis and rheotaxis in orientation of sharks toward chemical stimuli. *Comp. Biochem. Physiol.*, **42A**, 79–84.

Mearns, K.J. (1985) Response of Atlantic salmon (*Salmo salar* L.) yearlings to individual L-amino acids. *Aquaculture*, **48**, 253–9.

Mearns, K.J. (1986) Sensitivity of brown trout (*Salmo trutta* L.) and Atlantic salmon (*Salmo salar* L.) fry to amino acids at the start of exogenous feeding. *Aquaculture*, **55**, 191–200.

Mearns, K.J. (1989) Behavioural responses of salmonid fry to low amino acid concentrations. *J. Fish Biol.*, **34**, 223–32.

Murofushi, S. and Ina, K. (1981) Survey of feeding stimulants for the sea bream present in the dried pupae of silkworms. *Agric. biol. Chem.*, **45**, 1501–4.

Murofushi, S., Sano, A. and Ina, K. (1982) Structure–activity relationships of neutral amino acids as feeding stimulants for sea bream *Chrysophrys major*. *Nippon Nogeikagaku Kaishi*, **56**, 255–9. (In Japanese)

Neill, W.H. (1979) Mechanisms of fish distribution in heterothermal environments. *Am. Zool.*, **19**, 305–17.

Nelson, J.P. (1984) *Fishes of the World*, 2nd edn, Wiley, N.Y., 523 pp.

Ohsugi, T., Hidaka, I. and Ikeda, M. (1978) Taste receptor stimulation and feeding behaviour in the puffer, *Fugu pardalis*. II. Effects produced by mixtures of constituents of clam extracts. *Chem. Senses Flavour*, **3**, 355–68.

Olmsted, J.M.D. (1918) Experiments on the nature of the sense of smell in the common catfish, *Amiurus nebulosus* (LeSueur). *Am. J. Physiol.*, **46**, 443–58.

Olsén, K.H., Karlsson, L. and Helander, A. (1986) Food search behavior in Arctic charr, *Salvelinus alpinus* (L.), induced by food extracts and amino acids. *J. Chem. Ecol.*, **12**, 1987–98.

Oshima, K. and Gorbman, A. (1966a) Olfactory responses in the forebrain of goldfish and their modification by thyroxine treatment. *Gen. comp. Endocrinol.*, **7**, 398–409.

Oshima, K. and Gorbman, A. (1966b) Influence of thyroxine and steroid hormones on spontaneous and evoked unitary activity in the olfactory bulb of goldfish. *Gen. comp. Endocrinol.*, **7**, 482–91.

Parker, G.H. (1912) The directive influence of the sense of smell in the dogfish. *Bull. U.S. Bur. Fish.*, **32**, 35–46.

Pawson, M.G. (1977) Analysis of a natural chemical attractant for whiting *Merlangius merlangus* L. and cod *Gadus morhua* L. using a behavioural bioassay. *Comp. Biochem. Physiol.*, **56A**, 129–35.

Pearson, W.H., Miller, S.E. and Olla, B.L. (1980) Chemoreception in the food-searching and feeding behavior of the red hake, *Urophycis chuss* (Walbaum). *J. exp. mar. Biol. Ecol.*, **48**, 139–50.

Prochnow, J.A. (1989) Carrier for fish attractant. U.S. Patent no. 4,826, 691.

Rand, G., Kleerekoper, H. and Matis, J. (1975) Interaction of odour and flow perception and the effects of parathion in the locomotor orientation of the goldfish *Carassius auratus* L. *J. Fish Biol.*, **7**, 497–504.

Roberts, N.J. and Winn, H.E. (1962) Utilization of the senses in feeding behavior of the Johnny darter, *Etheostoma nigrum*. *Copeia*, **3**, 567–70.

Rottiers, D.V. and Lemm, C.A. (1985) Movement of underyearling walleyes in response to odor and visual cues. *Progve Fish Cult.*, **47**, 34–41.

Scarfe, A.D., Steele, C.W. and Rieke, G.K. (1985) Quantitative chemobehavior of fish: an improved methodology. *Env. Biol. Fishes*, **13**, 183–94.

Sosin, M. (1987) Call it macaroni. *Salt Wat. Sportsman*, **48**, 73–5.

Steven, D.M. (1959) Studies on the shoaling behaviour of fish. I. Responses of two species to changes of illumination and to olfactory stimuli. *J. exp. Biol.*, **35**, 261–80.

Sutterlin, A.M. (1975) Chemical attraction of some marine fish in their natural habitat. *J. Fish. Res. Bd Can.*, **32**, 729–38.

Sutterlin, A.M., Solemdal, P. and Tilseth, S. (1982) Baits in fisheries with emphasis on the North Atlantic cod fishing industry, in *Chemoreception in Fishes* (ed. T.J. Hara), Elsevier, Amsterdam, pp. 293–305.

Takaoka, O., Takii, K., Nakamura, M., Kumai, H. and Takeda, M. (1990) Identification of feeding stimulants for marbled rockfish. *Nippon Suisan Gakkaishi*, **56**, 345–51. (In Japanese)

Teichmann, H. (1959) Uber die Leistung des Geruchssinnes beim Aal (*Anguilla anguilla*). *Z. vergl. Physiol.*, **42**, 206–54.

Tester, A.L. (1963a) The role of olfaction in shark predation. *Pac. Sci.*, **17**, 145–70.

Tester, A.L. (1963b) Olfaction, gustation, and the common chemical sense in sharks, in *Sharks and Survival* (ed. P.W. Gilbert), D.C. Heath, Lexington, KY, pp. 255–82.

Tester, A.L., van Weel, P.B. and Naughton, J.J. (1955) Response of tuna to chemical stimuli. *Spec. scient. Rep. U.S. Fish. Wildl. Serv. Fish.*, **130**, 1–62.

Timms, A.M. and Kleerekoper, H. (1972) The locomotor responses of male *Ictalurus punctatus*, the channel catfish, to a pheromone released by the ripe female of the species. *Trans. Am. Fish. Soc.*, **101**, 302–10.

Tinsley, R. (1986) Invite a catfish to dinner. *Southern Outdoors*, **34**, 30–3.

U.S. Fish and Wildlife Service (1988) 1985 National Survey of Fishing, Hunting, and Wildlife Associated Recreation. U.S. Department of the Interior, 167 pp.

Wunder, W. (1927) Sinnesphysiologie Untersuchungen uber die Nahrungsaufnahme bei Verscheidenen Knochenfischarten. Z. vergl. Physiol., **6**, 67–98.

Chapter fifteen

Environmental contaminants and chemoreception in fishes

Dorthy A. Klaprat, Robert E. Evans and
Toshiaki J. Hara

15.1 INTRODUCTION

Anthropogenic wastes are continuously being discharged or leached into the aquatic environment. Dilution processes and seasonal inputs (e.g. snowmelt) subject aquatic organisms to fluctuating concentrations of chemicals. There is increasing concern regarding the ultimate, often unpredictable consequences of these changes in water quality. The short- and long-term effects of toxic substances on fish are two of the most important aspects of water pollution.

Fish chemoreceptive membranes directly exposed to the environment are predisposed to the deleterious effects of water-borne chemicals. Toxic substances may alter sensory perception via several modes of action: through cellular uptake and damage to organelles and enzyme systems, by direct interaction with membrane receptor sites, or by masking biologically important chemical signals through competitive binding.

Through behavioural and physiological adaptation, fish can survive and reproduce within a range of environmental conditions. How fish respond to different levels of contamination is important in assessing the ecological effects of pollution. Trace levels of contaminants could change the behaviour of a species and decrease its survival where it would otherwise be successful. Furthermore, avoidance of contaminants can alter fish distribution and movement, a change which may have repercussions throughout an ecosystem.

This chapter reviews investigations of the effects of aquatic contamination on chemoreception and physiological and behavioural responses, with emphasis on the most recent work. Reviews of previous work have been provided by Sutterlin (1974), Brown et al. (1982) and Hara et al. (1983).

15.2 HISTOLOGICAL EFFECTS OF CONTAMINANTS

Uptake and distribution

Chemoreceptive tissues, with the exception of the olfactory rosettes, have virtually been ignored with respect to uptake of toxins. The accumulation of certain metals has been examined in fish olfactory rosettes using atomic absorption spectrometry, gamma spectrometry (Brown et al., 1982) and whole-body autoradiography (Tjälve et al., 1986; Darnerud et al., 1989). The olfactory rosettes of rainbow trout, Oncorhynchus mykiss, held for 2 weeks in water containing 2.4 µM copper 1.3 µM cadmium or 0.75µM mercury accumulated a total metal load of 120–160 nmol g^{-1} wet weight (Brown et al., 1982). Cadmium was accumulated most rapidly, with maximal levels being reached within 24 h. Tjälve et al. (1986) exposed brown trout, Salmo trutta, to concentrations up to 0.09µM $^{109}Cd^{2+}$ for 1 or 2 weeks. After 2 weeks exposure at 9 nM, cadmium accumulated in the olfactory apparatus (no quantity provided), the gills (925 µg kg^{-1}) and the trunk kidney (2025 µg kg^{-1}). Autoradiography showed accumulation in the epithelium of the rosette, the olfactory nerve and the anterior part of the brain. It was suggested that cadmium was taken up in sensory cells in the olfactory epithelium and transported via the axon to the terminal in the bulb (Tjälve et al., 1986). Cadmium is a potent inducer of necrosis of the olfactory epithelial cells in fish (Stromberg et al., 1983). Accumulation of the metal in the olfactory nerve primordium of juvenile pike, Esox lucius, exposed for 40 days at concentrations > 4.5 µM led to alterations of olfactory nerve structure (Dedual, 1987). The metabolites of ^{14}C-labelled 1,2-dibromoethane, an agricultural nematocide, were also taken up by the olfactory rosettes of juvenile rainbow trout, as demonstrated by autoradiography (Darnerud et al., 1989). Irreversible binding of such reactive metabolites to cellular macromolecules is generally considered to be of importance for the toxicity of halogenated hydrocarbons (Darnerud et al., 1989). The accumulation of metals and other toxins by chemosensory cells, as in other tissues, may lead to impairment of cellular functions or to cell death.

Cellular effects of contaminants

Lesions associated with the chemosensory tissues provide convincing evidence that olfaction and taste systems can be highly vulnerable to water pollution. Most heavy metals are damaging to chemosensory tissues; the severity is generally related to concentration and duration of exposure, as well as to water conditions such as pH and hardness which affect contaminant solubility and speciation.

Olfactory tissue degeneration induced by zinc ions is especially pronounced. Irrigation of the olfactory organ of lamprey, *Entosphenus japonicus*, with 0.2 M zinc sulphate for a few seconds caused loss of cilia from the receptor cells, while higher concentrations (0.5 M) led to non-selective degeneration of the epithelium after 1–3 days (Suzuki and Suzuki, 1981). Low zinc concentrations of 10–170 µM produced degeneration in the olfactory receptor cells of channel catfish, *Ictalurus punctatus* (Cancalon, 1980). Furthermore, the olfactory receptor and sustentacular cell ratio decreased in catfish 4 days following several seconds of irrigation with zinc sulphate at concentrations > 30 mM, showing that the metal can act very specifically on sensory cells when contact with the mucosal tissue is brief (Cancalon, 1982). Zinc uptake by sensory cells and subsequent degeneration of various cellular components may be responsible for its destructive effects.

The degree of morphological change due to copper exposure depends on concentration and exposure time (Gardner, 1975). The olfactory sensory cells of the mummichog, *Fundulus heteroclitus*, and Atlantic silverside, *Menidia menidia*, were necrotized (irreversibly damaged), in 7.9 µM copper after 24 h and 6 h, respectively. The architecture of the olfactory organs in the silverside was essentially destroyed, and hyperplasia of sustentacular cells reduced the sensory surface area in the mummichog (Gardner and LaRoche, 1973). Further, Moran *et al.* (1987) demonstrated with transmission electron microscopy that copper ions were associated with selective loss of olfactory sensory cells in brown trout. Copper sulphate at 200 µM also caused damage to taste buds in goldfish, *Carassius auratus*, within a few hours (Vijayamadhavan and Iwai, 1975). Taste buds in the skin of brown bullhead, *Ictalurus nebulosus*, showed an advanced stage of degeneration, with pycnotic cells after 40 days exposure to 4.7 µM copper (Benedetti *et al.*, 1989).

Other metals have also been shown to be deleterious to chemosensory tissues. Stromberg *et al.* (1983) exposed fathead minnows, *Pimephales promelas*, to high cadmium concentrations of 107 µM for 96 h. Acute diffuse epithelial necrosis disrupting the normal architecture was observed through the full thickness of the olfactory epithelium. In chronic 40 day exposures of developing pike juveniles to 0.03 µM cadmium, the olfactory nerve was split and the fibrillar connection, which constitutes the frame of the olfactory nerve, between the nasal capsule and the olfactory lobe was destroyed (Dedual, 1987). Mercury at 2.5 µM and silver at 4.6 µM caused severe degenerative changes in neurosensory cells of the mummichog olfactory organs within 96 h (Gardner, 1975). The occurrence of a lesion in silver-exposed fish appeared to be an 'all-or-none' response for the concentration range 0.77 µM to 0.31 mM. Mercuric chloride at 100 µM destroyed taste buds and epithelial cells in the palatal organ of goldfish within 1 h (Vijayamadhavan and Iwai, 1975). Sensory cell disintegration was also found in the taste buds on the barbels and lips of catfish, *Ictalurus nebulosus* and tench, *Tinca tinca*, exposed to 50–100 μM lead for up to 6 months (Haider, 1975).

Histochemically stainable phospholipids are highly localized in the sensory neurones of the olfactory mucosa of several fish species (Evans and Hara, 1977). Alterations of the phospholipids coincide well with changes in electrophysiological activities (electroolfactogram, EOG) of the olfactory epithelium following axotomy (Evans et al., 1982). The effect of exposure to copper sulphate (2–4 μM) on the olfactory epithelium of whitefish, *Coregonus clupeaformis*, and rainbow trout is the elimination of stainable elements after 2 weeks but no discernible effect on sustentacular cells (Brown et al., 1982). Exposure to sublethal levels of cadmium and mercury, 0.8–2.4 μM, proved to be less destructive than these concentrations of copper after 2 weeks. In these fish there was only a slight reduction in the number of cells staining for phospholipid. Longer exposure time further diminished staining and electrophysiological bulbar responses (Brown et al., 1982).

Petroleum hydrocarbons can induce an array of damage to the chemosensory tissues of fish. Saturated and aromatic hydrocarbons, with their complex water-soluble fractions (WSF), are generally responsible for the toxicity of crude oils (Blumer, 1971). Exposure to 0.14 mg l^{-1} whole crude oil (WCO) for 7 days caused marked hyperplasia of the olfactory sustentacular cells and wide areas of cellular degeneration in the Atlantic silverside (Gardner, 1975). Similar results were found in the tidewater silverside, *Menidia beryllina*, exposed to 5 or 100 mg l^{-1} WCO for 21–30 days or 50% WSF for 7 days (Solangi and Overstreet, 1982). Olfactory mucosa of the hogchoker, *Trinectes maculatus*, exhibited severe necrosis of both neurosensory and sustentacular cells in 100 mg l^{-1} WCO after 38 days (Solangi and Overstreet, 1982). Olfactory lamellae also produced considerable mucus in response to hydrocarbon contamination. Seven-day exposure to water-soluble oil fractions at 0.167 ml l^{-1} induced metaplasia in Atlantic silverside, whereby sensory epithelium was replaced by a less differentiated cell type (Gardner, 1975). Larval sand sole, *Psettichthys melanostictus*, exposed to 800 ppb WSF for 8 days developed attenuated chemosensory cilia, and microridges on epithelial cells were lost (Hawkes, 1980). A saltwater-insoluble crude oil fraction at 0.58 mg l^{-1} induced lesions in Atlantic silverside similar to those observed following exposure to WCO (Gardner, 1975).

Specific water-soluble components of crude oil have also provided evidence of their deleterious effects on fish chemosensory tissues. Exposure of pink salmon, *Oncorhynchus gorbuscha*, fry to benzene for 12 days resulted in exhausted mucous cells at concentrations >0.15 mg l^{-1} and loss of cilia on olfactory lamellae surfaces at concentrations >0.51 mg l^{-1} (Babcock, 1985). Naphthalene, at concentrations of 0.02 mg l^{-1} for 15 days, caused olfactory sensory cell necrosis with lymphocytic infiltration and completely necrotic taste buds in the lips, mouth and pharynx in the mummichog (DiMichele and Taylor, 1978). Similarly, exposure to the water-soluble fraction of aviation fuel at 5 mg l^{-1} in seawater resulted in cell necrosis in the olfactory rosette of the fathead minnow after 72 h (Latendresse and Fisher, 1983).

Detergents have a pronounced detrimental effect on the neurosensory tissue because they reduce surface tension and act as fat solvents. Erosion of the taste buds was observed in yellow bullheads, *Ictalurus natalis*, exposed to concentrations of 0.5 mg l^{-1} linear or branched-chain alkylbenzene sulphonates (ABS) for up to 4 weeks. The olfactory epithelium was more resistant, requiring more than 4 ppm ABS to induce thickening of the borders of sensory cells in the olfactory lamellae, and to decrease the number of neurones (Bardach *et al.*, 1965). Cancalon (1983) found that the non-ionic detergent Triton X-100 at concentrations of 0.03–0.1% immediately severs microvilli and cilia of the receptor and sustentacular cells of the olfactory sensory areas of catfish. Higher detergent concentrations, 1–4%, extensively damaged and denuded the sensory and nonsensory areas of the lamellae.

Reports of the effects of low pH on the chemosensory tissues are few but consistent. Daye and Garside (1976), studying brook trout, *Salvelinus fontinalis*, fingerlings exposed to pH < 4.2 for 7 days, found mucous cells of increased size and number within the olfactory rosette. Similarly, there were large increases in the amount of mucus deposited on the olfactory rosette in fathead minnows exposed for 72 h to pH levels below 6.0 (Lemly and Smith, 1987), and increased numbers of mucus droplets over the olfactory epithelium in rainbow trout exposed to acidified water, pH 4.7, for 14 days (Klaprat *et al.*, 1988). Although the functional role of the mucus layer is not well understood, it has been suggested that it provides a protective layer through which odorants must diffuse before contacting receptor sites. In polluted water the increased mucus could trap and concentrate toxic substances and impair olfactory function. Extremely low pH, 4.0, completely destroyed the olfactory mucosa of fathead minnows (Cancalon, 1980). Alkaline conditions above pH 9.5 created changes in the nares of brook trout similar to that caused by acid stress, but in lesser degree (Daye and Garside, 1976). In nature, fish may be faced with acute and chronic exposure to dissolved and suspended metals such as aluminium, mobilized from soils by acidic precipitation (Driscoll *et al.*, 1980). Klaprat *et al.* (1988) have found that exposure of rainbow trout to pH 4.7 plus 5–20 µM aluminium resulted in irregularly shaped olfactory knobs, swellings on microridge cells, clumped microvilli and loss of receptor cell cilia.

Recovery from toxicant-induced lesions

Fish olfactory receptor cells possess the potential for turnover and for regeneration after injury (Evans *et al.*, 1982). Recovery from contaminant exposures has been shown for a number of fish species. Whitefish and rainbow trout olfactory epithelium exposed to 2.4 μM copper sulphate for 2 weeks showed gradual recovery of receptor cells by 12 weeks after removal from copper exposure (Hara et al., 1983). Wild brown trout experienced complete loss of their olfactory sensory cells after 2 days exposure in aquaria to elevated levels of copper, cadmium, cobalt and zinc; olfactory receptors regenerated within 8 days after the fish were returned to their home stream (Moran et al., 1987). The olfactory receptor cilia on lampreys, damaged by irrigation with 0.2 M zinc sulphate, grew back by 12 days (Suzuki and Suzuki, 1981). In these fish the binding activity of several amino acids was similar to control values after only 10 days. Following prolonged treatment with > 30 mM zinc sulphate, some regeneration was observed in limited areas of catfish olfactory lamellae, with receptor cell reappearing between days 45 and 55, and complete recovery requiring 60–65 days (Cancalon, 1982). Similarly, catfish showed significant receptor cell regeneration in extensively damaged sensory and non-sensory areas 2 months following exposure to 1–4% Triton X-100 detergent (Cancalon, 1983). However, regeneration of severed microvilli and cilia occurs after 4 days when catfish are exposed to lower concentrations of detergent (0.03–0.1%). Yellow bullhead chemoreceptors also damaged by detergent showed partial recovery after 6 weeks in clean water (Bardach et al., 1965). Depending on concentration used, partial to complete recovery was observed in the olfactory organ of tidewater silversides that had been exposed to whole crude oil or its water-soluble fraction for 20 days and then maintained in oil-free seawater for 17 days (Solangi and Overstreet, 1982). Cell renewal, presumably from the undifferentiated basal cells of chemosensory tissues, takes place once the contaminant is removed from the environment. The regenerative capacity of the olfactory neurones suggests that their continuous replacement may be an adaptation to injury from the environment during the normal life of the animal (Graziadei and Monti Graziadei, 1978a,b). However, morphological recovery does not imply a simultaneous functional recovery.

15.3 PHYSIOLOGICAL AND BIOCHEMICAL EFFECTS OF CONTAMINANTS

Olfactory electrophysiology

Preliminary studies indicated that metals such as mercury and copper depress olfactory responses (Sutterlin and Sutterlin, 1971; Hara, 1972). An electrophysiological method to monitor effects of sublethal concentrations of various toxicants on the olfactory responses of fishes was subsequently developed by Hara et al. (1976). The method consists of recording the olfactory bulbar electrical responses to a standard stimulant (food extract, amino acids, etc.) applied to the olfactory rosettes as they are perfused with a test toxicant of known concentration. Toxicant effects are generally determined by the degree of inhibition of responses to a standard stimulant over a predetermined period of time. When rainbow trout are exposed to >0.5 μM inorganic mercury or >0.13 μM copper, there is a depression of the olfactory response within 2 min. The degree of neurological depression increases with time of exposure and with metal concentration (Hara et al., 1976). Further studies with copper sulphate showed a biphasic bulbar response induced by stimuli: the effect of lower concentrations of copper is primarily inhibition (increasing with concentration up to 50 μM), whereas the presence of excess copper sulphate, >100 μM, augments the bulbar response during perfusion (Hara, 1981). Copper has also been shown to enhance taste responses in Atlantic salmon, *Salmo salar* (Sutterlin and Sutterlin, 1970). Although high concentrations of metals sometimes augmented responses during treatment, upon rinsing, the bulbar response to L-serine may be severely impaired. Brown et al. (1982) summarized the general patterns of depression of the olfactory bulbar response to 10 μM L-serine for various metals. Silver, mercury and copper were the most effective inhibitors, with IC50 (concentration which inhibited responses by 50%) values of 0.13–2.20 μM. The IC50 values for other metals were greater than 100 μM.

The concentration required to inhibit olfactory bulbar responses varies with individual metals and water conditions. The toxicity of natural lake waters contaminated with heavy metals from mining and smelting effluents has been investigated electrophysiologically in Arctic charr, *Salvelinus alpinus* (Thompson and Hara, 1977). The effect of the metals in lake water on the olfactory bulb response is less than that of an artificial lake water containing the same concentrations of the metals. Formation of metal–organic complexes are probably responsible for this amelioration of inhibitory effects. Brown et al. (1982) found that inhibition of bulbar electrical activity by cadmium, zinc, lead, cobalt and nickel occurred at levels that exceed lethal concentrations. High concentrations of cadmium applied to the olfactory epithelium of the carp, *Cyprinus carpio*, also inhibited total bulbar response (Takagi et al., 1978). Chronic exposure to 1.3 μM cadmium for 2 weeks resulted in no inhibition in trout, *Oncorhynchus mykiss* (Brown et al., 1982). Threshold concentrations of silver, copper and mercury causing minimal inactivation of bulbar responses, were estimated at 0.03, 0.2 and 0.4 μM, respectively. These values are 20 to 50% of the lethal levels of the metals for rainbow trout (MacLeod and Pessah, 1973; Davies et al., 1978; Miller and MacKay, 1980). Copper at 2.4 μM for 2 weeks completely inhibited bulbar responses to L-serine, and 0.75 μM mercury reduced the response by 40%. In general, chronic exposure to metals progressively inhibits the electrophysiological responsiveness of trout to standard stimulants (Brown et al., 1982).

Detergents have been shown to inhibit olfactory responses to standard stimulants. Sutterlin et al. (1971) demonstrated that some ionic detergents

(ABS, quaternary ammonium and imidazonium salts and diamines) block the neural discharges evoked by amino acids in the olfactory epithelium of Atlantic salmon parr at 1 mg l^{-1} or lower. Most blocking effects were found to be reversible following rinsing with water. Hara and Thompson (1978) observed decreased olfactory bulbar response to odorants when the nares of whitefish were perfused with sublethal levels of detergents. The response to a food stimulus was reduced almost immediately upon exposure to sodium lauryl sulphate (SLS) and the response was dose dependent over the range 0.1–10 mg l^{-1}. Irrigation of olfactory mucosa of catfish with 0.1% Triton X-100 removes membranous proteins and cellular processes from the receptor or sustentacular cells (Cancalon, 1980). The EOG response was abolished within an hour of treatment but returned to 50–60% of the original activity level during the next hour. Their ability to deplete the epithelial mucus layer, to dissolve chemosensory membranes and to denature receptor proteins renders detergents hazardous to fish (Tanford and Reynolds, 1976; Brown et al., 1982).

Low pH interferes with chemoreceptive processes. Hara (1976) and Thommesen (1983) noted that the electrophysiological response of the olfactory bulb of rainbow trout to different amino acids was highly dependent upon pH. Responses fell rapidly below pH 5.0 and were inhibited at pH <4.0 and >8.0 (Hara, 1976). Changes in pH probably influence the ionization of both stimulant molecules and receptor sites, thus limiting their binding for sensory transduction. The property of acidified water to dissolve metals in the environment poses an even more detrimental effect to fish chemoreception. For example, electrical responses recorded from the olfactory nerve in response to the amino acid L-serine were similar to, but smaller than, those in controls in rainbow trout exposed to acidified water (pH 4.7) for 2 weeks. When fish were exposed to acidified water (pH 4.7) and aluminium (20 µM), the response was significantly diminished (Klaprat et al., 1988).

Gustatory electrophysiology

Agents that effectively block olfactory electrophysiological responses may simultaneously affect the taste system. Although the concentrations used were not always realistic in that they exceeded lethal levels, electrical recordings from the taste system of carp (Hidaka and Yokota, 1967; Hidaka, 1970) and Atlantic salmon (Sutterlin and Sutterlin, 1970) showed that metals were effective blocking agents of gustatory responses. In particular, mercury and copper appear to have an inhibitory effect on all chemoreceptor types. The detergent ABS at concentrations as low as 1 mg l^{-1} impaired electrical responses from the taste nerve bundles innervating the barbel taste receptors of yellow bullhead in response to 0.01 M cysteine (Bardach et al., 1965). Increasing concentrations of hydrogen ion evoke electrical responses in taste system preparations of several fish species (Konishi et al., 1969; Hidaka, 1970, 1972; Sutterlin and Sutterlin, 1970; Yoshii et al., 1980; Marui et al., 1983; Yamashita et al., 1989). The evidence points to the existence of receptors for H^+ and for CO_2, which is released when acid is added to water containing carbonate salts. Studies dealing with toxic effects of pH on taste receptors are lacking.

Inhibition of receptor binding

Although the effects of contaminants on receptor–ligand binding (Chapter 7) is a promising area for research, studies with fish tissue are few (Cagan and Zeiger, 1978; Suzuki and Suzuki, 1981; Brown et al., 1982; Rehnberg and Schreck, 1986a). These studies have focused on metal-induced inhibition of chemostimulatory amino acid binding to olfactory cell membrane receptors. Brown et al. (1982) demonstrated that metals inhibited, in a dose-dependent fashion, L-serine binding to a sedimentable fraction isolated from the olfactory rosettes of rainbow trout. Mercury, silver and copper were the most effective inhibitors, with IC50 values of 1.0, 3.0 and 9.5 μM, respectively. Studies with rainbow trout (Cagan and Zeiger, 1978) and coho salmon, *Oncorhynchus kisutch* (Rehnberg and Schreck, 1986b) showed that some metals were essentially concentration independent and not especially inhibitory to binding. Copper, an effective inhibitor of L-serine binding in trout (Brown et al., 1982), was ineffective both in salmon (Rehnberg and Schreck, 1986b) and at high concentrations of 1 mM for L-alanine binding in rainbow trout (Cagan and Zeiger, 1978). Fish chronically exposed for 1 or 2 weeks to 2.4 μM copper showed a significant reduction in maximal binding capacity, but no change in binding affinity (Brown et al., 1982). In contrast, all workers agree that mercury is a most effective inhibitor of binding of chemostimulatory amino acids to fish olfactory preparations. In fact, Brown et al. (1982) suggested that the metal denatures the olfactory receptor for L-serine, because it irreversibly inhibits both binding and electrophysiological response. Metals could bind covalently to receptor sites for chemostimuli, thereby inhibiting binding almost irreversibly.

Besides producing inconsistent results, the receptor–ligand binding techniques appear to suffer from problematical sensitivity. For example Brown et al. (1982) showed binding of L-serine to be 10-fold more sensitive to inhibition by copper and cadmium than did Cagan and Zeiger (1978). The discrepancies point to differences in methodology and tissue preparations used. The subcellular fractions used in the binding studies do not represent pure fractions and so may not represent pure binding. Active transport, in addition to binding, may also be occurring (Brown and Hara, 1982). Although the inhibitory effects of metals on L-serine binding and electrophysiological responses show some parallelism, the recorded bulbar activity may be altered without comparable reduction in peripheral biochemical activity (Brown et al., 1982).

15. BEHAVIOURAL RESPONSES TO CONTAMINANTS

Responses by naïve fish

The most studied behavioural response of fish to metals is preference/avoidance behaviour, a sensitive indicator of sublethal exposure that can be objectively and rapidly quantified (reviews: Cherry and Cairns, 1982; Beitinger and Freeman, 1983; Giattina and Garton, 1983; Atchison et al., 1987). Avoidance or preference responses vary greatly with fish species and contaminant. Black and Birge (1980) noted that bluegills, *Lepomis macrochirus*, and largemouth bass, *Micropterus salmoides*, had lower response threshold concentrations than rainbow trout for cadmium, copper and zinc. Largemouth bass did not respond to 0.79 μM copper, which attracted goldfish and channel catfish (Timms et al., 1972). Rainbow trout displayed initial attraction to all lethal copper concentrations tested, with fish orientating towards the copper source (Pedder and Maly, 1985). The preference response was greatest at the highest copper concentrations tested, 0.05 and 0.06 mM, which led to high mortality followed in survivors by a belated avoidance response. Sprague (1968) attributed a difference in avoidance thresholds for rainbow trout (0.25 μM Cu) and Atlantic salmon (0.85 μM Cu) to a basic behavioural difference. Rainbow trout tend to swim more freely than salmon in the test trough and are thus more aware of any available choices between waters. Hartwell et al. (1989) studied the toxicity of five metals (arsenic, cadmium, chromium, copper and selenium) and the avoidance responses of golden shiner, *Notemigonus crysoleucas*. They concluded that the most toxic metals may not elicit an avoidance response. Selenium was not avoided at 44.2 μM, a concentration approaching one-half the 96-h LC50 (concentration which kills 50% of the fish in 96 h). Cadmium, the most toxic element tested in the golden shiner (96-h LC50 = 0.03 mM), was not avoided at 1.7 μM, whereas copper, the least toxic element tested (96-h LC50 = 1.33 mM), was avoided at 0.41 μM. McNicol and Scherer (1991) reported that whitefish (*Coregonus clupeaformis*) responded to concentrations of cadmium $\leqslant 0.01$ μM and $\geqslant 0.07$ μM but showed little response to those in between. In nature an avoidance response can elicit movement away from a contaminated area, resulting in an alteration of the aquatic community. A preference response can lead to exposure of the animal to a higher and perhaps lethal concentration of the contaminant (Giattina and Garton, 1983). Reproductive state can also influence the behavioural response of fish to metals. Sexually mature male fathead minnows avoided 4.34μM zinc in a test chamber when given a choice between dosed and clean water (Korver and Sprague, 1989). However, when breeding shelters were placed on one side of the chamber and territories became established, it took six times that concentration to cause an avoidance reaction.

Detergents affect the feeding and locomotor behaviour of fish. Exposure to sublethal concentrations of branched alkylbenzene sulphonates impaired feeding behaviour of yellow bullhead (Bardach et al., 1965) and flagfish, *Jordanella floridae* (Foster et al., 1966). After 6 weeks in detergent-free water, fish did not fully recover their food-searching behaviour. Rainbow trout exposed to ABS avoided concentrations of $0.4\,\text{mg}\,l^{-1}$ but not $10\,\text{mg}\,l^{-1}$ (Sprague and Drury, 1969). Ayu, *Plecoglossus altivelis*, avoided low concentrations but preferred high concentrations of linear alkylbenzene sulphonates (LAS) (Tatsukawa and Hidaka, 1978). Whitefish exposed to sodium lauryl sulphate preferred intermediate concentrations and did not respond to the lowest ($0.01\,\text{mg}\,l^{-1}$) and highest ($10\,\text{mg}\,l^{-1}$) concentrations tested (Hara and Thompson, 1978). Similarly, Hidaka and Tatsukawa (1989) observed that medaka, *Oryzias latipes*, avoided LAS at $10\text{–}30\,\mu\text{g}\,l^{-1}$ but not $>48\,\mu\text{g}\,l^{-1}$. They speculate that the responses were due to adaptation of chemoreceptors to the lower concentration and to deterioration of detection ability resulting from chemoreceptor damage at the higher concentration.

The freshwater spawning migration of salmonids is controlled to a great extent by olfactory cues (Hasler and Scholz, 1983). Several laboratory studies suggest that acidification of rivers may contribute to olfactory impairment during spawning migration. Royce-Malmgren and Watson (1987) monitored the response of juvenile Atlantic salmon to glycine and L-alanine while the pH of the test water was gradually reduced from 7.6 to 5.1. At pH 7.6 salmon were attracted to glycine and avoided L-alanine, whereas at pH 5.1 they were attracted to L-alanine and indifferent to glycine. Nakamura (1986) exposed the Japanese fat minnow, *Phoxinus lagowski*, to water at pH levels ranging from 1 to 5. He noted an immediate decrease in activity and a change in distribution as the minnows avoided the acidified water, although both returned to original state within 1–2 h, after exposure. Peterson et al. (1989) tested the preference/avoidance of 11 fish species in a continuous pH gradient from pH 4 to 10. They concluded that the pH avoidance thresholds were generally similar to the pH levels limiting natural distribution of the various species.

The range in avoidance concentrations determined for 18 fish species varies from 0.05 to $0.43\,\text{mg}\,l^{-1}$ total residual chlorine (TRC) (Cherry and Cairns, 1982). Salmonids are most sensitive, avoiding $0.05\text{–}0.10\,\text{mg}\,l^{-1}$ TRC; cyprinids as a group are more variable, showing avoidance thresholds as low as $0.05\text{–}0.2\,\text{mg}\,l^{-1}$ (rosyface shiner, *Notropis rubellus*) to as high as $0.21\text{–}0.43\,\text{mg}\,l^{-1}$ (golden shiner). Fish avoidance of low concentrations of chlorine (Hidaka and Tatsukawa, 1985) is influenced by environmental factors such as acclimation temperature, salinity, pH, light intensity and combinations of these variables (Giattina and Garton, 1983). Hose and

Stoffel (1980) reported a significant difference in avoidance behaviour of blacksmith, *Chromis punctipinnis*, to TRC that depended on the degree of starvation of the test animal. Starved fish did not show an avoidance response up to a concentration of $0.33 \, \text{mg} \, l^{-1}$, as opposed to an avoidance threshold concentration of $0.18 \, \text{mg} \, l^{-1}$ for satiated fish. When given a choice between chlorinated waters at different temperatures, shiner perch, *Cymatogaster aggregata* (Stober et al., 1980) and spotfin shiners, *Notropis spilopterus* (Giattina et al., 1981) were attracted to water heated to their preferred temperature and would tolerate twice the level of chlorine normally avoided at ambient temperature. Cherry et al. (1982) have reported similar results with other freshwater species.

Several studies indicate that fish will not necessarily avoid petroleum-contaminated water (Morrow, 1973; Bean, et al., 1974). Maynard and Weber (1981) found that >50% of presmolt salmon tested would not avoid a potentially toxic concentration of petroleum hydrocarbons, i.e. $3-4 \, \text{mg} \, l^{-1}$. However, avoidance was observed at lower concentrations, $0.2 \, \text{mg} \, l^{-1}$, of individual components of hydrocarbon mixtures. Also, feeding behaviour was found to be affected by exposure to hydrocarbons. Purdy (1989) exposed coho salmon to a mixture of seven aromatic hydrocarbons at two concentrations. At the low concentration, 0.08%, feeding behaviour was reduced, and at the higher concentration, 0.15%, feeding behaviour ceased for 3 days.

Studies on the effect of pesticides on fish behaviour deal primarily with avoidance of lethal concentrations and have been summarized by Sutterlin (1974), Giattina and Garton (1983) and Hara et al. (1983). Hidaka and Tatsukawa (1989) tested the avoidance response of medaka, with resected olfactory organs, to fenitrothion and confirmed that fish detect pesticides by olfaction. Folmar (1976) and Hidika et al. (1984) found that fish will avoid many pesticides at sublethal levels but will not avoid others, even at lethal concentrations. Hall et al. (1984) studied the avoidance response of striped bass, *Morone saxatilis*, and Atlantic menhaden, *Brevoorita tyrannus*, to an antifoulant material, bis(tri-N-butyltin) oxide. Atlantic menhaden and striped bass avoided concentrations of 5.5 and $14.7 \, \mu\text{g} \, l^{-1}$, respectively, which were much higher than the highest recorded environmental concentration ($0.92 \, \mu\text{g} \, l^{-1}$; Waldock, and Miller, 1983). Therefore it is unlikely that these fish would avoid pesticide concentrations that they would normally encounter in the environment.

Responses by pre-exposed fish

Pre-exposure to metals influences the subsequent avoidance responses of fish to the same metals. Anestis and Neufeld (1986) pre-exposed rainbow trout for 7–20 weeks to 0.06–0.10 μM potassium dichromate solutions, and noted that the avoidance thresholds of pre-exposed fish increased linearly with the level of pre-exposure. However, fish tended to prefer concentrations matching their pre-exposure level. Similarly, Hartwell et al. (1987) tested the effect of pre-exposure to metals on the preference/avoidance response of fathead minnows to a blend of metals (relative proportions: 1.00 copper, 0.54 chromium, 1.85 arsenic, 0.38 selenium). Although unexposed fish avoided 29 µg l^{-1} total metals, exposed fish preferred elevated metal concentrations equal to 3 times the holding exposure level (<294 µg l^{-1} total metals) and avoided concentrations 5 times the holding concentration (490 µg l^{-1} total metals) after 3 and 6 months pre-exposure, respectively. Fish were unresponsive to concentrations approaching 10 times the holding exposure level (980 µg l^{-1} total metals) after 9 months pre-exposure.

Pre-exposure to metals was also found to affect how fish respond to odorants. Adult female zebrafish, *Brachydanio rerio*, have a pheromonal sex attractant system that acts in attracting mature males to a spawning area (Bloom and Perlmutter, 1977). However, after 9 days exposure to 0.08 mM zinc, zebrafish failed to detect pheromone at concentrations that attracted control fish (Bloom et al., 1978). In another study, exposure of whitefish to 0.18 µM mercuric chloride for 1–2 weeks resulted in greatly reduced behavioural responses to food extract (Kamchen and Hara, 1980). Rehnberg and Schreck (1986b) used a two-choice Y-trough to test the effect of 0.1–10.0 µM mercury, copper or zinc on olfaction in salmonids, based on an avoidance reaction to L-serine. Avoidance behaviour was depressed by the addition of the metals to the test water.

Pre-exposure to 1–2 mg l^{-1} LAS significantly reduced chemoattraction of water conditioned by conspecifics to juvenile Arctic charr (Olsen and Höglund, 1985). Furthermore, Hidaka and Tatsukawa (1986) observed that fish size affected the avoidance response of medaka to LAS, with smaller fish showing avoidance at a lower concentration.

The attraction of Arctic charr to food extract was depressed after 14 days pre-exposure to pH 4.75 to 4.50 (Jones et al., 1985). Fathead minnows exposed to sublethal pH 6.0 for 72 h did not respond to a chemical stimulus that elicited a feeding response at higher pH levels (Lemly and Smith, 1985). This impairment was reversible: the response to feeding was restored in minnows allowed to recover in pH 8 water for 24 h (Lemly and Smith, 1987). Thus reduced feeding could contribute to the elimination of fathead minnows from natural waters when pH levels fall below <6.0.

In a laboratory study, Arctic charr accidentally exposed to >19 µg l^{-1} TRC for 6 days became hypoactive, more strongly thigmotactic and less responsive to a food extract (Jones and Hara, 1988). Post-exposure fish failed to respond positively to food extract for 20 days, and some fish displayed chlorine stress behaviour for at least 46 days.

Behavioural responses in the field

Disruption of fish migration and distribution patterns as a result of metal contamination has been demonstrated in field studies. Salmonids imprint to home-stream odours, and as adults use this olfactory information for their freshwater reproductive migration (Hasler and Scholz, 1983). The addition of 0.69 µM copper was shown to reduce the effectiveness of home-stream water as an attractant to Atlantic salmon (Sutterlin and Gray, 1973). Sprague et al. (1965) and Saunders and Sprague (1967) reported that 0.27–0.33 µM copper and 3.21–3.95 µM zinc caused downstream movements of migrating adult Atlantic salmon. They concluded that mixtures containing >0.60 µM copper and >7.34 µM zinc could block spawning runs. Geckler et al. (1976) added 1.89 µM copper to a stream and observed that several fish species avoided spawning in areas of high copper concentration. For example, the bluntnose minnow, *Pimephales notatus*, spawned only in areas where the copper concentration was <1.21 µM.

Beamish (1974) reported a decreased annual growth rate of white suckers, *Catastomus commersoni*, in an acidified lake despite an abundant food supply. This suggests that feeding intensity and/or food utilization were decreased as pH levels in the lake dropped. Also, there has been a decline in anadromous fish populations in rivers where the pH is above levels considered to be lethal (Haines, 1981; Muniz, 1981; Watt et al., 1983). Recruits may fail to reach maturity because they do not respond to food or to other stimuli, or as a result of spawning in areas unsuitable to the successful development of eggs, fry and juveniles. Johnson and Webster (1977) reported that brook trout avoided potential spawning areas when the pH of the water was reduced to 4.5.

Tsai (1973) surveyed 149 sewage treatment plants and found that the numbers of fish and of fish species were lowest near the discharge site, and gradually increased with distance from the site. These findings, plus observations from field studies, led Osborne et al. (1981) to conclude that fish move into potentially toxic chlorinated plumes to feed but move out before they are adversely affected. Therefore lethal or chronic effects may not occur in the field, provided that the discharge sites do not disturb migration routes and spawning areas. On the other hand, Weber et al. (1981) noted that mature salmon migrating upstream to spawn did not avoid a mixture of hydrocarbons at concentrations that juvenile salmon avoided in laboratory tests.

15.5 FUTURE RESEARCH

Olfaction and taste systems can be highly vulnerable to environmental contamination, as demonstrated by accumulation of toxins in chemosensory cells and lesions in chemosensory tissue. Behavioural responses of fish to adverse conditions, although important to survival, are variable and are not always related to lethal levels of pollutants. Physiological, biochemical and histochemical methods that are sensitive and specific to a chemical or group of chemicals should be developed to serve as early warning indicators, so that there is time for protective measures to be taken to avoid adverse effects to the ecosystem.

Biochemical studies in fish chemoreception have used ligand-binding studies to examine the interactions between amino acids and their receptors. The complex process that translates the initial recognition of ligand into a final cellular response is not completely understood (Chapter 7). More work is required to increase knowledge of the transduction mechanism for naturally occurring stimuli (e.g. pheromones) in order fully to understand the nature and mode of action of toxicants.

Recently the cytochrome P-450 monooxygenase (mixed-function oxidase) system has been identified as a detoxifying mechanism for many xenobiotics at the subcellular level. The olfactory epithelium of a number of fish species contains high levels of cytochrome P-450 which can be activated by exposure to hydrocarbons or heavy metals. Recent work suggests the presence of more than one form of inducible cytochrome P-450, and further research will determine whether or not they are toxicant specific. The basic biochemical and physiological functions of mixed-function oxidase systems must be understood before they can be used to predict the effect of contaminant-induced changes.

ACKNOWLEDGEMENTS

We thank Diane Malley and Rick McNicol for their constructive reviews of the manuscript.

REFERENCES

Anestis, I. and Neufeld, R.J. (1986) Avoidance–preference reactions of rainbow trout (*Salmo gairdneri*) after prolonged exposure to chromium (VI). *Water Res.*, **20**, 1233–41.

Atchison, G.J., Henry, M.G. and Sandheinrich, M.B. (1987) Effects of metals on fish behavior: a review. *Env. Biol. Fishes*, **18**, 11–25.

Babcock, M.M. (1985) Morphology of olfactory epithelium of pink salmon, *Oncorhynchus gorbuscha*, and changes following exposure to benzene: a scanning electron microscopy study, in *Marine Biology of Polar Regions and Effects of Stress on Marine Organisms* (eds J.S. Gray and M.E. Christiansen), Wiley, Chichester, pp. 259–67.

Bardach, J.E., Fujiya, M. and Holl, A. (1965) Detergents: effects on the chemical senses of the fish *Ictalurus natalis* (le Sueur). *Science*, Wash., D.C., **148**, 1605–7.

Beamish, R.J. (1974) Loss of fish populations from unexploited remote lakes in Ontario, Canada as a consequence of atmospheric fallout of acid. *Water Res.*, **8**, 85–95.

Bean, R.M., Vanderhorst, J.R. and Wilkinson, P. (1974) *Interdisciplinary study of the toxicity of petroleum to marine organisms*. Battelle Pacific Northwest Laboratories, Richmond, 31 pp.

Beitinger, T.L. and Freeman, L. (1983) Behavioral avoidance and selection responses of fishes to chemicals. *Resid. Rev.*, **90**, 35–55.

Benedetti, I., Albano, A.G. and Mola, L. (1989) Histomorphological changes in some organs of the brown bullhead, *Ictalurus nebulosus* Le Sueur, following short- and long-term exposure to copper. *J. Fish Biol.*, **34**, 273–80.

Black, J.A. and Birge, W.J. (1980) An avoidance response bioassay for aquatic pollutants. *Univ. Kentucky, Water Resources Res. Inst. Res. Rep.*, **123**, 1–34.

Bloom, H.D. and Perlmutter, A. (1977) A sexual aggregating pheromone system in the zebrafish, *Brachydanio rerio* (Hamilton-Buchanan). *J. exp. Zool.*, **199**, 215–26.

Bloom, H.D., Perlmutter, A. and Seeley, R.J. (1978) Effect of a sublethal concentration of zinc on an aggregating pheromone system in the zebrafish. *Brachydanio rerio* (Hamilton-Buchanan). *Environ. Pollut.*, **17**, 127–31.

Blumer, M. (1971) Scientific aspects of the oil spill problem, in *Environmental Affairs*, **1**, Woods Hole Oceanographic Institution, Woods Hole, pp. 54–73.

Brown, S.B. and Hara, T.J. (1982) Biochemical aspects of amino acid receptors in olfaction and taste, in *Chemoreception in Fishes* (ed. T.J. Hara), Elsevier, Amsterdam, pp. 159–80.

Brown, S.B., Evans, R.E., Thompson, B.E. and Hara, T.J. (1982) Chemoreception and aquatic pollutants, in *Chemoreception in Fishes* (ed. T.J. Hara), Elsevier, Amsterdam, pp. 363–93.

Cagan, R.H. and Zeiger, W.N. (1978) Biochemical studies of olfaction: binding specificity of radioactively labeled stimuli to an isolated olfactory preparation from rainbow trout (*Salmo gairdneri*). *Proc. natn. Acad. Sci. U.S.A.*, **75**, 4679–83.

Cancalon, P. (1980) Effects of salts, pH and detergents on the catfish olfactory mucosa, in *Olfaction and Taste VII* (ed. H. van der Starre), IRL Press, London, pp. 73–6.

Cancalon, P. (1982) Degeneration and regeneration of the olfactory cells induced by $ZnSO_4$ and other chemicals. *Tissue and Cell*, **14**, 717–33.

Cancalon, P. (1983) Influence of a detergent on the catfish olfactory mucosa. *Tissue and Cell*, **15**, 245–58.

Cherry, D.S., and Cairns, J. (1982) Biological monitoring. V. Preference and avoidance studies. *Water Res.*, **16**, 263–301.

Cherry, D.S., Larrick, S.R., Giattina, J.D., Cairns, J. and van Hassel, J. (1982) Influence of temperature selection upon chlorine avoidance of cold and warm-water fish. *Can. J. Fish. Aquat. Sci.*, **39**, 162–73.

Darnerud, P.O., Lund, B.-O., Brittebo, E.B. and Brandt, I. (1989) 1,2-Dibromoethane and chloroform in the rainbow trout (*Salmo gairdneri*): studies on the distribution of nonvolatile and irreversibly bound metabolites. *J. Toxicol. environ. Health*, **26**, 209–21.

Davies, P.H., Goettl, J.P. and Sindey, J.R. (1978) Toxicity of silver to rainbow trout (*Salmo gairdneri*). *Water Res.*, **12**, 113–17.

Daye, P.G. and Garside, E.T. (1976) Histopathologic changes in surficial tissues of brook trout, *Salvelinus fontinalis* (Mitchill), exposed to acute and chronic levels of pH. *Can. J. Zool.*, **54**, 2140–55.

Dedual, M. (1987) The effect of cadmium upon the structure of the olfactory nerve in the young pike. *Toxicol. environ. Chem.*, **4**, 227–31.

DiMichele, L. and Taylor, M.H. (1978) Histopathological and physiological responses of *Fundulus heteroclitus* to naphthalene exposure. *J. Fish. Res. Bd Can.*, **35**, 1060–6.

Driscoll, C.T., jun., Baker, J.P., Bisogni, J.J., jun., and Schofield, C.L. (1980) Effect of aluminum speciation on fish in dilute acidified waters. *Nature, Lond.*, **284**, 161–4.

Evans, R.E. and Hara, T.J. (1977) Histochemical localization of phospholipids in the olfactory epithelium of fish. *Can. J. Zool.*, **55**, 776–81.

Evans, R.E., Zielinski, B. and Hara, T.J. (1982) Development and regeneration of the olfactory organ in rainbow trout, in *Chemoreception in Fishes* (ed. T.J. Hara), Elsevier, Amsterdam, pp. 15–37.

Folmar, L.C. (1976) Overt avoidance reaction of rainbow trout fry to nine herbicides. *Bull. environ. Contam. Toxicol.*, **15**, 509–14.

Foster, N.R., Scheier, A. and Cairns, J. (1966) Effects of ABS on feeding behavior of flag fish, *Jordanella floridae*. *Trans. Am. Fish. Soc.*, **95**, 109–10.

Gardner, G.R. (1975) Chemically induced lesions in estuarine or marine teleosts, in *The Pathology of Fishes* (eds W.E. Ribelin and G. Migaki), Univ. Wisconsin Press, Madison, pp. 657–93.

Gardner, G.R. and LaRoche, G. (1973) Copper induced lesions in estuarine teleosts. *J. Fish. Res. Bd Can.*, **30**, 363–8.

Geckler, J.R., Horning, W.B., Neiheisel, T.M., Pickering, Q.H., Robinson, E.L. and Stephen, C.E. (1976) Validity of laboratory tests for predicting copper toxicity in streams. *U.S. Environmental Protection Agency, Ecological Research Series*, EPA-600/3-76-116, Duluth, Minn., 192 pp.

Giattina, J.D. and Garton, R.R. (1983) A review of the preference–avoidance responses of fishes to aquatic contaminants. *Residue Rev.*, **87**, 44–90.

Giattina, J.D., Cherry, D.S., Larrick, S.R. and Cairns, J., jun. (1981) Comparison of laboratory and field avoidance behavior of fish in heated chlorinated water. *Trans. Am. Fish. Soc.*, **110**, 526–35.

Graziadei, P.P.C. and Monti Graziadei, G.A. (1978a) The olfactory system: a model for the study of neurogenesis and axon regeneration in mammals, in *Neural Plasticity* (ed. W. Cotman), Raven Press, New York, pp. 131–53.

Graziadei, P.P.C. and Monti Graziadei, G.A. (1978b) Continuous nerve cell renewal in the olfactory system, in *Handbook of Sensory Physiology*. IX. *Development of Sensory Systems* (ed. M. Jacobson), Springer-Verlag, Heidelberg, pp. 55–83.

Haider, G. (1975) Die Wirkung subletaler Bleikonzentrationen auf die Chemorezeptoren zweier Susswasserfischarten. *Hydrobiologia*, **47**, 291–300.

Haines, T.A. (1981) Acidic precipitation and its consequences for aquatic ecosystems: a review. *Trans. Am. Fish. Soc.*, **111**, 669–707.

Hall, L.W., jun., Pinkney, A.E., Zeger, S., Burton, D.T. and Lenkevich, M.J. (1984) Behavioral responses to two estuarine fish species subjected to bis(tri-n-butyltin)oxide. *Water Res. Bull.*, **20**, 235–9.

Hara, T.J. (1972) Electrical responses of the olfactory bulb of Pacific salmon *Oncorhynchus nerka* and *Oncorhynchus kisutch*. *J. Fish. Res. Bd Can.*, **29**, 1351–5.

Hara, T.J. (1976) Effects of pH on the olfactory responses to amino acids in rainbow trout, *Salmo gairdneri*. *Comp. Biochem. Physiol.*, **54A**, 37–9.

Hara. T.J. (1981) Behavioral and electrophysiological studies of chemosensory reactions in fish, in *Brain Mechanisms of Behaviour in Lower Vertebrates* (ed. P.J. Laming), Cambridge Univ. Press, Cambridge, pp. 123–36.

Hara, T.J. and Thompson, B.E. (1978) The reaction of whitefish, *Coregonus clupeaformis*, to the anionic detergent sodium lauryl sulfate and its effects on their olfactory responses. *Water Res.*, **12**, 893–7.

Hara, T.J., Brown, S.B. and Evans, R.E. (1983) Pollutants and chemoreception in aquatic organisms, in *Aquatic Toxicology* (ed. J.O. Nriagu), Wiley, New York, pp. 247–306.

Hara, T.J., Law, Y.M.C. and Macdonald, S. (1976) Effects of mercury and copper on the olfactory response in rainbow trout. *J. Fish. Res. Bd Can.*, **33**, 1568–73.

Hartwell, S.I., Cherry, D.S. and Cairns, J. jun. (1987) Avoidance responses of schooling fathead minnows (*Pimephales promelas*) to a blend of metals during a 9-month exposure. *Environ. Tox. Chem.*, **6**, 177–87.

Hartwell, S.I., Jin, J.H., Cherry, D.S. and Cairns J. jun. (1989) Toxicity versus avoidance response of golden shiner, *Notemigonus crysoleucas*, to five metals, *J. Fish Biol.*, **35**, 447–56.

Hasler, A.D. and Scholz, A.T. (1983) *Olfactory Imprinting and Homing in Salmon*, Springer-Verlag, Berlin, 134 pp.

Hawkes, J.W. (1980) The effects of xenobiotics on fish tissues: morphological studies. *Fed. Proc.*, **39**, 3230–6.

Hidaka, H. and Tatsukawa, R. (1985) Avoidance test of a fish, medaka (*Oryzias latipes*), to aquatic contaminants, with special reference to monochloramine. *Arch. environ. Contam. Toxicol.*, **14**, 565–71.

Hidaka, H. and Tatsukawa, R. (1986) Variations by sex and body size in the avoidance tests of sodium linear laurylbenzene sulfonate and fenitrothion using a fish, medaka *Oryzias latipes*. *Bull. Jap. Soc. scient. Fish.*, **52**, 1753–7.

Hidaka, H. and Tatsukawa, R. (1989) Avoidance by olfaction in a fish, medaka (*Oryzias latipes*), to aquatic contaminants. *Environ. Pollut.*, **56**, 299–309.

Hidaka, H., Hattanda, M. and Tatsukawa, R. (1984) Avoidance of pesticides with medakas (*Oryzias latipes*). *Nippon Nogei Kagaku Kaishi*, **58**, 145–51.

Hidaka, I. (1970) The effects of transition metals on the palatal chemoreceptors of the carp. *Jap. J. Physiol.*, **20**, 599–609.

Hidaka, I. (1972) Stimulation of the palatal chemoreceptors of the carp by mixed solutions of acid and salt. *Jap. J. Physiol.*, **29**, 39–51.

Hidaka, I. and Yokota, S. (1967) Taste receptor stimulation by sweet tasting substances in the carp. *Jap. J. Physiol.*, **17**, 652–66.

Hose, J.E. and Stoffel, R.J. (1980) Avoidance response of juvenile *Chromis punctipinnis* to chlorinated seawater. *Bull. environ. Contam. Toxicol.*, **25**, 929–35.

Johnson, D.W. and Webster, D.A. (1977) Avoidance of low pH in selection of spawning sites by brook trout (*Salvelinus fontinalis*). *J. Fish. Res. Bd Can.*, **34**, 2215–18.

Jones, K.A. and Hara, T.J. (1988) Behavioral alterations in Arctic char (*Salvelinus alpinus*) briefly exposed to sublethal chlorine levels. *Can. J. Fish. Aquat. Sci.*, **45**, 749–53.

Jones, K.A., Hara, T.J. and Scherer, E. (1985) Behavioral modification in Arctic charr (*Salvelinus alpinus*) chronically exposed to sublethal pH. *Physiol. Zool.*, **58**, 400–12.

Kamchen, R. and Hara, T.J. (1980) Behavioral reactions of whitefish (*Coregonus clupeaformis*) to food extract: an application to sublethal toxicity bioassay. *Can. tech. rep. Fish. Aquat. Sci.*, **975**, 182–91.

Klaprat, D.A., Brown, S.B. and Hara, T.J. (1988) The effect of low pH and aluminum on the olfactory organ of rainbow trout, *Salmo gairdneri*. *Env. Biol. Fishes*, **22**, 69–77.

Konishi, J., Hidaka, I., Toyota, M. and Matsuda, H. (1969) High sensitivity of the palatal chemoreceptors of the carp to carbon dioxide. *Jap. J. Physiol.*, **19**, 327–41.

Korver, R.M. and Sprague, J.B. (1989) Zinc avoidance by fathead minnows (*Pimephales promelas*): computerized tracking and greater ecological relevance. *Can. J. Fish. Aquat. Sci.*, **46**, 494–502.

Latendresse, J.R., II, and Fisher, J.W. (1983) Histopathologic effects of JP-4 aviation fuel on fathead minnows (*Pimephales promelas*). *Bull. environ. Contam. Toxicol.*, **30**, 536–43.

Lemly, A.D. and Smith, R.J.F. (1985) Effect of acute exposure to acidified water on the behavioral response of fathead minnows, Pimephales promelas, to chemical feeding stimuli. *Aquat. Toxicol.*, **6**, 25–36.

Lemly, A.D. and Smith, R.J.F. (1987) Effects of chronic exposure to acidified water on chemoreception of feeding stimuli in fathead minnows (*Pimephales promelas*): mechanisms and ecological implications. *Environ. Contam. Toxicol.*, **6**, 225–38.

MacLeod, J.C. and Pessah, E. (1973) Temperature effects on mercury accumulation, toxicity, and metabolic rate in rainbow trout (*Salmo gairdneri*). *J. Fish. Res. Bd Can.*, **30**, 485–92.

Marui, T., Evans, R.E., Zielinski, B. and Hara, T.J. (1983) Gustatory responses of the rainbow trout (*Salmo gairdneri*) palate to amino acids and derivatives. *J. comp. Physiol.*, **153A**, 423–33.

Maynard, D.J. and Weber, D.D. (1981) Avoidance reactions of juvenile coho salmon (*Oncorhynchus kisutch*) to monocyclic aromatics. *Can. J. Fish. Aquat. Sci.*, **38**, 772–8.

McNicol, R.E. and Scherer, E. (1991) Behavioral responses of lake whitefish (*Coregonus clupeaformis*) to cadmium during preference–avoidance testing. *Environ. Contam. Toxicol.*, **10**, 225–34.

Miller, T.G. and MacKay, W.C. (1980) The effects of hardness, alkalinity, and pH of test water on the toxicity of copper to rainbow trout (*Salmo gairdneri*). *Water Res.*, **14**, 129–33.

Moran, D.T., Rowley, J.C. and Aiken, G. (1987) Trout olfactory receptors degenerate in response to water-borne ions: A potential bioassay for environmental neurotoxicology, in *Olfaction and Taste IX* (eds S.D. Roper and J. Atema), *Ann. N.Y. Acad. Sci.*, **510**, 509–11.

Morrow, J.E. (1973) Oil induced mortalities in juvenile coho and sockeye salmon. *J. mar. Res.*, **31**, 135–43.

Muniz, I.D. (1981) Acidification and the Norwegian Salmon, in *Acid Rain and the Atlantic Salmon*, (ed. L. Sochasky), International Atlantic Salmon Foundation, New York, pp. 65–72.

Nakamura, F. (1986) Avoidance behavior and swimming activity of fish to detect pH changes. *Bull. Environ. Contam. Toxicol.*, **37**, 808–15.

Olsén, K.H. and Höglund, L.B. (1985) Reduction by a surfactant of olfactory mediated attraction between juveniles of Arctic charr, *Salvelinus alpinus* (L.). *Aquat. Toxicol.*, **6**, 57–69.

Osborne, L.L., Iredale, D.R., Wrona, F.J. and Davies, R.W. (1981) Effects of chlorinated sewage effluent on fish in the Sheep River, Alberta. *Trans. Am. Fish. Soc.*, **110**, 536–40.

Pedder, S.J. and Maly, E.J. (1985) The effect of lethal copper solutions on the behavior of rainbow trout, *Salmo gairdneri*. *Arch. environ. Contam. Toxicol.*, **14**, 501–7.

Peterson, R.H., Coombs, K., Power, J. and Paim, U. (1989) Responses of several fish species to pH gradients. *Can. J. Zool.*, **67**, 1566–72.

Purdy, J.E. (1989) The effects of brief exposure to aromatic hydrocarbons on feeding and avoidance behaviour in coho salmon, *Oncorhynchus kisutch*. *J. Fish Biol.*, **34**, 621–9.

Rehnberg, B.G. and Schreck, C.B. (1986a) The olfactory L-serine receptor in coho salmon: biochemical specificity and behavioral response. *J. comp. Physiol.*, **159**, 61–7.

Rehnberg, B.C. and Schreck, C.B. (1986b) Acute metal toxicology of olfaction in coho salmon: behavior, receptors, and odor–metal complexation. *Bull. environ. Contam. Toxicol.*, **36**, 579–86.

Royce-Malmgren, C.H. and Watson, W.H., III (1987) Modification of olfactory-related behavior in juvenile Atlantic salmon by changes in pH. *J. Chem. Ecol.*, **13**, 533–46.

Saunders, R.L. and Sprague, J.B. (1967) Effects of copper–zinc mining pollution on a spawning migration of Atlantic salmon. *Water Res.*, **1**, 419–32.

Solangi, M.A. and Overstreet, R.M. (1982) Histopathological changes in two estuarine fishes, *Menidia beryllina* (Cope) and *Trinectes maculatus* (Bloch and Schneider), exposed to crude oil and its water-soluble fractions. *J. Fish Diseases*, **5**, 13–25.

Sprague, J.B. (1968). Avoidance reactions of rainbow trout to zinc sulphate solutions. *Water Res.*, **2**, 367–72.

Sprague, J.B. and Drury, D.E., (1969) Avoidance reactions of salmonid fish to representative pollutants. *Adv. Water Pollut. Res.*, **4**, 169–79.

Sprague, J.B., Elson, P.F. and Saunders, R.L. (1965) Sublethal copper–zinc pollution in a salmon river – a field and laboratory study. *Int. J. Air Water Pollut.*, **9**, 531–43.

Stober, Q.J., Dinnel, P.A., Hurlburt, D.F. and DiJulio, P.H. (1980) Acute toxicity and behavioral responses of coho salmon (*Oncorhynchus kisutch*) and shiner perch (*Cymatogaster aggregata*) to chlorine in heated sea-water. *Water Res.*, **14**, 347–54.

Stromberg, P.C., Ferrante, J.G. and Carter, S. (1983) Pathology of lethal and sublethal exposure of fathead minnows, *Pimephales promelas*, to cadmium: a model for aquatic toxicity assessment. *J. Toxicol. Environ. Health*, **11**, 247–59.

Sutterlin, A.M. (1974). Pollutants and the chemical senses of aquatic animals – perspective and review. *Chem. Senses Flavor*, **1**, 167–78.

Sutterlin, A.M. and Gray, R. (1973) Chemical basis for homing of Atlantic salmon (*Salmo salar*) to a hatchery. *J. Fish. Res. Bd Can.*, **30**, 985–9.

Sutterlin, A.M. and Sutterlin, N. (1970) Taste responses in Atlantic salmon (*Salmo salar*) parr. *J. Fish. Res. Bd Can.*, **27**, 1927–42.

Sutterlin, A.M. and Sutterlin, N. (1971) Electrical responses of the olfactory epithelium of Atlantic salmon (*Salmo salar*). *J. Fish. Res. Bd Can.*, **28**, 565–72.

Sutterlin, A.M., Sutterlin, N. and Rand S., (1971) The influence of synthetic surfactants on the functional properties of the olfactory epithelium of Atlantic salmon. *Fish. Res. Bd Can., Tech. Rep.*, **287**, 8 pp.

Suzuki, N. and Suzuki, Y. (1981) Changes in epithelial structure and binding of amino acids followed by zinc sulfate irrigation of lamprey olfactory epithelium. *Taste and Smell*, **15**, 48–51.

Takagi, S.F., Iino, M., Yarita, H. and Mori, K. (1978) Ionic stimulation of the olfactory epithelium in the bullfrog and the carp. *Jap. J. Physiol.*, **28**, 129–48.

Tanford, C., and Reynolds, J.A. (1976) Characterization of membrane proteins in detergent solutions. *Biochim. biophys. Acta*, **457**, 133–70.

Tatsukawa, R. and Hidaka, H. (1978) Avoidance of chemical substances by fish – avoidance of detergents by ayu (*Plecoglossus altivelis*). *Nippon Nogei Kagaku Kaishi*, **52**, 263–70.

Thommesen, G. (1983) Morphology, distribution, and specificity of olfactory receptor cells in salmonid fishes. *Acta physiol. scand.*, **117**, 241–9.

Thompson, B.E., and Hara, T.J. (1977) Chemosensory bioassay of toxicity of lake waters contaminated with heavy metals from mining effluents. *Water Pollut. Res. Can.*, **12**, 179–89.

Timms, A.M., Kleerekoper, H. and Matis, J. (1972) Locomotor response of goldfish, channel catfish, and largemouth bass to a 'copper-polluted' mass of water in an open field. *Water Resour. Res.*, **8**, 1574–80.

Tjälve, H., Gottofrey, J. and Björklund, I. (1986) Tissue disposition of $^{109}Cd^{2+}$ in the brown trout (*Salmo trutta*) studied by autoradiography and impulse counting. *Toxicol. environ. Chem.*, **12**, 31–45.

Tsai, C. (1973) Water quality and fish life below sewage outfalls. *Trans. Am. Fish. Soc.*, **102**, 281–92.

Vijayamadhavan, K.T. and Iwai, T. (1975) Histochemical observations on the permeation of heavy metals into taste buds of goldfish. *Bull. Jap. Soc. scient. Fish.*, **41**, 631–9.

Waldock, M.J. and Miller, D. (1983) The determination of total and tributyl tin in seawater and oysters in areas of high pleasure craft activity. *International Council for the Exploration of the Sea, Marine Environmental Quality Committee*, CM 1983/E:12, Essex, England. 17 pp.

Watt, W.D., Scott, C.D. and White, W.J. (1983) Evidence of acidification of some Nova Scotian rivers and its impact on Atlantic salmon, *Salmo salar*. *Can. J. Fish. Aquat. Sci.*, **40**, 462–73.

Weber, D.D., Maynard, D.J., Gronlund, W.D. and Konchin, V. (1981) Avoidance reactions of migrating adult salmon to petroleum hydrocarbons. *Can. J. Fish. Aquat. Sci.*, **38**, 779–81.

Yamashita, S., Evans, R.E. and Hara, T.J. (1989) Specificity of the gustatory chemoreceptors for CO_2 and H^+ in rainbow trout (*Oncorhynchus mykiss*). *Can. J. Fish. Aquat. Sci.*, **46**, 1730–4.

Yoshii, K., Kashiwayanagi, M., Kurihara, K. and Kobatake, Y. (1980) High sensitivity of eel palatine receptors to carbon dioxide. *Comp. Biochem. Physiol.*, **66A**, 327–30.

Author index

Abdel-Latif, A.A. 136–7, 145
Abe, Y. 316
Abraham, M. 122
Ache, B.W. 193–4
Adam, H. 123
Adamek, G.D. 154, 166
Adams, M.A. 223, 301, 305, 314, 317
Adrian, E.D. 49, 54, 155, 166
Adron, J.W. 187, 189, 190, 192, 196, 272, 285–6, 306, 318
Aida, K. 224
Aiken, G. 339
Akabas, M.H. 119, 121
Al-Awqati, Q. 121
Albano, A.G. 325
Albers, P.C.H. 225
Alkon, D.L. 139, 145
Allen, G.H. 262, 266
Allendorf, F.W. 247, 253, 266
Allison, A.C. 40–1, 54, 154, 166
Amouriq, L. 201, 221
Andres, K.H. 40–1, 43, 54
Anestis, I. 332, 335
Angevine, J.B. 83, 100
Anholt, R.R.H. 144–5, 148
Anrep, B. von 117, 121
Anthony, J. 16, 36
Appelget, J. 309, 314
Arai, S. 168, 194, 223, 246
Ariens-Kappers, C.U. 191–2
Armstrong, R.H. 266
Arnold, G.P. 194, 263, 266
Aronson, L.R. 53–4
Arzt, A.H. 124
Astrand, M. 236, 246
Atchison, G.J. 330, 335
Atema, J. 10, 28, 60, 75–6, 80, 84, 91, 100, 123, 172–3, 181, 190, 192, 194, 197, 203, 216–7, 221–2, 240, 245, 248, 271, 275, 285, 289, 291, 293, 296, 300, 301, 310, 314–5
Axel, R. 5, 10, 145, 146
Azuma, S. 97, 100
Atz, J.W. 16, 32

Baatrup, E. 115, 119, 121, 173, 192
Babcock, M.M. 324, 335
Bailey, R.M. 253, 269, 309, 315
Baker, J.P. 336
Bakhtin, E.K. 25, 32
Ball, A.K. 59
Bams, R.A. 230, 244, 256, 264, 266
Bannister, L.H. 23, 25, 27, 32
Bardach, J.E. 60, 76, 79, 80, 100, 114, 121, 172, 191–2, 197, 199, 200, 221, 248, 289, 292, 293, 296–7, 304–5, 307, 315, 325–6, 328, 331, 335
Bardele, C. 73, 78
Barlow, G.W. 243, 246
Barnett, C. 240, 243–4
Barrington, E.J.W. 283, 285
Barry, M.A. 81, 83, 85, 87, 100
Bartheld, C.S. von 40, 44, 45, 51, 53–4, 118, 121, 191–2
Bass, A.H. 40, 44, 51–2, 54, 56
Bateson, W. 288, 291, 296–7, 310, 314
Bauer, D.H. 55
Baum, R.M. 314–5
Bayley, D.L. 146–8
Beamish, R.J. 334–5
Beard, J. 30, 32
Beauchamp, G.K. 200–1
Becker, L.R. 225
Beidler, L.M. 28, 34, 36, 116–7, 121, 163, 166
Beitinger, T.L. 330, 335
Belghaug, R. 159, 166, 240, 244
Bell, W.J. 292, 315
Belousova, T.A. 117, 121
Belvedere, P.C. 222

Benedetti, I. 322, 325
Bentivegna, F. 222
Berg, M. 255, 266
Berghe, L. van den 22, 32
Berridge, M.J. 136–7, 145, 191–2
Bertelsen, E. 21, 32, 36
Bertmar, G. 16, 28, 31–2, 256–7, 266
Bethe, A. 199, 221
Bianchi, S. 33
Bijvank, G.J. 33, 124
Billard, R.E. 224
Birge, W.J. 330, 336
Bishai, H.M. 119, 121
Bisogni, J.J., jun. 336
Bjerring, H.G. 30, 33
Björklund, I. 340
Black, J.A. 330, 336
Blackmore, P.F. 148
Blackwell, J.F. 169
Blake, J.R. 37
Blanchi, D. 118, 121
Blaustein, A.R. 229, 231, 242, 244
Blaxter, J.H.S. 116, 121
Bloom, H.D. 326, 333
Blum, V. 224
Blumenthal, K.M. 285, 315
Blumer, M. 324, 336
Boeke, J. 60, 76
Boekhoff, I. 144–5, 221
Boeynaems, J.M. 157, 167
Bohman, A. 246
Bolte, L. 35
Bonaventura, J. 242, 247
Bone, Q. 16, 33
Bornfeld, N. 33
Bossert, W.H. 212, 228
Boudreau, J.C. 155, 167
Bouffard, R.E. 211–2, 221
Boyle, A.G. 138–9, 141–2, 144–8
Bradley, B. 284–5
Braford, M.R. 40–1, 50, 58
Brand, J.G. 4–5, 127–8, 135, 143, 145–9, 150, 152, 172, 182–4, 186, 192–3, 196–7, 218
Brandt, I. 336
Brannon, E.L. 242, 245, 255, 262, 266, 269
Breer, H. 145, 218, 221
Breipohl, W. 21, 23, 28, 33, 124
Brett, J.R. 240, 245, 301, 315
Breucker, H. 23, 25, 27, 29, 33, 37–9
Brightman, M.W. 25, 36

Brittebo, E.B. 336
Brodal, A. 36
Bromley, P.G. 271, 285
Bronshtein, A.A. 25, 33
Brown, R.E. 200, 221
Brown, S.B. 5, 10, 11, 130, 146, 261, 267, 316–7, 322, 324, 327–9, 330, 336–8
Bruch, R.C. 4, 5, 127, 129, 131–4, 136–9, 141, 144–7, 150, 152, 172, 182, 185–6, 193, 218
Bryant, B.P. 128, 130, 145–8, 177, 178, 192–3, 216, 221, 240, 245
Buck, L. 5, 10, 145–6
Buckland, F. 249, 255, 266
Bujan Varela, J. 34
Burd, G. 56
Burne, R.H. 17, 19, 21–2, 33
Burton, D.T. 337
Busack, C.A. 231, 242, 247
Bütschli, O. 14, 33
Byrd, R.P. 5, 10, 131, 147, 162, 167

Cadena-Roa, M. 286
Cagan, R.H. 5–6, 10, 36, 127–9, 130, 134–5, 142, 145–9, 188, 192–3, 197, 218, 225, 329, 336
Cahn, P.H. 296, 304, 316
Cairns, J. 330–1, 336–7
Cajal, S. Ramón y 43, 55
Camino, E. 121
Canario, A.V.M. 205, 208, 226
Cancalon, P. 25, 27–8, 33, 322, 325–6, 328, 336
Capoor, A.S. 35
Caprio, J. 5, 6, 10, 27, 34, 38, 60, 76, 79, 85–6, 100, 101, 115–7, 121, 122, 127, 129, 131, 137, 147, 155–9, 160, 162–3, 167, 169, 170, 172–4, 176–9, 183–5, 188–9, 190–1, 193–7, 217, 221
Carr, W.E.S. 149, 181, 187–8, 193–4, 241, 245, 275, 285, 296–7, 301, 306–7, 310, 314–5
Carter, S. 340
Caruso, J.H. 21, 31, 33
Case, J. 114, 121, 292, 296, 305, 307, 315
Case, J.F. 203, 227
Castelucci, M. 123
Cetta, F. 211, 221
Chamberlain, K.J. 226, 241, 245, 285, 307, 315

Chaney, T.B. 187, 193, 241, 245, 285, 307, 315
Chapman, D.W. 255, 269
Charest, R. 148
Chase, R. 55
Chaykovskay, A.V. 248
Chen, L.C. 204, 216, 221
Cherry, D.C. 330, 331–2, 336, 337
Chidester, F.E. 255, 266
Child, A.R. 230, 245
Chien, A.K. 216, 222
Chohan, K.S. 59
Christian, J.J. 246
Claas, B. 45, 57
Clemens, W.A. 256, 266
Closs, L.E. 223
Cobb, J.L. 173, 197
Cole, K.S. 215, 222
Cole, L.W. 117, 122
Collins, G.B. 267
Colombo, L. 199, 202–3, 220, 222
Concalon, P. 129, 131, 147
Connes, R. 104, 122
Coombs, K. 339
Cooper, J.C. 7, 10, 269
Cope, O.B. 253, 266
Corazza, L. 34
Cordier, R. 60, 76
Corrigan, A. 56
Cotman, C.W. 83, 100
Courtenay, S.C. 239, 242, 245
Cox, R.A. 36
Craigie, E.H. 255–6, 266
Crickmer, R. 100, 315
Crosby, E.C. 192
Crow, R.T. 201, 222
Crozier, W.J. 117, 122

Dahl, K. 255, 266
Dahlgren, U. 33
Darnerud, P.O. 322, 336
Darrelmann, C. 33
Darwin, C. 199, 222
Davenport, C.J. 173–4, 185, 190–1, 193
Davies, P.H. 327, 336
Davies, R.W. 339
Davis, R.E. 40, 50–1, 55, 58
Daye, P.G. 325, 336
deBruin, J.P.C. 53, 55, 226
Dedual, M. 322–3, 336
Delacy, A.C. 257–8, 263, 266
Delanoy, R. 315

Delfino, G. 27–8, 33
Dell, M.B. 259, 267
Dempsey, C.H. 274, 285, 300, 303, 315
Derby, C.D. 173, 181, 193–4
Demski, L.S. 45, 50, 53, 55, 217, 219, 222
Derivot, J.H. 16, 17, 25, 33
Derscheid, J.M. 17, 19, 33
Desgranges, J.C. 75–6
deSombre, E.R. 223
Dethier, S. 40, 45, 51, 57
Devitsina, G.V. 121
De Vries, M.C. 245
Diaz, J.P. 122
DiJulio, P.H. 340
DiMichele, L. 325, 336
Dinnel, P.A. 340
Dittman, A.H. 261, 263, 269
Dockray, G.J. 283, 285
Dodd, J. 121
Dodson, J.J. 11, 194, 202, 222, 225
Dogiel, A. 28, 34
Dohrn, A. 30, 34
Dolack, M.K. 36, 149
Donaldson, L.R. 262, 266
Dore, F.Y. 11, 225
Dotan, A. 122
Doty, R.L. 221
Döving, K.B. 22, 34, 40, 49, 52, 55, 115, 119, 121, 154, 157, 159, 165–7, 181–2, 187, 194, 197, 202–3, 216, 219, 222, 235, 240–2, 244–5, 247, 260, 263, 266, 296, 304, 314–5
Drickamer, L.C. 200, 222
Driscoll, C.T. 325, 336
Drongelen, W. van 154, 167
Drury, D.E. 331, 340
Dubois-Dauphin, M. 34, 45, 55
Dudek, J. 167, 193
Dudley, C.A. 223
Dulka, J.G. 40, 55–6, 59, 168–9, 205, 207, 219, 222, 224, 227–8
Dumont, J.E. 157, 167
Dunham, P.J. 203, 222
Dupé, M. 33, 300, 316

Easter, S.S., jun. 38
Ebbesson, S.O.E. 40, 51–2, 55, 59
Ebner, V. von 60, 76
Echteler, S.M. 50, 53, 55
Eguchi, A. 316

Ellingsen, O.F. 187, 194, 296, 304, 314–5
Ellis, D.M. 267
Elsdon, J.W.V. 269
Elson, P.F. 340
Elston, R. 29, 34
Emmanuel, M.E. 202, 222
Emmerling, M.R. 75–6
Enger, P.S. 55, 245
Ercolini, A. 33
Erickson, J.R. 27, 34, 129, 147, 158–9, 167
Etingof, R.N. 131–4, 148–9
Evans, H.E. 11, 77
Evans, R.E. 11–2, 25, 28–9, 34, 125, 153, 167–8, 194, 196, 223, 246, 316, 324–5, 336–7, 339, 341
Exton, J.H. 136, 147–8

Fange, R. 273, 285
Fänge, R. 36
Farbman, A.I. 73, 77
Farenholz, C. 103, 122
Fattal, B. 122
Felbeck, H. 186, 194
Fentress, J.C. 191–2
Fauconneau, B. 216, 225
Ferrante, J.G. 340
Fesenko, E.E. 11, 148
File, D.M. 225
Finger, T.E. 4, 11, 40, 51, 53, 55, 75, 77, 85–6, 88–9, 91–2, 94, 95, 97, 100–1, 111, 113–4, 119, 122, 124, 172, 176, 194, 197, 217
Fink, S.V. 216, 228
Fishelson, L. 108, 122
Fisher, J.W. 25, 34, 325, 338
Fisknes, B. 240, 245
FitzGerald, G.J. 243–5, 248
Fletcher, D.J. 229, 245, 272, 285, 300, 316
Folmar, L.C. 332, 336
Forey, P.L. 36
Forward, R.B. 242, 245
Foster, N.R. 331, 337
Fox, H. 108, 109, 122
Francis, M.G. 301, 318
Freeman, L. 330
Frisco, C. 222
Frost, W.E. 256, 266
Fuiman, L.A. 116, 121
Fujimoto, K. 286
Fujita, I. 40–2, 45, 49, 50–3, 55, 56, 58, 156, 165, 167, 218, 222
Fujita, T. 74, 77, 119, 122
Fujiwara, K. 159, 168
Fujiya, M. 192, 335
Fuke, S. 272, 285
Fukuda, A. 225
Fukuda, K. 314, 316
Fukuoka, T. 58
Funakoshi, M. 187, 191, 194, 196
Furuichi, M. 283, 285
Furukawa, A. 278, 285
Fusetani, N. 195–6, 316, 318

Gage, S.P. 92, 100
Gagnon, J. 197
Gagosian, R.B. 203, 221–2
Gandolfi, G. 201, 222
Gans, C. 31, 34, 36
Gardiner, B.G. 36
Gardner, G.R. 322–4, 337
Gardner, W.S. 157, 167
Garside, E.T. 325, 336
Garton, R.R. 330–2, 337
Gawrilenko, A. 29, 34
Geckler, J.R. 334, 337
Gee, J.H. 256, 258, 269
Gemne, G. 154, 165, 167
Gerald, J.W. 55
Gerking, S.D. 258, 267
Gesteland, R.C. 166
Getchell, M.L. 74, 77, 147
Getchell, T.V. 77, 127, 129, 144, 147–8, 153, 167
Giattina, J.D. 330–2, 336–7
Gibbs, J. 311, 316
Giduck, S.A. 273, 277, 283, 285
Gilman, A.G. 133–5, 147
Gleeson, R.A. 149, 193, 203, 223
Godet, R. 300, 316
Goettle, J.P. 336
Goetz, F.W. 59, 169, 211, 221, 223, 226–7
Goetz, R. 33, 205
Goh, Y. 159, 167–8, 176, 179, 187–9, 190, 194, 272, 285, 306, 316
Gold, G.H. 137, 141–2, 144, 146, 148, 152, 168
Goldschmid, A. 123
Gondek, A.R. 315
González Santander, R. 29, 34
Göppert, E. 35
Gorbman, A. 220, 225, 300, 319

Gordon, K.D. 173, 194
Gottofrey, J. 340
Granie-Prie, M. 122
Granstrom, E. 211, 223
Grandt, D. 33
Grant, P.T. 196, 286
Gray, R. 262, 269, 334, 340
Graziadei, P.P.C. 28, 34, 326, 337
Greene, C.W. 288, 291, 298, 316
Greene, G.L. 223
Grillo, M. 147
Grober, M.S. 45, 56
Groenix Van Zoelen, R.F.O. 225
Gronlund, W.D. 341
Groot, C. 239, 241, 245, 260, 267
Grove, D. 273, 285
Grover-Johnson, N. 73, 77
Groves, A.B. 256–8, 267
Gruber, D. 296, 304, 316, 318
Gruber, S.H. 39, 170
Gunnerød, T.B. 262, 267
Gunning, G.E. 258, 267
Guardabassi, A. 118, 121
Guselnikov, V.I. 45–6, 56
Guselnikova, K.G. 45–6, 56

Haider, G. 324, 337
Haines, T.A. 334, 337
Hall, L.W., jun. 332, 337
Haller, C.J. 119, 124
Halpren-Sebold, L. 226
Halpren-Sebold, L.R. 45, 56
Halvorsen, M. 256, 258, 262–3, 267
Hama, K. 40–1, 43, 56–7
Hamilton, W.D. 229, 231, 242, 245
Hammar, J, 236, 245
Hanyu, I. 224
Hara, T.J. 2, 5, 6, 9, 10–2, 18, 25, 29, 30, 34, 38–9, 40, 56, 59, 60, 77, 125, 129, 130, 146, 149, 153–9, 160–2, 164, 167–9, 170, 175–6, 178–9, 182–3, 190, 194–6, 204, 205, 208, 217–8, 220, 222–3, 226–7, 231, 240, 242, 245–8, 261, 267, 298, 300, 304, 307, 316–7, 322, 324, 326–9, 331–3, 336–9, 340–1
Harada, K. 181, 190, 194, 301–3, 305, 307, 314, 316
Harada, S. 173, 195, 196
Harcup, M.F. 256, 258, 267
Haredn Jones, F.R. 267
Harpaz, S. 147, 193

Harris, J.E. 108, 122
Harrison, H.M. 309, 315
Hartt, A.C. 259, 267
Hartwell, S.I. 330, 333, 337
Hasegawa, T. 169
Hashimoto, T. 122
Hashimoto, Y. 189, 195–6, 286, 303, 307, 316, 318
Hasler, A.D. 7, 11, 12, 216, 223, 230, 256–7, 259, 261, 267, 269, 270, 331, 334, 338
Hatt, H. 126, 127, 147, 173, 181, 195
Hattanda, M. 338
Haukkamaa, M. 218, 223
Havre, N. van 243–5, 248
Hawkes, J.W. 23, 34, 324, 338
Hayama, T. 85, 86, 100, 195
Haynes, K. 225
Haynes, L.J. 298, 301–2, 307, 316
Heggberrget, T. 267
Helander, A. 197, 247, 319
Helton, C.D. 34
Henrikson, R.C. 23, 34
Henry, M.G. 335
Hepher, B. 283, 285
Hernádi, L. 25, 34
Herrick, C.J. 3, 11, 60, 75, 77, 79, 80, 86, 88, 89, 91, 100, 191, 195
Hert, E. 243, 246
Hidaka, H. 331–3, 338, 340
Hidaka, I. 174–5, 177–8, 181, 187–9, 190, 196, 198, 275, 285, 306, 316, 318, 328–9, 338
Hilgers, H. 39
Hirata, K. 74, 77
Hiratsuka, T. 123
Hirsch, P.J. 7, 10
Hoagland, H. 172, 195
Hiyama, Y. 256–7, 267
Hoar, W.S. 230, 246
Hobden, B.R. 168
Hobson, E.S. 292–3, 297, 317
Hodgson, E.S. 289, 292–3, 301, 303, 307, 317–8
Hodgson, T. 56
Hoese, D. 289, 296, 298, 305, 317
Hoese, H.D. 289, 296, 298, 305, 317
Höglund, L.B. 232, 236, 246, 333, 339
Holl, A. 16, 17, 19, 21–2, 25, 28, 31, 34, 37, 108, 124, 192, 335
Holland, K. 80, 100, 289, 315, 317
Holland, K.N. 189, 195, 289, 301, 317
Holley, A. 29, 34, 55, 167

Holmberg, K. 159, 167
Holmes, W.G. 229, 246
Holmgren, N. 41, 43, 45, 50, 56
Honda, E. 124
Honda, H. 202, 216, 223
Hong-qi, Z. 314
Horall, R.M. 267, 269
Horning, W.B. 337
Hose, J.E. 331, 338
Hosokawa, H. 285
Houman, H. 122, 196
Huber, G.C. 192
Hughes, G.M. 108, 122
Huisman, E.A. 225
Hunt, S. 108, 122
Hunter, J.R. 216, 223
Huntsman, A.G. 264, 270
Huque, T. 136, 138, 144–8
Hurk, R. van den 169, 204, 216, 224–5, 227–8
Hurlburt, D.F. 340
Hutchings, J.A. 255, 267
Huxley, T.H. 16, 35
Hvidsten, N.A. 267

Ichikawa, M. 25, 27, 35, 40–1, 43, 45, 51, 56, 58
Idler, D.R. 224, 318
Iger, Y. 120, 122
Iino, M. 340
Ikeda, I. 272, 285
Ikeda, M. 318
Ikehara, W. 315
Ina, K. 272, 275, 286, 306, 307, 317–8
Ingersol, D.W. 202, 224
Inone, T. 318
Iredale, D.R. 339
Irvine, R. 137, 145
Ishida, H. 286
Ishida, Y. 174–5, 178, 181, 192, 195
Ishioka, K. 267
Ishiwata, N. 273, 286
Ito, H. 51, 53, 56, 58–9, 101
Itoh, Z. 283, 286
Iwai, T. 31, 35, 322, 324, 340
Iwanaga, T. 38

Jacquin, M.F. 198
Jahn, L.A. 29, 35, 242, 246, 256–7, 267
Jakubowski, M. 3, 11, 21, 22, 25, 35, 60, 73–4, 75, 77, 104–5, 107–9, 112, 116, 120, 122

Jennes, L. 57
Jensen, E.V. 218, 223
Jin, J.H. 337
Johannesson, B. 262, 267
Johansen, K. 35
Johansen, P.H. 14, 201, 223
Johnsen, P.B. 79, 100, 266, 293, 297, 301, 305, 314, 317
Johnson, D.W. 334, 338
Johnson, P.C. 310, 317
Johnston, J.A.A. 269
Jonas, R.E.E. 318
Jones, D.T. 133, 137, 147
Jones, J.W. 262, 267
Jones, K.A. 8, 9, 188, 195, 271, 297–9, 302, 304, 307, 317, 333, 338
Jonsson, B. 249, 250, 267
Jourdan, F. 34
Junger, H. 80, 101

Kaasa, H. 255, 267
Kah, O. 45, 56
Kajihara, T. 267
Kalinoski, D.L. 128–9, 131–6, 143, 144, 145–8, 186, 192–3
Kallius, E. 35
Kamchen, R. 333, 338
Kamekawa, S. 287
Kamo, N. 198
Kang, J. 188, 195
Kanno, T. 77, 122
Kanwal, J.S. 4, 85–6, 88–9, 91–2, 94–5, 97, 100–1, 172–4, 176, 189, 191, 195, 217
Kapoor, B.G. 2, 11, 60, 75, 77
Kapoor, A.S. 22, 35
Kare, M.R. 146, 285
Karlson, P. 7, 11, 200, 223
Karlsson, I. 197, 247, 319
Kasahara, Y. 101, 194
Kashiwayanagi, M. 154, 168, 341
Kasumyan, A.O. 8, 11
Kato, F. 255, 268
Kato, M. 58
Katsuki, Y. 118, 122–3
Katz, U. 117, 122
Kaufman, B. 55
Kawakita, K. 194
Kawamura, Y. 100, 102
Keenleyside, M.H.A. 243, 246
Kelley, M.J. 218, 223
Kendall, J.I. 122

Kendall, R.L. 268
Kimura, I. 286
Kindahl, H. 211, 223
Kirschenblatt, J. 200, 223
Kishi, K. 56
Kitamura, S. 159, 168
Kitoh, J. 122, 195
Kittredge, J.S. 202–3, 219, 223–4
Kiyohara, S. 101, 116, 118, 122, 173, 175–9, 181, 183–8, 191, 195–8
Klaprat, D.A. 9, 60, 77, 168, 298, 325, 328, 338
Kleerekoper, H. 1, 11–5, 22, 29, 35, 40, 56, 291–3, 295, 298, 300, 303, 307, 309, 317–9, 340
Klumpp, S. 149
Knigge, K.M. 53, 55
Knudsen, E.I. 86, 101
Kobatake, Y. 198, 341
Kobayashi, H. 159, 168, 194
Kobayashi, M. 205, 224
Kobayashi, S. 77, 122
Kobayashi, K. 316
Kohbara, J. 185–6, 196
Kolliker, A. 103, 123
Kolmer, W.C. 60, 77
Komatsu, K. 123
Konchin, V. 341
Konev, S.V. 149
Konishi, J. 172, 196, 329, 338
Konishi, M. 86, 101
Konosu, S. 187, 195–6, 275, 285–6, 307, 316, 318
Kosaka, T. 40–1, 43, 56–7
Kotrschal, K. 11, 80, 101, 104, 107–9, 112–5, 117–8, 121–5
Korver, R.M. 330, 338
Koyama, T. 310
Koyama, Y. 52–3, 56
Krapivinskaya, L.D. 11, 148
Krautgartner, W.D. 11, 123–4, 148
Krueger, J.M. 128, 148
Kruse, K.C. 301, 318
Kubomatsu, T. 195, 316
Kuhme, W. 243, 246
Kumai, H. 276–7, 286, 319
Kumazawa, T. 146, 149, 184, 196–7
Kunysz, E. 21, 22, 35
Kurihara, K. 134, 148, 168, 169, 198, 341
Kurosaki, Y. 316
Kusunoki, M. 58
Kux, J. 22, 35, 39

Kyle, A.L. 49, 52–3, 57, 59, 165, 168–9, 211, 213, 218–9, 224, 227
Kyle, H.M. 21, 35

LaBar, G.W. 256–7, 268
Labarca, P. 144, 148
Lambert, J.G.D. 204, 216, 224–8
LaMorte, V.J. 148
Lancet, D. 127, 129, 137, 144, 148, 218–9, 224
Lane, E.B. 103–5, 108–9, 111–6, 122–3, 125
LaRoche, G. 322, 337
Laraña Solé, A. 34
Larrick, S.R. 336–7
Lasker, R. 224
Latendresse, J.R. II 325, 338
Laugwitz, H.J. 33
Laumen, J. 216, 224
Law, Y.M.C. 168, 337
Lee, C.T. 202, 224
Lee, G.F. 157, 167
Leftheris, K. 177–8, 193
Le Magnen, J. 34
Lemly, A.D. 298, 318, 325, 333, 338
Lemm, C.A. 306–7, 319
Lenkevich, M.J. 337
Leonard, C.M. 45, 59, 79, 101
Leuko, A.V. 149
Levenick, C.V. 76
Levine, R.L. 40, 45, 51, 57
Leydig, F. 79, 101
Li, W. 21, 31, 37
Lieramann, K. 21–2, 35
Liley, N.R. 7, 11, 40, 57, 199, 200–1, 216, 222, 224–5, 227
Limberger, D. 243, 248
Lindsey, C.C. 255–6, 268
Lindsey, J. 224
Liron, N. 37
Littleton, J.T. 181, 196
Little, E.E. 189, 196, 297, 301, 318
Lockle, D.M. 244, 246
Lorz, H.W. 255–7, 268
Lowe, G.A. 41, 43, 45, 57
Lubosch, W. 31, 35
Luciano, L. 119, 123
Ludwig, C. 5, 10, 155, 166
Lund, B.-Q. 336
Luscher, M. 7, 11, 200, 223
Lynch, C.J. 136, 148
Lythgoe, J.N. 220, 225

Macdonald, S. 168, 194, 223, 240, 245–6, 261, 267, 337
Machidori, S. 255, 268
MacKay, W.C. 327, 339
Mackie, A.M. 187–9, 190, 192, 196, 271–2, 285–6, 306, 318
MacKinnon, D. 301, 315
Mackay-Sim, A. 33
MacLeod, N.K. 41, 43, 45, 57
Madison, D.M. 246
Maeda, H. 316
Mair, R.G. 166
Maiwa, T. 267
Maller, J.L. 219, 224
Maly, E.J. 330, 339
Malyukina, G.A. 121
Maneckjee, A. 219, 224
Marc, R.E. 27, 36, 129, 148
Marcanto, A. 222
Marchaterre, M.A. 56
Margolis, F.L. 127, 147–8
Margolis-Nunno, H. 226
Marshall, A.M. 30, 35
Marshall, N.B. 16, 21, 31, 33, 35
Martíinez Cuadrado, G. 34
Marui, T. 6, 38, 85–6, 101, 122, 127, 168, 173, 175–9, 186–7, 189, 190–1, 194, 196–7, 223, 246, 329, 339
Marusov, Y.A. 289, 296, 318
Masai, H. 101
Mason, J.R. 124
Matis, J. 318–9, 340
Matoltsy, A.G. 23, 34
Matsuda, H. 338
Matsui, H. 272, 286, 306–7, 317
Matsui, K. 197, 287
Matsui, T. 12
Matsuo, R. 102
Matsushima, 58
Matsushima, T. 169
Matsushita, T. 286
Mattei, X. 33
Matthes, E. 17, 35
Mattie, D.R. 34
Matthewson, R.F. 289, 292–3, 301, 303, 307, 317–8
Mayers, S.P. 250–1, 284, 286
Maynard, D.J. 332, 339, 341
McBride, J.R. 291, 296, 300, 311, 318
McCleave, G.P. 194
McCleave, J.D. 256–7, 268
McKaye, K.R. 243, 246

McKinnon, J.S. 216, 224
McLeod, J.C. 327, 339
McLeod, P. 34
McIsaac, D.O. 264, 268
McMahon, T.E. 291, 296, 318
McNally, B. 308, 318
McNicol, R.E. 330, 339
Mearns, K.J. 188, 197, 289, 304, 318
Mees, F. 21, 35
Meinel, M. 34
Meis, S. 120, 124
Melinkat, R. 19, 22, 33, 35, 39
Melrose, D.R. 204, 224
Meng, Q. 15, 36
Meredith, M. 2, 12, 165, 170
Merkel, F. 79, 101
Meteiller, R. 274, 286
Meyer, D.L. 45, 54–5, 118, 121, 191–2
Meyer, J.H. 201, 224
Meyer, M. 103, 112, 123
Michel, W. 178, 181, 186, 196–7
Michener, C.D. 229, 245
Miller, D. 332, 341
Miller, M.G. 198
Miller, R.B. 256–7, 268
Miller, S.E. 319
Miller, T.G. 327, 339
Millot, J. 16, 36
Mills, D. 264, 268
Milstead, M.L. 193
Mimura, T. 287
Minami, T. 225
Misono, M. 318
Mitchell, A.I. 271–2, 286, 306, 318
Miyamoto, T. 148
Miyoshi, M. 38
Miyoshi, S. 38
Mizuno, Y. 225
Mizushima, U. 280, 286
Moate, 109
Mola, L. 325
Möller, P.C. 28, 36
Monti Graziadei, G.A. 28, 34, 326, 337
Moore, A. 173, 197, 220, 225
Moran, D.T. 25, 27, 37, 322, 326, 339
Morgensen, J. 300, 303, 307, 317
Mori, K. 43, 57, 340
Mori, Y. 196
Morin, P.P. 8, 11, 220, 225
Morita, Y. 56, 58, 82, 85, 87, 89, 91, 94, 100–1
Morrill, A.D. 103, 123
Morris, J. 55

Morrow, J.E. 332, 339
Moss, R.L. 223
Moulton, D.G. 28, 36, 49, 57, 221
Mozell, M.M. 49, 57
Mrowka, W. 243, 246
Mugford, A. 221
Muller, J.F. 27, 36, 129, 148
Muniz, I.D. 334, 339
Munshi, J.S. 108, 122
Munz, H. 45, 57
Murai, T. 246
Murakami, T. 40, 51, 53, 58, 101
Murofushi, S. 306-7, 318
Murray, R.G. 60, 73-4, 77
Myers, G.S. 268
Myrberg, A.A., jun. 243, 246

Nabekura, J. 218, 225
Nada, O. 74, 77-8, 119, 123
Nadji, M. 223
Nagahama, Y. 219, 225
Naidu, P. 226
Nakahara, S. 124
Nakahura, H. 38
Nakamura, F. 331, 339
Nakamura, I. 31, 35
Nakamura, M. 12, 286, 319
Nakamura, T. 141-2, 144, 148, 152, 168
Nakano, Y. 38
Nakao, Y. 274, 275, 287
Naughton, J.J. 319
Neiheisel, T.M. 337
Neill, W.H. 194, 292, 318
Nelson, J.P. 318
Nelson, K. 216, 225
Netherton, J.C. 285, 315
Neufeld, R.J. 332, 335
Nevitt, G.A. 22, 36
Nieuwenhuys, R. 40-1, 50, 58, 154, 168
Nickum, J.G. 34
Nielsen, J.G. 21, 32, 36
Nilsson, A. 283, 286
Nilsson, N.-A. 246
Nordeng, H. 8, 11, 55, 230, 245-6, 249, 250-6, 259, 260, 268
Norgren, R. 79, 101
Norman, J.R. 31, 36
Northcote, T.G. 255-7, 268
Northcutt, R.G. 31, 34, 36, 40-1, 44, 50, 55, 58, 81, 102, 165, 167, 217, 222
Norton, L.E. 81, 87, 100

Nose, T. 195-6, 246, 316, 318
Notenboom, C.D. 225
Novoselov, V.I. 5, 11, 131-2, 134, 148
Nunez-Rodriguez, J. 56
Nyu, N. 195

Oakley, B. 166, 245
O'Connor, J.F. 256, 268
Oda, S. 195
Oelschläger, H.A. 22, 36
Ogata, H. 159, 168, 240, 246
Oggolter, H. 33
O'Hara, R.K. 229, 231, 242, 244
Ohno, T. 159, 162, 169
Ohsugi, T. 195, 306, 316, 318
Ojha, P.P. 22, 35
Ojima, H. 56
Oka, Y. 40-1, 43, 45, 52, 57-9
Okumoto, N. 58, 169
Olbe, L. 273, 278, 286
Olla, B.L. 319
Olmstead, J.M.D. 289, 297, 298, 301-2, 319
Olsén, K.H. 7, 115, 123, 181, 197, 231-3, 235-8, 241, 246-7, 262, 307, 319, 333, 339
Ono, R.D. 108, 123
Onoda, N. 118, 123
Oomura, Y. 225
Oordt, P.W.J. van 169, 224, 225-6
Oppel, A. 60, 77
Orkand, P.M. 36, 149
Orton, H. 262, 267
Osawaa, Y. 35
Osborne, L.L. 334, 339
Oshima, K. 200, 225, 300, 319
Osse, J.W. 102
Ostretsova, I.G. 149
Ottoson, D. 153-5, 169
Overstreet, R.M. 324, 326, 339

Pain, U. 339
Panchen, A.L. 31, 36
Pankhurst, N.E. 220, 225-6
Pappas, G.D. 75, 77
Parfenova, E.V. 131-2, 134, 148
Paris, J. 122
Park, Y.S. 145
Parker, G.H. 4, 11, 75, 77, 79, 101, 115-8, 123, 292, 298, 308, 319
Parrett, T. 38
Partridge, B.L. 201-2, 209, 225
Partridge, L.R. 36

Patterson, C. 36
Patterson, R.L.S. 224
Paulos, L.M. 34
Pavlov, D.S. 8, 11
Pawson, M.G. 241, 247, 289, 292, 304, 307, 319
Pearson, W.H. 296, 319
Peddar, S.J. 330, 339
Pellegrini, V. 36
Perlmutter, A. 333, 336
Pern, U. 224
Pessah, E. 327, 339
Peter, R.E. 52–3, 56–7, 168, 211, 219, 224–5, 227
Peters, R.C. 4, 11, 115, 119, 123–4, 169, 225
Peterson, R.H. 331, 339
Pevsner, J. 149
Pevzner, R.A. 11, 77, 109, 124
Pfefferkorn, G.E. 33
Pfeiffer, W. 16, 31, 36
Pickering, Q.H. 337
Pinkney, A.E. 337
Pipping, M. 17, 22, 36
Pollack, E.I. 217, 225
Powell, T.P.S. 43, 58
Power, G. 256, 268
Power, J. 339
Powers, J.B. 50, 58
Preston, R.L. 186, 197
Price, J.L. 43, 58
Price, S. 134, 148
Prochnow, J.A. 311, 319
Purdy, J.E. 332, 339
Putney, J.W. 139, 148
Puzdrowski, R.L. 118, 124
Pyatkina, G.A. 25, 27, 29, 36

Quinn, T.P. 231, 239, 242, 245, 247, 260–4, 266–9

Rabinowitz, J.L. 136, 146–8
Raderman-Little, R. 74, 77
Rahamin, E. 122
Raisman, G. 50, 58
Raleigh, R.F. 255, 269
Rand, G. 298, 319
Rand, S. 340
Rao, Ch. V. 219, 225
Rasmussen, H. 139, 145
Reale, E. 123
Reed, H.C.B. 224
Reed, R.R. 133, 137, 147

Reese, T.S. 25, 36
Rehn, B. 33
Rehnberg, B.G. 131–2, 148, 329, 333, 339
Reinke, W. 29, 36
Resink, J.W. 159, 160, 169, 204–5, 217–8, 220, 224–5
Restrepo, D. 141, 144, 146, 148
Reutter, K. 3, 11, 60, 73–5, 77–8, 108–9, 113, 116, 120, 124–5
Reynolds, J.A. 328, 340
Rhein, L.D. 27, 36, 129, 130, 149, 218, 225
Ribbink, A.J. 243, 247
Richards, I.S. 240, 247
Ricker, W.E. 230, 247, 251, 269
Riehl, R. 23, 37
Rieke, G.K. 319
Ritchie, D.G. 36
Rittschof, D. 242, 245, 247
Rivier, J.E. 56
Roberts, N.J. 296, 319
Robinson, E.L. 337
Robinson, J.J. 167, 193
Rogers, D.C. 119, 124
Röhlich, P. 25, 34
Rolls, E.T. 102
Roper, S.D. 60, 72, 75, 78, 113, 120, 124, 127, 149
Rose, J.E. 76
Rosen, D.E. 31, 36
Rosenblum, P.M. 218, 225
Ross, L.S. 14, 38
Rossi, A.C. 216, 225
Roth, A. 109, 124
Rottiers, D.V. 306–7, 319
Rowley, J.C., III 25, 27, 37, 339
Royce-Malgren, C.H. 331, 339
Rubio Sáez, M. 34
Rückert, K. 33
Rulli, R.D. 131, 137–8, 146
Ruska, H. 123
Russell, I.J. 118, 124
Ruyetruyet, J.P. 286
Ryman, N. 230, 247, 253, 266, 269

Saglio, P. 216, 225
Saidel, W.M. 51, 55
Sakamoto, N. 56
Sandheinrich, M.B. 335
Sangster, A.W. 316
Sano, A. 318
Saruya, R. 318

Sato, R. 267
Sato, Y. 155, 169
Satou, M. 3, 40–1, 43–9, 50, 52–3, 55–9, 156, 165, 169, 170, 217–8
Saunders, R.L. 253, 256, 258, 269, 334, 339, 340
Savage, G.E. 53, 58
Scalia, F. 40, 51, 59
Scarfe, A.D. 291, 303, 319
Scheer, B.T. 251, 255, 269
Scheich, H. 55
Scheier, A. 337
Scherer, E. 317, 330, 338–9
Schiffman, S.S. 173, 197
Schmalhausen, I.I. 31, 37
Schofield, C.L. 336
Scholz, A.T. 7, 11, 220, 223, 259, 261, 267, 269, 331, 334, 338
Scholz, K. 56
Schoonen, W.G.E.J. 204, 224–6, 228
Schreck, C.B. 131–2, 148, 329, 333, 339
Schreibman, M.P. 45, 56, 220, 226
Schreiner, K.E. 104, 124
Schulte, E. 13, 17, 23, 25, 27–8, 34, 37, 108, 124
Schultz, E. 149
Schultze, M. 37
Schwanzel-Fukuda, M. 45, 55
Scott, A.P. 205, 208, 220, 225–6
Scott, C.D. 341
Scott, G.I. 76
Seeley, R.J. 336
Segaar, J. 53, 59, 217, 226
Segil, N. 56
Seifert, P. 203, 226
Sellick, P.M. 118, 124
Selset, R. 49, 55, 165, 167, 182, 194, 197, 219, 222, 240–1, 245, 247–8, 260, 269
Sennewald, K. 197
Sewertzoff, A.N. 29, 37
Shapovalov, L. 256, 269
Shaulsky, G. 147
Sheiko, L.M. 128, 149
Sheldon, R.E. 41, 43, 45–6, 50, 59, 116, 124, 165, 169
Shepherd, G.M. 28, 37, 43, 50, 59
Shepherd, R. 247
Sherman, P.W. 229, 246
Shibata, T. 318
Shiga, T. 51
Shimamura, A. 78

Shimamura, S. 124
Shimeno, S. 285, 287
Shiraishi, A. 169
Shoji, T. 168
Sibbing, F.A. 82, 87, 89, 102
Sienkiew, Z.J. 102
Silver, W.L. 4, 11, 38, 111, 114, 117, 119, 124, 153, 155, 159, 169, 217, 226
Simon, S.A. 148
Sindey, J.R. 336
Sklar, P.B. 149
Slapnick, S.M. 76
Sleigh, M.A. 29, 37
Smith, A.B. 223
Smith, L.L. 309, 314
Smith, L.S. 271, 286
Smith, R.J.F. 7, 11–2, 215, 222, 298, 318, 325, 333, 338
Smith, R.S. 244, 248
Snyder, S.H. 137, 144, 149
Sobkowicz, 76
Solangi, M.A. 324, 326, 339
Solemdal, P. 197, 319
Solomon, D.J. 199, 226
Sorensen, P.W. 3, 5–7, 40, 49, 55–6, 59, 157–9, 160–1, 163–5, 167–9, 200, 202, 205–8, 211–3, 215–9, 220, 222, 224–8
Sosin, M. 309, 319
Spencer, M. 203, 227
Splechtna, H. 39
Sprague, J.B. 330–1, 334, 338–9, 340
Springer, A.D. 45, 59
Stabell, O.B. 7, 8, 230, 236, 239, 240, 242, 248, 255, 256, 258, 260–1, 262–3, 267, 269
Stacey, N.E. 40, 49, 55–7, 59, 165, 167–9, 199, 200, 205, 208, 210–1, 213, 215–7, 219, 222, 224–8
Ståhl, G. 247, 253, 266, 269
Starling, E.H. 200, 227
Steele, C.W. 319
Steenderen, G.W. van 123
Steinlen, S. 139, 149
Stell, W.K. 45, 50, 59
Stephen, C.E. 337
Steven, D.M. 291–2, 296, 307, 316, 319
Stober, Q.J. 332, 340
Stoffel, R.J. 332, 338
Stone, B.M. 301, 318
Stott, B. 258, 269

Strahan, R. 14, 35
Striedter, G. 81, 93, 95–6, 102
Stromberg, P.C. 322–3, 340
Strottman, J. 145
Stuart, T.A. 255–6, 261–2, 269
Stumpf, W.E. 57
Su, J. 21, 31, 37
Suga, N. 86, 102
Sugimoto, K. 149
Sugiyama, T. 316
Sutterlin, A.M. 5, 12, 156, 159, 162, 169, 172, 176, 179, 197, 240, 248, 262, 269, 293, 296, 302, 304–7, 310, 319, 322, 326–9, 332, 334, 340
Sutterlin, N. 5, 12, 156, 159, 162, 169, 172, 176, 179, 197, 240, 248, 326–9, 340
Suzuki, N. 5, 12, 151–3, 155–7, 159, 169, 170, 172, 197, 322, 326, 329, 340
Suzuki, Y. 28, 37, 322, 326, 329, 340
Suyama, K. 267
Sveinsson, T. 5, 12, 153, 159, 160–1, 168, 170
Sykes, P.A. 271, 285
Sze, P.Y. 218, 227

Taborsky, M. 243, 248
Taft, A.C. 256, 269
Takada, O. 272, 286
Takagi, S.F. 58, 327, 340
Takahashi, F.T. 203, 223–4
Takahashi, S. 38
Takaoka, O. 305, 319
Takashima, R. 286
Takeda, M. 9, 28, 37, 190, 197, 271, 272–4, 285–7, 319
Takei, K. 169
Takeuchi, H. 169
Takii, K. 9, 197, 274–6, 278–9, 280–2, 286–7, 319
Tamura, T. 159, 167–8, 176, 179, 187–9, 190, 194–5, 272, 285, 306, 316
Tanaka, S. 255, 270
Tanford, C. 328, 340
Taniuchi, T. 267
Tareilus, E. 145, 221
Tash, J.C. 291, 296, 318
Tate, S. 147
Tatsukawa, R. 331–3, 338, 340
Tavolga, W.N. 200–1, 226

Taylor, G. 317
Taylor, M.H. 325, 336
Teeter, J.H. 79, 100, 131–2, 137–9, 141–4, 146–9, 184, 186, 189, 195–7, 289, 293, 297, 301, 317
Teichmann, H. 17, 29, 37, 292, 319
Terloun, A. 102
Terry, M. 224
Tester, A.L. 15, 37, 288, 291, 296–8, 301–2, 307, 311, 319
Theisen, B. 14–6, 18–9, 21–3, 25, 27–9, 31, 37, 39
Thornhill, R.A. 25, 28, 38
Thomas, L. 2, 12
Thomas, S. 316
Thommesen, G. 22, 27, 34, 38, 49, 59, 129, 149, 154–5, 157, 162, 164, 167, 170, 194, 218, 222, 227, 245, 328, 340
Thompson, B.E. 327–8, 331, 336–7, 340
Thompson, H.W. 181, 193
Thorpe, J.E. 261, 270
Threatte, R.M. 285
Tilseth, S. 197, 319
Tilzey, R.D.J. 255–6, 270
Timms, A.M. 309, 319, 330, 340
Tinsley, R. 309, 319
Tjälve, H. 322, 340
Tobin, T.R. 292, 315
Todd, J.H. 100, 190, 197, 199, 200, 221, 240, 248, 315
Toft, R. 256, 257, 266, 270
Tolson, G.M. 239, 247, 260, 269
Tomlinson, N. 318
Torii, K. 188, 197
Towle, A.C. 218, 227
Toyoshima, K. 74, 78, 114, 124
Toyota, M. 338
Trapido-Rosenthal, H.G. 127, 149
Trefethen, P.S. 267
Trujillo-Cenoz, O. 13, 38
Tsai, C. 334, 340
Tscharntke, H. 109, 124
Tsukahara, H. 278, 285
Tucker, D. 5, 12, 60, 78, 116, 124, 153, 156–7, 159, 169, 170, 183, 197, 261, 270

Uchida, M. 196
Ueda, K. 18, 21, 25, 27–8, 31, 35, 38, 41, 45, 51, 55–9, 155, 169, 170
Ueda, S. 56

Author index

Uehara, K. 23, 38
Ui, M. 149
Urade, R. 147
Urayasu, M. 274
Uskov, I.A. 248
Uskova, Y.T. 240, 248
Utter, F.M. 266

Vale, W.W. 56
Van Der Kraak, G.J. 55, 59, 205, 222, 227–8
van der Meche-Jacobi, M.E. 226
van Erkel, G.A. 14, 15, 35
Van Zoelen, G.A. 228
Vanegas, H. 53, 56, 59
Vijayamadhavan, K.T. 322, 324, 340
Vodyanoy, V. 141–2, 149
Vogel, W.O.P. 109, 124
Volotvskii, I.D. 149
Voorthuis, P.K. 169, 225

Waldman, B. 229, 242, 244, 248
Waldoek, M.J. 332, 341
Walker, S.E. 59
Wantanabe, K. 202, 215, 219, 228
Watson, W.H., III 331, 339
Watt, W.D. 334, 341
Waxman, S.G. 75, 77
Weber, D.D. 332, 334, 339, 341
Weber, E.H. 79, 102
Webster, D.A. 334, 338
Weel, P.B. van 319
Wegert, S. 177, 184, 197
Weisbart, M. 224
Weitzman, S.H. 216, 228
Wendelaar Bonga, S.E. 120, 124
Westerberg, H. 263, 266, 270
Whitear, M. 3, 4, 11, 23, 38, 60, 73–5, 77, 103–4, 107–9, 111–9, 120–3, 125
White, H.C. 256, 264, 270
White, W.J. 341
Whoriskey, F.G., jun. 244, 248
Wiedersheim, R. 19, 38
Wiley, S. 186, 194
Wilkins, N.P. 264, 270
Williams, L.G. 247
Williams, R. 267
Williamson, J.R. 137, 149

Wilson, E.O. 212, 228
Wilson, S.W. 29, 38
Wilton, R. 317
Winans, S.S. 50, 58
Winn, H.E. 202, 226, 296, 319
Wirsig, C.R. 45, 59
Wisby, W.J. 7, 12, 256–7, 259, 267, 270
Witt, M. 74, 78, 116, 125
Wrona, F.J. 339
Wullimann, M.F. 81, 94, 102
Wunder, W. 289, 320
Wysocki, C.J. 2, 12, 165, 170

Yamada, J. 23, 38
Yamagishi, M. 28, 38
Yamaguchi, K. 58, 155, 170
Yamamori, K. 6, 12
Yamamoto, M. 13, 18, 21, 27–8, 31, 38, 117, 125, 129, 149, 154, 170
Yamamoto, T. 97, 100, 102
Yamamoto, Y. 195
Yamashita, S. 6, 12, 118, 122, 125, 184, 195–6, 198, 329, 341
Yamazaki, F. 202, 215, 219, 228
Yanagisawa, K. 118, 122
Yarita, H. 340
Yaxley, S. 97, 102
Yin, M. 15, 36
Yokota, S. 328, 338
Yokouchi, C. 198
Yonezawa, H. 196
Yoshii, K. 169, 173, 175, 177–9, 181, 183–4, 187–9, 198, 329, 341
Yugari, Y. 287
Yumikari, K. 284, 287

Zala, C.A. 240, 245
Zeger, S. 337
Zeiger, W.N. 5, 6, 10, 130, 147, 329, 336
Zeiske, E. 2, 15, 17, 19, 21–3, 25, 28–9, 31, 33, 35, 37, 39, 41, 129, 159, 164, 170, 217, 228
Zielinski, B. 2, 6, 11, 12, 25, 29, 30, 34, 39, 77, 129, 149, 154, 164, 168, 170, 178, 194, 196, 242, 248, 336, 339
Zippel, H.P. 33

Species index

Acipenseridae
  Acipenser 296
  Acipenser ruthenus (sterlet) 104, 107
Actinopterygii 17, 31
African catfish, *see* Clariidae
African cichlid, *see* Cichlidae
African lungfish, *see* Protopteridae
Agonidae
  Agonus cataphractus (hooknose, pogge) 108
Air breathing catfish, *see* Clariidae
Amago salmon, *see* Salmonidae
Amberjack, *see* Carangidae
Ameiurus, *see* Ictalurus
American eel, *see* Anguillidae
Amiidae
  Amia calva (bowfin) 100, 108
Ammodytidae
  Ammodytes personatus (sand lance) 21, 274
Anabantidae
  Anabas testudineus (climbing perch) 108, 122
Angelfish, *see* Cichlidae
Anglerfishes 21
Anguillidae
  Anguilla anguilla (European eel) 108, 225, 285
  Anguilla japonica (Japanese eel) 9, 173, 175, 176, 178, 179, 183, 184, 187–9, 190, 272, 274–6, 278, 279, 280, 281, 284, 303
  Anguilla rostrata (American eel) 56, 159, 169, 226
Antarctic dragonfish, *see* Bathydraconidae
Arctic charr, *see* Salmonidae
Ariidae
  Arius felis (sea catfish) 19, 21, 22, 178, 186

Atherinidae
  Bedotia geayi 19
  Menidia beryllina (tidewater silverside) 324, 326
  Menidia menidia (silverside) 302, 305, 323, 324
Atlantic menhaden, *see* Clupeidae
Atlantic salmon, *see* Salmonidae
Australian lungfish, *see* Ceratodontidae
Ayu, *see* Plecoglossidae

Baltic sea trout (brown trout), *see* Salmonidae
Bass, *see* Serranidae
Bathydraconidae
Belonidae
  Belone belone (needlefish) 22, 25, 29
  Gymnodraco acuticeps (Antarctic dragonfish) 21
Belontiidae
  Colisa 216, 225
  Colisa labiosa 216
  Colisa lalia 216
  Macropodus opercularis 51, 55
  Trichogaster 224
  Trichogaster trichopterus (blue gourami) 217, 225
Black gobies, *see* Gobiidae
Blacksmith, *see* Pomacentridae
Blenniidae
  Blennius pavo 224
  Blennius pholis (shanny) 108
  Blennius tentacularis 108, 124
Blue gourami, *see* Belontiidae
Bluegill, *see* Centrarchidae
Bonnethead shark, *see* Sphyrnidae
Bothidae
  Scophthalmus maximus (turbot) 190, 272, 285, 286, 306, 318
  Scophthalmus rhombus (brill) 272

Bowfin, see Amiidae
Brachiopterygii 17, 31
Brill, see Bothidae
Brook charr or trout, see Salmonidae
Brook lamprey, see Petromyzontidae
Brown catfish, see Ictaluridae
Brown trout (Baltic sea trout), see Salmonidae
Bullhead catfishes, see Ictaluridae
Burbot, see Gadidae

Carangidae
  Seriola dumerili (amberjack) 175, 181
  Seriola quinqueradiata (yellowtail) 159, 168, 175, 181, 190, 272, 274, 275, 283, 305, 316
  Trachurus japonica (jack mackerel) 174, 179, 181, 272, 285
Carcharhinidae
  Mustelus canis (dogfish) 292, 296, 319
  Negaprion brevirostris (lemon shark) 159, 293, 303
  Rhizoprionodon porosus 16
Caribbean sharpnose shark, see Carcharhinidae
Carp, see Cyprinidae
Catfish, see Clariidae
Catfish, see Ictaluridae
Catfish, see Plotosidae
Catfish, see Siluridae
Catostomidae
  Catostomus commersoni (white sucker) 334
Centrarchidae
  Lepomis cyanellus (green sunfish) 81, 94, 102, 223
  Lepomis mactochirus (bluegill) 55, 330
  Lepomis megalotis megalotis (longear sunfish) 258
  Micropterus salmoides (largemouth bass) 301, 318, 330
Ceratodontidae
  Neoceratodus forsteri (Australian lungfish) 17, 25
Channel catfish, see Ictaluridae
Characidae
  Paracheirodon innesi (neon tetra) 108
  Serrasalmus nattereri (piranha) 51, 53, 55
Charr, see Salmonidae

Chimaeridae
  Chimaera monstrosa 15, 286
Chinook salmon, see Salmonidae
Chub mackerel, see Scombridae
Chum salmon, see Salmonidae
Cichlidae
  Cichlasoma citrinellum (Midas cichlid) 243
  Clarias batrachus 107
  Clarias gariepinus (African catfish) 159, 169, 204, 205, 216, 217, 225, 226
  Haplochromis burtoni 107
  Hemichromis bimaculatus (jewal fish) 243
  Oreochromis aureus 108
  Pterophyllum scalare (angelfish) 222
  Sarotherodon mossambicus (Mozambique cichlid) 124
  Tilapia nilotica 173, 174, 179, 182
  Tilapia zillii 305, 314, 315
Clariidae
  Clarias batrachus (air breathing catfish) 107
  Clarias gariepinus (African catfish) 159, 169
Clearnose skate, see Rajiidae
Climbing perch, see Anabantidae
Clingfish, see Gobiesocidae
Clupeidae
  Alosa (shad) 311
  Brevoorita tyrannus (menhaden) 311, 332
  Clupea harengus (herring) 105, 108, 115, 274, 300, 303, 311, 315
  Dorosoma petenense (threadfin shad) 318
  Sprattus sprattus (spratt) 108, 111
Cobitididae
  Cobitis 104, 109
  Cobitis taenia (spined loach) 105, 108
  Misgurnus anguillicaudatus (Oriental weatherfish) 159, 168, 181, 190, 223, 303, 316
Cod, see Gadidae
Coho salmon, see Salmonidae
Conger eel, see Congridae
Congridae
  Conger myriaster (conger eel) 43, 57, 159, 296
Coregonidae
  Coregonus clupeaformis (whitefish) 324, 326, 328, 330, 331, 333

Crossopterygii 31
Cyclostomes 14
　Coregonus clupeaformis (whitefish) 159, 168
Cyprinidae 19, 171
　Brachydanio rerio (zebrafish) 29, 204, 216, 221, 227, 228, 303, 333
　Carassius auratus (goldfish) 3, 6, 7, 43, 49, 50–9, 80, 81, 83, 84, 87–9, 91, 94, 100, 101, 102, 108, 124, 156, 158, 159, 160, 164, 167–9, 199, 205–16, 218–28, 297, 298, 300, 316, 317, 319, 323, 324
　Carassius carassius (crucian carp) 80, 87, 92, 94, 101
　Carassius carassius grandoculis (gold crucian carp) 155, 169
　Cyprinus carpio (carp) 41, 42, 45–9, 50–3, 55, 56, 58, 59, 83, 87–9, 102, 117, 122, 131, 132, 134, 154, 155, 159, 167, 168, 170, 173, 175, 176, 179, 180, 182–4, 191, 283, 327, 328
　Notemigonus crysoleucus (golden shiner) 330
　Notropis rubellus (rosyface shiner) 331
　Notropis spilopterus (spotfin shiner) 332
　Notropis umbratilis (redfin shiner) 223
　Phoxinus 107
　Phoxinus eos (northern redbelly dace) 104
　Phoxinus lagowski (Japanese fat minnow) 331
　Phoxinus phoxinus (minow) 104, 105, 107
　Pimephales notatus (bluntnose minnow) 334
　Pimephales promelas (fathead minnow) 222, 298, 318, 323, 325, 333
　Pseudorasbora parva (topmouth minnow) 173, 174, 176
　Rhodeus ocellatus ocellatus 223
　Rutilus rutilus (roach) 107
　Tinca tinca (tench) 22, 45, 55, 155, 324
Cyprinodontidae
　Fundulus heteroclitus (killifish) 25, 27, 29, 302, 304, 316, 323, 325

　Jordanella floridae (flagfish) 331

Darters, see Percidae
Dasyatidae
　Dasyatis pastinaca (skate) 11, 24, 131–4, 148
　Dasyatis sabina (stingray) 159, 169
Dogfish, see Carcharhinidae
Dolly Varden charr, see Salmonidae
Dover sole, see Soleidae

Esocidae
　Esox lucius (pike) 21, 22, 322
Elasmobranchs 15, 31
Embiotocidae
　Cymatogaster aggregata (shiner perch) 332
Exocoetoidae 21
Fathead minnow, see Cyprinidae
Flagfish, see Cyprinodontidae
Flatfish 31
Flounder, see Pleuronectidae

Gadidae 73
　Ciliata 114, 115, 123
　Ciliata mustela (five bearded rockling) 4, 105, 123, 124
　Gadus morhua (cod) 49, 55, 57, 117, 165, 218, 219, 222, 285, 292, 296, 304, 315, 319
　Gaidropsarus 114, 123
　Gaidropsarus mediterraneus 107, 123
　Lota lota (burbot) 165, 167
　Pollachius virens (saithe, pollock) 108
　Urophycis chuss (red hake, squirrel hake) 121, 296, 315, 319
Gasterosteidae
　Gasterosteus 104, 111
　Gasterosteus aculeatus (threespine stickleback) 59, 105, 114, 217, 226, 243, 244
　Spinachia spinachia (fifteenspine stickleback) 19, 21, 108
Giant moray, see Muraenidae
Gobiidae
　Bathygobius soporator 201, 227
　Gobiosoma bosci (goby) 305, 317
　Gobius jozo (black gobies) 203, 222
　Periopthalmus koelreuteri (mudskipper) 108
　Pomatoschistus 104, 112, 113
　Pomatoschistus minutus (sand goby) 104, 105, 107, 108, 111

Species index

Gobiesocidae 21
Goby, see Gobiidae
Gold crucian carp, see Cyprinidae
Golden shiner, see Cyprinidae
Goldfish, see Cyprinidae
Green sunfish, see Centrarchidae
Grunt, see Haemulidae
Gulper eel, see Saccopharyngidae
Guppy, see Poeciliidae

Haemulidae
  Orthopristis chrysopterus (pigfish) 272, 285, 301, 306, 315
  Parapristipoma trilineatum (grunt) 174, 179, 181
Hagfish, see Myxinidae
Hake, see Gadidae or Merlucciidae
Hammerhead shark, see Sphyrnidae
Hardhead sea catfish, see Ariidae
Hatchetfish, see Sternoptychidae
Hawaiian goatfish, see Mullidae
Herring, see Clupeidae
Hogchoker, see Soleidae
Hooknose, see Agonidae
Hyborhynchus notatus (bluntnose minnow) 259
Holocentrus ascensionis 56

Ictaluridae 171, 186, 296, 309, 317, 319
  Ameiurus nebulosus 319
  Ictalurus 73, 74, 75, 113, 116, 118, 123
  Ictalurus catus (white catfish) 5, 12, 155, 156, 159, 172
  Ictalurus lacustris punctatus (southern channel catfish) 314, 315
  Ictalurus melas (black bullhead) 104, 105, 114, 155, 289, 290
  Ictalurus natalis (yellow bullhead) 172, 292, 293, 304, 314, 325, 326, 328, 331
  Ictalurus nebulosus (brown catfish) 11, 51, 55, 80, 84, 100, 107, 127, 128, 172, 190, 240, 304, 323, 324
  Ictalurus punctatus (channel catfish) 6, 51, 53–5, 80, 81, 85, 88, 89, 92, 97, 100–2, 127–9, 130, 131, 134–7, 139, 141, 142, 144, 145, 147, 148, 158, 159, 162, 167, 172–4, 176, 181, 182, 183–6, 188, 189, 191, 309, 314, 317, 318, 319, 323, 326
  Ictalurus serracanthus 159

Jack mackerel, see Carangidae
Jacks, see Carangidae
Japanese eel, see Anguillidae
Japanese fat minnow, see Cyprinidae
Jewal fish, see Cichlidae

Killifish, see Cyprinodontidae
Knifefish, see Notopteridae

Lake charr or trout, see Salmonidae
Lampreys, see Petromyzontidae
Lampridae
  Neolamprologus (formerly *Lamprologus*) *brichardi* 243
Largemouth bass, see Centrarchidae
Lemon shark, see Carcharhinidae
Lepidosirenidae
  Lepidosiren 16
Latimeridae 16
Loaches, see Cobitididae
Longear sunfish, see Centrarchidae
Loricariidae
  Ancistrus cirrhosus (vieja) 108
Lungfishes, see Lepidosirenidae or Protopteridae

Masu salmon, see Salmonidae
Melanotaeniidae
  Melanotaenia maccullochi (rainbow fish) 25, 29
Marbled rockfish, see Scorpaenidae
Medaka or ricefish, see Oryziidae
Menhaden, see Clupeidae
Merluccidae
  Merluccius (hake) 293, 296
Midas cichlid, see Cichlidae
Minnows, see Cyprinidae
Monognathidae 21
Moray eel, see Muraenidae
Mozambique cichlid, see Cichlidae
Mudskipper, see Gobiidae
Mugilidae
  Mugil cephalus 159, 174, 179
Mullets, see Mugilidae
Mullidae
  Parupeneus porphyreus (Hawaiian goatfish) 100, 317
  Parupenus multifaciatus 100

Mummichog (killifish), see Cyprinodontidae
Muraenidae
  Gymnothorax funebris (moray eel) 51, 59
  Gumnothorax javanicus (giant moray) 19
  Rhinomuraena ambonensis (moray eel) 19
Myxinidae
  Myxine glutinosa (hagfish) 14, 15, 124, 159, 167, 286, 307
  Polistotrema stouti 291, 316

Needlefish, see Belonidae
Neon tetra, see Characidae
Notopteridae
  Notopterus notopterus (knifefish) 108
Nurse shark, see Orectolobidae

Odacidae
  Siphonostoma typhle 19, 21
Orectolobidae
  Ginglymostoma cirratum (nurse shark) 292–5, 303, 318
Oriental weatherfish, see Cobitididae
Oryziidae
  Oryzias latipes (medaka or ricefish) 331

Parrotfish, see Tetraodontidae
Percichthyidae
  Morone saxatilis (striped bass) 332
Percidae
  Etheostoma nigrum (darter) 296, 319
  Stizostedian lucioperca (zander, pike perch) 107
  Stizostedian vitreum (walleye) 306, 319
Petromyzontidae
  Entosphenus japonicus 151, 323, 326
  Lampetra fluviatilis (brook or river lamprey) 25, 28, 173
Pigfish, see Haemulidae
Pike, see Esocidae
Pink salmon, see Salmonidae
Pipefish, see Odacidae
Piranha, see Characidae
Plecoglossidae
  Plecoglossus altivelis (ayu) 223, 331
Pleuronectidae
  Pseudopleuronectes americanus (flounder) 293, 306
  Psettichthys metanostictus (sand sole) 324
Plotosidae
  Plotosus lineatus (formerly *Plotosus anguillaris*; sea catfish) 60, 61, 65, 68, 69, 70, 71, 73, 75, 85, 101, 123, 172, 173, 182, 184, 186, 191
Poeciliidae
  Mollienesia(=*Poecilia*) *sphenosp* 217, 228
  Poecilia reticulata (guppy) 105, 108, 201, 222, 223, 224
  Xiphophorus helleri (swordtail) 21, 108, 124
  Xiphophorus maculatus 56
Pogges, see Agonidae
Polyteridae
  Erpetoichthys calabaricus (reedfish) 18, 28, 108
Pomacentridae
  Chromis punctipinnis (blacksmith) 332
Porgies, see Sparidae
Protopteridae
  Protopterus 16, 17, 109, 112, 113, 124
  Protopterus aethiopicus (lungfish) 105, 108, 111
  Protopterus amphibius 108
  Protopterus annectens 25, 297, 299, 316
Pteromyzontidae
  Lampetra fluviatilis (lamprey) 103
  Lampetra planeri (brook lamprey) 103, 104, 105, 107, 111, 121
  Petromyzon marinus (sea lamprey) 291, 303, 317
Puffers, see Tetraodontidae

Rabbitfish, see Siganidae
Rainbow fish, see Melanotaeniidae
Rainbow trout, see Salmonidae
Rajiidae
  Raja clavata (roker, thornback ray) 109, 124
  Raja eglentaria (clearnose skate) 100
Red sea bream, see Sparidae
Redfin shiner, see Cyprinidae
Reedfish, see Polypteridae
River lamprey, see Petromyzontidae
Roaches, see Cyprinidae

Species index

Rocklings, see Gadidae
Rosyface shiner, see Cyprinidae

Saccopharyngidae 21
Saithe, see Gadidae
Salmonidae
  Oncorhynchus clarki (cutthroat trout) 250, 255, 257, 258
  Oncorhynchus gorbuscha (pink salmon) 230, 250–3, 255, 324
  Oncorhynchus keta (chum salmon) 230, 250–3, 255, 318
  Oncorhynchus kisutch (coho or silver salmon) 7, 131, 159, 168, 231, 239, 242, 250, 251, 255, 257, 315, 329
  Oncorhynchus masou (masu salmon) 233, 250–3, 255
  Oncorhynchus mykiss (formerly *Salmo gairdneri*; rainbow trout) 10–2, 52, 57, 58, 59, 118, 125, 130, 146, 147, 149, 154, 159, 160–2, 164, 167, 168, 170, 173, 175, 176, 182, 183, 189, 190, 218, 219, 222, 223, 225, 240, 250, 255, 272, 285, 302, 304, 317, 322, 325–7, 329, 330–2
  Oncorhynchus nerka (sockeye salmon) 27, 29, 156, 159, 168, 169, 239, 241, 250–3, 255, 256, 262, 300, 318
  Orcorhynchus rhodorus (amago salmon) 223
  Oncorhynchus tshawytscha (chinook or spring salmon) 250, 251, 252, 253, 255, 315
  Salmo alpinus, see *Salvelinus alpinus*
  Salmo gairdneri, see *Oncorhynchus mykiss*
  Salmo malma (Dolly Varden charr) 250, 255
  Salmo salar (Atlantic salmon) 5, 8, 11, 12, 122, 156, 159, 169, 172, 176, 179, 220, 225, 236, 250–3, 255, 257, 259, 263, 264, 304, 318, 327, 328, 330–4
  Salmo trutta (brown trout or Baltic sea trout) 28, 122, 250, 255, 257–9, 304, 318, 322, 326
  Salvelinus 6
  Salvelinus alpinus (Arctic charr) 12, 55, 123, 159, 160–2, 166, 170, 175, 227, 231, 235–7, 240–2, 250, 259, 298, 299, 304, 317, 319, 327, 333
  Salvelinus fontinalis (brook charr or trout) 159, 168, 175, 179, 221, 224, 250, 255, 325, 334
  Salvelinus namaycush (lake charr or trout) 159, 160, 175, 250
Sand eel, see Ammodytidae
Sand lance, see Ammodytidae
Sand sole, see Pleuronectidae
Scombridae
  Euthynnus affinisa (tuna) 31, 296, 318, 319
  Scomber japonicus (chub mackerel) 174, 181
  Thunnus albacares (yellowfin tuna) 333, 315
Scorpaenidae
  Sebasticus marmoratus (marbled rockfish) 51, 53, 58, 272, 286, 305, 319
Sea catfish, see Ariidae and Plotosidae
Sea eel, see Congridae
Sea robin, see Triglidae
Serranidae
  Dicentrarchus labrax (bass) 104, 122
Shad, see Clupeidae
Sharks, see also Carcharhinidae 15
Shanny, see Blenniidae
Shiner perch, see Embiotocidae
Siganidae
  Siganus fuscescens (rabbitfish) 174, 179, 181
Siluridae 171
  Silurus glanis (European catfish) 21, 22, 125
Silver salmon, see Salmonidae
Silversides, see Atherinidae
Skates, see Dasyatidae or Rajiidae
Snakefish, see Synodontidae
Sockeye salmon, see Salmonidae
Soleidae
  Solea (sole) 291, 306, 296
  Trinectes maculatus (hogchoker) 324
  Solea solea (Dover sole) 272, 274, 285, 286, 318
South American lungfish, see Lepidosirenidae
Sparidae
  Chrysophyrys major (red sea bream) 159, 174, 176, 188, 189, 306, 307, 316, 317, 318

Sparidae—Contd.
  Pagrus major (porgy) 159, 272, 283, 285, 286
Sphyrnidae
  Sphyrna (hammerhead shark) 293
  Sphyrna tiburo (bonnethead shark) 297, 317
Spotfin shiner, see Cyprinidae
Sprat, see Clupeidae
Spring salmon, see Salmonidae
Stargazer, see Uranoscopidae
Sterlet, see Acipenseridae
Sternoptychidae 19
Stingray, see Dasyatidae
Striped bass, see Percichthyidae
Sturgeons, see also Acipenseridae 27, 29
Swamp eel, see Synbranchidae
Swordtail, see Poeciliidae
Synbranchidae
  Monopterus cuchia (swamp eel) 108
Synodontidae
  Trachinocephalus myops (snakefish) 21

Tench, see Cyprinidae
Teraponidae
  Therapon oxyrhynchus (tiger fish) 174, 181
Tetraodontidae
  Arothron nigropunctatus (parrotfish) 19
  Fugu pardalis (puffer) 173, 175, 181, 183–5, 188, 189, 190, 191, 272, 285, 286, 306, 316, 318

Fugu rubripes (tiger puffer) 276, 277
Threadfin shad, see Clupeidae
Threespine stickleback, see Gasterosteidae
Tiger fish, see Teraponidae
Tiger puffer, see Tetraodontidae
Tomcod, see Gadidae
Topmouth minnow, see Cyprinidae
Triglidae
  Prionotus 119, 123
  Prionotus carolinus (sea robin) 4, 11, 111, 121, 122, 124, 292, 305, 315
  Prionotus evolans 121, 305, 315
  Trigla lucerna (sapphirine gurnard) 105
Tuna, see Scombridae
Turbot, see Bothidae

Uranoscopidae
  Astroscopus (stargazer) 21
  Ichthyoscopus (stargazer) 21

Vieja, see Loricariidae

Walleye, see Percidae
White catfish, see Ictaluridae
White sucker, see Catostomidae
Whitefish, see Coregonidae

Yellow bullhead, see Ictaluridae
Yellowtail, see Carangidae

Zander, see Percidae
Zebrafish, see Cyprinidae

Subject index

Acetylcholinesterase 113
Acidic waters 325, 328, 334
Action potentials 151–2
Adaptation 30–1
Adenyl cyclase, see Transduction
Alanine 127, 128
L-Alanine 127–8, 241, 271–8
D-Alanine-best fibres 186
L-Alanine-best fibres 185
Aliphatic acids 179–81
Altruistic behaviour 229
Amino acids 5–6, 115, 127, 130, 135, 142, 156, 157, 160–3, 164, 172, 207, 240, 241, 302–7, 314, 326
   dose-response relationship 176–7
   feeding behaviour 188–90
   feeding stimuli 271–2, 275, 276, 277
   gustatory responses 172, 174–5, 176, 177–9, 192, 328–29
   intraspecific chemical cues 240
   ligand binding 329
   mixtures 187
   olfactory responses 176, 327–8
   olfactory specificity 160–1
   receptor site 182–6, 192
   stereospecificity 177–9
L-α-Amino acids 160, 162
Ammonia 240, 241
Amphidromous, see also Migration system 250
Anadromous, see also Migration system 250, 252, 258
Anadromous populations 230
Androgen 204, 215
Anosmic fish 202, 257, 263
Apical processes 65, 104, 107, 109, 111, 112
Appetite 300
Aquaculture 271, 274, 275, 284
Aquatic environment 1, 156, 321

Aquatic odorant 156, 157
Arginine 127, 132
L-Arginine 128–9, 143–4
L-Arginine-best fibres 185
Arousal phase 289–91
Axonemal complex 25, 27
Axon, see also Olfactory organ 23, 24, 25, 29
Axotomy 324

Baits 308–11, 314
Bait shy 301
Barbels 61, 63, 81, 83, 172
Basal cell
   gustatory 63, 67, 69, 70, 71, 73, 74, 75, 120
   olfactory 24, 23, 28
Basal lamina, see Taste buds
Behavioural response
   contaminant effects 330–4
Benthic search behaviour 296
Betaine, see also Quarternary ammonium bases 272, 274, 307
Bile 235, 241
Bile salts 6, 157, 158, 161, 163, 241
   gustatory stimuli 181–2
   olfactory stimuli 157, 158, 161, 163
Binding
   activity 326
   affinity 157
   amino acids 130
   capacity 132
   competition 131
   cross-adaptation 129, 132
   dissociation constant 128–32
   metal effect 329
   gustatory 127–9
   contaminant inhibition 329–30
   lectin 131
   ligand 5
   olfactory 129–33

Binding—Contd.
  pheromones 218
  profile 127
  receptor 5
  transport 130
Binding site 127, 128
  alanine 127, 132, 134
  arginine 127, 132, 134
  basic amino acids 128
  cilia 129–33
  G protein 132
  neutral amino acids 128, 131
Bipolar neurone 25
Bisexual pheromones, see also Pheromones 213, 215
Blood glucose 282
Brush cells 119

Catadromous, see also Migration system 251
Cell renewal 74
Central nervous system 41, 75, 171 176
Centrifugal fibre 43, 44, 45, 46, 49 52
Centrioles 27, 66, 67
Cephalic reflex 273, 277, 278, 280–4
Chemical cues
  ammonia 240, 241
  bile 235, 241
  free amino acids 240, 241
  intestinal contents 235, 236, 241
  intraspecific odours 230
  skin mucus 241
  urine 241
Chemical feeding stimuli 301
Chemical lure Ch.14
  attractant 310–1
  future development 311
Chemical search image 300–1
Chemical signal 1, 7
Chemical stimuli 1
Chemoreceptive tissues
  metal accumulation 322
  degeneration 323
  disintegration 324
  erosion 325
  hyperplasia 324
  loss of cilia 323, 324
  metaplasia 324
  sustentacular cell decrease 323
  toxin uptake 322
Chemosensory organ 2

Choana 31
Cholecystokinin (CCK) 283
Cholera toxin 133
Chumming 308–9
Cilia 25, 27, 153, 154, 166, 326
Ciliated nonsensory cell 24, 28, 29
Ciliated receptor cell, see also Olfactory receptor cells 24, 25–7, 28, 29–30
Claviform bodies 113
Commercial fishing 288, 307–8, 310
Common chemical sense 4, 72, 75, 116–9
Communication 200, 203, 213
Concentration-response curve, see also Concentration-response relationship 157, 162–3, 154
Concentration-response relationship 5, 156–7
Consummation phase 290, 297
Contaminants 321
  behavioural effect 321, 330–4
  biochemical effect 326–30
  chemoreception Ch. 15
  fish distribution 321
  histological effects 322–6
  physiological effect 326–30
Convergence 154
Corium papilla 63, 66, 67, 72
Courtships 165, 201
Cranial nerves 75
Cross-adaptation 182–4, 187–8, 208, 212
  gustatory 182–4, 187–8
  olfactory 5, 161, 162–4, 208, 212
Cross-reactivity 163
Crustecdysone 202–3
Cyclic AMP, see also Second messengers 134, 135, 137–9
Cyclic GMP, see also Second messengers 139
Cyclic-nucleotide-gated conductance 141–4
Cyclosmate 22
Cytochrome P-450 monooxygenase 335

D-R relations, see Dose-response relationships
Dark sensory cell 61, 63, 65–6, 67, 70, 71, 73, 76
Deciliation 154
Degeneration 25, 323–4

Subject index

Dermal papilla, see Corium papilla
Development
  gustatory 61–6
  olfactory 29–31
Diadromous, see also Migration system 251
Diencephalic projection 95–7
Dietary extract 274–6, 284
Dietary nutritive value 273
Digestive mechanism 273
17α,20β-Dihydroxy-4-pregnen-3-one (17,20P or 17α,20βP) 6, 59–60, 158–61, 205–15, 218–9, 220
17,20P, see 17α,20β-dihydroxy-4-pregnen-3-one
17α,20βP, see 17α,20β-dihydroxy-4-pregnen-3-one
Dip baits 310
Dissociation constant 128–32, 163
DNA recombinant 10
Dose-response relationships, see also Concentration-response relationships 176–7
Downstream migration 253
Dynamic range, see also Concentration-response curves 157
Dynein arms 25, 29

Ectohormone 199
EGTA 136, 152
Electroceptor 109, 118
Electroencephalogram (EEG) 220
Electroencephalographic (EEG) response 154–5
Electroolfactogram (EOG) 151, 152–3, 155, 157, 158, 160–1, 163, 207, 208, 209, 211, 212, 217–8
  contaminant effect 324
  phasic response 153
  tonic response 153
Endohormone 199
Enhancement 187–8
Environmental contaminants Ch.15
Epidermis 109
Epithelial cell 23–4
Etiocholanolone 203
Eversion 92
Evolution, see also Olfactory organ 202–3
Excitatory postsynaptic potential (EPSP) 46–8
Exponential curve 157
Exteroceptor 298

Extracellular recording 153
Extraoral gustatory system 80, 81

Facial lobe 80, 81, 82, 83, 84, 85, 86, 87, 88–9, 91, 97
Facial nerve 75, 83, 113–4, 117, 119, 171, 173, 176, 185
Facial taste system 176
Feed efficiency 276
Feeding
  food search 84, 91
  stimulus localization 81, 83, 86, 87
Feeding activity 272, 273, 275
Feeding attractant 207
Feeding behaviour 188–90, 297
  amino acids 188–90
  gustatory 188–90
  nucleosides 190
  nucleotides 190
Feeding deterrents 297
Feeding frenzy 288, 298
Feeding stimulants Ch.13, 8–9
  absorption 273
  dietary application 273–6
  digestion 273, 278, 280
  effects on nutrition 277–84
  energy retention 276
  enzyme activity 278, 279
  feeding activity 275
  nutrient metabolism 280
  physical properties 272
  protein efficiency 276
  protein retention 276
  quaternary ammonium bases 271, 272
Field potentials 46–7
Flask cell 109
Flavoured diet 273–7
  survival rate 274
  weight gain 274
Fluviarium 231, 232, 233, 235, 236
Food extract 295, 299, 301
Food intake 273, 274
Food odour 289, 291, 296
Food search behaviour Ch.14
  chemical stimulants 303–6
  digging behaviour 297
  experience 300–1
  gradient search 292, 293
  looping behaviour 297
  modifiers 297–301
Food search inducer 301–7
  amino acids 302–7

Food search inducer—*Contd.*
  chemical nature 302–7
  natural materials 301–2
Food search pattern 288
Foskolin 135
Free nerve ending 113, 116, 117–8

G$_{olf}$ *see* Transduction
G protein, *see also* Second messengers 4, 126, 133, 138, 185
Gastronintestinal digesta 278–9
Generator potentials 152
Genetic homogeneity 265
Genetic pollution 264
Giant output neurons 89
Gill arches 61
Glial cell 68
Glossophalyngeal nerve 75, 83, 171, 173, 176
Glycine 271–8
Glycocalyx 67, 109, 112, 116
Glycogen granules 65
Glycoprotein 131, 132
Goblet cell 24
Golgi complex, *see also* Golgi system 25, 28
Golgi system 65–6, 71, 76, 111–2
Gonadotrophin (GtH) 205, 208, 209–10, 213–4
Gonadotrophin hormone-releasing hormone (GnRH) 217
Granule cell 43–7, 49
Grower diets 275–7
Growth performance 273, 274–7, 283, 284
Guanine nucleotide, *see* Transduction
Guanyl cyclase, *see* Transduction
Gustation, *see also* Taste Ch.9, Ch.13, 2, 217
  historical background 172–7
  receptor event 127
  receptor site types 182–6
Gustatory cell 61, 63, 73
Gustatory inputs 84–8
Gustatory nerve 6
Gustatory organ Ch.4, 3
Gustatory receptor 7
Gustatory receptor cell 3
Gustatory response
  aliphatic acids 179–81
  amino acid mixtures 187
  amino acids 172, 174–5, 177–9, 192,

  contaminant effect 328–9
  D-amino acids 178, 186
  enhancement 187–8
  facial taste fibres 185–6
  L-α amino acids 178
  limited response range 175, 178–9, 192
  nucleotides 181
  peptides 178
  specificity 174–5
  wide response range 174, 178–9, 192
Gustatory system
  central projection Ch.5
  central representation Ch.5
  descending gustatory projections 88–91
  facial gustatory centre 80
  gross morphology 83–4
  medullary networks 88–91, 94
  supramedullary gustatory pathways 91–7, 98
  telencephalic connections 94–7

Hemidesmosomes 67, 71
Hepatic enzyme activity 280–3, 284
Hillock, *see* Taste buds
Histological effects
  toxic substance 322–6
Home-area preference 258
Homeward migration 254
Homing *see also* Migration 7, 230–1, 255–9
Homing behaviour 249, 255–9
  ecological implication 264
  imprinting hypothesis 259, 261
  kin recognition 262
  non-reproductive 258
  olfactory control 263–4
  olfactory sense 258
  pheromone hypothesis 259–60, 261, 264
  pheromone profile 262
  reproductive 255–8, 265
  scent marking 262
  sensory mechanism 255
Hormonal metabolites 203, 216
Hormonal pheromone 200–3, 205, 215–6
  endocrine state 200–2
  evolution 202–3
  hormone metabolism 203–5
  model 205–15

Subject index

Horseradish peroxidase (HRP) 41, 42, 85, 87, 94
Hyperpolarization 152
Hyperplasia 324
Hypophysectomy 201

Imprinting 7, 220, 243, 261
 thyroid activity 220
Imprinting hypothesis 259
Induced waves 154
Inferior lobe 92, 99
Inhibitory postsynaptic potential (IPSP) 46–8, 60
Inositol trisphosphate (IP_3), see also Second messengers 136, 137–9
Interdigitation 71
Intermediate cell 63, 66, 74
Interneurone 74
Intestinal contents 235, 236, 241
Intracellular recording 151
Intraoral gustatory system 80
Interspecific interactions 115
Intraspecific odours
 chemical cues 230, 240–1
Intraspecific population 265
Intratelencephalic fibre 51
Inversion 92
Ion channels 4, 133, 140–4
 cyclic-nucleotide-gated 141–2
 stimulus-activated 140–1, 142–4
Isosmate 22
Isthmic gustatory centre 92–4

Junctional complex 23, 28

Kin recognition Ch.11
 familiarity and previous association 229
 learning 235, 240, 242
 phenotypic matching 229, 230
 preference 231, 236
 recognition allelles 230
 self-matching 235
 shoaling 231, 242, 244
 sibling recognition 231–5
 spatial distribution 229, 230
Kineses 292
Kinetic analysis 162–3
Kinocilia 17, 22, 29
Klinotaxis 292

Lakeward migration 255
Lateral olfactory system 50

Lateral Olfactory tract (LOT), see also olfactory tract 45, 48, 49, 51, 52, 219
LC 50 329, 330
Lectin 74, 131
Light sensory cell 61, 63, 65, 70, 71, 73
Light-dark cells
 nomenclature 72–3
Locomotor activity 289, 291

Macromolecular receptor 129
Map
 gustatory inputs 85–7
 oropharyngeal 86
Marginal cell, see Taste buds
Mass action law 163
Mechanoreceptive activity, see also Tactile inputs 155–6, 172, 191
Medial olfactory system 50, 52
Medial olfactory tract (MOT), see also Olfactory tract 45, 48, 49, 52, 205, 219
Membrane resistance, see also Olfactory neurone 151
Membrane conductance, see also Olfactory neurone 152
Membrane potential, see also Olfactory neurone 151, 153
Merkel cell 3, 74, 113, 115
Metals
 behavioural response 330
 effects on binding 329
 gustatory response 328
 ligand binding 329
 low pH 325
 olfactory response 326–8
Metaplasia 324
Microridge 23
Microtubules 25, 65, 66, 67, 72,111
Microvilli 326
 gustatory cell 65, 66, 73, 76
 olfactory 165
Solitary chemoreceptors Ch.6, 109
Microvillous receptor cell 24, 25–7, 30
Midline lobe 87
Migration
 contaminant effect 334
 salmon homing 7–8
Migratory cycle 253
Migratory system
 amphidromous 250

Migratory system—*Contd.*
  anadromous 250, 252, 258
  catadromous 251
  diadromous 251
  land-locked stock 250, 254
  oceanodromous 251
  potamodromous 250, 254–5, 258
Milt 158, 205–7, 210, 213, 214
Mitochondria 65, 66, 68, 69, 111
Mitral cell, *see also* Olfactory bulb 155
Mixed-function oxidase, *see* Cytochrome P-450 monooxygenase
Modifiers of food search 297–301
Molecular mechanism
  chemoreception Ch.7
Morpholine 261
Mucous cell 28
Mucus 74, 109, 115–6, 119, 241, 325
Multiunit activity 176
Multivillous cell 107, 115
Myelinated nerve fibres 72

Nasal
  bridge 14
  duct 15
  flaps 15, 16, 19, 21, 22
  tube 16, 17, 18
Nasopharyngeal duct 14, 15
Nasopharyngeal pouch 14
Necrosis 323, 324, 325
Negative rheotaxis 263
Nerve fibre plexus 63, 68–9
Nerve twig recording (NTR)
  gustatory 176
  olfactory 153, 162
Neural coding 164–6
Neuromast 108, 118
Non-linear regression model fitting 162–3
Non-sibling 231, 233, 235
Nonsensory cells 27–9
Nonsensory cilia, *see also* Kinocilia 16
Nonsensory epithelium 22, 23, 24, 26
  histology 22–9
Nostrils 14, 15, 16, 17, 18, 19, 21, 22, 29
Nucleosides 181, 190
  feeding stimuli 271, 272
Nucleotides 307
  feeding stimuli 271, 272
  gustatory stimuli 181, 190
Nucleus ambiguous 89, 91
Nutrition Ch.13, 8

Nutritional requirement 273

Oceanodromous, *see also* Migratory system 251
Odour mixtures, *see also* Amino acid mixtures 216
Olfaction 2, 249
  freshwater migration 257
  home stream detection 257
Olfactory
  chamber 13, 15, 17, 18, 21, 31
  cilia 25, 129–33
  epithelium 13, 14, 24, 27, 28
  groove 29
  knob 25
  lamellae 14, 15, 16, 17, 18, 21, 22, 27
  papilla 21, 22, 29
  placode 29, 31
  receptor cell 14, 23, 25
  rosette 14, 16, 21
  ventilation 14, 16, 21, 22, 23
  ventilation sac 14, 21, 22, 31
Olfactory ablation 201, 204, 217
Olfactory bulb 3, 154, 219
  anatomy 41–5
  central projection Ch.3
  concentric layer 45
  electrical activity 154–5
  electrophysiology 45–50
  field potentials 46, 47
  glomerular layer 41
  glomeruli 154
  granule cell 43, 44, 45, 46, 47, 49, 155
  internal cell layer 41
  mitral cell 41–43, 44, 45, 46, 47, 48, 49, 52, 154, 155
  mitral cell layer 41
  olfactory nerve layer 41
  ruffed cell 43, 44
  synaptic organization Ch.3
Olfactory bulb response
  contaminant effect 327–8
Olfactory epithelium
  contaminant effect 323–5
Olfactory glomerulus, *see also* Olfactory bulb 154
Olfactory hypothesis, *see also* Homing behaviour 259–63
Olfactory mechanisms Ch.8
Olfactory nerve 2, 13, 29, 41, 42, 43, 49

Olfactory neurone 150, 151–2
  action potential 151, 153
  conductance 152
  electrical characteristics 151–2
  hyperpolarization 152
  resistance 151
  resting membrane potential 151
Olfactory organ Ch.2
  adaptation 30–1
  central 2
  development 29–30
  evolution 30–1
  gross morphology 13–22
  histology 22–9
  peripheral 2
  sixual dimorphism 21, 31
Olfactory receptor 164, 218–9
  population 164
Olfactory receptor cells 2, 14, 129, 164–5, 323–4
  ciliated 2, 129, 164–5
  microvillar 2, 129, 164–5
Olfactory receptor sites 130–3
Olfactory response 5, 157
  contaminant effect 326–8
  dynamic range 157
  sensitivity 150
  specificity 150
Olfactory rosette
  contaminant effect 322–3
  structure 20–1
Olfactory system
  functional separation 165
  lateral 50
  medial 50, 52
  subdivision 3, 165
Olfactory tract 41, 42, 155–6, 165, 205, 219
  electrical activity 155–6
  lateral (LOT) 45–9, 51
  medial (MOT) 45–9, 51
Olfactory ventilation sac 14, 21, 22, 31
Oligovillous cell 107, 109, 111, 112, 113, 115, 119,
Oocyte maturation 205, 206
Orobranchial motoneurones 89
Outward migration 254
Ovalian fluid 203–4
Oviparous 201
Ovoviviparous 201
Ovulation, *see also* Pheromones 202, 209, 213

Palatability 272, 273
Palatal organ 82, 84, 87–8, 90
Paraneurones 74, 119–20
Patch clamp 10, 152
Peameability 152
Perinuclear cisternae 65, 67
Peripheral gustatory organ, *see also* Taste buds 61
Pertussis toxin, *see* Transduction
PGFs, *see* Prostaglandins
pH 179, 325, 328, 334
Pharynx 61, 325
Phenotypic matching, *see* Kin recognition
Pheromone hypothesis 259, 264
Pheromones 7, 8, 49, 50, 199–200, 309
  17,20P 165
  bisexual 213, 215
  central mechanism 219
  chemical ecology 264
  definition 199–200
  hormone 203–3
  long-distance signalling 216
  odour mixture 216
  olfactory stimulation 207–15
  ovulation 208, 211, 220
  population-specific 259, 265
  postovulatory 160
  preovulatory 163
  primer 212, 215
  prostaglandin F_{2a} 165
  releaser 212–3, 215
  sex 161
  short-range signalling 216
  social cues 216
  strain-specific substance 259, 265
  trails 260
Phospholipase D 136
Phospholipid stain 324
Phospholipase A1 136
Phylogeny 29–30
Plasmalemma 67
Population hybrid 261
Population structure 250–1
Population-specific chemical cues 231, 235–41, 243
Population-specific odours 239, 241
Positive rheotaxis 263
Postovulatory pheromone, *see also* Pheromones 160
Postsynaptic cisternae 75
Potadromous population 230

Potamodromous *see also* Migratory system 250, 254, 255, 258
Power function 157
Pre-exposure 332–3
Preference behaviour, *see also* Kin recognition 231, 236
Preference-avoidance behaviour 330–3
Preovulatory pheromone *see also* Pheromones 163
L-Proline 6, 161, 186, 271–8
Prostaglandins (PGFs) 5, 49, 158–60, 161–2, 164, 206, 211–5, 218
  metabolites 212, 219
  olfactory specificity 161
  olfactory thresholds 158, 159, 160
Protein kinase C 136

Quarternary ammonium bases, *see also* Betaine 272

Radiolabelled ligands 211
Raphe 14, 15, 16, 17, 18, 21, 22
Receptor cell, *see also* Olfactory neurone 25–7
  axon 23, 24, 25, 29
  ciliated 25, 26, 27
  dentrite 23, 25, 27
  gustatory 61, 63, 73
  microvillous 25, 26, 27
  olfactory 150
Receptor field 63, 65, 66–7, 73, 76
Receptor site 150
Receptor site type
  classification 183
  gustatory 182–4
Receptor subtypes, *see also* Receptor site type and receptor types 164
Receptor types
  acidic amino acids 162
  affinity 163
  basic amino acids 162
  cross-apaptation 162
  kinetic analysis 162–3
  long-chained neutral amino acids 162
  multiplicity 162
  non-linear model fitting 162–3
  short-chained neutral amino acids 162
Reciprocal synapse 43–4
Regeneration 25

Regenerative cell 74
Reproductive behaviour 202, 212, 213, 217
Resting membrane potentials, *see also* Olfactory neurone 151
Reticular formation 91
Reticulospinal system 91
Rheotaxis 263, 264, 293, 296
Ribosome 25, 65
Rod cell 27
Rough endoplasmic reticulum (rER) 25, 65, 66, 112
Ruffed cell 43, 44

Salmonid ecology 265
Saturation 160
Scanning electron microscopy 26, 60, 62–3, 109
Schwann cell 68
Search phase 290, 291–7
Seaward migration 263
Second messengers 4, 185, 218
  calcium regulation 139
  cyclic AMP 134, 135, 137–9
  cyclic GMP 139
  diacylyglycerol (DAG) 136
  fate 139–41
  G protein 4, 126, 133, 138, 185
  gustation 134–7
  inositol trisphosphate (IP_3) 136, 137–9
  olfaction 134–9
Secondary gustatory nucleus 92, 93, 99
Secondary olfactory areas 50–2
Seeding 308, 309
Seminal vesicle 204
Sensitivity
  gustatory 6
  olfactory 5–6, 156–60
  receptor 5
Sensory cells 25–6, 61, 63, 66, 70
Sensory epithelium 22–3, 24, 26, 31, 61 63–6
  histology 22–9
  receptor cell 23, 25–7, 28
  supporting cell 23, 25, 27
L-Serine 158, 160, 207, 209
Sex hormone 300
Sex pheromones Ch.10, 200–3, 205–15, 217, 309
  17,20P 205–15, 218, 220
  androgen 204, 215

conjugated steroids 204, 208
crustecdysone 202–3
etiocholanolone glucuronide 203
glucuronated steroids 204
neural response 217–20
oestradiol 201
olfactory sensitivity 202, 204
ovarian fluid 203, 209
PGF metabolites 211, 212
postovulatory 209
preovulatory 209
progestational steroids 208
seminal vesicle 204, 205
testicular extracts 204
testosterone 202, 219, 220
urine 203
Sex steriods, 6, 158, 203–20
Sexual arousal 160
Sexual receptivity 205–6, 211
Shoaling, see also Kin recognition 231, 242, 244
Sibling-specific chemical cues 231
Siblings, see also kin recognition 230, 231, 233, 235
Sigmoidal curve 157, 163
Signal transmission 4, 154
Single fibre activity
   gustatory 185
Sight feeder 296
Smooth endoplasmic reticulum (sER) 25, 65, 66, 67, 112
Social interaction 242
Solitary chemosensory cell Ch.6, 4
   apical process 104, 107, 109, 112
   cytology 109–12
   distinction 120
   distribution 104–9
   nerve innervation 113–4
   physiology 114–6
   transmitter 113
Solitary chemosensitivity 115
Spawning 205, 209, 213
   contaminant effect 334
Spawning synchrony 205, 206
Species-specificity 216, 315
Specificity, see also Structure-activity relationships
   gustatory response 5–6
   olfactory response 5–6
Spinal nerve 113–4, 116
Spines 23, 74
Sports fishing 288, 307–8, 310, 312–4

Starter diet 273–5
Stationary behaviour 258
Stem cell 28, 74
Stereoselectivity 131
Stereospecificity 311, 314
   gustatory 177–9
   olfactory 160–2
Steroid glucuronides 160
Steroidal metabolites 203, 205
Steroids 5, 157–60, 161, 163–4
   17,20P 6, 158–61
   bile salts 6, 157, 158, 161, 163
   EOG response 158
   olfactory specificity 161
   olfactory stimuli 157
   olfactory thresholds 157–60
Stock characteristics 251–3
Stock-specific chemical cues 230
Stock-specific odours 239
Stratified squamous epithelium 22, 28, 71
Stress 298, 299
Structure-activity relationship, see also Stereospecificity 5, 131, 314
   gustatory 177–9
   olfactory 160–2
Sublethal effect 326
Superior secondary gustatory 94
Supplemental diets 274–7, 284
Supporting cell
   gustatory 61, 63, 73, 116, 120
   olfactory 23, 24, 25, 27, 28
Sustentacular cells, see also Supporting cell 323
Sympatric population 236, 237
Synapse
   asymmetrical 43
   excitatory 46
   organization 44
   solitary chemoreceptor 108, 113
   taste bud 63, 69–70, 71, 72, 73, 75
Synergism 187–8, 192
Synthetic chemical lures 311–4

Tactile cues 296, 298
Tactile function 114, 115
Tactile inputs 82
Tactile receptor 114–5, 117
Tactile sensitivity 191
Taste activity enhancement 187–8, 192
Taste buds Ch.4, 3, 79, 83, 90, 103, 104, 107, 109, 112, 113, 114, 115, 116

basal cell 120
bipolar cell 103, 104
distribution 173
gustatory cell 104, 107, 109, 115, 116, 119, 120
sensory cell 103, 104, 114
supporting cell 116, 120
basal cell 63, 67, 69, 70, 71, 73, 74, 75
basal lamina 63, 67, 72, 75
contaminant effects 325
degenerating cell 63, 66, 74
distribution 61–3
fine structure 60, 63–72,
gross anatomy 60, 61–9
hillock 63
marginal cell 63, 71–2
nerve fibre plexus 63, 68–9
receptor field 63, 65, 66–7, 73, 76
Taste cortex 97
Taste pore, see also Receptor field 116
Taste receptor sites
 L-alanine 127–8
 L-arginine 143–4, 128–9
Taxes 292
TB see Taste buds
Telencephalon 50–3
Telencephalic connections 94–7
 centrifugal fibres 51, 52–3
 eversion 50
 evolution 50, 51
 inversion 50
 non-olfactory sensory inputs 53
 secondary olfactory areas 51–2
 terminal regions 51
Terminal nerve (TN) 45, 50, 156, 165 217, 218
Testosterone 202, 203
 hormonal influence 219–20
Thalamic taste nucleus 92
Thermosensitivity 155–6
Threshold concentration 327
Thresholds
 amino acids 173, 177
 gustatory 172–3, 177, 179, 181, 182
 olfactory 156–60
 toxicant effect 327
Thyroxine 300
Tight junction, see also Junctional complex 71
TN cell, see Terminal nerve
Tonofilaments 28, 65, 66
Torus semicircularis 92

Toxic substance
 1,2-dibromoethane 322
 aviation fuel 325
 benzene 324
 crude oil 324
 destructive effects 323
 detergents 325, 326, 327–8, 331
 halogenated hydrocarbons 322
 hydrocarbon 324
 low pH 325, 328, 334
 metals 322–4, 326, 327, 329, 330
 petroleum hydrocarbons 324
Toxicant effect
 behavioural response 330–4
 gustatory 328–9
 migration 330, 334
 olfactory 326–8
 recovery 325–6
 thresholds 327
Toxicant-induced lesions, see also Chemoreceptive tissues 325–6
Transduction Ch.7, 119, 164
 adenyl cyclase 133, 137, 138
 cholera toxin substrate 133
 cyclic AMP 134, 135, 137–9
 G protein 133
 G_{olf} 137
 guanine nucleotide 133
 guanyl cyclase 139
 gustation Ch.7, 185–6
 molecular mechanisms Ch.7
 olfaction Ch.7, 152, 153–4, 164
 pertussis toxin substance 133
 pheromone 218
 receptor occupancy 137
 receptor-mediated 127
 second messenger 133
 sensory 151–3
 gustatory sequence 143
 olfactory sequence 144
 signal 4, 5
Transmission electron microscopy 13, 26, 60, 72
Transmitters 109
Transplantation 260–1
Transport see also Binding 130
Trigeminal nerve 75, 172, 191
Trigeminal receptor 115–7
Triton X-100, 154, 326, 328
Tropotaxis 292

Unmyelinated nerve fibres 72
Unit recording 162

Urine 203, 241

Vagal lobe 81, 82, 84, 86, 87, 88, 89, 90, 91, 97, 99, 114
Vagus nerve 75, 83, 171, 173
  sensorimotor circuit 89
Ventroposterior diencephalon 95
Vesicles 65–6, 67, 69–70, 76, 111–2

Vibratile fin 105, 109, 113, 114
Viscerosensory brain-stem 80
Vomeronasal organ 2, 165

Water pollution Ch.15, 9

Y-maze 231, 233